Problems and Solutions in Electricity and Magnetism

Online at: https://doi.org/10.1088/978-0-7503-6477-5

Problems and Solutions in Electricity and Magnetism

Pradeep Kumar Sharma MSc, MBA (IIT), Mtech, MPhil, PhD, AMIE, CEng(I) MIET, MInstP, SMIEEE
Consultant physicist and researcher, (India)

IOP Publishing, Bristol, UK

ISBN   978-0-7503-6477-5 (ebook)
ISBN   978-0-7503-6473-7 (print)
ISBN   978-0-7503-6474-4 (myPrint)
ISBN   978-0-7503-6476-8 (mobi)

DOI   10.1088/978-0-7503-6477-5

Version: 20250901

IOP ebooks

British Library Cataloguing-in-Publication Data: A catalogue record for this book is available from the British Library.

Published by IOP Publishing, wholly owned by The Institute of Physics, London

IOP Publishing, No.2 The Distillery, Glassfields, Avon Street, Bristol, BS2 0GR, UK

US Office: IOP Publishing, Inc., 190 North Independence Mall West, Suite 601, Philadelphia, PA 19106, USA

This book is dedicated to my mother Mrs Mukta Manjari Sharma and my mother-in-law Mrs Annapurna Dash for their constant support, motivation and blessings.

Contents

4 Current, resistance and electromotive force 4-1

Preface

Overview of the series

From my studenthood to date I have been in search of a series of books on College Physics that carries every concept in depth along with the scholarly problems to cater for the needs of potential students preparing for top entrance examinations. I found that students and scholars were spending hundreds of hours gathering information from different sources, websites and books on each concept and chapter. This practice still continues today. This is due to the lack of a desired college physics book series. We have a lot of books in the market; each has its own merits and limitations. Practically, it is not possible to get everything in a single book. However, a potential student expects everything in a book such as varieties of adequate problems with detailed theories and examples and solved problems.

Students are confused by the scattering and variety of sources for the problems, as well as their provenance and validity of solutions. So, not just helpless students but also teachers need concise and precise books where maximum concepts are covered systematically. The proposed series of books is a sincere effort to minimize all the above limitations of existing books and resources and maximise the potential of the books in terms of content, quality, rigor, and depth in theories, questions, and problems.

There are no such books available in the market balanced with problems, theories, quality, and quantity of problems. In this series, we have attempted to balance the books with theories, examples, and problems using a systematic approach for concept building. By virtue of my experience and expertise, as well as suggestions and recommendations of my colleagues and hundreds of my gifted students, the proposed series of books is an outcome of my strong desire to transform physics-phobia into a physics-loving attitude for the students.

Readership

At present, sophomores prepare for entrance examinations to get into premier universities like Oxford, Cambridge, etc, in the United Kingdom; MIT, Princeton, etc, in the Unites States of America. Millions of students all over the world appear for Physics Olympiads (both national and international) and other international physics examinations such as Physics GRE, etc. In India, millions of students appear for national-level examinations for prestigious Indian Institutes of Technology (IITs) through a toughest Joint Entrance Examination called IIT-JEE with the lowest success rate in the world.

Moreover, this book is also useful for students preparing for the PhD qualifying examinations of top universities and Physics GRE examinations. In India, students preparing for UPSC examinations and semester examinations and physics majors in BSc and BEng (or any branch of engineering study) can use this book as a handbook of physics problems. This book will be ultimately useful for Indian students preparing for JEE (Mains and Advanced).

To educate potential students, teachers should be strong in concepts and problemsolving ability. I strongly hope that this text could be a valuable reference book for JEE-educators in India due to its content and quality, such as familiar problems, theories, creative problems, and problems with detailed solutions along with impressive coloured diagrams throughout. I have handpicked many problems from various sources. Each problem is observed being asked in different ways in various examinations all over the world. I spent adequate time in analysing each problem's most original version and then put it in my series. Furthermore, I prefer to put all possible questions linked with the original problems in one place so that the student does not have to waste their time thinking too much and wandering.

About this book

This book contains ten chapters: (1) Electric field and potential, (2) Properties of conductors and dielectrics, (3) Capacitance, (4) Current, resistance and electromotive force (EMF), (5) Direct current circuits, (6) Magnetic field, (7) Magnetic force and energy, (8) Electromagnetic induction, (9) Inductance, and (10) Alternating current.

Chapter 1 introduces the concept of electrostatic force by Coulomb's law. The electric force defines electric field and the concept of work-energy defines electric potential. Then the calculation of electric field is done by using both Coulomb's law and Gauss' law for some known charge configurations. By using the field ideas, the calculation of electric potential is done for different charge configurations. At the end of the electrostatics, the field and potential calculation of an electric dipoles is discussed. Furthermore, I explained the interaction of a dipole with an external electric field.

In chapter 2, I define the concepts of conductors, dielectric and insulators and discuss their properties. Using these ideas of conductors and dielectric, the first passive electric circuit element, that is, *capacitance* (*C*), is explained in chapter 3. Thus, first three chapters deals with electrostatics.

Chapters 4 and 5 deal with current electricity. In chapter 4, I define the electric current. The second passive circuit element, that is, *resistance* (*R*) and the active circuit element *electromotive force* (emf) that causes a current, are explained. The relation between voltage (potential difference in rough sense) and current is established in the form of Ohm's law. In chapter 5, Kirchoff's voltage and current laws and their applications for different direct current circuits are discussed. At the end of this chapter, RC circuits and their response to DC are also discussed. Furthermore, different measuring instruments such as voltmeter, ammeter, meter bridge, potentiometer etc, are explained.

Magnetism is dealt in chapters 6 and 7. In chapter 6, magnetic field is defined and by using Biot–Savart's law and Ampere circuital law, the calculation of magnetic field is explained. In chapter 7, the interaction of a magnet or current carrying conductor with an external magnetic field or another current carrying system is discussed by Ampère's force law, in the form of force, energy etc. At the end of the magnetostatics, the field and potential calculation of a magnetic dipole is discussed.

The concept of Lorentz force is explained at the end of the chapter by taking the example of cyclotron, motion of a point charge in a uniform crossed electric and magnetic fields. Furthermore, I explained the interaction of a magnetic dipole with an external magnetic field of permanent magnets or electromagnets.

Chapter 8 deals with the generation of electricity governed by Faraday's law of electromagnetic induction and its application such as electric generators, transformers etc. In chapter 9, the third passive circuit element, that is, *inductance* (L) is defined. Calculation of self-inductance and mutual-inductance for different current carrying systems is discussed. The response of RL circuit with DC is also discussed at the end of this chapter. In chapter 10, I define the alternating current (AC) that is obtained by an alternating current generator discussed in chapter 9. I explained the response of the electric passive circuit elements such as resistance (R), capacitance (C) and inductance (L) to an AC supply in the form of reactance and impedance. Series and parallel RLC circuits are dealt with in their resonance conditions.

How to use this book

Students should first complete the given theory with the examples and then try to solve the problems in their own way. Sometimes students find better methods for solving the problems and should not rely solely on the methods provided. However, if a student falters after repeated attempts, they can refer to the given solutions. Although I tried to create error-free text, some unavoidable mistakes might appear in some subtle form each time. So there will always be scope for improvement of the content. I request my readers to critically analyze my work and give their valuable comments and suggestions for the overall improvement of this book to make it error-free in due course.

Acknowledgments

It goes without saying that a teacher is worth millions of books. An ideal teacher is built by the association of potential students. I acquired my experience and expertise by the association of gifted students of premier institutes. So, first of all, I would like to express my gratitude to the directors of all the institutes where I worked. Most illustrious are Brilliant Tutorials Private Limited, Madras (Chennai), FIIT-JEE Ltd, New-Delhi, and Narayana Group of Educational Institutions, Hyderabad.

My high school science and mathematics teacher Mr Ramesh Chandra Behera imparted an ever-lasting style of solving problems and presented the theory in a handy and student-friendly way. I am also indebted to my professors Professor Sarat Mohapatra and Professor P K Dhir for imparting their deepest knowledge of electricity and magnetism and invaluable suggestions regarding my previous books on electromagnetism which helped me to improve the standard of this book. All the attributes of my gifted teachers (many of them have left their mortal bodies) and students (who are now global leaders in their own fields) are reflected in these books. So, I express my sincere thanks to all my revered teachers (gurus) and past students.

I am thankful to Professor T Surya Kumar, Professor Bhanumati and Professor G N Subramanian of BT who was well-versed with both physics and mathematics helped me to solve the standard problems in mechanics.

In FIIT-JEE Limited, New-Delhi, I thank my seniors Er Srikant Kumar and Er P K Mishra for imparting a standard subject knowledge enhancement programme that had a great impact on this book.

I thank the late Mr Prakash Chand Bathla for offering me to write five books under his nationally leading publication house (G R Bathla & Sons). Based upon my experience as a national level author and educator, I could attain a chance to write for an international publishing house (IOPP).

While I headed the department of Physics in Narayana Group of educational institutions, I would like to thank all my previous top students who edited my books. Furthermore, I express my deepest gratitude to the principals and deans of Nayayana Group, Hyderabad especially Mr Krishna Reddy and Mr Ramalinga Reddy, with whom I worked major portion of my professional career, for giving me operational freedom, status, stability and respect. There, I could complete some theories and examples of the present series in rudimentary form. My sincere thanks and admiration to some leading physics educators such as Er Aditya Sachan, Er L N Prusty, Er Sekhar Somnath, Mr Monoj Pandey and Mr S K Singh. I am also thankful to Professor Kundal Rao and Professor Raghunath of Narayana IIT and Professor Srinivasa Chary of Sri-Mega for their suggestions and inspirations for my publishing works.

I would like to express my profound gratitude to my wife Usha in supporting and bearing me in the pandemic in 2020 when I started conceptualizing this book-series. I remain obliged to the commissioning editor of IOP Publishing Mr John Navas for his insightful comments, suggestions, and expertise, which have enhanced the rigor and depth of this work. Furthermore, I thank Mr David McDade and Phoebe Hooper, who streamlined the publication work of this book.

I express my gratitude to my Ex-publisher Mr Monoj Bathla who suggested me to accept the offer of IOPP realising the suitability of the publisher with my work.

I sincerely thank Er. Bismay Parida (Readers institute, Balasore, Odisha State) for his continuous effort in typing the manuscript in time.

I thank Professor Peter Dobson (Oxford University) who taught me how to do the things with perfection and I am applying this idea in the present publication. I am deeply indebted to Dr Benjamin Hourahine, Professor Yu Chen of University of Strathclyde for imparting a standard knowledge of nanoscience so that I could include this fastest growing field in my problem book series.

One of my notable friends is Mr Rajinder Sehra, director of S&RJ Ltd and Foot Print Media Production near Glasgow. His constant motivation for writing this series is also praiseworthy.

Furthermore, I would like to thank a potential Physics educator Mr Mithilesh for finding time to review some of the problems in my book. At last, my sincere thanks to Er K K Khandelwal (a graduate from IIT, an ex-baeurocrat and a gifted senior Physics educator) for reviewing some of the controversial problems in my book.

I especially thank my son Mr Hayagriva Sharma for looking after me during the last editing of this book.

This book would not have been possible without the collective contributions and support of all those mentioned above. Their guidance and encouragement have been instrumental in the completion of this significant milestone in my authorship. Taking this as a blessing of the Almighty, I pray for the attainment of knowledge that would be an ultimate solution to all problems of human being and other living entities.

Pradeep Kumar Sharma
20 April 2025

Author biography

Pradeep Kumar Sharma

He is a well-known physics educator in India possessing more than three decades of experience in physics education and research in training the aspirants of the joint entrance examination conducted by prestigious Indian Institutes of Technology, popularly known as IIT-JEE. Many of his students also won gold and silver medals in national and international physics Olympiads. His vast experience as a potential teacher, team leader and head of the department in some premier institutes like Brilliant tutorials, (Chennai), FIIT-JEE Ltd (New Delhi), Narayana Group (Andhra and Telangana) etc., made him extend his service as a consultant physicist to mentor both students and teachers of reputed groups in India. He has authored bestselling study materials and five books known as GRB Understanding Physics for the entrance examinations. He has been associating as a research scholar of physics education, nanoscience, metaphysics and management in some Indian and foreign universities such as Oxford University, Strathclyde University, Sofia University, Indian Institutes of Technology, Patna etc. Furthermore, he is continuing his research while affiliated with various national and international organizations such as IEEE (USA), IET (UK), IE(I), IOP(UK) etc. He has published dozens of papers in national and international journals like IEEE-Scopus journals and journals published by Institute of Physics (UK). He is currently busy in completing the problems and solutions of a series of six books which will be ready to publish very shortly. Also, he is planning to design a unique interactive study material in the mode of Active Teaching and Active Learning (ATAL) that will make the physics easier for the students to learn.

Foreword

It is my pleasure to write a foreword for this book authored by Dr Pradeep Kumar Sharma. The author of this book knew me while he was a lecturer of Physics in Brilliant Tutorials (BT), Chennai and I was a professor of Physics in Indian Institute of Technology (IIT), Madras. I also know the author for his five books—*GRB Understanding Physics*. The author gained a vast experience and expertise in training the best students for various competitive examinations such as IIT-JEE, Olympiads etc. In 2020, the author personally visited my house in Chennai and requested me to edit this book. Although I could not edit, I spent some time in reviewing the matter and gave some suggestions to improve the overall quality and content of this book.

This book is well-written and well-balanced with the theories and problems. All concepts are covered in-depth. The explanations have been presented in detail in a simple and lucid style. Most appropriate examples are given by the author based on his vast experience of teaching physics in reputed institutes across the country such as BT, FIIT-JEE etc. The book has an impressive layout and excellent quality of its production. In each chapter, the author has included systematic theories with best examples and scholarly set solved problems. This will assist both teachers and students in building and strengthening the concepts of physics.

This book will be immensely useful to all college students preparing for entrance examinations such as IIT-JEE, Physics Olympiads etc. Furthermore, this book could be useful for Physics GRE and PhD Qualifying examinations of top universities across the world. The teachers can also use this book for enhancing their subject knowledge so as to impart a better physics education. Each topic is dealt scrupulously so as to enhance the student's understanding of the subject to a great extent. I thank the author for this splendid work as the result of his hard work and determination. I wish him all the best for his forthcoming books.

Dr Jagabandhu Majhi
Professor of Physics (Retired)
Indian Institute of Technology, Madras

IOP Publishing

Problems and Solutions in Electricity and Magnetism

Pradeep Kumar Sharma

Chapter 1

Electric field and potential

1.1 Introduction

In the standard model of particle physics electromagnetic force has a second place among the four fundamental forces of nature: gravitation, electromagnetism, nuclear and weak forces. Coulomb was the first person to measure the electric force between two point charges and formulate a law for attractive and repulsive electrostatic force between the point charges. The knowledge of electricity dates back to ancient Egyptian civilization (2750 BCE) to glorify the electric fish of the Nile river as 'Thunderer of the Nile'. The Greek philosopher Thales of Miletus discovered the attraction between amber and fur. In the 17th century Gilbert, and in the 18th century Benjamin Franklin, Galvani and Volta made significant contributions in practical electricity. The major developments occurred in the 19th century with Michel Faraday, Carl Friedrich Gauss, André-Marie Ampère, Thomas Edison, James Clerk Maxwell etc.

1.2 Charge and matter

Matter is composed of atoms that has electrons, protons and neutrons. Each electron and proton carry an equal quantity of electric charge. If we call the electron negatively charged, the proton is positively charged. Since each atom has equal numbers of electrons and protons, an atom is electrically neutral. The charge of an electron is $q_e = -e = -1.6 \times 10^{-19}$ Coulomb and the charge of a proton is $q_p = +e = +1.6 \times 10^{-19}$ Coulomb. As the electrons are attracted by the nucleus by coulomb's force which is much less than the nuclear force, it is much easier to strip an electron from an atom than a proton from a nucleus. So, we can remove the electrons from an object easily by different ways such as rubbing, heating, irradiating etc. By removing an electron, neutral matter becomes electron lacking and gets a positive charge of 1.6×10^{-19} C. If uncharged matter receives an electron, it becomes electron rich and gets a negative charge of -1.6×10^{-19} C. This means that by losing or gaining electrons, matter becomes electrically charged. If a neutral

doi:10.1088/978-0-7503-6477-5ch1

body loses n electrons, its charge will be $q = +ne$; if it gains electrons, its charge will be $q = -ne$. So, a charged object has an integral multiple of electrons; this means that charge is quantized. The electrons lost by one object will be gained by the other objects. Then, the charge is conserved. An electron and its anti-particle (positron) will annihilate by emitting two gamma photons in opposite directions such that the conservation of charge holds good. Let us now tabulate the properties of charges:

1. The total charge is conserved in a closed system.
2. The total charge is quantized in a closed system.
3. The charges of an electron or proton etc, remain invariant, whereas their mass can vary with velocity as

$$m = \frac{m_0}{\sqrt{1 - \frac{v^2}{c^2}}}$$

4. Like charges repel and unlike charges attract each other.

So, *the charge is a property of matter by virtue of which it can experience an electric and magnetic force when placed in an electromagnetic field.*

Example 1 Find the (a) charge of a copper sphere of mass $m = 6.5$ kg when each copper atom loses two electrons, (b) mass lost by the copper sphere.
   Solution

(a) If $n =$ number of electrons lost per atom, the charge of an atom will be

$$q = +ne \tag{1.1}$$

In m kg of copper, the numbers of atoms is

$$N = (m/Z)N_0, \tag{1.2}$$

where $Z =$ atomic mass $= (63.5/1000)$ kg and $N_0 =$ Avogadro's number then, the total charge of the copper sphere is

$$Q = Nq \tag{1.3}$$

Using the last three equations, we have

$$Q = Ne = (mne/Z)N_0 = (6.5)(2)(1.6 \times 10^{-19})(6.02 \times 10^{23})/(63.5/1000) = 1.98 \times 10^7 \text{ C Ans.}$$

(b) The total electrons lost by the copper sphere is

$$N' = Nn \tag{1.4}$$

Using equations (1.2) and (1.4), we have

$$N' = Nn = \{(m/Z)N_0\}n = mnN_0/Z \tag{1.5}$$

If m_e = mass of each electron, the extra mass lost by the copper sphere is

$$m_{\text{extra}} = N'm_e \tag{1.6}$$

using the last two equations, we have

$$m_{\text{extra}} = N'm_e = mm_e nN_0/Z = (6.5)(9.1 \times 10^{-31})(2)(6.02 \times 10^{23})/(63.5/1000)$$
$$= 11.215 \times 10^{-5}\,\text{kg Ans.}$$

1.3 Methods of charging

1.3.1 Frictional

As we know, generally matter is neutral. Due to the electron transfer between the uncharged bodies, one will be electron lacking by losing electrons and becomes positively charged, and the other will be electron rich and negatively charged. This transfer of electrons can happen by rubbing the materials. In the process of rubbing, energy is supplied in the form of heat which will be utilized to remove the surface electrons from an object and enter into the other object. This depends upon the nature of the material. For instance, when we rub resin with glass, glass gets a positive charge and the resin becomes negatively charged. After rubbing a balloon on dry hair, combing the dry hair can produce static electricity. This process is used to charge the insulators.

1.3.2 Conduction

In this process, charge (electron) flows in the conducting wire when we connect between charged and uncharged conducting bodies. For instance, when a charged conductor having a charge, $+Q$ say, touches an uncharged conductor by a conducting wire or by direct physical touch, due to the difference in potential, the charge flows between the conductors. As a result, the neutral conductor will get charged. We will talk more about it in the next chapter. In this process a physical connection between the charged bodies is done either by a conducting wire or by just touching them against each other. This method is used to charge conductors.

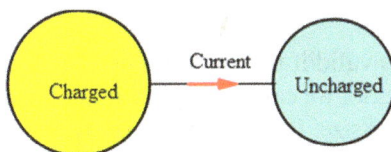

Neutral conductor is charged by the process of conduction of electrons from a charged conductor by a conducting wire.

1.3.3 Induction

This process is used for charging a conductor. This process does not require physical contact between the bodies. Let us take a neutral conductor A and a charged body B (it need not be a conductor). Let the charge of B be $+Q$. Just hold the charged body B near A. Due to the effect of electrostatic induction (as discussed in the next chapter in detail) in the neutral body A, a charge $(-q,$ say) opposite to that of the charged body appears at the side of the neutral body which is nearer to the charged body B. As body A is neutral, an equal and opposite charge $+q$ will be induced at the side of A farthest from body B. If we connect body A with Earth, due to conduction the free induced charge $+q$ will be transferred to Earth leaving behind the conductor A charged to $-q$. We will talk in detail about it in the next chapter.

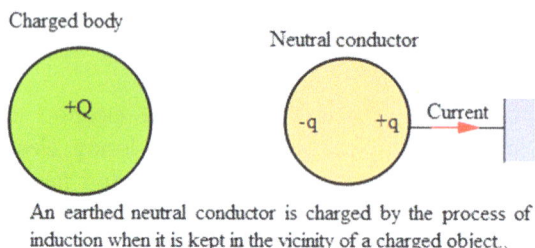

Charged body

Neutral conductor

+Q

-q   +q   Current

An earthed neutral conductor is charged by the process of induction when it is kept in the vicinity of a charged object.

1.4 Coulomb's law

In 1683, Newton published his universal law of gravitation between two-point masses. In 1770, English scientist Cavendish experimentally speculated on the action of the principle of inverse square law between two-point charges without publishing his results. In 1785, French scientist Charls Augustine De Coulomb, using torsion balance, published the papers regarding the electric force (attraction and repulsion) between point charges. The concept of Coulomb's electrostatic force was based upon the inverse square law of Newton's law of universal gravitation.

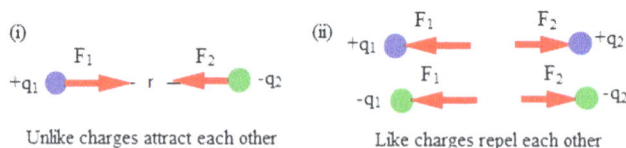

(i)

F_1 F_2

$+q_1$ r $-q_2$

Unlike charges attract each other

(ii)

F_1 F_2

$+q_1$ $+q_2$

F_1 F_2

$-q_1$ $-q_2$

Like charges repel each other

The inverse square law is valid if the charge is either point-like or its distribution possess spherical symmetry. So, we cannot apply Coulomb's law directly for charged bodies of arbitrary shape and size. If the point charges coincide, Coulomb's law does not hold good. Finally, the charge particles must be at rest to avoid the radiation of energy. If the charge particles move, a magnetic force and nonconservative induced electric force may arise apart from Coulomb's force.

Let us take a system of two-point charges. If both the charges are like (either +ve or −ve), they repel each other; if the charges are unlike (one is +ve and the other is −ve), they attract each other. The force of interaction (attraction or repulsion) between any two charged particles q_1 and q_2 varies linearly with the magnitude of charges and inversely proportional to the square of the distance between them.

According to Newton's 3rd law, each particle experiences equal and opposite force;

$$\vec{F_1} = -\vec{F_2}$$

So, $\vec{F_1}$ and $\vec{F_2}$ are action–reaction pairs. The line of action of Coulombic (electrostatic) forces $\vec{F_1}$ and $\vec{F_2}$ passes through the line joining the particles q_1 and q_2.

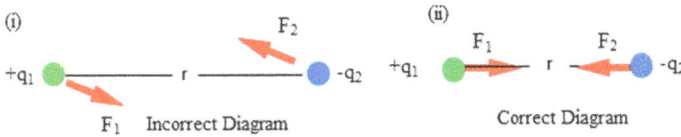

(i)

F_2

$+q_1$ —————— r —————— $-q_2$

F_1 Incorrect Diagram

(ii)

F_1 F_2

$+q_1$ — r — $-q_2$

Correct Diagram

Since, $|\vec{F}|_1 = |\vec{F_2}| = F$ (say), according to Coulomb's law

$$F \propto q_1 q_2, \text{ so also } F \propto \frac{1}{r^2}$$

Then, combining the last two expressions, we can write

$$F \propto \frac{q_1 q_2}{r^2}$$

Equalizing both sides by a constant 'K', we have

$$F = K\frac{q_1 q_2}{r^2}, \text{ where}$$

F_1 F_2

$+q_1$ — r — $-q_2$

The forces are equal, opposite and collinear (pass through the line of separation of the particles)

It was experimentally verified that the Coulomb's constant $K = 1/(4\pi\epsilon_o) = 9 \times 10^9$ Nm^2 C^{-2}; $\epsilon_o = 8.8542 \times 10^{-12}$ F m^{-1}, is known as absolute permittivity of vacuum. It can be calculated by the formula

$$\varepsilon_o \mu_o = \frac{1}{c^2}, \text{ where } \mu_o = 4\pi \times 10^{-7} \text{ H } m^{-1} \text{ and } c = \text{speed of flight in free space} = 3 \times 10^8 \text{ m s}^{-1}$$

So, the force of interaction *between* any two charge particles does not depend on any material medium and the nature of bodies. However, the force acting on each charge due to the combined effect of the other charges and the polarization of the surrounding medium will be less than that without them (in vacuum), which will be explained in later sections.

Thus, Coulomb's law states that:

Each charged particle attracts or repels the other particle along the line joining them. The electrostatic force of interaction between any two charged particles is directly proportional to the product of the magnitude of the charges and inversely proportional to the square of the distance of separation between the particles.

The Coulombic (electrostatic) force acting on q_2 due to q_1 is given as

$$\vec{F_1} = -\vec{F_2} = \frac{Kq_1q_2}{r^2}\hat{r}_{12}, \quad \text{where } \hat{r}_{21} = \frac{\vec{r}_{21}}{r}$$

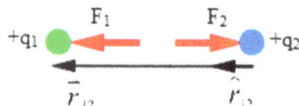

Example 2 A point charge Q is divided in two parts so that a they repel each other with a maximum force F_{max} at a given distance of separation R. Find the value of Q.

Solution
Let the charge be divided in to two parts x and $Q - x$. The coulombic force F at a given distance R is

$$F = \frac{Q(Q - x)}{4\pi\varepsilon_0 R^2} \tag{1.7}$$

For the force to be maximum at the distance R, $dF/dx = 0$; so we have

$$\frac{d}{dx}\{Q(Q - x)\} = 0 \tag{1.8}$$

$$\Rightarrow Q(2x - Q) = 0$$

Since $Q \neq 0$, we have

$$2x - Q = 0$$

This gives us

$$x = Q/2 \tag{1.9}$$

Using equations (1.8) and (1.9), we have

$$F\frac{Q(Q - Q/2)}{4\pi\varepsilon_0 R^2} \frac{Q^2}{8\pi\varepsilon_0 R^2}_{max}$$

$$\Rightarrow Q = \left(\sqrt{8\pi\varepsilon_0 F_{max}}()\right) \quad \text{Ans.}$$

1.5 Electric field and field intensity, superposition of electric field

Field idea: After the discovery of Newton's law of universal gravitation, one aspect surprised everyone, namely how can a body pull another body without touching it?

Newton himself could not give any satisfactory explanation for this which was called action-at-a distance. Long after Newton, in the third decade of the 19th century, Michael Faraday adopted the idea of field to explain the interaction between the charged particles. According to this model, every body has its own field of force. For instance, when we release a stone from any place above the Earth's surface, it accelerates. This means that at any point in the space surrounding the Earth, a force acts on an object. Hence, the space surrounding the Earth is called a region (or field or space) of a gravitational force. In short, we call it the *gravitational field*.

As discussed in the last book, *Problems and Solutions in Many-Particle Systems*, the gravitational field is the region where a gravitational force acts on an object. The gravitational force field is a vector field because 'force' is a vector. Any particle sets up its own gravitational field in the surrounding space by virtue of which it pulls the surrounding objects.

This field idea can be extended to describe the force acting on a point charge due to another charged object. The electrical field is the region where an electric (or Coulombic) force acts on a particle. Electrical force field is a vector field because 'force' is a vector. Any charged particle sets up its own electric field in the surrounding space by virtue of which it pulls or pushes the surrounding charged particles.

Field strength: As explained in the last book, to find the strength or intensity of a gravitational field of any point-object of mass M at any point, we need to put a test mass at that point. Then we measure the attraction force F_g acting on m which is directly proportional to the test mass m. But the ratio of force F_g and test mass m remains constant at a point, which is defined as gravitational field strength at the given point. As the gravitational force is always attractive, the gravitational field strength points towards the mass M.

Likewise, the electric field strength or field density due to a charge Q, say, at a given point is directly proportional to the force acting on the test charge q; a test charge is a very small positive charge. But the ratio of force and charge remains the same at the given point, which is defined as electric field strength at that point. As the electric force can be either attractive, or repulsive, the field strength due to the point charge Q either points towards or away from the charge.

Symbolically,

$$|\vec{F}| \propto m \text{ for gravitational field and } |\vec{F}| \propto q \text{ for electric field.}$$

Hence, $\frac{|\vec{F}|}{m}$ is a constant quantity at any point of a gravitational field; $\frac{|\vec{F}|}{q}$ is a constant quantity at any point of an electric field. This means that, $\frac{\vec{F}}{m} = \text{Constant} = \vec{E}$ (say) which is called the gravitational field strength; $\frac{\vec{F}}{q} = \text{constant} = \vec{E}$, is the electric field strength. We have not put any subscript for gravitational or electric field, just denoted it by a symbol $\vec{E}$.

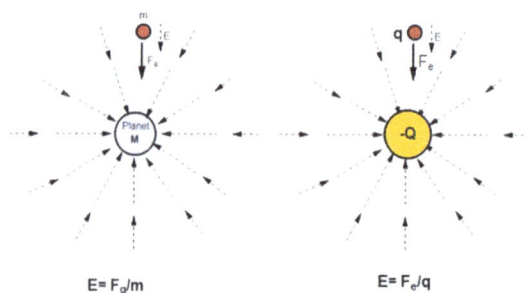

$$E = F_g/m \qquad\qquad E = F_e/q$$

It should be noted that gravitational field intensity $= \dfrac{F}{m} = \vec{a}$ (acceleration of the particle) whose unit is m s^{-2} or Newton kg^{-1} in SI units.

Hence, gravitational field strength (or intensity) at any point is equal to the acceleration of a particle placed at that point due to that gravitational field. However, electric field strength $\vec{E}$ at any point is defined as the force acting on a unit point charge 'q' placed at that point; $\vec{E} = \dfrac{F}{q}$; $\vec{E}$ is a vector quantity that points in the direction of electric force $\vec{F}$.

Superposition of $\vec{E}$: If there are many charged objects, each produces its own electric field.

Hence, the net force acting on a particle of charge q is given by

$$\vec{F} = \sum \vec{F_i},$$

where $\vec{F_i}$ is the electric force due to the ith charge of the system.

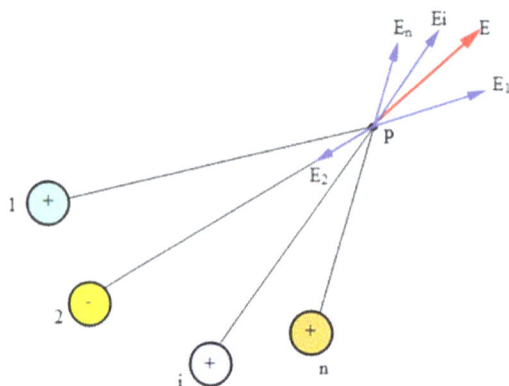

The net electric field E at any point P due to a group of point charges is equal to the vector sum or resultant of all individual electric fields. In other words, the superposition of all individual electric fields produces a net elctric field at the point P.

Dividing both sides by charge q of the particle P, we have

$$\frac{\vec{F}}{q} = \sum \frac{\vec{F_i}}{q}$$

1-8

Substituting $\frac{\vec{F_i}}{q} = \sum \vec{E_i}$ (field strength due to the ith system at P), we have

$\frac{\vec{F}}{q} = \sum \vec{E_i}$, where $\frac{\vec{F}}{q} =$ Total (net) field strength at $P = \vec{E}$ (say)

Then, the net field is

$$\vec{E} = \sum \vec{E_i}$$

The net electric field strength at any point is equal to the vector sum of the field strengths contributed by all charged bodies.

This is known as the principle of superposition of electric field.

1.6 Calculation of electric field intensity

In this section we will find the strength of electric field of different types of charge distribution. For this purpose, we need to use the basic formula

$$\vec{E} = \sum \frac{\vec{F}}{q},$$

where $\vec{F} =$ force acting as a test charge q placed at the point where we want to measure the field intensity.

Point charge: Let us take a point charge Q at O, (say) and try to find its electric field strength at a point P, at a distance r, say, from O. For this, we need to put a test charge q at P. The force acting on q due to Q is given as

$$\vec{F} = -\frac{KQq}{r^2}\hat{r} \tag{1.10}$$

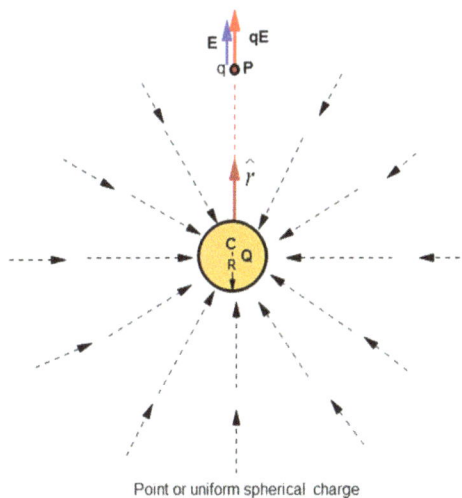

Point or uniform spherical charge

The strength of electric field of charge Q at P is given as

$$\vec{E} = \frac{\vec{F}}{q} \tag{1.11}$$

Substituting $\vec{F}$ from equation (1.10) in equation (1.11), we have

$$\vec{E} = \frac{\frac{KQq}{r^2}\hat{r}}{q}$$

$$\Rightarrow \vec{E} = \frac{KQ}{r^2}\hat{r}$$

The above equation tells us that

The magnitude of strength of electric field due to a point charge Q at point P is directly proportional to the charge q and inversely proportional to the square of distance of separation r between the point charge and the point P where you want to find the field strength. The direction of the field strength $\vec{E}$ is given by '$\hat{r}$', which tells us that $\vec{E}$ points along the line joining the point charge and the point P under consideration. If the charge Q is positive, it points radially outward and if the charge is negative, field strength points radially inward. But the gravitational field intensity is always radially inward due to gravitational attraction.

Discrete charge distribution: Now you can use expression $\vec{E} = \frac{KQ}{r^2}\hat{r}$ as a basic formula for field strength due to a point charge. Let us use it in the following example of discrete charge distribution.

Example 3 Two identical particles each of charge $+Q$ are separated by a distance $2R$. Find the electric field strength at P situated at a distance from O on the perpendicular bisector OP.

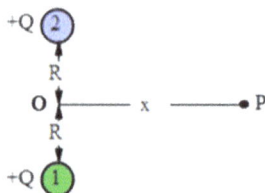

Solution

Since, P is situated at equal distance from both particles, the magnitude of $\vec{E}$, that is, $|\vec{E}|$ (or E) is the same for both particles, at P. If $\vec{E_1}$ and $\vec{E_2}$ are the field strengths due to the particles 1 and 2 at P, respectively, we have

$$E_1 = E_2 = \frac{KQ}{r^2} \tag{1.12}$$

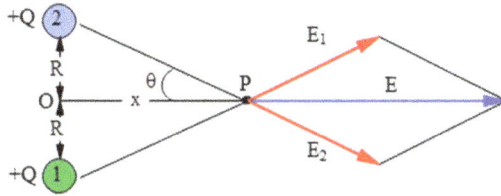

Resolving $\vec{E_1}$ and $\vec{E_2}$ along x- and y-axis, we have

$$E_x = E_1 \cos\theta + E_2 \cos\theta \tag{1.13}$$

$$E_y = E_1 \sin\theta - E_2 \sin\theta \tag{1.14}$$

Using equations (1.12), (1.13) and (1.14), the magnitude of total field intensity is,

$$E = E_x = \frac{2KQ}{r^2} \cos\theta,$$

where $\cos\theta = \frac{x}{r} = \frac{x}{\sqrt{R^2 + x^2}}$ and $r = \sqrt{R^2 + x^2}$

$$\Rightarrow E = \frac{2KQx}{(R^2 + x^2)^{\frac{3}{2}}} \text{ Ans.}$$

N.B: By putting $dE/dx = 0$, you can find that at $x =$ value of $x = \frac{R}{\sqrt{2}}$ E is maximum and putting this value, we have maximum value of $E = \frac{4KQ}{3\sqrt{3}R^2}$. From the above example let us conclude the following points:

1. Since $E_y = 0$, the net field strength is directed away from the mid-point O (of the line joining the particles) along the x-axis. This means that a negatively charged particle placed at any point on the perpendicular bisector (x-axis) will accelerate towards the origin possessing a stable equilibrium. But a positive point charge will accelerate away from O along the x-axis, possessing an unstable equilibrium. Hence, the origin is called 'stable equilibrium position' for a negative charge but unstable equilibrium for a positive point charge q. We will show that the negative point charge will oscillate simple harmonically along the x-axis, about O. In other words, v is maximum and a will be zero at O, which is the condition of stable equilibrium.
2. You can also show that a +ve point charge experiences a stable equilibrium and a −ve point charge experiences an unstable equilibrium at O along the y-axis.
3. Since, $|\vec{E}| = \frac{2KQx}{(R^2 + x^2)^{\frac{3}{2}}}$, when $x = 0$, $E = 0$; when $x \to \infty$, $E \to 0$. This means that the modulus or magnitude of E attains a maximum value at $x = \frac{R}{\sqrt{2}}$ because $\frac{dE}{dx} = 0$ and $\frac{d^2E}{dx^2} < 0$.

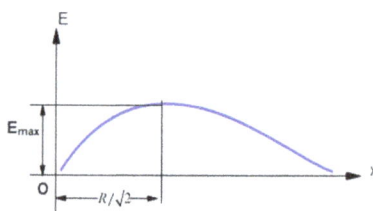

4. Referring to the above equation, the graph $E = f(x)$ is an asymptote.

Now, by using the formula $E = \frac{KQ}{r^2}$ for a point charge, let us find the field intensity due to continuous charge distribution.

Continuous charge distribution: In continuous mass distribution (extended object), we cannot distinguish one particle from the others because the particles are distributed continuously. In this case, take an element of mass dq in the extended object. It behaves as a point charge. Then using the basic formula, the field intensity $\overrightarrow{dE}$ due to the point charge at the given point P is

$$d\,\overrightarrow{E} = \frac{Kdq}{r^2}\hat{r}$$

Here, you need to check the directions of field intensities due to other elements of the body. If all elementary field intensities have the same direction, we can sum up (integrate) $\overrightarrow{dE}$ directly to obtain the net field intensity, which can be given by $\overrightarrow{E} = \hat{r}K\int\frac{dq}{r^2}$

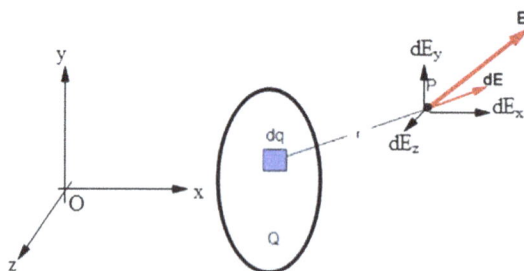

In volume charge distribution, take an elementary charge dq, find its electric field dE, resolve it in to componets dE_x, dE_y and dE_z; then integrate them to get the net electric field in x,y and z directions. The resultant of these components is equal to the net field at the given point P.

If the directions of field intensities at P are different for different elements of the body, we must resolve dE into components dE_x, dE_y and dE_z along x-, y- and z-axes, respectively. Then we integrate the corresponding elementary field intensities to obtain E_x, E_y and E_z. Hence, the net field intensity of the body at P can be given as

$$\overrightarrow{E} = E_x\hat{i} + E_y\hat{j} + E_z\hat{k}$$

In each case, dq must be expressed in terms of r as follows. For linear charge, the linear charge density at any point P can be given as $\lambda = \frac{dq}{dx}$,

$$\Rightarrow dq = \lambda \, dx, \quad \text{where} \quad \lambda = f(x)$$

For surface (area) charge distribution, areal charge density is given as

$$\sigma = \frac{dq}{dA}$$

$$\Rightarrow dq = \sigma \, dA, \quad \text{where} \quad \sigma = f(r)$$

Similarly, for volume charge distribution volume charge density is given as

$$\rho = \frac{dq}{dV}$$

$$\Rightarrow dq = \rho dV, \quad \text{where} \quad \rho = f(r)$$

In the case-to-case basis, we can express dq in terms of r (or any linear dimension)
Let us summarize the above explanation:
To find $\vec{E}$ for an extended object
1. Take an elementary charge dq
2. Find the field intensity $d\vec{E}$ due to dq

$$d\vec{E} = \frac{Kdq}{r^2}\hat{r}$$

3. Resolve $d\vec{E}$ along x-, y- and z-axes to find dE_x, dE_y and dE_z, respectively.
4. Express, $dq = \lambda \, dx$ (for linear charge distribution)

$$= \sigma dA \text{ (for areal charge distribution)}$$

$$= \rho dV \text{ (volume charge distribution)}$$

5. Use the given relations of mass densities λ, σ and ρ as the function of (linear dimensions);

$$\lambda = f(x), \quad \sigma = f(r) \quad \text{and} \quad \rho = f(r)$$

6. Integrating dE_x, dE_y and dE_z, obtain E_x, E_y and E_z, respectively.
7. Finally, the net field intensity is given as: $\vec{E} = E_x\hat{i} + E_y\hat{j} + E_z\hat{k}$

Now you can use the above ideas in finding field intensities of linear, surface and volume charge distribution through the following examples.
   Linear charge distribution: Under this type, we have wires (straight and bent). Let us take the example of a straight thin wire.

Example 4 Find an expression for field intensity due to the finite straight wire of charge Q and length l, at the point P. Assume that the charge is uniformly distributed.

Solution

Let us take an elementary charge dq at a distance r from the point P. The field intensity of charge dq at P is

$$d\vec{E} = \frac{Kdq}{r^2}\hat{i}, \tag{1.15}$$

where

$$dq = \lambda\,dr \tag{1.16}$$

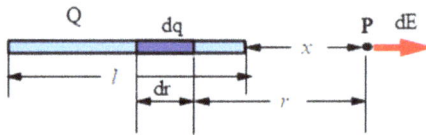

Since, the field intensities due to each element of the straight wire have the same direction (towards left), we can directly integrate dE.

Then, we have

$$\vec{E} = \int d\vec{E} \tag{1.17}$$

Using equations (1.15), (1.16) and (1.17), we have

$$\vec{E} = K\hat{i}\int\frac{\lambda\,dr}{r^2}$$

Since the wire is uniform, λ is constant. Hence, you can take λ out of the integral to obtain

$$\vec{E} = -K\lambda\hat{i}\int\frac{dr}{r^2}$$

As the wire extends from x to $x + l$, we have

$$\vec{E} = K\lambda\hat{i}\int_x^{x+l}\frac{dx}{r^2} = K\lambda\hat{i}\left[\frac{1}{r}\right]_x^{x+l}$$

$$= k\lambda\hat{i}\left(\frac{1}{x} - \frac{1}{x+l}\right) = \frac{k\lambda l}{x(x+l)}\hat{i}\ \text{Ans.}$$

In the previous example, we found E at a familiar position. Let us derive a general expression of E due to finite wire in the following example.

Example 5 (Uniformly charged straight thin rod) Find the electric field strength due to a uniform thin wire AB of linear charge density λ, at a point P which is situated at a perpendicular distance x from the wire. The point P subtends internal angles θ_1 and θ_2 at the ends of the wire as shown in the figure.

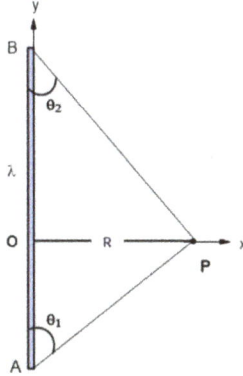

Solution

Drop perpendicular from P onto the wire. The point of intersection O of the perpendicular OX and the wire is assumed as the origin. Now, let us take an element of mass dq of the wire at distance y from the origin. If the elementary length of dq is dy,

$$dq = \lambda \, dy$$

Then, the field intensity of dq at P is,

$$dE = \frac{Kdq}{r^2}$$

Substituting $dq = \lambda \, dy$, we have

$$dE = K\lambda \frac{dy}{r^2}$$

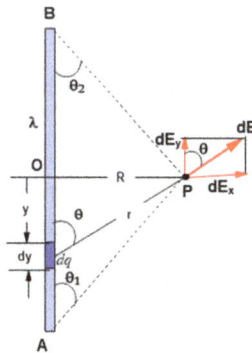

Since, the direction of $\overrightarrow{dE}$-differs from element to element, we need to resolve dE along the x- and y-axes to obtain

$$dE_x = dE \sin \theta \ \text{ and } \ dE_y = dE \cos \theta$$

Then, the net field intensities in the x- and y-directions are given as:

$$E_x \int dE \sin \theta \ \text{ and } \ E_y = \int dE \cos \theta,$$

where $dE = K\lambda \frac{dy}{r^2}$. $\sin \theta = \frac{R}{r}$ and $\cos \theta = \frac{y}{r}$

Then we obtain the following two equations

$$E_x = K\lambda x \int \frac{dy}{r^3} \ \text{ and } \ E_y = K\lambda \int \frac{y\,dy}{r^3}$$

Let us now evaluate the integrals by putting

$$y = R \cot \theta, \ r = R \operatorname{cosec} \theta$$

and

$$dy = -R \operatorname{cosec}^2 \theta d\theta$$

Then, we have

$$E_x = -\frac{K\lambda}{R} \int_{\theta_1}^{\pi-\theta_2} \sin \theta \ d\theta,$$

where the upper limit of the angle is given as $\theta = \pi - \theta_2$ because $\theta = $ angle between $+y$-direction and the position vector of the point P.

Then, we have

$$E_x = \frac{K\lambda}{R}(\cos \theta_1 + \cos \theta_2) \quad \text{Ans.}$$

Similarly, we have

$$E_y = -\frac{K\lambda}{R} \int_{\theta_1}^{(\pi-\theta_2)} \cos \theta \ d\theta$$

This gives

$$E_y = \frac{K\lambda}{R}(\sin \theta_2 - \sin \theta_1) \quad \text{Ans.}$$

N.B: Student task

    1. At any point on the perpendicular bisector of the wire at a distance x, by putting $\theta_1 = \theta_2$, in the following expression

$$E_x = \frac{K\lambda}{R}(\cos \theta_1 + \cos \theta_2),$$

we have

$$E = E_x = \frac{2K\theta}{R} \cos \theta,$$

where $x = R$ and

$$\cos \theta = \frac{l/2}{\sqrt{R^2 + (l/2)^2}}$$

Putting $\lambda l = Q$, we have

$$E = \frac{2KQ}{R^2\sqrt{4R^2 + l^2}}$$

2. For infinite long straight wire, putting $\theta_1 = \theta_2 = 0$ in the expression

$$E_x = \frac{K\lambda}{R}(\cos \theta_1 + \cos \theta_2),$$

we have

$$E_x = \frac{2K\lambda}{R} \leftarrow \quad \text{and} \quad E_y = 0$$

3. In general, if the coordinates of A, B and P are known from the given set, the corresponding angles θ_1 and θ_2 are given as $\sin \theta_1 = \frac{R}{\sqrt{R^2 + y^2}}$, $\cos \theta_1 = \frac{y}{\sqrt{R^2 + y^2}}$, $\sin \theta_2 = \frac{R}{\sqrt{R^2 + (l-y)}}$ and $\cos \theta_2 = \frac{l-y}{\sqrt{R^2 + (l-y)^2}}$

Then substituting the values of θ_1 and θ_2, we can find the E_x and E_y. The direction of E_x is always directed towards the wire AB along OP; the direction of E_y is positive (↑) when $\sin \theta_2 > \sin \theta_1$ (or $\theta_2 > \theta_1$) and vice versa. When $\theta_1 = \theta_2$ in the case of perpendicular bisector, $E_y = 0$.

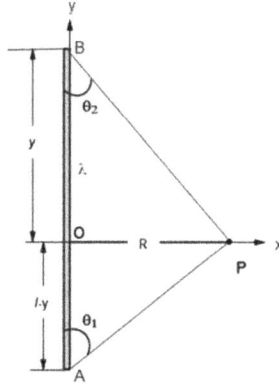

Let us now bend the wire in the form of a circular arc and try to find the field intensity at the centre curvature of the circle in the following example.

Example 6 (Uniformly charged circular arc) Find the electric field intensity of a thin circular arc having linear charge density λ, at the centre C of the curvature. The arc subtends an angle β at the centre C.

Solution

Let us take an element of charge dq at an angle θ. The field intensity of dq at C is, $dE = \frac{K\,dq}{R^2}$, where $dq = \lambda\,dl$

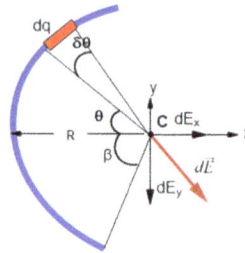

Substituting $dl = R\,d\theta$, we have

$$dE = \frac{K\lambda}{R}d\theta \tag{1.18}$$

Since, the angle θ of $d\vec{E}$ changes, we need to resolve $d\vec{E}$ in the x- and y-axes to obtain

$$dE_x = dE\cos\theta \quad \text{and} \quad dE_y = dE\sin\theta$$

As the arc is symmetrical about the x-axis, the strength of field of the upper and lower halves of the arc get cancelled. Hence, the net field intensity is directed along the x-axis (towards left)

Then, we have

$$E_{\text{net}} = E_x \int dE_x,$$

where $dE_x = dE\cos\theta$

$$\Rightarrow E_{\text{net}} = \int dE\cos\theta \tag{1.19}$$

Using equations (1.18) and (1.19), we have

$$E_{\text{net}} = \frac{K\lambda}{R}\int_{-\beta/2}^{\beta/2}\cos\theta\,d\theta$$

$$= \frac{2K\lambda}{R}\int_{0}^{\beta/2}\cos\theta\,d\theta$$

$$\Rightarrow E_{net} = \frac{2K\lambda}{R} \frac{\sin \frac{\beta}{2}}{\beta} \quad \text{Ans.}$$

N.B:

1. Using the formula $E_{net} = \frac{2K\lambda}{R} \frac{\sin \frac{\beta}{2}}{\beta}$, find the field strength (intensity) of a semi-circular wire of charge q and length l. can be found as

$$E = \frac{2Kq}{l^2}$$

2. In the above example, if you write $E = \int dE$, where $dE = \frac{K\lambda}{R} d\theta$ you will get $E = \frac{K\lambda\beta}{R}$ which will lead to a wrong result. Since, the directions of $\overrightarrow{dE}$ are different for different elements of the arc, you cannot write $|\overrightarrow{E}| = \int |\overrightarrow{dE}|$. Hence, we need to resolve $\overrightarrow{dE}$ into components and then integrate them.

3. In the formula $E = \frac{4K}{R\beta} \sin \frac{\beta}{2}$, if we put $\beta = 0$, we will get $E = 0$. This means that the field strength due to a circular wire at its centre is zero.

Let us now find the field strength at the axial points of a circular wire in the following example.

Example 7 (Uniformly charged ring) Find the expression for the field intensity at any point on the axis of a circular wire of charge Q and radius R.

Solution

The field due to the element of charge dq is

$$dE = \frac{Kdq}{r^2} \tag{1.20}$$

The axial component of dE is

$$dE_x = dE \cos \theta$$

Integrating the axial component, we have

$$E_x = \int dE_x = \int dE \cos \theta \tag{1.21}$$

Using equations (1.20) and (1.21), we have $E_x = K \int \frac{dq}{r^2} \cos \theta$

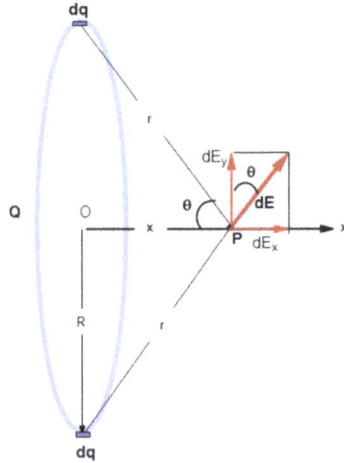

Since, each element has same distance r from P, we can take the constant quantities $\cos\theta$ and r out of the integral.

Then, we have

$$E_x = \frac{K\cos\theta}{r^2}\int dq \qquad (1.22)$$

Since, we integrate the field of all elements, we can write

$$\int dq = Q,$$

where Q = mass of the ring.

Substituting $\int dq = Q$ in equation (1.22), we have

$$\cos\theta = \frac{x}{\sqrt{R^2 + x^2}} \quad \text{and} \quad r = \sqrt{R^2 + x^2}$$

Then, we have

$$E_x = \frac{KQx}{(R^2 + x^2)^{3/2}}$$

The radial component of dE is

$$dE_r = dE\sin\theta$$

Since, the ring is symmetrical about the axis, the net radial field intensity is zero.

Hence, the net field is axially directed towards the centre of the circular ring, whose magnitude is given as:

$$E_{net} = E_x = \frac{KQx}{(R^2 + x^2)^{3/2}} \quad \text{Ans.}$$

Since $E_{net} = 0$ when $x = 0$ and $E \to \infty$ when $E \to \infty$, the graph of E versus x is an asymptote. E is maximum when $x = \dfrac{R}{\sqrt{2}}$

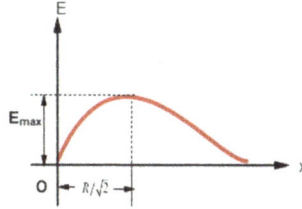

Surface charge distribution: Using the formula for field intensity of a ring, we can find the field intensity at the axial points of a disc because the disc is a combination of thin concentric rings.

Example 8 (Uniformly charged disc) Derive an expression of electric field strength of a uniform disc of radius R, surface charge density σ, at an axial point situated at a distance x from the centre of the disc.

Solution

Let us take a thin ring of radius r and thickness dr. The mass of the ring can be given as

$$q_{ring} = \sigma dA,$$

where dA = area of the strip (shaded portion) of the ring = $2\pi r dr$ and σ = surface mass density.

Hence,

$$q_{ring} = 2\pi \sigma r \, dr \tag{1.23}$$

As we derived in the previous example, the field due to the ring is,

$$E_{ring} = \frac{K q_{ring} x}{(r^2 + x^2)^{3/2}} \tag{1.24}$$

Substituting q_{ring} from equation (1.23) in equation (1.24), we have

$$E_{ring} = 2\pi \sigma K x \frac{r \, dr}{(r^2 + x^2)^{3/2}} \tag{1.25}$$

Since the disc is a combination of concentric rings of radii ranging from $r = 0$ to $r = R$ and field due to each ring has same direction (towards O) the field due to the disc can be given by;

$$E_{disc} = \int E_{ring} \tag{1.26}$$

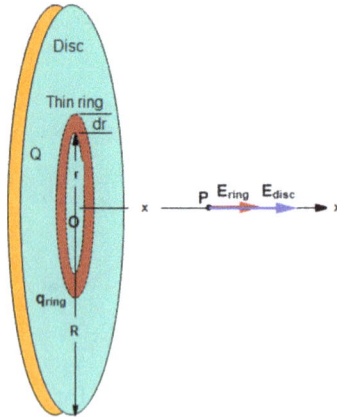

Substituting E_{ring} from equation (1.25) in equation (1.26), we have

$$E_{\text{disc}} = 2\pi\sigma Kx \int_0^R \frac{r\,dr}{(r^2 + x^2)^{3/2}} \Rightarrow E_{\text{disc}} = 2\pi\sigma K\left(1 - \frac{x}{\sqrt{R^2 + x^2}}\right)$$

$$\Rightarrow E_{\text{disc}} = \frac{\sigma}{2\varepsilon_o}\left(1 - \frac{x}{\sqrt{R^2 + x^2}}\right) \quad \text{Ans.}$$

N.B:

1. In the foregoing example, we cannot take σ out of the integral if it is a function of radial distance r, that is, non-uniform surface charge distribution.
2. In the function $E = 2\pi\sigma K(1 - \frac{x}{\sqrt{R^2 + x^2}})$ if we put $x = 0$, we get $E = 2\pi\sigma K$. If $x \to 0$, $E \to 0$. This means that E is nearly uniform at very close points of a finite disc.

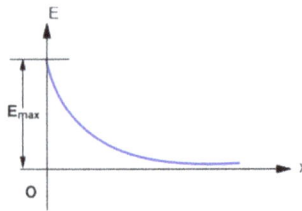

3. If we put $R \to \infty$, we will get $E = 2\pi\sigma K$. This tells us that $E = $ constant at any point in space, due to a large sheet.

Let us now take a closed surface and field intensity.

Example 9 (Thin spherical shell) Derive an expression for intensity due to a thin sphere of charge q and radius R.

Solution

Let us take a point P inside the spherical shell. Let the elementary patches of areas dA_1 and dA_2 at both sides of the point P subtend a solid angle $d\Omega$ at P.

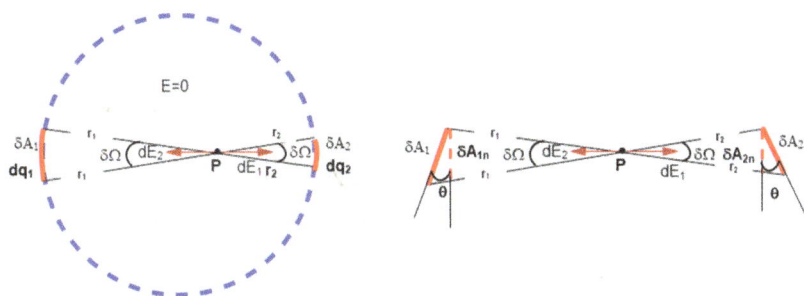

The charges of the elementary patches are $dq_1 = \sigma dA_1$ and $dq_2 = \sigma dA_2$. The fields due to the elements at P are

$$dE_1 = \frac{Kdq_1}{r_1^2} = \frac{K\sigma dA_1}{r_1^2}$$

and

$$dE_2 = \frac{Kdq_2}{r_2^2} = \frac{K\sigma dA_2}{r_2^2}$$

Since, $d\vec{E_1}$ and $d\vec{E_2}$ oppose each other, the net field is,

$$dE = |dE_1 - dE_2| = \left|\frac{K\sigma dA_1}{r_1^2} - \frac{K\sigma dA_2}{r_2^2}\right| = K\sigma\left|\frac{dA_1}{r_1^2} - \frac{dA_2}{r_2^2}\right| \qquad (1.27)$$

Since, $\theta_1 = \theta_2 (=\theta$, say) by the properties of tangents at the ends of a chord, where θ_1 and θ_2 are the angles between the surface elements dA_1 and dA_2 and the corresponding perpendiculars to the chords, we can write

$$dA_{1n} = dA_1 \cos\theta_1 = dA_1 \cos\theta \text{ and } dA_{2n} = dA_2 \cos\theta_2 = dA_2 \cos\theta.$$

where dA_{1n} and dA_{2n} are the areas perpendicular to the line passing through P.

Then, substituting $dA_1 = \frac{dA_{1n}}{\cos\theta}$ and $dA_2 = \frac{dA_{2n}}{\cos\theta}$ in equation (i), we have

$$dE = \frac{K\sigma}{\cos\theta}\left(\frac{dA_{1n}}{r_1^2} - \frac{dA_{2n}}{r_2^2}\right)$$

We know that the opposite solid angles are equal, given as

$$\frac{dA_{1n}}{r_1^2} = \frac{dA_{2n}}{r_2^2} = d\Omega \text{ (solid angle) according to solid geometry.}$$

Using the last two equations, we have

$$dE = 0$$

This means that the net electrostatic field intensity inside a thin spherical shell of uniform charge distribution is zero.

In other words,

The electrostatic field strength inside a thin uniformly charged spherical shell is zero.

The field strength at any point outside the spherical shell situated at a radial distance r can be calculated as follows.

Take a thin ring at an angular position θ. The charge of the ring is

$$q_r = \sigma dA,$$

where dA = area of the strip of the ring

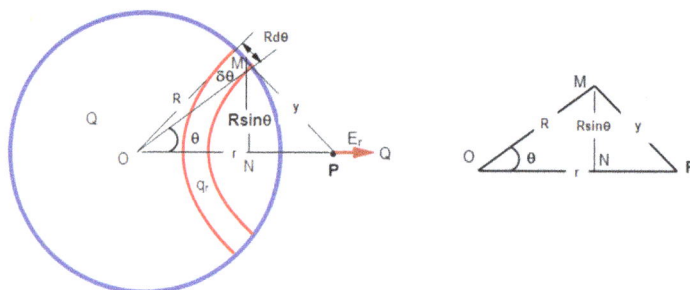

$$= 2\pi(R \sin \theta)(Rd\theta) = 2\pi R^2 \sin \theta d\theta$$

This gives

$$q_r = 2\pi\sigma R^2 \sin \theta d\theta \qquad (1.28)$$

Then, the field due to the ring at P is,

$$E_r = \frac{Kq_r x}{y^3} \qquad (1.29)$$

In the $\triangle OMP$, we have

$$y^2 = r^2 + R^2 - 2rR \cos \theta \qquad (1.30)$$

Differentiating equation (1.30), we have

$$y\, dy = rR \sin \theta d\theta \qquad (1.31)$$

In $\triangle MNP$ $x = r - R \cos \theta$, where $R \cos \theta = \frac{-y^2 + R^2 + r^2}{2r}$ from equation (1.30)

This gives

$$x = \frac{r^2 - R^2 + y^2}{2r} \qquad (1.32)$$

Substituting q_r from equation (1.28), $\sin\theta d\theta$ from equation (1.31) and x from equation (1.32), in equation (1.29) to obtain

$$E_r = \frac{\pi\sigma KR}{r^2}\left(\frac{r^2 - R^2}{y^2} + 1\right)dy$$

Finally, integrating E_r obtain

$$E_{\text{sph}} = \int E_{\text{ring}}$$

Substituting E_r, we have

$$E_{\text{sph}} = \frac{\pi\sigma KR}{r^2}\int_{r-R}^{r+r}\left(\frac{r^2 - R^2}{y^2} + 1\right)dy$$

After evaluating the integration, we have

$$E_{\text{sph}} = \frac{4\pi K\sigma R^2}{r^2},$$

where $4\pi R^2\sigma = Q$ (charge of the shell)

$$\Rightarrow E_{\text{sph}} = \frac{KQ}{r^2}.$$

N.B: We note the following points from the foregoing discussions.
1. $E = 0$; $r < R$; field inside the shell is zero.
2. $E = \frac{KQ}{r^2}$; $r \geqslant R$; field varies obeying inverse square law, outside the shell.

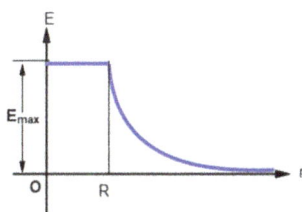

3. The E–r graph tells us that E decreases from $\frac{KQ}{R^2}$ to zero obeying the inverse square law.
4. Since, $E = \frac{KQ}{r^2}$ for any outside point of a symmetrical spherical charge distribution, it behaves as a point charge. However, for any other charge distribution (linear, areal and other non-uniform distribution of charge), we cannot substitute the object by a point charge.
5. For any inside point P of the spherical shell, we have

$$E_{\text{sph}} = \frac{\pi\sigma KR}{r^2}\int_{R-r}^{R+r}\left(\frac{r^2 - R^2}{y^2} + 1\right) = 0$$

Volume charge distribution: Let us now use the expression of field intensity due to a thin spherical shell to find the field due to a solid sphere.

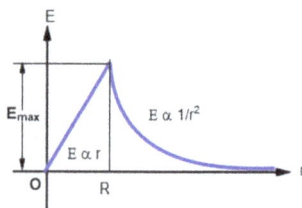

Example 10 Derive an expression for the field strength due to a uniform solid sphere of charge q and radius R.

Solution

Let us take a point P at a radial distance r from the centre O of the sphere. Draw a sphere of radius r passing through P having its centre at O, shown as a shaded region in the figure below.

Since, point P is lying inside hollow sphere 2, the field at P due to the hollow sphere is zero. Then, the field at P is solely due to the solid sphere 1. Since, point P is situated just outside the uniform sphere 1, using the concept of spherical symmetry, field intensity due to the sphere at P can be given by

$$E = \frac{Kq}{r^2},$$

where q = charge of the sphere of radius r given as

$$q = \frac{4}{3}\pi r^3 \rho; \quad \rho = \text{volume charge density.}$$

Then, we have

$$E = \frac{4}{3}K\pi\rho r; \quad r \leqslant R \text{ Ans.}$$

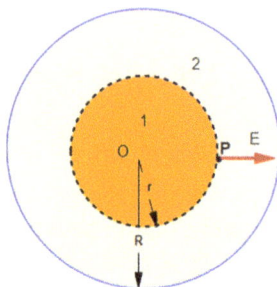

If we consider the point outside or at the surface of the sphere, following the concept of spherical symmetry, we have

$$E = \frac{KQ}{r^2}; r \geqslant R,$$

where Q = total charge of the sphere. Ans.

1.7 Work done by an electric field

We know that electrostatic field is conservative as it obeys inverse square law. If we release a particle of charge q at a point P in an electrostatic field, the electrostatic force acting on the particle is,

$$\vec{F_{el}} = q\vec{E},$$

where $\vec{E}$ = electrostatic field strength at P. If the particle moves in any arbitrary curve, for an elementary displacement $\vec{ds}$, the work done by gravity is

$$dW_{el} = \vec{F_{el}} \cdot d\vec{s},$$

where

$$\vec{F_{el}} = q\vec{E}.$$

This gives

$$dW_{el} = q\vec{E} \cdot d\vec{s}$$

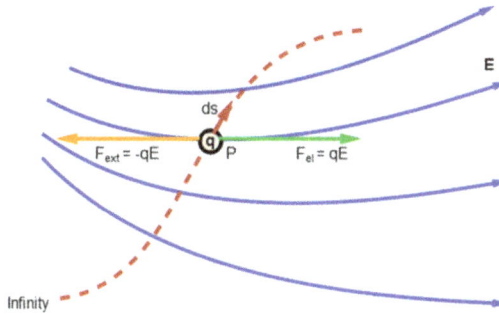

When the particle undergoes displacement from point 1 to point 2, integrating the elementary work done, the total work done by the electric field is,

$$W_{el} = \int dW_{el},$$

where $dW_{el} = q\vec{E} \cdot d\vec{s}$ -

$$\Rightarrow W_{el} = \int_1^2 q\vec{E} \cdot d\vec{s}$$

$$\Rightarrow W_{el} = q \int_1^2 \vec{E} \cdot d\vec{s}$$

Example 11 A fixed sphere of radius R and charge Q is uniformly distributed throughout its volume. Find the work done by the electric field of the sphere when a particle of charge q is displaced from the centre of the sphere through a diametrical chute to the (a) surface of the sphere (b) infinity. Assume $R =$ radius of the sphere.

Solution

(a) As we have derived earlier,

$$E = \frac{KQ}{R^3}r; \quad r \leqslant R$$

Hence, the work done by gravity is,

$$W_{el} = q \int_0^R \vec{E} \cdot d\vec{s},$$

where $\vec{E} = \frac{KQ}{R^3}\hat{r}$ and $d\vec{s} = dr\,\hat{r}$.

$$\Rightarrow W_{gr} = \frac{KQq}{R^3} \int_0^R r\,dr$$

After evaluating the integral, we have

$$W_{gr} = \frac{KQq}{2R} \quad \text{Ans.}$$

(b) As we have derived earlier,

$$E = \frac{KQ}{r^2}; \quad r \geqslant R$$

The work done by the external agent in slowly removing the particle from surface to infinity is

$$W'_{el} = \int_R^\infty E\,dr = \int_R^\infty \frac{KQq}{r^2}\,dr$$

$$= \frac{KQq}{R}$$

Adding last two works done, the work done by the external agent in slowly removing the particle from the centre to infinity is

$$W_{el} = \frac{KQq}{2R} + \frac{KQq}{R} = \frac{3KQq}{2R}$$

1.8 Electrostatic potential energy between two particles

You know that the potential energy of interaction between the two particles of masses m_1 and m_2 is

$$U = -\frac{Gm_1m_2}{r}$$

You can transform by writing $G = K$, $m_1 = q_1$, $m_2 = q_2$ and change the $-$ve sign with positive sign because here the electric force is repulsion between two positive charges, whereas gravitational force is attraction. So, for a system of two charged particles of masses q_1 and q_2 separated by a distance r, the potential energy of interaction between them is given as:

$$U = \frac{Kq_1q_2}{r}$$

But the above process is not the derivation. Let us start deriving from the basics. For the sake of simplicity, let us assume that $q_1 = Q$ and $q_2 = q$ and place Q fixed and move the charge q from infinity slowly in any arbitrary path (shown as a dotted line). While doing so, the electrostatic force will push the charge q radially away while the arbitrary path of shifting the charge q need not be radial. Then, any elementary displacement can have one radial and one normal component.

$$\vec{ds} = \vec{ds_n} + \vec{ds_r}$$

While shifting the charge q from infinity to point P, the work done by the electric field is given as

$$W_{el} = q \int_{\infty}^{P} \vec{E} \cdot d\vec{s}$$

Using the last two equations, we have

$$W_{el} = q \int_{\infty}^{P} \vec{E} \cdot d\vec{s} = q \int_{\infty}^{P} \vec{E} \cdot (d\vec{s_n} + d\vec{s_r}) = q \int_{\infty}^{P} \vec{E} \cdot d\vec{s_n} + q \int_{\infty}^{P} \vec{E} \cdot d\vec{s_r}$$

Since $\vec{E}$ is perpendicular to $d\vec{s_n}$, $\vec{E} \cdot d\vec{s_n} = 0$; so, we have

$$W_{el} = q \int_{\infty}^{P} \vec{E} \cdot d\vec{s_r}$$

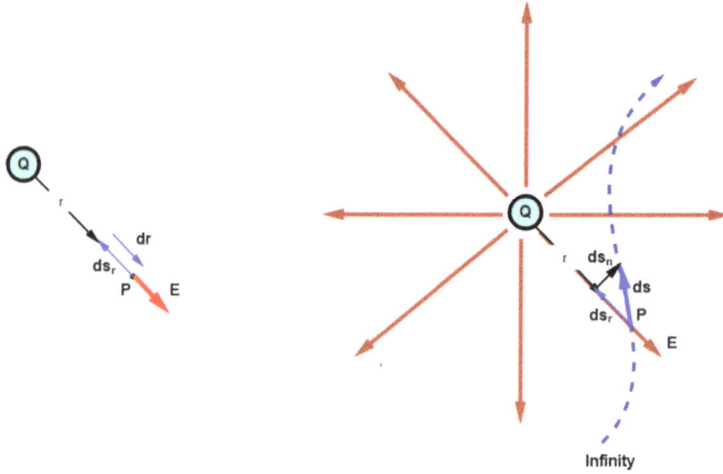

Infinity

Since $\vec{E}$ and $\vec{ds_r}$ are opposite, we can write

$$W_{el} = -q \int_{\infty}^{P} E ds_r$$

Since the radius vector decreases as we are coming from infinity towards with the charge q, we can write

$$ds_r = -dr$$

$$W_{el} = q \int_{\infty}^{P} E dr$$

Now putting the position of P as r, we have

$$W_{el} = q \int_{\infty}^{r} E.dr$$

Putting $E = \frac{KQ}{r^2}$, we have

$$W_{el} = q \int_{\infty}^{r} \frac{KQ}{r^2} \cdot dr = KQq \int_{\infty}^{r} \frac{dr}{r^2}$$

Integrating, we have

$$W_{el} = -\frac{KQq}{r}$$

According to work–energy theorem, the work done by a conservative force is equal to the loss (negative of the change) of the potential energy, which can be symbolically given as

$$W_{con} = -\Delta U$$

Here the conservative force is the electrostatic force. So, using the last two equations we can write

$$W_{el} = -\Delta U = \frac{KQq}{r}$$

$$\Rightarrow -(U_r - U_\infty) = -\frac{KQq}{r}$$

At infinity, the electrical force is zero; so, there is no potential energy there. Then, by putting $U_\infty = 0$ in the last expression and simplifying the factors, we have

$$U_r = \frac{KQq}{r}$$

N.B: If you put $Q = q_1$ and $q = q_2$ as assumes at the beginning of the derivation, we can write

$$U_r = \frac{Kq_1q_2}{r}$$

The above expression tells us the following.

Electric potential energy of interaction between two point charges is directly proportional to the product of their charges and inversely proportional to the distance of separation between the particles. Electric potential energy can be positive if both the charges are either positive or negative; it can be negative if one charge is positive and the other is negative. In other words, electrostatic potential energy between the charged particles can be positive if the electrostatic force of interaction between the charged particles is repulsive and *vice versa*. However, gravitational potential energy is always negative due to attraction force.

The work done by electric field in bringing the particles from infinity to the given configuration is given as

$$W_{el} = -\frac{KQq}{r}$$

In other words, the work done by the external agent in slowly bringing the particles from infinity to the given configuration can be given by

$$W_{ext} = -W_{el} = \frac{KQq}{r}$$

Hence, the potential energy is given as

$$U = -W_{el} = W_{ext}$$

The above discussion states that

The electric potential energy of interaction between two particles can be defined as the negative work done by their mutual gravitational forces or minimum work done by an external agent in bringing (assembling) the particles from the infinity to the given configuration. For minimum work done, the external agent must shift all particles slowly from infinity.

Example 12 The total energy of an electron in Bohr's hydrogen atom is known to be -13.6 eV in ground state. Find the radius of Bohr's hydrogen atom in ground state.

Solution

We know that the electron possesses a kinetic energy KE as it revolves in a circular path around the nucleus. Due to the Coulombic attraction the electron possesses an electrostatic potential energy U. Then, the total energy of an electron is the sum of its kinetic and potential energy, given as

$$E = KE + U \tag{1.33}$$

It is known that the electrostatic attraction between the electron and nucleus is

$$F_{el} = \frac{Ke^2}{r^2} \tag{1.34}$$

This force provides the necessary centripetal acceleration

$$a_{cp} = \frac{v^2}{r} \tag{1.35}$$

Applying Newton's 2nd law on the electron,

$$F_{el} = ma_{cp} \tag{1.36}$$

Using last three equations, we have

$$\frac{Kq^2}{r^2} = m\frac{v^2}{r}$$

Then, the KE of the electron is

$$KE = \frac{mv^2}{2} = \frac{Ke^2}{2r} \tag{1.37}$$

The electrostatic potential energy of the electron relative to the nucleus is

$$U = -\frac{Ke^2}{r} \tag{1.38}$$

Using the equations (1.33), (1.37) and (1.38), the total energy of the electron is

$$E = KE + U = \frac{Ke^2}{2r} + \left(-\frac{Ke^2}{r} \right) = -\frac{Ke^2}{2r} \tag{1.39}$$

Putting $E = -13.6e$ in equation (1.39) we have

$$E = -\frac{Ke^2}{2r} = -13.6e$$

So, Bohr's radius of ground state of a hydrogen atom is

$$r = \frac{Ke}{2 \times 13.6} = \frac{(9 \times 10^9)(1.6 \times 10^{-16})}{2 \times 13.6} = 0.529 \times 10^{-10}\, m = 0.529\, \text{Å Ans.}$$

1.9 Electric potential

For the sake of simplicity, let us take a fixed charged body. You may call it source charge. It produces its electric field around it. We have shown an arbitrary pattern of an electric field due to a source charge. Then bring a test charge q *slowly* from infinity (or any other reference point of zero-force) to a point P following an arbitrary path (shown as a dotted line). As discussed earlier, the distance much greater than the size of the charged body can be considered as an infinite distance. A test charge is imagined as a point charge because we want to find the field and potential of the field at a point. Furthermore, the test charge must be extremely small compared to the source charge such that it cannot distort its charge distribution. 'Slow-shifting' means moving the test charge nearly with nearly zero velocity and zero acceleration. So, you have to push the test charge with a force $\overrightarrow{F_{ext}}$ such that this will counteract the electrical repulsive force F_{el} acting on the test charge.

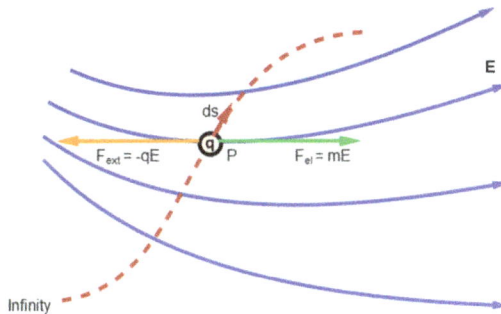

As the acceleration is zero, the net force is zero; so, we can write

$$\overrightarrow{F_{net}} = \overrightarrow{F_{ext}} + \overrightarrow{F_{gr}} = m\overrightarrow{a} = 0$$

Then, we have

$$\overrightarrow{F_{ext}} = -\overrightarrow{F_{gr}} \tag{1.40}$$

That is why we stated that the external force will be equal in magnitude and oppositely in direction of electrical force. While the particle undergoes an elementary displacement $d\,\overrightarrow{s}$, the elementary work done by the external agent is

$$dW_{ext} = \overrightarrow{F_{ext}} \cdot d\,\overrightarrow{s} \tag{1.41}$$

Substituting $\overrightarrow{F_{ext}} = -\overrightarrow{F_{el}}$ from equation (1.40) in equation (1.41), we have

$$dW_{ext} = -\overrightarrow{F_{el}} \cdot d\overrightarrow{s}$$

Integrating the elementary works, the external work done in bringing the particle from infinity to point P is

$$W_{ext} = \int dW_{ext} = -\int_{\infty}^{P} \overrightarrow{F_{el}} \cdot d\overrightarrow{s}, \tag{1.42}$$

where $\overrightarrow{F_{el}} = q\overrightarrow{E}; \overrightarrow{E} =$ Electric field intensity.

$$\Rightarrow W_{ext} = -q \int_{\infty}^{P} \overrightarrow{E} \cdot d\overrightarrow{s}$$

The electric field due to the static charges, the electrostatic field, is conservative, $q\int_{\infty}^{P} \overrightarrow{E} \cdot d\overrightarrow{s}$, that is, total work done by the electric field between infinity and point P does not depend on the path followed by the particle. In other words, $\int_{\infty}^{P} \overrightarrow{E} \cdot d\overrightarrow{s}$, is a constant quantity at a point P of an electrostatic field. This constant quantity is treated as a property of the electric field at any point, known as electric potential.

Definition of potential: This means that minimum work is done, W_{ext}, by an external agent in *slowly* shifting the particle from infinity will be different for different point charges. However, in each case, the ratio of work done W_{ext} and charge q, that is, $\frac{W_{ext}}{q}$ is a constant quantity for point P. We call it electric potential at point P, denoted by the letter V. Symbolically, the potential at P is given as

$$V_P = \frac{W_{ext}}{q} = -\int_{\infty}^{P} \overrightarrow{E} \cdot d\overrightarrow{s}$$

It is standard practice in physics to define some quantities by putting the denominator as *one* (unity). Following this rule, if you put $q = 1$, we have

$$V_P = W_{ext}$$

The above expression tells us the following.

The work done by an external agent in slowly shifting a test charge of one coulomb (unit charge) from infinity to any point in an electrostatic field of an object (source charge) is termed as 'potential' of the field (or the object creating the field) at that point. The term potential is derived from the terms of work done, *potential* shows the stamina of the field. The potential of a point P can also be defined as the work done per unit charge in shifting it slowly from infinity to the point. The relation between the field strength (intensity) and potential is given as

$$V = -\int_{\infty}^{P\rightarrow} \overrightarrow{E} \cdot d\overrightarrow{s}$$

Physical significance of potential: If an external agent does more work in shifting a particle of charge 1 C from infinity to any point, we can say that the potential of that point is more. Hence, it is related with the work done by (or stamina of) the external agent or the electric field. Thus, in contrary to the field strength $\overrightarrow{E}$, potential V is a scalar; $\vec{E}$ is the force per unit charge and V is the work done per unit charge. The unit of gravitation potential is joule/kg and it is a negative quantity, whereas electrostatic potential can be positive and negative. In some books you may find the potential being defined as 'potential energy per unit charge'. If you follow this definition, you must be more careful to understand the meaning of 'potential energy per unit charge'. If you follow this definition, you must be careful to understand the meaning of 'potential energy'. Here, the word potential energy means the potential energy of interaction between the particle and the object (or system), but not the total potential energy of interaction between all particles of the object. Recapitulating, potential signifies energy or work, that is *stamina* in crude language; field intensity tells us about the force or *strength*.

N.B: Every point of an electrostatic field is characterized by field strength (or intensity) $\vec{E}$ and potential V;

$$\vec{E} = \frac{\vec{F}}{q}; \quad V = \frac{W_{ext}(\infty \to P)}{q} = -\int_{\infty}^{P \to} \vec{E} \cdot d\vec{s}$$

Some authors express the electric field as $\vec{E} = \underset{q \to 0}{\text{Lim}} \frac{\vec{F}}{q}$; $\text{Lim}(q \to 0)$ means the charge q must be vanishingly small so as not to disturb the charge distribution of the source charge. Thus, electric field of the source charge remains the same. This *tight* definition will be very useful in understanding the conductor properties because the presence of a tiny change or a very small electric field can alter the surface charge distribution of the conductor. So, if you do not like this expression for mathematical reasons, it does not matter. Already we have assumed that our test charge is very small (point-like and has extremely small charge). Use of limit is just a mathematical treatment to idealize the situation of defining the electric field of any charged system without disturbing its charge configuration. In dielectrics, things are more ideal but in conductors, we must stick to the ideal definition of a test charge q *tends to zero* (in both volume and magnitude of charge).

1.10 Calculation of electric potential

Discrete charge distribution: We will find electric potential of discrete and continuous charge distribution as we did for the gravitational case. The basic formula, that is $V = \frac{Kq}{r}$, gives us the potential due to a point charge. For a group of particles, we can find the potential due to each and then add them algebraically to obtain the total potential, at any point. Hence, the potential at P due to discrete charge distribution of n particles of masses $q_1, q_2, ..., q_n$ situated at distances $r_1, r_2, ..., r_n$ from point P can be given as

$$V = V_1 + V_2 + \cdots + V_n$$

where $V_1 = \frac{Kq_1}{r_1}, V_2 = \frac{Kq_2}{r_2} \ldots.. V_n = \frac{Kq_n}{r_n}$

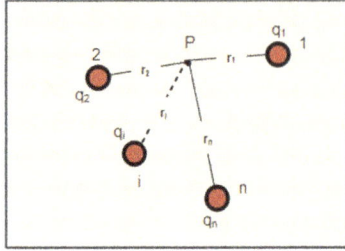

$$\Rightarrow V = K\left(\frac{q_1}{r_1} + \frac{q_2}{r_2} + \cdots + \frac{q_n}{r_n}\right)$$

In a nutshell, we can write the above expression as:

$$V = K\sum_{i=1}^{i=n}\frac{q_i}{r_i}$$

let us use the above formula in the following example.

Example 13 (Discrete point charges) Find the electric potential of a system of two particles of charges q_1 and q_2 separated by a distance r, at a point where the field intensity of the system is zero.

Solution

Let the field at P be zero.

Then, we have

$$\overrightarrow{E}_{\text{net}} = \overrightarrow{E}_1 + \overrightarrow{E}_2 = 0$$

$$\Rightarrow E_1 = E_2$$

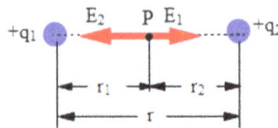

$$\Rightarrow \frac{Kq_1}{r_1^2} = \frac{Kq_2}{r_2^2}$$

$$\Rightarrow \frac{r_1}{r_2} = \left(\frac{q_1}{q_2}\right)^{1/2} \tag{1.43}$$

As per the given system

$$r_1 + r_2 = r \tag{1.44}$$

Using equations (1.43) and (1.44), we have

$$r_1 = \frac{r\sqrt{q_1}}{\sqrt{q_1} + \sqrt{q_2}} \qquad (1.45)$$

$$r_2 = \frac{r\sqrt{q_2}}{\sqrt{q_1} + \sqrt{q_2}} \qquad (1.46)$$

The total potential at P is

$$V = K\left(\frac{q_1}{r_1} + \frac{q_2}{r_2}\right) \qquad (1.47)$$

Putting the values of r_1 and r_2 from equations (1.45) and (1.46) in the equation (1.47), we have

$$V = \frac{K(\sqrt{q_1} + \sqrt{q_2})^2}{r} \quad \text{Ans.}$$

Continuous mass distribution: For any extended object, first of all we consider an element of mass dm. Then, the potential due to the extended object is,

$$dV = K\frac{dq}{r}$$

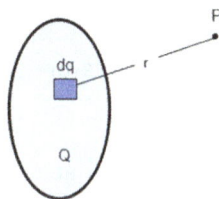

The total potential due to the extended object is

$$V = \int dV = K\int \frac{dq}{r},$$

where $dq = \lambda dx$ for linear mass distribution

$\qquad = \sigma\, dA$ for surface mass distribution
$\qquad = \rho\, dV$ for volume mass distribution.

Example 14 (Uniformly charged thin rod) Find the electric potential at P of a uniform thin rod of linear charge density λ and length l, as shown in the figure.

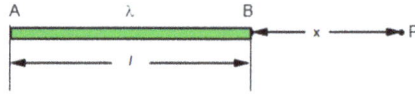

Solution
Potential due to the element of charge dq at P is, $dV = K\frac{dq}{r}$, where $dq = \lambda\, dr$

$$\Rightarrow dV = K\lambda\frac{dr}{r}$$

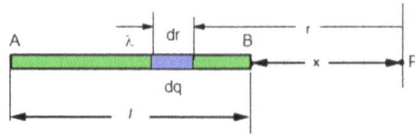

Integrating the elementary potentials, total potential due to the rod is

$$V = \int dV = K\lambda \int_x^{l+x} \frac{dr}{r}$$

$$V = K\lambda \ln \frac{l+x}{x} \text{ Ans.}$$

Example 15 (Uniformly charged circular arc) Find the electric potential of a thin circular arc of charge Q and radius of curvature R, at the centre of curvature.
Solution

(a) The potential due to the element dq at C is $dV = K\frac{dq}{R}$
Then, the total potential due to the arc is

$$V = \int dV = \int_0^q K\frac{dq}{R}$$

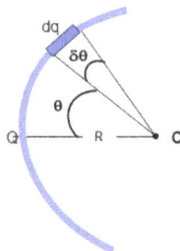

Since, K and R are constant, taking them out of the integral, we have

$$V = \frac{K}{R} \int_0^Q dq$$

$$\Rightarrow V = \frac{KQ}{R} \text{ Ans.}$$

N.B: Since each element of a ring is equidistant from the centre C of the ring, the potential at the centre of a ring can be given as $V = \frac{KQ}{R}$ which does not depend upon the charge distribution.

Example 16 (Uniformly charged circular ring) Find the electric potential of a thin circular ring of charge Q and radius R, at a point P on the axis of the ring located at a distance x from the centre of the ring.

Solution

The potential of the element dq of the ring at P is given by,

$$dV = \frac{Kdq}{r}$$

Since, each element dq of the ring is situated at a distance r from the axial point P, the potential of the ring is given by,

$$V = \int dV = \int_0^Q K\frac{dq}{r}$$

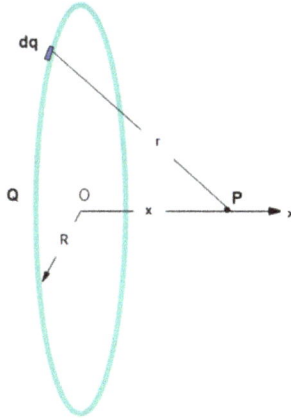

$$\Rightarrow V = \frac{KQ}{r} \left(\text{but not, } \frac{KQ}{R} \right),$$

where $r = \sqrt{R^2 + x^2}$.

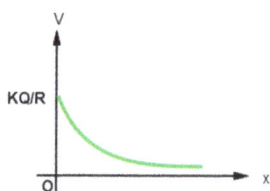

$$\Rightarrow V = \frac{Kq_{\text{ring}}}{\sqrt{R_{\text{ring}}^2 + x^2}} = \frac{KQ}{\sqrt{R^2 + x^2}}$$

At $x = 0$, $V = \frac{KQ}{R}$; If $x \to \infty$, $V \to 0$

This means that the potential varies with with axial distance x. Ans.

Example 17 (Uniformly charged disc) Using the formula of the electrical potential obtained for the ring in the previous example, derive an expression for electrical potential at the axial point which is situated at a distance x from the centre of a disc of radius R and surface charge density σ.

Solution

Let us take a thin ring of radius r and thickness dr. Then, the charge of the ring is, $q_{\text{ring}} = \sigma(2\pi r \, dr)$, as explained earlier.

As derived in the previous example the potential due to the ring is,

$$V_{\text{ring}} = -\frac{Kq_{\text{ring}}}{\sqrt{R_{\text{ring}}^2 + x^2}}$$

where $q_{\text{ring}} = 2\pi\sigma r \, dr$ and $R_{\text{ring}} = r \Rightarrow V_{\text{ring}} = 2\pi\sigma K \frac{r \, dr}{\sqrt{r^2 + x^2}}$

Integrating V_{ring}, we have

$$V_{\text{disc}} = \int V_{\text{ring}}$$

$$= -2\pi\sigma K \int_0^R \frac{r \, dr}{\sqrt{r^2 + x^2}}$$

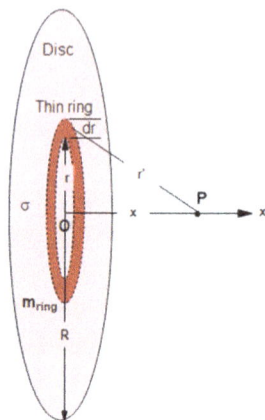

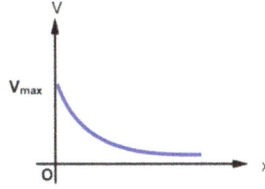

Let $\sqrt{r^2 + x^2} = t$; then, $r\, dr = t\, dt$ when $r = 0$, $t = x$; when $r = R$,

$$t = \sqrt{R^2 + x^2}$$

$$\Rightarrow V_{\text{disc}} = \pi\sigma K \int_R^{\sqrt{R^2 + x^2}} \frac{t\, dt}{t}$$

After evaluating the integral, we have

$$V_{\text{disc}} = \pi\sigma K(\sqrt{R^2 + x^2} - x) \text{ Ans.}$$

N.B:
1. At $x = 0$, $V = \pi K\sigma R = \sigma R/4\varepsilon_o = V_{\text{max}}$ and at $x \to \infty$, $V \to 0$.
2. If $R \to \infty$, $V \to \infty$. This means that V is infinite for a large sheet.

Example 18 (Uniformly charged thin spherical shell) Derive an expression for the electric potential at a point having a radial distance r, outside of a thin spherical shell of surface charge density σ and radius R.

Solution
Take a thin ring (portion between two red arcs) at an angular position θ. As explained earlier, the charge of the ring is

$$q_{\text{ring}} = \sigma(2\pi R^2 \sin \theta\, d\theta)$$

$$\Rightarrow q_{\text{ring}} = 2\pi\sigma R^2 \sin \theta\, d\theta \tag{1.48}$$

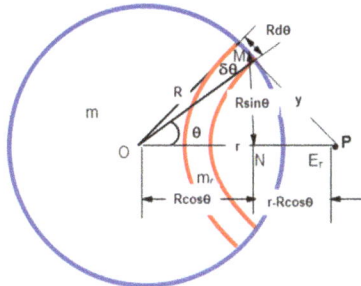

The potential due to the ring at P is

$$V_{\text{ring}} = \frac{Kq_{\text{ring}}}{y} \qquad (1.49)$$

Using last two equations,

$$V_{\text{ring}} = -\frac{2\pi\sigma R^2 K \sin\theta \, d\theta}{y} \qquad (1.50)$$

In the triangle OMP, applying the cosine rule,

$$R^2 + r^2 - 2Rr\cos\theta = y^2$$

Differentiating both sides,

$$2Rr\sin\theta \, d\theta = 2y \, dy$$

$$\Rightarrow \sin\theta \, d\theta = \frac{y \, dy}{Rr} \qquad (1.51)$$

Substituting $\sin\theta \, d\theta$ from equation (1.51) in equation (1.50), we have

$$V_{\text{ring}} = \frac{2\pi\sigma KR}{r} dy \qquad (1.52)$$

The thin spherical shell is the combination of coaxial rings of angular positions θ ranging from $0°$ to $180°$. Then, the value of y must vary from $(r - R)$ to $(R + r)$. Integrating V_{ring}. we have

$$V_{\text{sph}} = \int V_{\text{ring}} \qquad (1.53)$$

Using the last two equations,

$$V_{\text{sph}} = \frac{2\pi\sigma KR}{r} \int_{r-R}^{r+R} dy$$

$$V_{\text{sph}} = \frac{4\pi\sigma KR^2}{r} = \frac{KQ}{r} (\because 4\pi\sigma R^2 = Q) \text{ Ans.}$$

To find the potential inside the spherical shell, we take the point P inside the shell. Then, the lower limit becomes $R - r$ and the upper limit of the integral remains the same $(=R + r)$

So, the aforementioned integral can be written as

$$V_{\text{sph}} = \frac{2\pi\sigma KR}{r} \int_{R-r}^{R+r} dy.$$

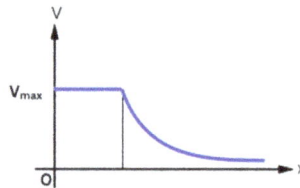

Substituting $\sigma(4\pi R^2) = Q$(mass of the spherical shell), we have

$$V = \frac{KQ}{r}; r \geqslant R$$
$$= \frac{KQ}{R}; r \leqslant R$$

This means that the potential of the sphere at any inside point is equal to a $\frac{KQ}{R}$ and it decreases hyperbolically to zero (or V decreases from the maximum $\frac{KQ}{R}$ inside to zero as $r \to \infty$ (we go radially away). Ans.

Example 19 (Uniformly charged solid sphere) Find the electric potential of a uniform solid sphere of radius R and volume charge density ρ at any point having a radial distance r.

Solution

We divide the sphere into parts: sphere 1 (shaded portion) of radius r containing point P and the hollow sphere 2 of inner radius r and outer radius R. If we can find potential due to each sphere at P and then sum then up, we will get the potential due to the given sphere at P. Let us proceed with this idea.

Let the potentials due to sphere 1 and hollow sphere 2 at P be V_1 and V_2, respectively. Since sphere 1 is composed of thin spherical shells of radii ranging from 0 to r and point P is situated outside of each thin sphere (spherical shells), the potential due to 1 at P is equal to the sum of potentials due to each thin shell. Taking a thin spherical shell of charge dq, the potential at P is,

$$dV = K\frac{dq}{r}$$

Integrating dV, the total potential due to the sphere of radius r at P is,

$$V_1 = \int dV = \frac{K}{r} \int_0^{q'} dq = \frac{Kq'}{r},$$

where q' = charge of the sphere 1 $= \frac{4}{3}\pi r^3 \rho$

$$\Rightarrow V_1 = \frac{4\pi K\rho r^2}{3} = \frac{\rho r^2}{3\varepsilon_o} \tag{1.54}$$

Let us now calculate V_2. For this let us take a spherical shell of mass dm and radius x. Since point P lies inside the shell, its potential at P is given as

$$dV = K\frac{dq}{x}$$

as discussed earlier.

Integrating dV, the potential due to hollow sphere 2 at P is,

$$V_2 = \int dV = \int_r^R K\frac{dq}{x}, \quad \text{where } dq = \rho(4\pi x^2 dx)$$

$$\Rightarrow V_2 = 4\pi\rho K \int_r^R x \, dx$$

$$\Rightarrow V_2 = 2\pi\rho K(R^2 - r^2) \tag{1.55}$$

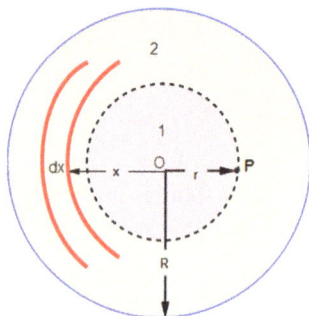

Adding V_1 and V_2, we have the total potential at P given by

$$V = V_1 + V_2 \tag{1.56}$$

Substituting V_1 from equation (1.54) from equation (1.55) in equation (1.56), we have

$$V = \frac{4\pi K\rho}{6}(3R^2 - r^2); \ r \leqslant R$$
$$= \frac{\rho}{6\varepsilon_0}(3R^2 - r^2); \ r \leqslant R \text{ Ans.}$$

As discussed earlier, for any outside point ($r > R$), we have

$$V = \frac{Kq'}{r},$$

where $q' =$ mass of the total sphere. Because we have taken point P outside of the given sphere,

$$q' = \frac{4}{3}\pi R^3 \rho$$

Using the last two equations,

$$V = \frac{4\pi R^3 \rho K}{3r}; \ r \geqslant R$$
$$= \frac{R^3 \rho}{3\varepsilon_0 r}; \ r \geqslant R \text{ Ans.}$$

Alternative solution:
We can also use the expression

$$V = -\int_\infty^r E \, dr$$

to obtain the same result.

As derived earlier,

$$E = \frac{KQ}{R^3}r; \, r \leqslant R$$

$$V = -\int_{\infty}^{R} E \, dr - \int_{R}^{r} E \, dr$$

Since, $V = -\frac{4\pi K\rho}{6}(3R^2 - r^2)$, putting $\frac{4\pi R^3}{3}\rho = Q$, we have $V = \frac{KQ}{2R^3}(3R^2 - r^2)$, for $r \leqslant R$.

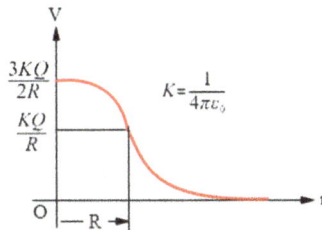

This means that potential decreases from $\frac{3KQ}{2R}$ to $\frac{KQ}{R}$ parabolically from the centre to the surface of the sphere.

As the sphere possesses spherical symmetry, it behaves as a point charge for the points outside the sphere. Hence, potential $V = \frac{KQ}{r}$ for $r \geqslant R$. This means that potential varies hyperbolically outside the sphere like a thin spherical shell or point charge.

1.11 Electrostatic potential energy

1.11.1 Discrete particle system—a system of a group of charged particles

We have derived the expression of electric potential energy by using the concept of *work* and the *work–energy (W–E) theorem* for a two-particle system. Let us apply the *W–E* theorem for a many-particle system to develop a general formula for electric potential energy of interaction of a group of particles. As discussed earlier, potential energy of a two-particle system of charges q_1 and q_2 can be given as the minimum work done by an external agent (or negative of work done by the electric field) in assembling the system (particles) by slowly bringing each charge from infinity to the given configuration. Let us assume that we have n particles of charge $q_1, q_2, q_3, \ldots, q_n$. If we bring q_1 slowly from infinity and place it at position (point) 1, we will not have to do any work because no external field was present.

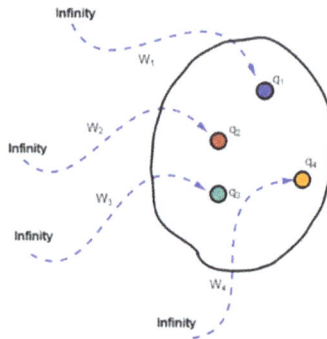

The energy of interaction of a system of point charges is
equal to the work done by an external agent to bring the
charges slowly from infinity to form the given configuration.

$$W_1 = 0 \qquad (1.57)$$

Now bring the particle q_2 slowly from infinity and place it at point 2. In this process
the work done against the Coulombic force of q_1 is

$$W_2 = \frac{Kq_1q_2}{r_{21}} \qquad (1.58)$$

Similarly, when you bring particle q_3 slowly from infinity to point 3, you have to do
work $\frac{Kq_1q_2}{r_{31}}$ against the electrostatic force of q_1, $\frac{Kq_2q_3}{r_{32}}$ against the gravity of q_2.

Hence, the total work done in shifting q_3 is

$$W_3 = \frac{Kq_1q_3}{r_{31}} + \frac{Kq_2q_3}{r_{32}} \qquad (1.59)$$

Likewise, the total external work done in bringing the particle q_4 slowly from infinity
to point 4 is

$$W_4 = \frac{Kq_1q_4}{r_{41}} + \frac{Kq_2q_4}{r_{42}} + \frac{Kq_3q_4}{r_{43}} \qquad (1.60)$$

Adding equations (1.57), (1.58), (1.59) and (1.60), we have the total work

$$W = \frac{Kq_1q_2}{r_{21}} + \frac{Kq_1q_3}{r_{31}} + \frac{Kq_2q_3}{r_{32}} + \frac{Kq_1q_4}{r_{41}} + \frac{Kq_2q_4}{r_{42}} + \frac{Kq_3q_4}{r_{43}} \qquad (1.61)$$

Let us now do the same thing but in reverse order. First of all, you bring q_4 from
infinity and place it at point 4. The work done is zero because no field was there.
Hence,

$$W'_1 = 0 \qquad (1.62)$$

Then bring q_3 slowly from infinity and place it at point 3. The external work done is

$$W'_2 = \frac{Kq_3q_4}{r_{34}} \qquad (1.63)$$

Similarly total work done in bringing q_2 from infinity to point 2

$$W'_3 = \frac{Kq_2q_4}{r_{24}} + \frac{Kq_2q_3}{r_{23}}$$ (1.64)

The external work done in bringing q_1 from infinity to point 1

$$W'_4 = \frac{Kq_1q_4}{r_{14}} + \frac{Kq_1q_3}{r_{13}} + \frac{Kq_1q_2}{r_{12}}$$ (1.65)

Adding equations (1.62), (1.63), (1.64) and (1.65), we have

$$W' = \frac{Kq_3q_4}{r_{34}} + \frac{Kq_2q_4}{r_{24}} + \frac{Kq_2q_3}{r_{23}} \\ \frac{Kq_1q_4}{r_{14}} + \frac{Kq_1q_3}{r_{13}} + \frac{Kq_1q_2}{r_{12}}$$ (1.66)

In both cases we have exactly same configurations of the particles. Hence, the external work done must be the same in both cases. Then

$$W + W' = 2W$$ (1.67)

Substituting W from equation (1.61) and W' from equation (1.66) in equation (1.67) and grouping the terms, we have

$$2W = q_1\left(\frac{Kq_2}{r_{12}} + \frac{Kq_3}{r_{13}} + \frac{Kq_4}{r_{14}}\right) + q_2\left(\frac{Kq_1}{r_{21}} + \frac{Kq_3}{r_{23}} + \frac{Kq_4}{r_{24}}\right) \\ + q_3\left(\frac{Kq_1}{r_{31}} + \frac{Kq_2}{r_{32}} + \frac{Kq_4}{r_{34}}\right) + q_4\left(\frac{Kq_1}{r_{41}} + \frac{Kq_2}{r_{42}} + \frac{Kq_3}{r_{43}}\right)$$ (1.68)

Substituting $\left(\frac{Kq_2}{r_{12}} + \frac{Kq_3}{r_{13}} + \frac{Kq_4}{r_{14}}\right) = V_1$ (potential at 1 due to all particles except q_1), $\left(\frac{Kq_1}{r_{21}} + \frac{Kq_3}{r_{23}} + \frac{Kq_4}{r_{24}}\right) = V_2$ potential due to all except q_3, $\left(\frac{Kq_1}{r_{31}} + \frac{Kq_2}{r_{32}} + \frac{Kq_4}{r_{34}}\right) = V_3$ potential due to all except q_3 and $\left(\frac{Kq_1}{r_{41}} + \frac{Kq_2}{r_{42}} + \frac{Kq_3}{r_{43}}\right) = V_4$ potential due to all except q_4 in equation (1.68), we have

$$2W = q_1V_1 + q_2V_2 + q_3V_3 + q_4V_4$$

Following the above procedure for n particles, we have

$$2W = q_1V_1 + q_2V_2 + \ldots + q_nV_n$$

In a nutshell, we can write

$$2W = \sum_{i=1}^{i=n} q_iV_i$$

where V_i = potential at ith point due to all particles except q_1.

$$\Rightarrow W = W_{ext} = \frac{1}{2}\sum_{i=1}^{i=n} q_iV_i$$

As we know,

$$W_{ext} = U - U_\infty$$

and $U_\infty = 0$ (because $F = 0$, at infinity). Then, we have

$$U = \frac{1}{2}\sum_{i=1}^{i=n} q_i V_i$$

Example 20 Eight particles each of charge q are situated at the vertices of a cube of side l. Find the electric potential energy of interaction of the system of eight particles.

Solution

In a cube we have four corners (1, 3, 6 and 7) each having charge $+q$ and the other four corners (2, 4, 5 and 8) each having charge $-q$. Hence, the first four similar points must have same potential; $V_1 = V_3 = V_6 = V_7 = V_+$ (say). Similarly, the second four similar points must have same potential $V_2 = V_4 = V_5 = V_8 = V_-$ (say). Furthermore, put $q_1 = q_3 = q_6 = q_7 = +q$ and $q_2 = q_4 = q_5 = q_8 = -q$. Putting all these values in the expression

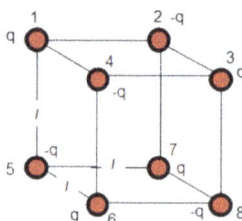

$$U = \frac{1}{2}\sum_{i=1}^{i=n} q_i V_i,$$

we have

$$U = \left(\frac{1}{2}(+q)(V_+)\right)(4) + \left(\frac{1}{2}(-q)(V_-)\right)(4) = 2q(V_+ - V_-) \qquad (1.69)$$

The potential at the sites of positive charge is given as

$$\text{where } V_+ = \frac{3K(-q)}{l} + \frac{3K(+q)}{l\sqrt{2}} + \frac{K(-q)}{l\sqrt{3}} = \frac{Kq}{l}\left(-3 + \frac{3}{\sqrt{2}} + \frac{-1}{\sqrt{3}}\right),$$

Similarly, you can prove that

$$V_- = \frac{3K(+q)}{l} + \frac{3K(-q)}{l\sqrt{2}} + \frac{K(+q)}{l\sqrt{3}} = -\frac{Kq}{l}\left(-3 + \frac{3}{\sqrt{2}} + \frac{-1}{\sqrt{3}}\right)$$

This means that $U = 2q(V_+ - V_-) = \frac{4Kq^2}{l}\left(-3 + \frac{3}{\sqrt{2}} + \frac{-1}{\sqrt{3}}\right)$ Ans.

N.B: At first sight, students may get confused to think $V_+ = V_-$, and tempted to write the answer as zero. But, if you proceed carefully, you can find that $V_+ = -V_-$; so, $U = =2q(V_+ - V_-) = 4qV_+$, is a non-zero answer. Some students may misinterpret it as since the net charge is zero, the potential energy should be zero, which is again a misconception.

1.11.2 Potential energy due to continuous charge distribution:

Let us take the expression $U = \frac{1}{2}\sum q_i V_i$ for discrete mass distribution. In continuous charge distribution we cannot distinguish one particle from the others. Hence, we write the ith charge as dq, that is, $q_i = dq$. Furthermore, the potential at the point where the ith charge is present can be written as simply V, that is, $V_i = V$; then replacing the sign '$\sum$' by internal sign '$\int$' in the above formula, we have

$$U = \frac{1}{2}\int V \, dq$$

where $V =$ potential at point P and
$\quad dq = \lambda \, dx$ (for linear charge distribution)
$\quad = \sigma \, dA$ (for surface charge distribution)
$\quad = \rho \, dV$ (for volume charge distribution)

In the above formula, $V =$ electric potential of any point inside (not outside) the object

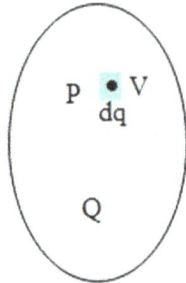

An element of charge dq at P is taken in a continuous charge distribution of charge Q; if V=potential of the body at point P, the electric potential energy of the body is given as $U = \frac{1}{2}\int Vdq$.

N.B: The above formula is presented without any proof, which can be proved by using vector calculus which is beyond the scope of this book.
Let us use the above formula.

Example 21 Find the electric potential energy possessed by a thin spherical shell of charge q and radius R.
Solution
Let us take an elementary charge dq at P. The potential at P is

$$V = -\frac{KQ}{R}$$

Substituting $V = \frac{KQ}{R}$ in the formula,

$$U = \frac{1}{2} \int V \, dq,$$

we have

$$U = \frac{1}{2} \int_0^Q \left(\frac{KQ}{R}\right) dq$$

Since, $\frac{KQ}{R}$ is constant, take it out of the integral to obtain

$$U = \frac{KQ}{2R} \int_0^Q dq$$

$$\Rightarrow U = \frac{KQ^2}{2R} \text{ Ans.}$$

N.B: In the previous example, if you write $dq = dQ$ and do not take it out of the integral, you will get wrong result as follows:

$$U = \frac{1}{2} \int_0^Q \frac{KQ}{R} \cdot dQ = \frac{KQ^2}{4R}$$

1.12 Earth's electric field

You may think that Earth is neutral; so, it does not have any electric field. But this is not true. It is interesting to note that our Earth has an electric field in addition to its well-known gravitational field that we have described in the chapter of 'gravitation' in the last book. Earth holds all of us due to its *gravity*.

The gas molecules in the ionosphere become ionized by the solar wind forming a spherical shell of positive charges. This induces equal negative charge on the surface of Earth with an average surface charge density of 2.65×10^{-9} C m^{-2}. Although the Earth's surface is 300 000 V negative relative to the ionosphere, as a whole, the Earth along with its atmosphere is neutral.

The electric field near Earth's surface practically varies from 100 to 300 volts per meter. This means that you have few hundred volts between your head and toes. But it is interesting to note that we do not get an electric shock and remain quite safe in this electric field. We will discuss more about this in the next chapter in the context of properties of a conductor.

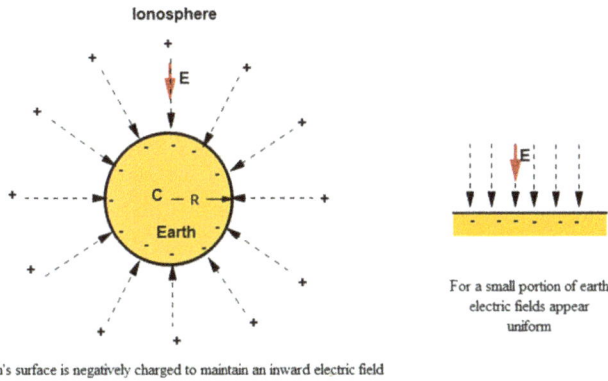

Earth's surface is negatively charged to maintain an inward electric field

Example 22

(a) Draw the graph of electric field and potential versus radial distance for a uniformly charged solid sphere of charge Q and radius R.

(b) Find the electric potential energy possessed by it.

Solution

(a) As the sphere has uniform volume charge density, and it possesses spherical symmetry, the electric field strength E due to the sphere can be given as:

$$E = \frac{KQ}{R^3}r; \ r \leqslant R$$
$$= \frac{KQ}{r^2}; \ r \geqslant R \quad \text{Ans.}$$

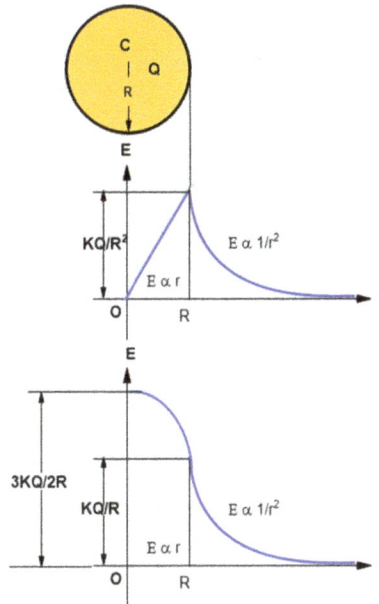

For any inside point, the potential is

$$V = \frac{KQ}{2R}(3R^2 - r^2); \; r \leqslant R$$

For any outside point, the potential is

$$V = \frac{KQ}{r}; \; r \geqslant R$$

We can see that both graphs $E = f(r)$ and $V = f(r)$ are continuous. Ans.

(b) Put $V = \frac{KQ}{2R^3}(3R^2 - r^2)$ and $dq = 4\pi\rho r^2 dr$ in the expression $U = \frac{1}{2}\int V\, dq$, where '$r$' ranges from zero to R to obtain

$$U = \frac{1}{2}\int V\, dq = \frac{1}{2}\int_0^R \frac{KQ}{2R^3}(3R^2 - r^2)(4\pi\rho r^2 dr)$$

$$= \frac{\pi KQ\rho}{R^3}\int_0^R (3R^2 - r^2)(r^2 dr)$$

$$= \frac{\pi KQ\rho}{R^3}\{3R^2 R^3/3 - r^5/5\} = \frac{\pi KQ\rho}{R^3}\frac{4R^5}{5}$$

Putting $4\pi K = (\frac{1}{\varepsilon_o})$ and $\rho = Q/(\frac{4\pi R^3}{3})$ and simplifying the factors, we have

$$U = \frac{3Q^2}{20\pi\varepsilon_o R} \; \text{Ans.}$$

N.B: You can also obtain the same expression when you find the energy possessed by a thin spherical shell of radius r and then integrate it.

1.13 Relation between field and potential

As we learnt, field strength is equal to force per unit charge and potential is equal to work done per unit charge. So, the relation between the field and potential is basically the relation between force and work done. When a test charge q at a point P undergoes an elementary displacement $\vec{ds}$, the work done by the external agent is equal to the negative of the work done by the electric field:

$$dW_{\text{ext}} = -dW_{el} = -q\,\vec{E}\cdot\vec{ds} = -q\,E\cdot ds\,\cos\theta = -q\,(E\,\cos\theta)ds,$$

where $\vec{E}$ is an electric field at point P. You know that $E\cos\theta$ is the projection or component of electric field along the s-axis. So, we can write

$$E\cos\theta = E_s$$

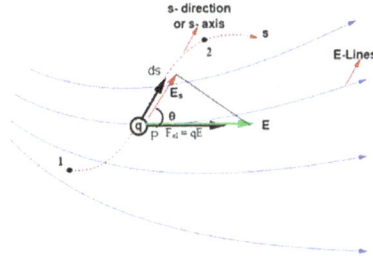

When a test charge q undergoes a small displacement ds, its component along the electric field is responsible for doing external work dW_{ext}; this work done upon the test charge is defined as the change in potential; $dV = dW_{ext}/q$.

Using the last two equations, we can write,

$$dW_{el} = -q\, E_s\, ds; \text{ or, } dW_{el}/q = -E_s\, ds$$

As we explained in section 1.7, $dW_{el}/q = dV =$ elementary potential difference along the s-direction/axis. Then, we have $dV = -E_s\, ds$; so, finally, we have the differential relation between $\vec{E}$ in and V, given as

$$E_s = -\frac{dV}{ds}$$

Here the s-axis is a generalized curve which you can call curvilinear coordinate system that can be categorized as special coordinate systems such as cartesian, cylindrical, spherical etc. In the Cartesian coordinate system, s-axis can be given as x-, y- and z-axes.

This tells us that, the electric field strength E_s along any s-direction (x, y etc) is numerically equal to the negative of the potential gradient along that direction.

In the Cartesian or rectangular coordinate system,

$$\vec{E_x} = -\frac{\partial V}{\partial x}\hat{i}, \vec{E_y} = -\frac{\partial V}{\partial y}\hat{j}, \vec{E_z} = -\frac{\partial V}{\partial z}\hat{k}$$

Summing of these three components, the net field is given as

$$\vec{E} = \vec{E_x} + \vec{E_y} + \vec{E_z} = -\frac{\partial V}{\partial x}\hat{i} - \frac{\partial V}{\partial y}\hat{j} - \frac{\partial V}{\partial z}\hat{k}$$

$$\vec{E} = -\left(\frac{\partial V}{\partial x}\hat{i} + \frac{\partial V}{\partial y}\hat{j} + \frac{\partial V}{\partial z}\hat{k}\right) = -\left(\frac{\partial}{\partial x}\hat{i} + \frac{\partial}{\partial y}\hat{j} + \frac{\partial}{\partial z}\hat{k}\right)V,$$

where $\left(\frac{\partial}{\partial x}\hat{i} + \frac{\partial}{\partial y}\hat{j} + \frac{\partial}{\partial z}\hat{k}\right)$ is known as Del operator, denoted as the symbol '∇'. So, we can write it in a handy form as

$$\vec{E} = -\nabla V,$$

where

$$\nabla = \left(\frac{\partial}{\partial x}\hat{i} + \frac{\partial}{\partial y}\hat{j} + \frac{\partial}{\partial z}\hat{k}\right)$$

N.B: We can also write

$$\frac{dV}{ds} = -E \cos \theta,$$

This is the derivative of potential V along a particular direction (s-direction or s-axis), which is given by the factor $\cos \theta$, where θ = angle between the displacement and electric field (E-lines). So, the above derivative is called directional derivative. This means that along any given direction (s, say), we cannot get the total value of the electric field; we just get the field partially, that is, $E \cos \theta$. So, by using the partial derivative, we write it in terms of partial (directional) derivative as

$$\frac{dV}{ds} = \frac{\partial V}{\partial s} = -E \cos \theta$$

If we chose $\cos \theta = -1$, we have the maximum potential gradient, given as

$$\left(\frac{dV}{ds}\right)_{\text{max}} = -E$$

This tells us that the electric field strength at any point is numerically equal to the negative of the maximum potential gradient at that point. In other words, electric field points in the direction in which the potential decreases most rapidly with distance, that is, with maximum potential gradient. So, potential decreases most sharply with distance in the direction of the electric field.

1.14 Equipotential line

When the potential does not change along a line, this is called equipotential or the V-line. To, understand it better, let the potential change by δV in a certain direction while undergoing a small displacement δs. By the concept of work, $\delta V = -E \delta s \cos \theta$. If potential remains the same, put $\delta V = 0$. So, $\cos \theta = 0$, Then, both V and E-lines will cross at right angles to each other. This is the first criteria for drawing equipotential lines if E-lines are given or vice versa. Let us recast the expression

$$\vec{E} = -\left(\frac{dV}{ds}\right)_{\text{max}} \hat{n}$$

The right side gives us the field strength which is directly proportional to the density of E-lines; the left side term is the maximum potential gradient that signifies the density of V-lines. So, we can conclude that the density of V-lines is directly proportional to the density of E-lines. In other words, where the densities of E-lines are greater, the densities of V-lines must be greater. So, it is simple to understand that the V-lines must be more crowded at the strong electric fields and vice versa. This is the second criteria for drawing equipotential lines if E-lines are given or vice versa.

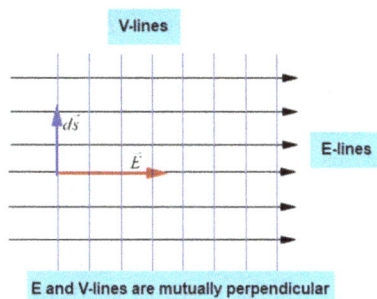

V-lines

$\vec{ds}$

$\vec{E}$

E-lines

E and V-lines are mutually perpendicular

You can see that so long as you are moving normal to the field lines, you do not have to perform work in transporting a test charge. If work done is zero, we can say that the potential difference is zero; the potential remains the same along that line. Zero work means that the force is perpendicular to the displacement $\vec{ds}$; since the force $\vec{F}$ is proportional to field strength $\vec{E}$, ($\vec{F} = q\,\vec{E}$), we can say that you must always move perpendicular to the field lines to maintain a potential by performing zero work. So, along an equipotential line the work done is always zero between any two points. Equipotential lines can also be termed as V-lines just like the electric lines of forces are called E-lines.

In practice, you can imagine a bunch of lines of force (or E-field lines) in a three-dimensional pattern in space. So, the equipotential points can form innumerable equipotential lines forming an equipotential surface. In other words, you can draw any line called an equipotential line in an equipotential plane or surface. The equipotential surfaces can be flat (plane) and curved. For instance, an infinite (very large) charged sheet has a uniform electric field that can form a flat equipotential surface in space. An infinite (very long) uniform line charge forms a cylindrical equipotential surface. A point charge or a uniformly charged sphere forms a spherical equipotential surface around the charge. On a sheet of paper or teaching board, the cylindrical and spherical equipotential surfaces can be drawn as a set of concentric circles. If we move a test charge slowly in an equipotential surface (or line), the work done will be zero; if we move a test charge slowly between any two equipotential surfaces following different paths, every time same work will be done. This means that the work done does not depend upon the path followed between any two equipotential surfaces. In other words, each equipotential surface or line has a specific potential energy per unit charge. This justifies the conservativeness of an electrostatic or gravitational field.

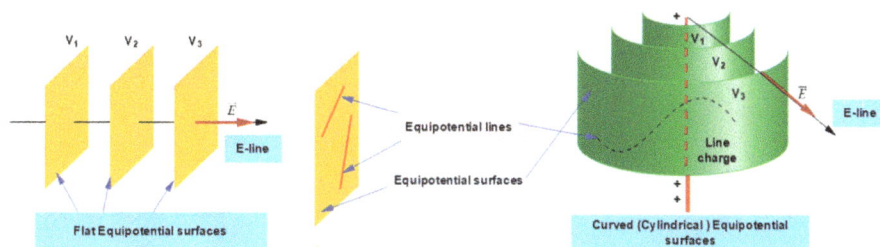

V_1 V_2 V_3

E

E-line

Flat Equipotential surfaces

Equipotential lines

Equipotential surfaces

V_1
V_2
V_3

$\vec{F}$

E-line

Line charge

Curved (Cylindrical) Equipotential surfaces

For a uniform electric field, the spacings of E-field lines must remain uniform. Therefore, the spacings of V-lines also remain uniform. However, for a non-uniform electrostatic field, the spacings of the equipotential surface will vary. We need to be careful in deciding the variations of the spacings of V-lines. First, you should see whether the potential increases or decreases in a given direction. You can also use the idea that the potential increases most rapidly (with distance but not with time) in the direction of the electric field and vice versa. Second, you will have to draw the V-lines perpendicular to E-lines at every point of space. Third, you must see that the density or crowding of V-lines will be more at a stronger electric field. If the field strength increases in any direction, the potential must decrease in that direction. Decreasing potential means less crowding of V-lines.

For the non-uniform field, equipotential lines are dense at point 1 where the field is strong and vice versa at point 2. We must be careful in understanding two factors such as variation (increasing or decreasing) of the spacings of equipotential lines and variation of potential. It is clear that potential decreases most rapidly with distance in the direction of (electric) E-field lines.

N.B: These points must be taken in to account while drawing E- and V-pattern.

1. E- and V-lines must be normal to each other.
2. The density of equipotential lines shows the maximum potential gradient which is numerically equal to an electric field.
3. The density of E-lines is proportional to an electric field strength.
4. The density of V-lines is proportional to the density of E-lines.

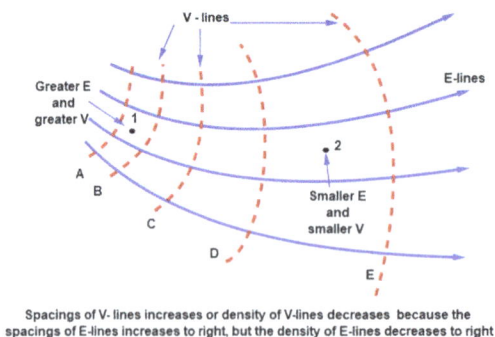

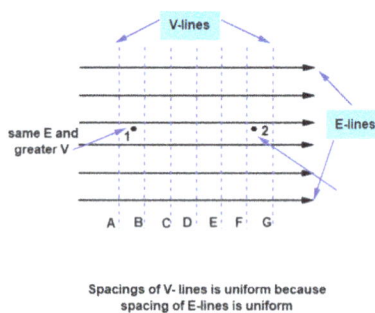

Spacings of V- lines increases or density of V-lines decreases because the spacings of E-lines increases to right, but the density of E-lines decreases to right

Spacings of V- lines is uniform because spacing of E-lines is uniform

Example 23 Draw the field pattern for the pattern of equipotential lines.

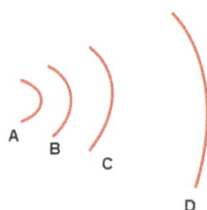

Solution

First of all, since the density of *V*-lines decreases radially away, we can say that the potential gradient decreases accordingly. As the electric field is directly proportional to the maximum potential gradient, the field strength also decreases radially away. So, the density of lines of force must decrease with radial distance. The second criteria is that always *E*- and *V*-lines must intersect at right angles at each point of space. So the patterns of *E*-lines must be radially out, as shown in the figure.

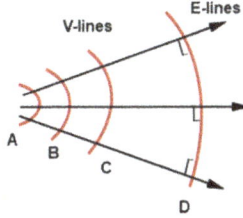

Here we can understand that the *E* lines are radial, but equipotential lines or *V* lines) are circular in a flat surface (2D) and concentric spheres in a space (3D). Ans.

Example 24 A charge of $q = 1.5$ micro coulomb is transferred from an equipotential surface of $V_1 = -10$ V to another having $V_2 = 5$ V. Find the work done by the (a) external agent (b) electric field.
   Solution

(a) The work done by the external agent is given as

$$W_{\text{ext}} = q(V_2 - V_1) = (1.5 \times 10^{-6})\{(5) - (-10)\} = 2.25 \times 10^{-5}\,\text{J Ans.}$$

(b) The work done by the electric field is given as

$$W_{\text{ext}} = -W_{\text{ext}} = -2.25 \times 10^{-5}\,\text{J Ans.}$$

1.15 Electric dipole

An electric dipole has two equal and opposite point charges separated by a very small distance. For instance, all polar molecules like HCl water etc, are the practical examples of dipoles.

1.15.1 Dipole moment

Let the particles *A* and *B* of charge $+q$ and $-q$ have positions $\vec{r_1}$ and $\vec{r_2}$, respectively relative to origin *O*. The moments of the charges relative to *O* are given as $+q\vec{r_1}$ and $-q\vec{r_2}$, respectively. Adding these two monopole moments vectorially, we have the net moment of the dipole, that is, dipole moment, given as

$$\vec{p} = +q\vec{r_1} - q\vec{r_2} = q(\vec{r_1} - \vec{r_2}); \text{ by putting } (\vec{r_1} - \vec{r_2}) = \vec{r},$$

we have

$$\vec{p} = q \ \vec{r},$$

where $\vec{r}$ = position vector of the point charge $-q$ relative to $+q$ is $\vec{r}$; so, the dipole moment is a vector quantity pointing from $+q$ to $-q$ and it depends upon the charge of the particles and the distance of separation between the particles. Its unit is coulomb meter or Debye; 1 Debye = 3.335×10^{-30} coulomb meter.

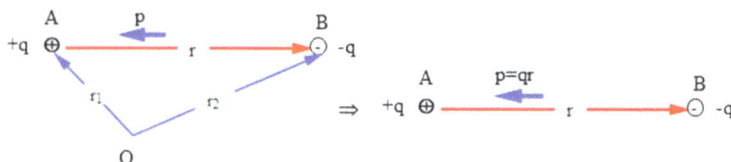

The dipole moment is equal to the vector sum of mono-pole moments of two equal and opposite charges;it does not depend upon any reference frame in contrary to monopole moment.

Example 25 Find the dipole moment of (a) water (b) carbon dioxide molecule.
  Solution

(a) Due to the presence of lone pairs of electrons in the oxygen molecules, the shape of a water molecule is a bent structure. The bond angle is $\theta = 104.5°$. The individual bond moment of each hydrogen atom relative to the oxygen atom is $p = 1.5$ Deybe. Now we have two dipoles each of moment p making an angle $\theta = 104.5°$. By applying parallelogram law of vectors, or component method of vector addition, the resultant dipole moment is given as

$$p_r = 2p \cos (\theta/2) = 2(1.5D) \cos(104.5°/2) = 1.837 \text{ D Ans.}$$

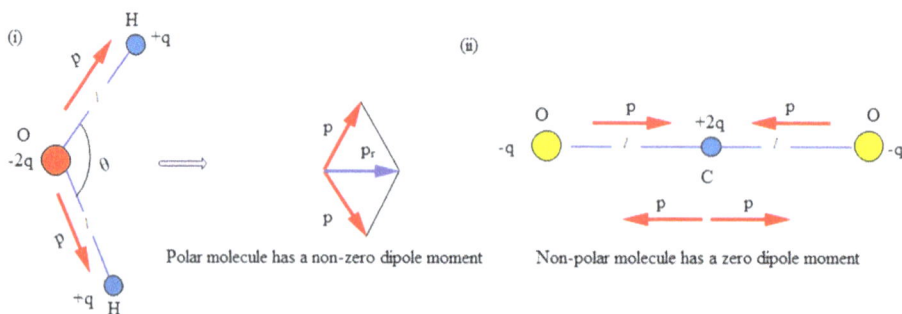

Polar molecule has a non-zero dipole moment   Non-polar molecule has a zero dipole moment

(b) Carbon dioxide is a linear molecule due to the absence of lone pairs. So, the bond moments of each oxygen atom with carbon atom are oppositely pointed and the net moment is zero. Ans.

1.15.2 Potential due to a dipole

The potential V due to a dipole at a point P having a distance r which is much greater than the length l of the dipole can be given as the sum of the potential due to each charge of the dipole at P. So, we can write

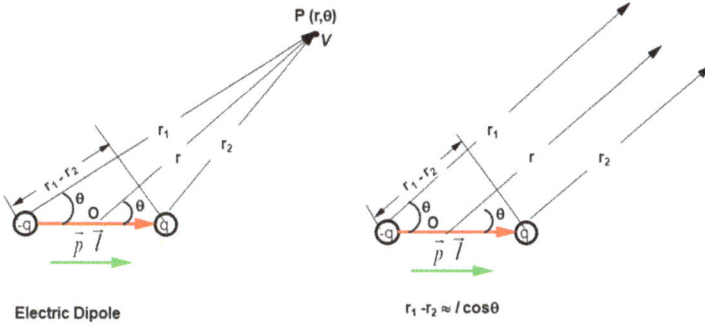

Electric Dipole

$r_1 - r_2 \approx l \cos\theta$

$$V = V_+ + V_- = \frac{Kq}{r_2} + \frac{K(-q)}{r_1} = Kq\left(\frac{1}{r_2} - \frac{1}{r_1}\right)$$

$$\Rightarrow V = Kq\left(\frac{1}{r_2} - \frac{1}{r_1}\right) = Kq\left(\frac{r_1 - r_2}{r_1 r_2}\right) \tag{1.70}$$

Since $r \gg l$, we can write $r_1 \cong r_2 \cong r$ as the three lines are approximately parallel an $r_1 - r_2 \approx l \cos\theta$. Putting these values in equation (1.70), we have

$$\Rightarrow V = Kq\left(\frac{r_1 - r_2}{r_1 r_2}\right) \simeq Kq\left(\frac{l \cos\theta}{r \cdot r}\right) = Kql\left(\frac{\cos\theta}{r^2}\right),$$

where $K = \frac{1}{4\pi\varepsilon_0}$ and $ql = p$. Then, we have

$$V = \frac{p \cos\theta}{4\pi\varepsilon_0 r^2} \tag{1.71}$$

The above expression tells us that *the potential due to an electric dipole is inversely proportional to the square of the distance of the point from the dipole and directly proportional to the* $\cos\theta$, *where* θ = *angle between the dipole moment and the position vector of point P relative to the dipole.* Please note that the dipole is point-like compared to the distance r.

1.15.3 Electric field due to a dipole

As the potential varies with both r and θ, using the relation $E = -(\frac{dV}{ds})_s$, we find the components of electric field in r and θ directions. Let us keep θ constant and move a little bit radially away by a small distance δr. If the change in potential is δV. Then, the negative of the potential gradient, that is, $-\delta V/\delta r$ is numerically equal to the component of electric field in r-direction, that is, E_r. It can be given as

$$E_r = -\left(\frac{\delta V}{\delta r}\right)_{\theta=C} = -\left(\frac{\partial V}{\partial r}\right) \tag{1.72}$$

Putting V from equation (1.71) in equation (1.72), we have

$$E_r = -\left(\frac{\partial V}{\partial r}\right)_{\theta=C} = -\frac{\partial}{\partial r}\left(\frac{p\cos\theta}{4\pi\varepsilon_o r^2}\right) = -\frac{p\cos\theta}{4\pi\varepsilon_o r^2}\frac{\partial}{\partial r}\left(\frac{1}{r^2}\right) = \frac{2p\cos\theta}{4\pi\varepsilon_o r^3}$$

$$E_r = \frac{2p\cos\theta}{4\pi\varepsilon_o r^3} \qquad (1.73)$$

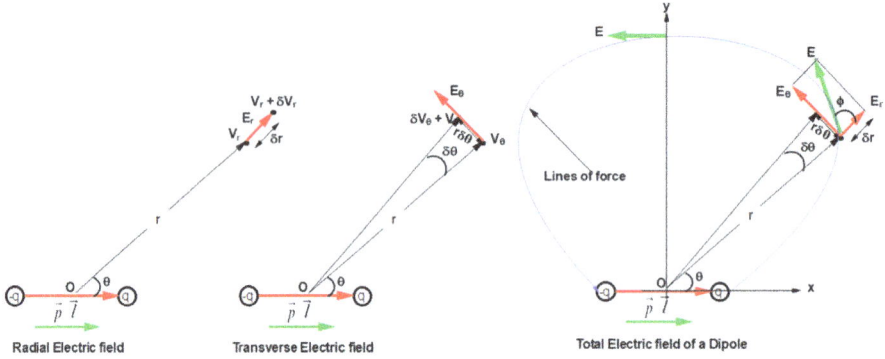

Radial Electric field   Transverse Electric field   Total Electric field of a Dipole

Let us keep r constant and move a little bit in θ-directions in the increasing order of θ by a small distance $\delta s_\theta = r\,\delta\theta$. If the change in potential is δV. Then, the negative of the potential gradient, that is, $-\delta V/r\delta\theta$ is numerically equal to the component of electric field in θ-direction, that is, E_θ. It can be given as

$$E_\theta = -\left(\frac{\delta V}{r\delta\theta}\right)_{r=C} = -\frac{\partial V}{r\partial\theta} \qquad (1.74)$$

Putting V from equation (1.71) in equation (1.74), we have

$$E_\theta = -\left(\frac{\partial V}{r\partial\theta}\right) = -\frac{\partial}{r\partial\theta}\left(\frac{p\cos\theta}{4\pi\varepsilon_o r^2}\right) = -\frac{p}{4\pi\varepsilon_o r^2}\frac{\partial}{\partial\theta}(\cos\theta)$$

Then, we have

$$E_\theta = \frac{p\sin\theta}{4\pi\varepsilon_o r^3} \qquad (1.75)$$

Combining these two components from equations (1.73) and (1.75), the net electric field at P is

$$\vec{E} = \vec{E_r} + \vec{E_\theta} = E_r\hat{r} + E_\theta\hat{\theta} = \frac{2p\cos\theta}{4\pi\varepsilon_o r^3}\hat{r} + \frac{p\sin\theta}{4\pi\varepsilon_o r^3}\hat{\theta}$$

$$\Rightarrow \vec{E} = \frac{p}{4\pi\varepsilon_o r^3}(2\cos\theta\hat{r} + p\sin\theta\hat{\theta}) \qquad (1.76)$$

The magnitude of the electric field is

$$\Rightarrow E = \frac{p}{4\pi\varepsilon_o r^3}\sqrt{1 + 3\cos^2\theta} \qquad (1.78)$$

The angle made by $\vec{E}$ with $\vec{r}$ is given as

$$\phi = \tan^{-1}(E_\theta/E_r) = \tan^{-1}\left(\frac{p\sin\theta}{4\pi\varepsilon_o r^3}\Big/\frac{2p\cos\theta}{4\pi\varepsilon_o r^3}\right) = \tan^{-1}(\tan\theta/2) \qquad (1.79)$$

1.15.4 *E* and *V* Pattern of the dipole

In a tiny dipole (shown as a tiny red arrow) the *E*-field lines emanate from the +ve charge and terminate on the −ve charge of the dipole. So, *E*-lines are not closed and are given by arrows showing a directional nature of the field strength. However, the *V*-lines or equipotential lines are closed (circles having shifted centres). You can see that the *E*- and *V*-lines cross each other at right angles. The tiny black arrow is the dipole moment that points from −ve to +ve charge of the dipole. The perpendicular bisector of the dipole is called zero-potential line or flat surface because any point on this line or flat surface is equidistant from both positive and negative charges. The field lines are parallel to the dipole moment while passing through the zero-potential line which is a flat surface in three-dimensional space.

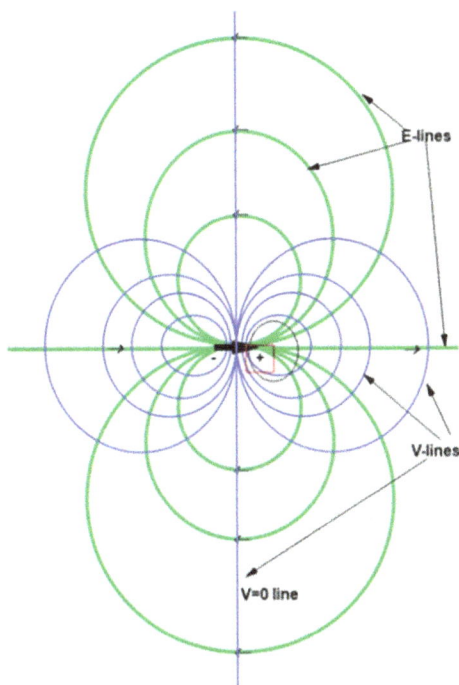

E and V pattern of a Tiny Dipole (Black arrow)

1.16 Field ideas in hydrodynamics

The idea of field lines was conceptualized by Micheal Faraday. It is borrowed from hydrodynamics where it is used as a term *streamline*. We have talked about it in the chapter of fluid dynamics. A streamline is a path followed by an element of fluid over a period of time. Two such elements in a streamline flow do not collide. This means that two streamlines will never intersect. The rate of flow of the fluid is called flux of the streamlines. If it is volume of liquid flowing per second, it is volume flux; if it is mass of liquid flowing per second, we call it mass flux. The volume flux of liquid in a streamline is given as

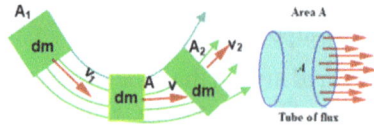

$$\phi = \frac{dV}{dt} = \vec{v} \cdot \vec{A} = vA \cos \theta$$

Since a streamline is characterized by the velocity of the fluid elements, we can say that the flux is the velocity flux or $\vec{v}$-flux, which can be given as

$$\phi_v = \vec{v} \cdot \vec{A}$$

1.16.1 Flux of $\vec{v}$-field (ϕ_v)

Flux is a word used in science and engineering bearing different meanings: mass flux, volume flux, energy flux, momentum flux etc. For the vector fields, we have $\vec{v}$-flux, $\vec{E}$-flux, $\vec{B}$-flux, $\vec{j}$-flux, v-flux etc. Here, we use the word 'flux' as 'rate of flow'; if it is rate of flow of mass, it is called mass flux; if it is rate of flow of volume, we call it volume flux; if it is the flow of energy, we call it energy flux etc. The word 'flux' of anything roughly represents the 'time rate of flow of that thing', denoted by rate R or ϕ. The volume flux of flowing air can be given as rate of flow of air; mathematically, the flux of air flow is given as

$$\phi = \frac{dV}{dt} = A_n v = Av \cos \theta = \vec{A} \cdot \vec{v},$$

where A_n= area perpendicular to the direction of flow (or velocity). Since the flow of fluid is characterized by the velocity vector, the flux is called flux of v-field, given as

$$\phi_v = \vec{A} \cdot \vec{v}$$

Physical significance of ϕ_v: When the rate of flow is more, we can say that flux of liquid is more. We can increase the flux in three different ways; either by increasing the speed v keeping the cross-section constant or by increasing cross-section A keeping flow velocity constant, or by increasing both A and v.

It means $\phi_v \propto A_n$ and $\phi_v \propto v$

More ϕ_v means more rapid flow of fluid. Accordingly, you need to imagine more streamlines or lines of flux (or force) of fluid. It means $\phi_v \propto$ Number of streamlines

In other words, $\phi \equiv$ Number of lines of forces (stream or v-lines)

The above expression relates the physical quantity 'flux' to the geometrical (mathematical) quantity 'number of lines of force'.

Flux passing through a flat surface: If you ask 'how much air flows, it has no precise meaning. If you ask 'what is the rate of flow through any loop of area A, it makes a better sense. The rate of flow (flux) passing through this surface (loop) is given as $\phi_v = \vec{v} \cdot \vec{A} = vA \cos \theta = v_n A = vA_n$, where θ = angle made by the outward normal of the area with the velocity $\vec{v}$ of the fluid, $v_n (= v \cos \theta)$ is the component of velocity of liquid particles perpendicular to the surface and projection of the area of the loop perpendicular to the flow velocity.

When $\theta = 90°$ as shown in the case (2) of the flowing figure, no air flows through the loop. In this case the air just sweeps the area. When the angle of orientation $\theta = 0$, maximum air flows per second through the loop as shown in case (2) of the following figure; so, it is practically evident that how much fluid passes through the area per second, that is, rate of flow or flux is governed by the orientation angle θ of the area under consideration for a constant flow velocity.

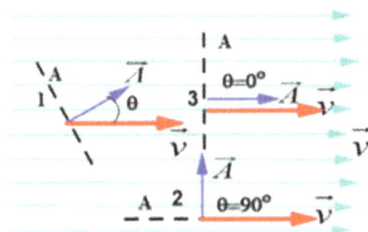

Streamline air flow.

Since the flux of the fluid is directly proportional to the numbers of $\vec{v}$-field lines or streamlines, we can write the following.

Flux passing through a curved surface: If the surface is curved, take an elementary area $d\vec{A}$. The flux passing through the elementary area is

$$d\phi = \vec{v} \cdot d\vec{A} = v dA \cos\theta.$$

Then, the total flux is $\phi = \int d\phi$, where $d\phi = \vec{v} \cdot d\vec{A}$

This gives $\phi = \int \vec{v} \cdot d\vec{A}$.

The above expression tells us that the surface integral of velocity gives the $\vec{v}$-field flux passing through the curved surface.

Flux passing through a closed surface: If you want to find the total flux passing through a closed surface, consider an elementary patch (area) on the surface. Find the flux $d\phi$ passing through the elementary area. Then, the total passing through the closed surface is

$$\phi = \oint d\phi$$

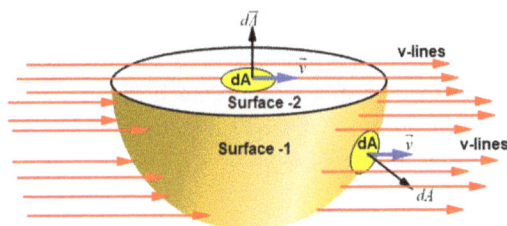

Substituting $d\phi = \vec{v} \cdot d\vec{A}$, we have

$$\phi = \oint \vec{v} \cdot d\vec{A},$$

where $\oint$ is called the closed integral.

1-63

The closed surface integral of velocity gives us the total flux passing through the closed surface.

Algebraic nature of flux: When the net flux ϕ passing through a closed surface is positive, we must think of a 'source' of generating the flux. This signifies that the outward flux is greater than inward flux. In other words, amount of liquid going out of the closed surface is greater than amount of liquid coming into (entering) the closed surface. Similarly, if the flux ϕ is negative the net flux is inward. This means, the inward flux is greater than the outward flux. In that case, something must be there to suck the liquid. This is known as 'sink'. If ϕ is zero, then the inward and outward fluxes are equal. It means, there is no *source* and *sink* in the region enclosed by the curved surface.

1. In a closed surface, if $\phi > 0$, net flux is outward; hence there is a source in the region bounded by the closed surface.
2. If $\phi < 0$, net flux is inward; hence there is a sink in the region bounded by a closed surface.
3. If $\phi = 0$, no net flux passes; hence there is no source and no sink.

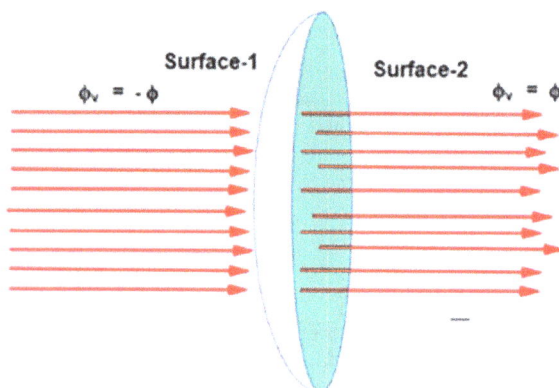

The total number of v-lines crossing the surface-1(inward) is equal to that crossing the surface-2(outward). Since the number of v-lines represent the v-flux (amount of liquid flowing per second or volume rate of flow),we can say that the net flux passing through the closed surface(sum of two surfaces) is zero.

Surface-1 (curved) has inward flux $-\varnothing$ and surface-2 (flat) has outward flux $+\varnothing$; so, the total flux passing through the closed surface (surface-1 plus surface-2) is zero. This is due to the algebraic nature of flux.

1.16.2 Flux density of $\vec{v}$-field

The ratio of flux and component of area A_n is given as

$$\frac{\phi}{A \cos \theta} = v,$$

where $A \cos \theta = A_n$, component of A normal to velocity of flow. The ratio of flux and normal area is defined as density of flux or *flux density*. So, the flux density of $\vec{v}$-field can be given as

$$\frac{\phi_v}{A_n} = v$$

This states that density of $\vec{v}$-field or flux is numerically equal to the velocity vector $\vec{v}$ of the $\vec{v}$-field.

1.16.3 Flux density at a point

Flux density of $\vec{v}$-field is numerically equal to the velocity of the flow. Then, the velocity of the fluid at a point is given by the flux density at a point. For this purpose, we consider a small area ΔA_n perpendicular to a bundle of streamlines. Then we find the flux $\Delta \phi$ passing through this area. This will give you an average flux density. If we reduce the area ΔA_n to a point, the flux $\Delta \phi$ will tend to decrease. Then, the ratio of $\frac{\Delta \phi}{\Delta A_n}$ when $\Delta A_n \to 0$ will give us the flux density at a point, which can be stated as

$$\lim_{\Delta A_n \to 0} \frac{\Delta \phi}{\Delta A_n} = \frac{d\phi}{dA_n} = v$$

Physical significance of flux density: As you know, the flow of liquid is associated with moving liquid particles. Hence, a moving liquid is characterized by velocity $\vec{v}$. It means, a moving liquid is a field of velocity vector (or $\vec{v}$-field). In this way, the streamlines can also be termed as v-flux lines of v-lines. Hence, the density of v-flux at any point is defined as velocity of liquid at that point. When you say flux density of v-flux is more, it means that v is more. Hence, the lines of forces are more crowded. Please remember that the flux density of v-field is none other than 'velocity' itself.

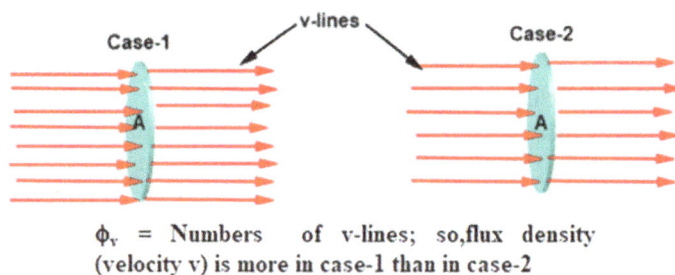

ϕ_v = Numbers of v-lines; so, flux density (velocity v) is more in case-1 than in case-2

Streamlines are more crowded in case-1 than in case-2; so the flux density of v-flux lines, that is velocity v is more in case-1 than in case-2. Physically, the liquid moves faster through the area A in case-1 than in case-2.

1.17 Field ideas in electrostatics

1.17.1 Flux of $\vec{E}$-field

In electrostatics, we have two types of charges, namely +ve and −ve charges; a positive point charge pushes a test charge away and a negative point charge pulls the test charge towards it. The line along which a test charge experiences a force is called electric lines of force. So, the lines of forces emanate from a positive charge and terminate on a negative charge. It is just like streamlines diverge from a point called source and converge to a point called a sink. So, a positive charge behaves as a source and a negative charge behaves as a sink.

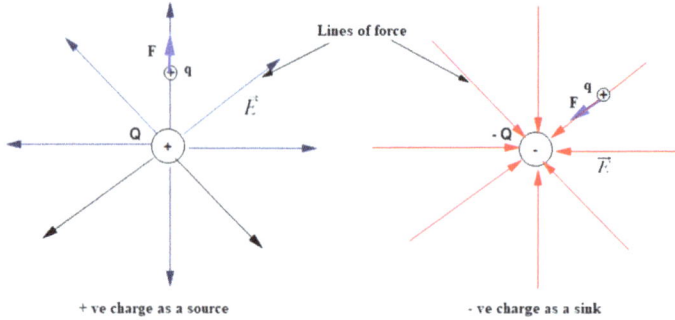

+ ve charge as a source   - ve charge as a sink

You may ask, what is flowing from or towards the charges? Here nothing flows like a liquid, but we can practically feel that a constant outward thrust (force) acts on the test charge due to the presence of the fixed charge $+Q$. Similarly, the test charge q experiences an inward pull due to the fixed charge $-Q$. The generation of the attraction or repulsion force in electrostatics can be explained by using the concept of virtual photons following the hydrodynamical method or model. We cannot perceive this force field with our eyes just like gravitational and magnetic force fields. However, these fields are very real because they are physically present in imparting the respective forces. So, these fields are not abstract, they are as real as the v-fields in fluid dynamics. The physical interpretation of the fields is Faraday's original ideas that changed the Newtonian concept of 'action at a distance' and revolutionized the way of thinking on Nature in terms of fields and particles. The electrostatic field is characterized by the field strength $\vec{E}$. It is called a static field because the field strength at any point does not vary with time. It is made equivalent to the $\vec{v}$-field of fluid dynamics where the velocity does not change with time at a point. In this case $\vec{v}$-field is static or potential. For the potential flow, the field density $\vec{v}$ can be expressed as a *gradient* of a potential φ of $\vec{v}$-field. In the same way, the electric field strength $\vec{E}$ can be expressed as a *gradient* of a potential V (or φ) of $\vec{E}$-field. There is a close resemblance between these two potential fields such as $\vec{v}$-field in hydro-dynamics and $\vec{E}$-field in electrostatics. In both cases, φ can be denoted as a potential function. The gradient of this potential defines the field vector. The gradient of φ in $\vec{v}$-field along any direction defines $\vec{v}$ along that direction; similarly, the gradient of φ in $\vec{E}$-field defines $\vec{E}$ along that direction. So, the static or potential $\vec{v}$-field and $\vec{E}$-field can be treated by the same concepts of divergence, curl, gradient and Laplacian, which are four operators in vector calculus. Much of the development was done by the brilliant self-taught British electrical engineer Oliver Heaviside in the last part of the 19th century. You should not get confused with the term *dynamics* of fluid and statics of electric charges. In both cases, the value of the field vectors $\vec{v}$ and $\vec{E}$ do not change with time—remaining stationary or static. In hydrodynamics the fluid moves giving us a velocity, but the $\vec{v}$-lines can be made static by maintaining the flux constant.

The flux $\vec{v}$-field over a closed surface can be given as

$$\phi_v = \oint \vec{v} \cdot d\vec{A}$$

Similarly, the flux $\vec{E}$-field over a closed surface can be given as

$$\phi_E = \oint \vec{E} \cdot d\vec{A}$$

1.18 Gauss law in electrostatics

Let us consider a fixed positive point charge $+Q$ and imagine a surface enclosing the charge. The charge spreads its radially outward electric fields to infinity. It is a matter of common sense that the flux of the electric field produced by the charge passing through a closed surface does not depend upon the shape and size of the surface. So, for the sake of mathematical simplicity we can imagine a spherical Gaussian surface as shown as a dotted circle of radius r, centered at the charge. The flux passing through is the Gaussian surface is

$$\phi_E = \oint \vec{E} \cdot d\vec{A}$$

Because the direction of the elementary patch $d\vec{A}$ of the Gaussian surface and electric field $\vec{E}$ are both radially outward,

$$\phi_E = \oint \vec{E} \cdot d\vec{A} = \oint E \cdot dA \cos 0 = \oint E \cdot dA$$

Since the electric field has the same magnitude at each element of the Gaussian surface (sphere), we can take it out of the integral.

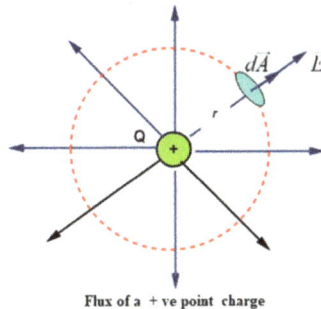

Flux of a +ve point charge

So, we have $\oint \vec{E} \cdot d\vec{A} = E \oint dA$, where $E = Q/(4\pi\epsilon_0 r^2)$ and $\oint dA = $ Area of the sphere $= 4\pi r^2$. Then, the flux out is

$$\phi_E = \oint \vec{E} \cdot d\vec{A} = Q/\varepsilon_0$$

This states that *the flux electric field* $\overrightarrow{E}$ *out of the Gaussian surface is numerically equal to the charge Q enclosed by the Gaussian surface divided by the absolute permittivity of the vacuum. In short, we can say that 'flux out is equal to charge enclosed divided by* ϵ_o. This is the statement of Gauss's law.

Since the Gauss law is valid for a point charge, it must be valid for a group of discrete or continuous charges.

Summarizing the above facts, we can conclude that:

1. The flux of a liquid passing through a flat surface area A with a velocity v is given as $\phi = vA \cos \theta =$ where $\theta =$ angle between $\overrightarrow{v}$ and $\overrightarrow{A}$ and $\overrightarrow{A}$ is area vector directed along outward normal.

2. Since $\cos \theta$ can be positive, negative and zero, we can have positive, negative and zero flux. Hence, flux is an 'algebraic scalar quantity'. When ϕ is +ve, it represents an outward flux; if ϕ is −ve, it represents inward flux. Outward and inward direction is decided by the location of the observer.

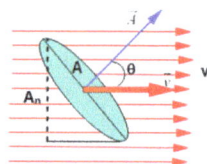

1.19 Applications of Gauss's law

Gauss's law is valid for all types of charge configuration but may not be useful for unsymmetrical charge distribution. However, for some symmetrical charge distribution, it will be mathematically simpler to find the electric field.

Example 26 *Spherical symmetry*: Find the electric field due to a charged sphere of radius R with a uniform volume charge distribution. The volume charge density at a the radial distance r is $\rho(r) = \rho =$ constant for uniform volume charge distribution.
   Solution

   (a) Inside the sphere; $r \leqslant R$
      Flux out of the spherical Gaussian surface of radius r is

$$\phi_E = \oint \overrightarrow{E} \cdot d\overrightarrow{A} = \oint E \cdot dA \cos 0° = E \oint dA = EA, \quad \text{where } A = 4\pi r^2$$

   Then,

$$\phi_E = 4\pi r^2 E \qquad (1.80)$$

The charge enclosed by the spherical Gaussian surface is

$$q_{en} = \int \rho dV$$

$$= \int_0^r \rho \, 4\pi r^2 dr, \quad \text{where} \ \rho = \text{constant} \tag{1.81}$$

$$= 4\pi\rho \int_0^r r^2 dr = 4\pi\rho r^3/3$$

Applying Gauss's law,

$$\phi_E = \frac{q_{en}}{\varepsilon_0} \tag{1.82}$$

Using the last three equations, we have

$$\Rightarrow E 4\pi r^2 = \frac{\rho(4\pi r^2/3)}{\varepsilon_0}$$

$$\Rightarrow E = \frac{\rho r}{3\varepsilon_0}; \ r \leqslant R \ \text{Ans.}$$

(b) Outside the sphere; $r \leqslant R$

Flux out of the spherical Gaussian surface of radius r is

$$\phi_E = EA, \quad \text{where} \ A = 4\pi r^2$$

Then,

$$\phi_E = 4\pi r^2 E \tag{1.83}$$

Since the spherical Gaussian surface is bigger than the charged sphere, the charge enclosed by the Gaussian surface will be equal to the total charge of the sphere. So, the charge enclosed is

$$q_{en} = Q_{sphere} = \rho \, (4\pi r^3/3)$$

$$4\pi\rho R^3/3 \tag{1.84}$$

Applying Gauss's law,

$$\phi_E = \frac{q_{en}}{\varepsilon_0} \tag{1.85}$$

Using the last three equations, we have

$$\Rightarrow E 4\pi r^2 = \frac{\rho(4\pi R^2/3)}{\varepsilon_0}$$

$$\Rightarrow E = \frac{\rho R^2}{3\varepsilon_0 r^2}; \ r \geqslant R \ \text{Ans.}$$

Example 27 A cup of radius R is placed in a uniform vertical electric field of strength $\vec{E}$. Fine the flux passing through the curved surface.

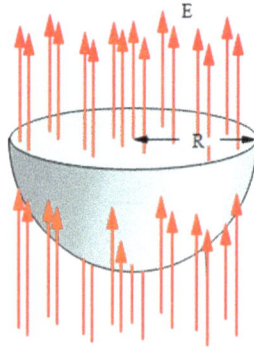

Solution

Let us cover the cup by a plate equal to the area of the circle of radius R. Let the curved surface of the cup be numbered as 1 and covered plate, that is, flat surface be numbered as 2. As the closed surface $(1 + 2)$ does not enclose any charge, the net flux out must be zero according to Gauss's law. Let the flux coming towards the curved surface (inward flux) be ϕ_1 and the flux going out (outward flux) of the flat surface be ϕ_2. Then the net flux passing through the curved surface is $\phi = -\phi_1 + (+\phi_2) = q_{en}/\varepsilon_0 = 0$; so, we have $\phi_1 = \phi_2$, where $\phi_2 = E(\pi R^2)$ because both the outward normal to the flat surface and the electric field have the same direction. Hence, the flux passing through the curved surface is $\phi_1 = \pi R^2 E$ Ans.

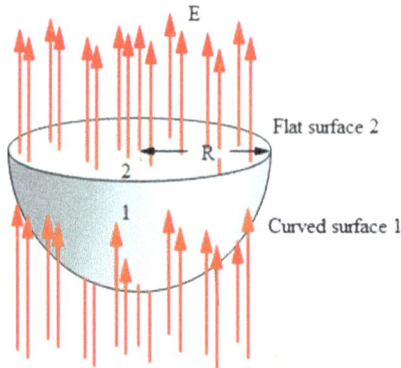

Example 28 *Cylindrical symmetry:* Find the electric field due to a straight infinite thread of linear charge density λ.

Solution

For mathematical simplicity, we choose a cylindrical Gaussian surface having top, bottom flat surface and side curved surface. As the thread is positively charged, the electric field is radially outward. We can see that the elementary patch of area $\overrightarrow{dA}$ and electric field $\vec{E}$ of top and bottom surface are perpendicular to each other. So $\vec{E} \cdot d\vec{A} = 0$; this means that no flux passes through the top and bottom surfaces as the electric field lines sweep these surfaces.

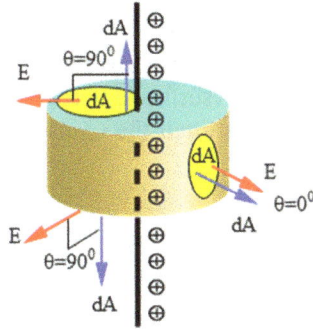

But, an elementary patch of area $\overrightarrow{dA}$ and electric field $\vec{E}$ of the side (curved) surface are parallel to each other. So the flux passing through the area $\overrightarrow{dA}$ is $\vec{E} \cdot d\vec{A} = EdA$. Integrating, the net flux out is

$$\phi_E = \oint \vec{E} \cdot d\vec{A} = \oint E \cdot dA \cos 0° = E \oint dA = EA,$$

where $A =$ area of the curved surface $= 2\pi rl$

Then,

$$\phi_E = 2\pi rlE \tag{1.86}$$

The charge enclosed by the cylindrical Gaussian surface is

$$q_{en} = \lambda l \tag{1.87}$$

Applying Gauss's law,

$$\phi_E = \frac{q_{en}}{\varepsilon_0} \tag{1.88}$$

Using the last three equations, we have

$$2\pi rlE = \lambda l/\varepsilon_0$$

Then, we have

$$E = \frac{\lambda}{2\pi\varepsilon_0 r} \text{ Ans.}$$

Example 29 *Plane symmetry:* Find the electric field due to a thin infinite flat sheet of surface charge density σ.

Solution

For mathematical simplicity we choose a cylindrical Gaussian surface. It has top and bottom flat surfaces and one side with a curved surface. As the large flat sheet is positively charged, the electric field is uniform pointing vertically up for the top surface and vertically down for the bottom surface of the sheet. Then, the elementary patch of area $d\vec{A}$ and electric field $\vec{E}$ of top and bottom surfaces are parallel to each other. So, the net flux passing out through both top and bottom surface is $\phi_{top} + \phi_{top} = \int EdA + \int EdA = 2\int EdA = 2E\int dA = 2EA$ where A = area of top or bottom of the Gaussian surface. However, an elementary patch of area $d\vec{A}$ and electric field $\vec{E}$ of the side (curved) surface are perpendicular to each other. As the electric field lines sweep the curved surfaces, zero flux passes through the curved surface. Then, the net flux passing through the Gaussian surface is

$$\phi_E = \oint \vec{E} \cdot d\vec{A} = \phi_{top} + \phi_{top} + \phi_{side} = 2EA \tag{1.89}$$

Since the sheet is extremely thin, the same charge belongs to both upper and lower face. So the charge enclosed by the cylindrical Gaussian surface is

$$q_{en} = \sigma A \text{ (but not } 2\sigma A) \tag{1.90}$$

Gauss's law is given as

$$\phi_E = \frac{q_{en}}{\varepsilon_0} \tag{1.91}$$

Using the last three equations, we have

$$2EA = \frac{\sigma A}{\varepsilon_0}$$

Finally, we have

$$E = \frac{\sigma}{2\varepsilon_0} \text{ Ans.}$$

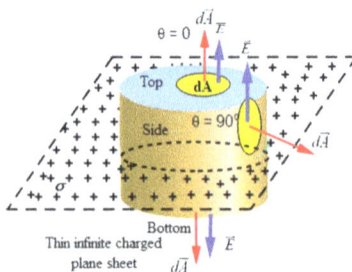

Thin infinite charged plane sheet

Problem 1 At what value of (a) q/m (b) distance between two point charges of mass m and change q, will the net force between them be zero?

Solution

(a) The attraction force due to gravity between the charge particles each of mass m is

$$F_{gr} = G\frac{m^2}{r^2} \tag{1.92}$$

The electrostatic repulsion between the charges is

$$F_{el} = \frac{q^2}{4\pi\varepsilon_0 r^2} \tag{1.93}$$

Since the net force on each is zero,

$$F_{el} = F_{gr} \tag{1.94}$$

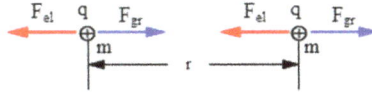

Using all the above three equations, we have

$$\frac{q^2}{4\pi\varepsilon_0 r^2} = \frac{Gm^2}{r^2}$$

$$\Rightarrow \left(\frac{q}{m}\right)^2 = 4\pi\varepsilon_0 G$$

$$\Rightarrow \frac{q}{m} = \sqrt{4\pi\varepsilon_0 G}$$

$$= \sqrt{\frac{1}{9 \times 10^9} \times 6.67 \times 10^{-11}}$$

$$= 0.862 \times 10^{-10} \text{ C kg}^{-1} \quad \text{Ans.}$$

(b) Since each force obeys inverse square law and opposes each other, the factor r^2 will get cancelled from both sides. The only factor is that the value of specific charge '$\frac{q}{m}$' must be equal to $\sqrt{4\pi\varepsilon_0 G}$ which is a constant quality. So irrespective of the separation distance the net force on each change particle will be zero for the above value of $\frac{q}{m}$. Ans.

Problem 2 Two positive point charges Q and q are placed separated by a distance ℓ, (a) where will a third charge (q', say) be placed so that all charges will be stationary, (b) what is the value of q'?
   Solution

(a) The net force on q' ($-$ve) is zero. So, we can write:

$$F_1 = F_2$$

$$\Rightarrow K\frac{Qq'}{x^2} = \frac{Kqq'}{(\ell - x)^2}$$

$$\Rightarrow \frac{x}{\ell - x} = \sqrt{\frac{Q}{q}}$$

$$\Rightarrow \frac{x}{\ell} = \frac{\sqrt{Q}}{\sqrt{Q} + \sqrt{q}}$$

$$\Rightarrow x = \frac{\sqrt{Q}}{\sqrt{Q} + \sqrt{q}} \ell \; \text{Ans.}$$

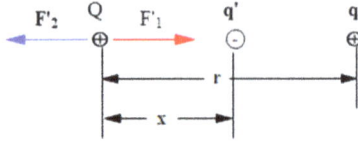

(b) The net force acting on Q is zero. So, we have

$$F'_1 = F'_2$$

$$\Rightarrow K \frac{Qq'}{x^2} = \frac{KQq}{\ell^2}$$

$$\Rightarrow q' = \frac{qx^2}{\ell^2}$$

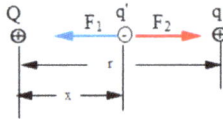

Putting the obtained value of x in the last expression,

$$q' = \frac{q}{\ell^2} \left(\frac{\sqrt{Q}}{\sqrt{Q} + \sqrt{q}} \right)^2 \ell^2$$

$$q' = \frac{Qq}{(\sqrt{Q} + \sqrt{q})^2} \; \text{Ans.}$$

Problem 3 Two identical charged bobs each of mass m and charge Q are suspended from a fixed point by two silk threads each of length ℓ in a liquid of relative permittivity ε_r. Assume σ and ρ are the density of the bob and liquid, respectively.

(a) If the angle between the threads is θ, find the electrical force of repulsion.
(b) Find the value of Q.
(c) Find the value of ε_r if the orientation of each string remains the same after the removal of the liquid.

Solution

(a) For equilibrium of the charge Q,

$$T \cos \theta + F_b = mg \qquad (1.95)$$

$$T \sin \theta = F, \qquad (1.96)$$

where $F_b=$ buoyant force

$$= v\rho g = \left(\frac{m}{\sigma}\right)\rho g$$

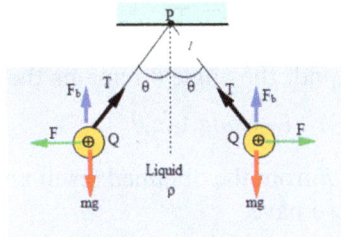

$$\Rightarrow F_b = mg\frac{\rho}{\sigma} \qquad (1.97)$$

Using equations (1.95) and (1.97),

$$T \cos \theta = mg\left(1 - \frac{\rho}{\sigma}\right) \qquad (1.98)$$

Using equations (1.96) and (1.98)

$$\tan \theta = \frac{F}{mg\left(1 - \frac{\rho}{\sigma}\right)}$$

$$\Rightarrow F = mg(1 - \frac{\rho}{\sigma})\tan \theta \quad \text{Ans.}$$

(b) The force of repulsion is given by Coulomb's law as follows

$$F = \frac{Q^2}{4\pi\varepsilon_0\varepsilon_r r^2}, \quad \text{where } r = 2\ell \sin \theta$$

$$\Rightarrow F = \frac{Q^2}{4\pi\varepsilon_0\varepsilon_r\left(4\ell^2\sin^2\theta\right)}$$

$$= \frac{Q^2}{10\pi t_0\varepsilon_r\ell^2\sin^2\theta}$$

Putting the obtained value of F, we have

$$mg\left(1 - \frac{\rho}{\sigma}\right)\tan\theta = \frac{Q^2}{16\pi\varepsilon_0\varepsilon_r\ell^2\sin^2\theta}$$

$$\Rightarrow Q = \sqrt{16\pi\varepsilon_0\varepsilon_r mg\ell^2(1 - \frac{\rho}{\sigma})\frac{\sin^3\theta}{\cos\theta}} \quad \text{Ans.}$$

(c) The value of relative permittivity is

$$\varepsilon_r = \frac{F_0}{F} \tag{1.99}$$

If we remove the liquid, the angle θ remains the same. So,

$$F_0 = mg\tan\theta \tag{1.100}$$

Putting the value of F from the obtained result and F_0 from equation (1.100) in equation (1.99), we have

$$\varepsilon_r = \frac{mg\tan\theta}{mg\tan\theta\left(1 - \frac{\rho}{\sigma}\right)}$$

$$\Rightarrow \varepsilon_r = \frac{\sigma}{\sigma - \rho} \quad \text{Ans.}$$

Problem 4 Four identical bobs each of mass m and charge Q (say) are suspended by four silk threads each of length ℓ. If each string makes an angle of 30° with the vertical, find the value of Q.

Solution

The repulsive force acting on Q due to the change at 1 and 2 are

$$F_1 = \frac{KQ^2}{r^2} \tag{1.101}$$

The repulsive force acting on Q due to the change at point 3 is

$$F_2 = \frac{KQ^2}{(\sqrt{2}r)^2} \tag{1.102}$$

Then, the net electric force acting on Q is

$$F = \sqrt{2}\,F_1 + F_2 \qquad (1.103)$$

Using the last three equations,

$$F = \frac{KQ^2}{r^2}\left[\sqrt{2} + \frac{1}{2}\right] \qquad (1.104)$$

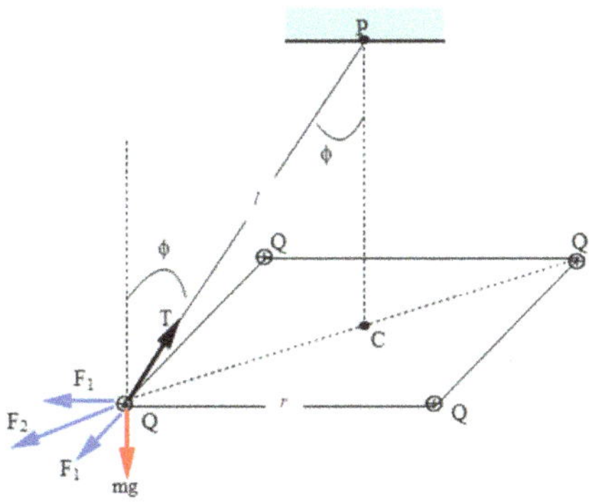

The vector triangle gives,

$$\frac{F_{\text{net}}}{mg} = \tan\phi$$

$$\Rightarrow F_{\text{net}} = mg\tan\phi \qquad (1.105)$$

Geometrically, we can write

$$\ell\sin\phi = \frac{r}{\sqrt{2}} \qquad (1.106)$$

From equations (1.104) and (1.106)

$$F_{\text{net}} = \frac{KQ^2}{2\ell^2\sin^2\phi}\left(\sqrt{2} + \frac{1}{2}\right) \qquad (1.107)$$

From equations (1.105) and (1.107)

$$\frac{KQ^2}{2\ell^2\sin^2\phi}\left(\sqrt{2} + \frac{1}{2}\right) = mg\tan\phi$$

Putting $\phi = 30^0$,

$$\frac{KQ^2}{2\ell^2\left(\frac{1}{2}\right)^2}\left(\sqrt{2} + \frac{1}{2}\right) = mg\frac{1}{\sqrt{3}}, \quad \text{where } K = \frac{1}{4\pi\epsilon_0}$$

$$\Rightarrow Q = \sqrt{\frac{4\pi \, \epsilon_0 \, mg\ell^2}{(2\sqrt{2} + 1)\sqrt{3}}} \quad \text{Ans.}$$

Problem 5 A charged bob A of mass m is suspended by a silk thread from P. It pushes another charged bob B fitted with a spring of stiffness k. If the system $(A + B)$ is equilibrium, find the
  (a) compression of the spring;
  (b) tension in the string;
  (c) if the string is cut, find the acceleration of A and B just after cutting the string.

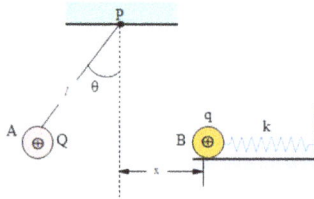

Solution

  (i)
      (a) For equilibrium of the charge q

$$F_s = kx = F_{el} \tag{1.108}$$

      For equilibrium of the charge Q

$$F_{el} = mg \tan \theta \tag{1.109}$$

      Using equations (1.108) and (1.109),

$$kx = mg \tan \theta$$

$$x = \frac{mg \tan \theta}{k} \quad \text{Ans.}$$

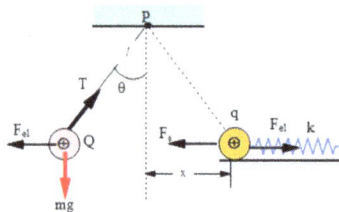

(b) The vertical force acting on the charge Q is

$$F_y = T \cos \theta - mg = 0$$

$$\Rightarrow T = mg \sec\theta \quad \text{Ans.}$$

(c) If the string is cut, tension will vanish immediately; $T = 0$. Then the net force acting on 'Q' is

$$\overrightarrow{F}_{\text{net}} = -F_{e\ell}\hat{i} - mg\hat{j} \tag{1.110}$$

Putting $F_{el} = mg \tan \theta$ from equation (1.109) in equation (1.110), we have

$$\overrightarrow{F}_{\text{net}} = -mg \tan \theta\hat{i} - mg\hat{j}$$

$$\Rightarrow m\overrightarrow{a} = -mg \tan \theta\hat{i} - mg\hat{j}$$

$$\Rightarrow \overrightarrow{a} = -g(\tan \theta\hat{i} + \hat{j})$$

or

$$|\overrightarrow{a}| = g\sqrt{1 + \tan^2 \theta} = \frac{g}{\cos \theta} \quad \text{Ans.}$$

The direction of $\overrightarrow{a}$ is opposite to the tension T; downward along the string.

Since the spring force does not change immediately after cutting the string, the acceleration of B will zero. Ans.

Problem 6 A flexible circular thread with uniform charge density λ and radius R is placed on an insulated surface. If a point charge Q is placed at the centre of the circle, find the tension developed in the string.

Solution

(a) The forces acting on an element of length $\delta\ell = R\delta Q$ at tension T acts at both ends tangentially and δF_{el} acts radially outward on the element.

As the element is at rest, the net radial force acting on the element is

$$\delta F_{e\ell} - T \sin \frac{\delta\theta}{2} - T \sin \frac{\delta\theta}{2} = 0$$

$$\Rightarrow \delta F_{el} = 2T \sin \frac{\delta\theta}{2}$$

$$\Rightarrow \delta F_{e\ell} = T\delta\theta \tag{1.111}$$

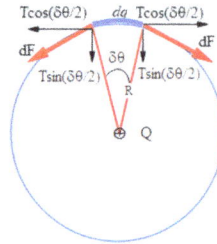

The force of interaction between Q and dq is

$$\delta F = \frac{Q\,dq}{4\pi\varepsilon_0 R^2} \tag{1.112}$$

Using equations (1.111) and (1.112),

$$\frac{Q\,dq}{4\pi\varepsilon_0 R^2} = T\delta Q$$

$$\Rightarrow T = \frac{Q}{4\pi\varepsilon_0 R}\left(\frac{dq}{Rd\theta}\right)$$

Since $\frac{dq}{Rd\theta} = \frac{dq}{d\ell} = \lambda$, we have

$$T = \frac{Q\lambda}{4\pi\varepsilon_0 R} \quad \text{Ans}$$

Problem 7 A heavy semi-infinite wire is placed along the x-axis having a uniform linear charge density λ. A flexible circular loop of uniform charge density λ' and mass m is placed in yz (vertical) plane as shown in the figure below. Find
    (a) the tension in the loop;
    (b) the net force acting on the loop;
    (c) the acceleration of the loop, if $m =$ mass of the loop. Neglect the tension of the string due to its own charge. Also neglect gravity.

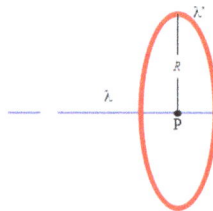

Solution

(a) The x and y component of electric field due to a semi-infinite charged wire at the perimeter of the ring is

$$E_x = \frac{\lambda}{4\pi\varepsilon_0 R} \text{ (to right)} \tag{1.113}$$

$$E_y = \frac{\lambda}{4\pi\varepsilon_0 R} \text{ (vertically up)} \tag{1.114}$$

The radially outward force acting on an elementary charge dq of the ring due to E_y is

$$\delta F = (\delta q) E_y \tag{1.115}$$

The radially inward force acting on the elementary charge dq of the ring is

$$\delta F_{in} = 2T \sin(\delta\theta/2) \simeq T\delta\theta, \tag{1.116}$$

where $T =$ tension in the string.

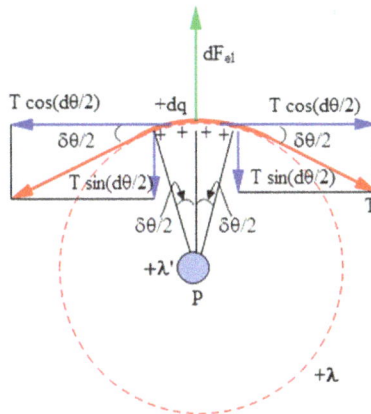

Then, using last two equations and referring to the previous problem, we have

$$(\delta q)\left(\frac{\lambda L}{4\pi\varepsilon_0 R}\right) = T d\theta \tag{1.117}$$

$$\Rightarrow T = \frac{\lambda}{4\pi t_0}\left(\frac{dq}{R d\theta}\right)$$

Putting $\frac{dq}{Rd\theta} = \lambda'$, we have

$$T = \frac{\lambda\lambda'}{4\pi\varepsilon_0} \text{ Ans.}$$

(b) Due to the parallel component of electric field E_x, we have the net force acting on the element of the ring is

$$\delta F_x = E_x dq$$

So, the net force acting on the loop is

$$F = \int F_x = F_x$$
$$= E_x \int dq = E_x q$$

Putting $E_x = \frac{\lambda}{4\pi\varepsilon_0 R}$, we have

$$F = \frac{\lambda q}{4\pi\varepsilon_0 R}, \quad \text{where } q = \lambda' 2\pi R$$

$$\Rightarrow F = \frac{\lambda\lambda'}{2\varepsilon_0} \text{ Ans.}$$

(c) The acceleration of the loop is

$$a = \frac{F}{m} = \frac{\lambda\lambda'}{2m\varepsilon_0} \text{ towards right.} \quad \text{Ans.}$$

Problem 8 A rod OP of length ℓ has a linear charge distribution which varies as cube of the distance x from O. Find the electric field at point O if the rod has a charge q.

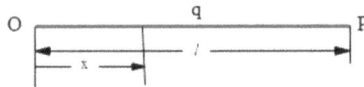

Solution

The electric field due to an element dq at O is

$$dE = \frac{dq}{4\pi\varepsilon_0 x^2}, \quad \text{where} \ \ dq = \lambda dx$$

$$\Rightarrow dE = \frac{\lambda \, dx}{4\pi\varepsilon_0 x^2} \tag{1.118}$$

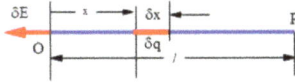

The net electric field at O is

$$E = \int dE = \frac{1}{4\pi\varepsilon_0} \int \frac{\lambda \, dx}{x^2}$$

If $\lambda = ax^3$, we have

$$E = \frac{1}{4\pi\varepsilon_0} \int \frac{ax^3 dx}{x^2}$$

$$= \frac{a}{4\pi\varepsilon_0} \int_0^{\ell} x \, dx$$

$$\Rightarrow E = \frac{a\ell^2}{8\pi\varepsilon_0} \tag{1.119}$$

The charge of the rod is given as

$$q = \int_0^{\ell} \lambda \, dx, \quad \text{where} \ \ \lambda = ax^3$$

$$\Rightarrow q = \int_0^{\ell} ax^3 \, dx = \frac{a\ell^4}{4}$$

$$\Rightarrow a = \frac{4q}{\ell^4} \tag{1.120}$$

Eliminating 'a' from equation (1.119) by equation (1.120),

$$E = \left(\frac{4q}{\ell^4}\right)\frac{\ell^2}{8\pi\varepsilon_0}$$

$$\Rightarrow E = \frac{q}{2\pi\varepsilon_0 \ell^2} \quad \text{Ans.}$$

Problem 9 Find the electric field due to a semi-infinite thin cylinder of uniform linear charge density at the centre of its end.

Solution

Take a thin ring of thickness dx at a distance of x from the end O. Then the charge of the ring is

$$q_{ring}(=dq) = \lambda dx$$

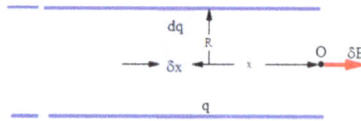

The electric field due to the ring at O is

$$E_{ring}(=dE) = \frac{q_{ring} x}{4\pi\varepsilon_0 (R^2 + x^2)^{\frac{3}{2}}} \tag{1.121}$$

The electric field due to the semi-infinite cylindrical shell is,

$$E = \int E_{ring} = \int_0^\infty \frac{q_{ring} x}{4\pi\varepsilon_0 (R^2 + x^2)^{\frac{3}{2}}}$$

Putting $q_{ring} = \lambda \, dx$, we have

$$E = \int_0^\infty \frac{(\lambda \, dx)x}{4\pi\varepsilon_0 \left(R^2 + x^2\right)^{\frac{3}{2}}}$$

$$= \frac{\lambda}{4\pi\varepsilon_0} \int_0^\infty \frac{x \, dx}{\left(R^2 + x^2\right)^{\frac{3}{2}}}$$

Putting $R^2 + x^2 = y$ and $2x dx = dy$ in the integration,

$$I = \int \frac{x dx}{(R^2 + x^2)^{\frac{3}{2}}},$$

we have

$$I = \int \frac{\frac{dy}{2}}{y^{\frac{3}{2}}} = \frac{1}{2} \int y^{-\frac{3}{2}} dy$$

$$= \frac{1}{2} \frac{y^{-\frac{3}{2}+1}}{-\frac{3}{2}+1} = -\frac{1}{\sqrt{y}}$$

$$\Rightarrow I = -\frac{1}{\sqrt{R^2 + x^2}} \Big|_0^\infty = \frac{1}{R}$$

Then, the total electric field is

$$E = \frac{\lambda}{4\pi\varepsilon_0}I, \quad \text{where } I = \frac{1}{R}$$

$$\Rightarrow E = \frac{\lambda}{4\pi\varepsilon_0 R} \quad \text{Ans.}$$

Problem 10 Two identical particles each of charge 'Q' are placed on the y-axis equidistant from the origin. Find the
  (a) expression for the electric field on x-axis;
  (b) Draw the
      (i) $E - x$ graph and measure the maximum electric field;
  (c) Describe the motion of a charge particle of charge '$-q$' placed very close to the origin along the x-axis.

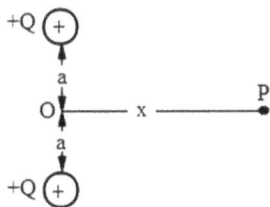

Solution

  (a) The electric field due to each charge is

$$E' = \frac{KQ}{r^2}$$

  The net electric field is pointing along $+x$-axis, given as

$$E = 2E' \cos\theta = 2\frac{KQ}{r^2}\frac{x}{r} = \frac{Qx}{2\pi\varepsilon_0 r^3}$$

$$\Rightarrow E = \frac{Qx}{2\pi\varepsilon_0(a^2 + x^2)^{\frac{3}{2}}} \quad \text{Ans.}$$

  (b)
      If $x \to 0$, $E \to 0$,
        If $x \to \infty$, $E \to 0$.
        Therefore, E is maximum at x, say, which can be obtained by putting

$$\frac{dE}{dx} = 0$$

$$\Rightarrow \frac{d}{dx}\left\{\frac{Qx}{2\pi\varepsilon_0(a^2 + x^2)^{\frac{3}{2}}}\right\} = 0$$

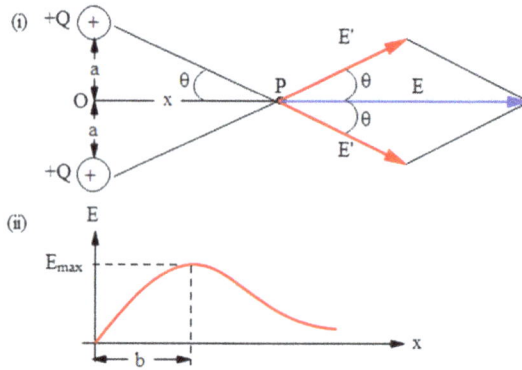

$$\Rightarrow x = \frac{a}{\sqrt{2}}$$

Then, putting the obtained value of q, we have

$$E_{max} = \frac{Q\left(\frac{a}{\sqrt{2}}\right)}{2\pi\varepsilon_0\left(a^2 + \frac{a^2}{2}\right)^{\frac{3}{2}}}$$

$$= \frac{Q\left(\frac{a}{\sqrt{2}}\right)}{2\pi\varepsilon_0\left(\frac{3a^2}{2}\right)^{\frac{3}{2}}}$$

$$\Rightarrow E_{max} = \frac{Q}{3\sqrt{3}\,\pi\varepsilon_0 a^2} \quad \text{Ans.}$$

(c) The field on x-axis is

$$E = \frac{Qx}{2\pi\varepsilon_0(a^2 + x^2)^{\frac{3}{2}}}$$

For small x, $a^2 + x^2 \simeq a^2$

$$\Rightarrow E \simeq \frac{Qx}{2\pi\varepsilon_0 a^3}$$

Then, the alternation for an q is

$$F = -Eq = -\left(\frac{Qx}{2\pi\varepsilon_0 a^3}\right)q$$

Compare the last equation with Hooke's law $F = -kx$

Then, we have

$$k = \frac{Qq}{2\pi\varepsilon_0 a^3}$$

So, the motion is oscillating with an angular frequency

$$\omega = \sqrt{\frac{k}{m}} = \sqrt{\frac{Qq}{2\pi\varepsilon_0 ma^3}} \text{ Ans.}$$

N.B: Along the x-axis, the charged particle $(-q)$ is in stable equilibrium and along the y-axis, it is in unstable equilibrium. For a positive charge $(+q)$, the reverse will occur.

Problem 11 Find the electric field due to the uniformly charge spherical cup that subtends plane angle of $\theta = 60°$ at its centre O. Assume that the surface charge density $= \sigma$.

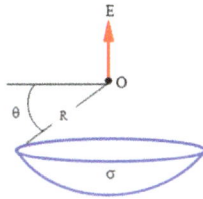

Solution
Take a thin horizontal ring of thickness $Rd\theta$ and radius $r = R\cos\theta$. So, the charge of the ring is

$$q_{\text{ring}} = \sigma dA = \sigma(2\pi r)Rd\theta$$
$$= 2\pi R\sigma r d\theta = 2\pi R\sigma(R\cos\theta)d\theta$$
$$= q_{\text{ring}} = 2\pi R^2\sigma\cos\theta d\theta$$

The electric field due to the ring at C is

$$E_{\text{ring}} = \frac{q_{\text{ring}}x}{4\pi\varepsilon_0(r^2 + x^2)^{\frac{3}{2}}}, \quad \text{where } r^2 + x^2 = R^2$$

$$\Rightarrow E_{\text{ring}} = \frac{2\pi R^2\sigma \cdot \cos\theta d\theta. x}{24\pi\varepsilon_0 R^3} = \frac{\sigma x\cos\theta d\theta}{2\varepsilon_0 R}$$

where $x = R\sin\theta$

$$\Rightarrow E_{\text{ring}} = \frac{\sigma}{2}\varepsilon_0\sin\theta\cos\theta d\theta$$

$$\Rightarrow E_{\text{ring}} = \frac{\sigma}{4\varepsilon_0} \sin 2\theta \qquad (1.122)$$

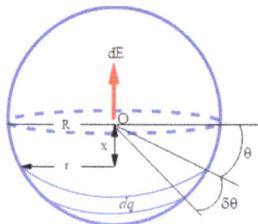

The electric field due to the cup is

$$E = E_{\text{ring}} = \frac{\sigma}{4\varepsilon_0} \int_{\theta_1 = 60°}^{\theta = 90°} \sin 2\theta \, d\theta$$

$$= \frac{\sigma}{4\varepsilon_0}(-\frac{1}{2} \cos 2\theta)|_{60°}^{90°}$$

$$= -\frac{\sigma}{8\varepsilon_0}(\cos 180 - \cos 120)$$

$$= -\frac{\sigma}{8\varepsilon_0}\{-1 - (-\frac{1}{2})\}$$

$$= \frac{\sigma}{8\varepsilon_0} \times \frac{3}{2} = \frac{3\sigma}{16\varepsilon_0} \text{ Ans.}$$

Problem 12 A circular patch of radius R is removed from a thin infinite plate of surface change density σ. Find the (a) electric field as the function of the axial distance x. (b) Frequency of oscillation of a particle of mass m and charge $-q$ placed at the centre of a circular patch. Neglect gravity.

Solution

(a) Take a thin ring of radius r and thickness dr. This has a charge $q_{\text{ring}} = \sigma dA$
$= \sigma \, 2\pi r dr$

$$\Rightarrow q_{\text{ring}} = 2\pi\sigma \, r dr \qquad (1.123)$$

Its electric field at any axial point P at a distance x from O is

$$dE = \frac{q_{\text{ring}} \, x}{4\pi\varepsilon_0(x^2 + r^2)^{\frac{3}{2}}}$$

Then the net field at P is

$$E = \int dE$$

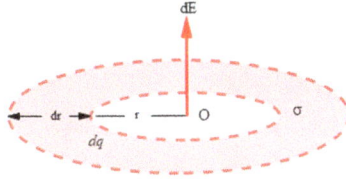

$$E = \frac{1}{4\pi\varepsilon_0} \int_R^\infty \frac{q_{\text{ring}} x}{(x^2 + r^2)^{\frac{3}{2}}} \qquad (1.124)$$

By using equations (1.123) and (1.124)

$$E = \frac{1}{4\pi\varepsilon_0} \int \frac{(2\pi\sigma r dr)x}{(x^2 + r^2)^{\frac{3}{2}}}$$

$$E = \frac{\sigma x}{2\varepsilon_0} \int_R^\infty \frac{r\,dr}{\left(r^2 + x^2\right)^{\frac{3}{2}}}$$

$$= \frac{\sigma x}{2\varepsilon_0} I,$$

where

$$I = \int_R^\infty \frac{r\,dr}{(r^2 + x^2)^{\frac{3}{2}}}$$

Put $r^2 + x^2 = u^2$

$$r\,dr = u\,du$$

$$\Rightarrow I = \int \frac{u\,du}{u^3} = \int u^{-2} du$$

$$= -\frac{1}{u}\Big| = -\frac{1}{\sqrt{r^2 + x^2}}\Big|_R^\infty$$

$$\Rightarrow I = \frac{1}{\sqrt{R^2 + x^2}}$$

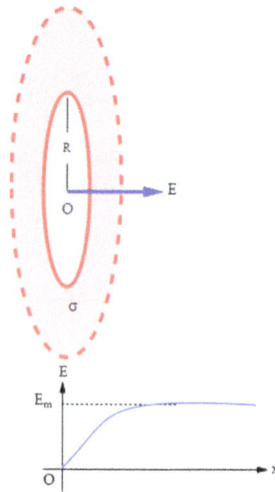

Then, the variation of electric field along the x-axis is given as

$$E = \frac{\sigma x}{2\varepsilon_0 \sqrt{R^2 + x^2}} \quad \text{Ans.}$$

(b) If $x \to 0$, $E \to 0$

$$\text{If } x \to \infty, \quad E \to \frac{\sigma}{2\varepsilon_0}$$

Then the graph of $E - x$ reaches a constant value of $\frac{\sigma}{2t_0}$ as x is large. However, near the centre of the hole

$$E \simeq \frac{\sigma x}{2\varepsilon_0 R} \quad \text{because } x \ll R$$

Then the point change can experience a linear attraction force,

$$F = -qE = \frac{-\sigma q x}{2\varepsilon_0 R} \tag{1.125}$$

Comparing the equation (1.125) equation

$$F = -k_{\text{eff}} x, \tag{1.126}$$

we have

$$k_{\text{eff}} = \frac{\sigma q}{2\varepsilon_0 R}$$

Then, the particle will oscillate with a frequency

$$F_{\text{ose}} = \frac{\omega}{2\pi} = \frac{1}{2\pi}\sqrt{\frac{k_{\text{eff}}}{m}}$$

$$\Rightarrow f_{\text{ose}} = \frac{1}{2\pi} \sqrt{\frac{\sigma q}{2\varepsilon_0 mR}} \quad \text{Ans.}$$

Problem 13 A thin plate of uniform charge density σ is placed on the horizontal (xz) plane. Find the electric field at P and Q.

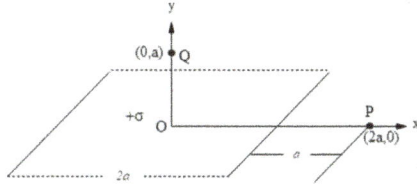

Take a thin strip of thickness dx. The linear charge density is

$$\lambda = \frac{\sigma \, d\ell. \, dx}{d\ell} = \sigma \, dx \qquad (1.127)$$

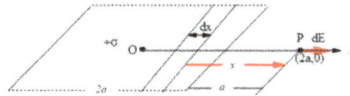

The electric field due to the infinite line charge at P is

$$dE = \frac{\lambda}{2\pi\varepsilon_0 x} \qquad (1.128)$$

Then, the net electric field due to plate at point P is

$$E = \int dE = \int \frac{\lambda}{2\pi\varepsilon_0 x}, \quad \text{where} \ \lambda = \sigma \, dx$$

$$\Rightarrow E = \frac{\sigma}{2\pi\varepsilon_0} \int_a^{\ell+a} \frac{dx}{x}$$

$$\Rightarrow E = \frac{\sigma}{2\pi\varepsilon_0} \ln\left(\frac{\ell + a}{a}\right) \quad \text{Ans.}$$

If $E'=$ electric field due to two symmetrical thin line charge, the net field at Q is

$$E_1 = 2E' \cos \theta$$

Then, the net field due to the string is

$$E = \int E_1 = \int 2E' \cos \theta \qquad (1.129)$$

where $E' = \frac{\lambda}{2\pi\varepsilon_0 r} = \frac{\lambda}{2\pi\varepsilon_0\sqrt{y^2 + x^2}}$ and $\cos\theta = \frac{y}{r}$

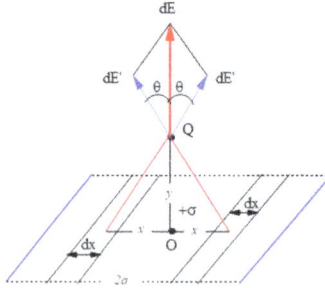

$$\Rightarrow E = \int 2\left\{\frac{\lambda y}{2\pi\varepsilon_0(y^2 + x^2)}\right\},$$

where $\lambda = \sigma\, dx$

$$\Rightarrow E = \frac{\sigma y}{\pi\varepsilon_0}\int\frac{dx}{(y^2 + x^2)}$$

$$= \frac{\sigma y}{\pi\varepsilon_0}\frac{1}{y}\tan^{-1}\frac{x}{y}\Big|_0^a$$

$$= \frac{\sigma}{\pi\varepsilon_0}[\tan^{-1}\frac{a}{y}]$$

Putting $y = a$ in the last expresssion, we have

$$E = \frac{\sigma}{4\varepsilon_0}\ \text{Ans.}$$

Problem 14 A very long thin strip of breadth b is bent into a semi-cylindrical shape of radius R. If $\sigma=$ surface charge density of the sheet, find the electric field at C.

Solution

Take a thin line of breadth $R\, d\theta$ and linear charge density

$$\lambda = dq, \quad \text{where}\quad dq = \sigma(Rd\theta)d\ell$$

$$\Rightarrow \lambda = \sigma R d\theta \tag{1.130}$$

The electric field due to their line charge is

$$dE' = \frac{\lambda}{2\pi\varepsilon_0 R} \tag{1.131}$$

Its vertical component is

$$dE_y = dE'\cos\theta = \frac{\lambda}{2\pi\varepsilon_0 R}\cos\theta \tag{1.132}$$

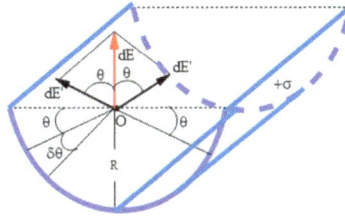

Then, the net field due to the sector of the long tube is

$$E = \int_0^{\frac{\phi}{2}} dE = 2 \int_0^{\frac{\phi}{2}} dE_y \qquad (1.133)$$

Using equations (1.132) and (1.133),

$$E = 2 \int \frac{\lambda}{2\pi\varepsilon_0 R} \cos\theta,$$

where $\lambda = \sigma R d\theta$ (equation (1.130))

$$\Rightarrow E = \int \frac{\lambda \cos\theta}{\pi\varepsilon_0 R} = \int \frac{\sigma R d\theta \cos\theta}{\pi\varepsilon_0 R}$$

$$\Rightarrow E = \frac{\sigma}{\pi\varepsilon_0} \int_0^{\frac{\phi}{2}} \cos\theta \, d\theta$$

$$= \frac{\sigma}{\pi\varepsilon_0} \sin\frac{\phi}{2},$$

where

$$\phi = \frac{b}{R}$$

$$\Rightarrow E = \frac{\sigma}{\pi\varepsilon_0} \sin\frac{b}{2R} \quad \text{Ans.}$$

Problem 15 A very long thin strip of breadth l is placed in the xz plane. If $\sigma =$ surface charge density of the sheet, find the electric field at P.

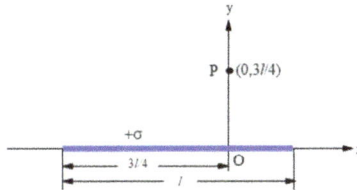

Solution
The electric field at P due to the thin strip is

$$dE = \frac{\lambda}{2\pi\varepsilon_0 r}$$

Then, the horizontal electric field is

$$dE_x = dE \sin\theta$$

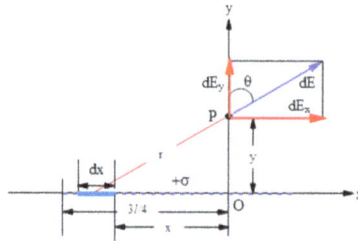

The net horizontal field due to the string is

$$E_x = \int dE_x = \int dE \sin \theta$$

$$= \int \frac{\lambda}{2\pi\varepsilon_0 r} \frac{x}{r} = \frac{\sigma}{2\pi\varepsilon_0} \int \frac{l\,dx}{r^2}$$

where $\lambda = \sigma\,dx$

$$\Rightarrow E_x = \frac{\sigma}{2\pi\varepsilon_0} \int_{x_1}^{x_2} \frac{x\,dx}{r^2}$$

$$= \frac{\sigma}{2\pi\varepsilon_0} \int_{x_1}^{x_2} \frac{x\,dx}{\left(x^2 + y^2\right)}$$

$$= \frac{\sigma}{4\pi\varepsilon_0} \int_{x_1}^{x_2} \frac{2x\,dx}{x^2 + y^2}$$

$$\Rightarrow E_x = \frac{\sigma}{4\pi\varepsilon_0} \ln(x^2 + y^2)|_{x_1}^{x_2}$$

$$\Rightarrow E_x = \frac{\sigma}{4\pi\varepsilon_0} \left| \ln\left(\frac{x_2^2 + y^2}{x_1^2 + y^2}\right) \right|$$

After evaluation, we have

$$E_x = \frac{\alpha}{4\pi\varepsilon_0} \ln 5 \qquad (1.134)$$

The vertical component of dE is

$$dE_y = dE \cos \theta$$

Then, the net vertical force is

$$E_y = \int dE_y = \int dE \cos \theta$$

$$= \int \frac{\lambda}{2\pi\varepsilon_0 r} \cdot \frac{y}{r},$$

where $\lambda = \sigma\,dx$

$$\Rightarrow E_y = \frac{\sigma y}{2\pi t_0} \int_{x_1}^{x_2} \frac{dx}{x^2 + y^2}$$

$$= \frac{\sigma y}{2\pi\varepsilon_0} \frac{1}{y} \tan^{-1} \frac{x}{y}|_{x_1}^{x_2}$$

$$= \frac{\sigma}{2\pi\varepsilon_0} \left| \tan^{-1}\left(\frac{x_2}{y}\right) + \tan^{-1}\left(\frac{x_1}{y}\right) \right|$$

$$= \frac{\sigma}{2\pi\varepsilon_0} \left| \tan^{-1}\left(\frac{\frac{3\ell}{4}}{\frac{\ell}{4}}\right) + \tan^{-1}\left(\frac{\frac{\ell}{4}}{\frac{\ell}{4}}\right) \right|$$

$$= \frac{\sigma}{2\pi\varepsilon_0} \left[\tan^{-1} 3 + \frac{\pi}{4} \right] \qquad (1.135)$$

Using the last two equations, the net field is

$$\vec{E} = E_x \hat{i} + E_y \hat{i}$$

$$= \frac{\sigma}{4\pi\varepsilon_0}\left[2\left(\tan^{-1}3 + \frac{\pi}{4}\right)\hat{j} + \ln 5\hat{i}\right]$$

$$\Rightarrow \vec{E} = \frac{\sigma}{4\pi\varepsilon_0}[\ln 5\,\hat{i} + (\frac{\pi}{2} + 2\tan^{-1}3)\hat{j}]\ \text{Ans.}$$

Problem 16

(a) Two dielectric spheres of volume charge density $+\rho$ and $-\rho$ are super-imposed such that the distance between their centre is a. (i) Find the electric field at any point of the common portion of the spheres. (ii) Draw the E-field pattern in the common portion of the spheres.

(b) If we remove a spherical portion from the sphere of positive charge, find the electric field inside the cavity formed.

(c) If the spheres in (a) have the same radius and their centres are separated by a small distance a, find the surface charge density of any point of either sphere having a position vector that makes an angle θ with the horizontal.

(d) If the surface charge density of a thin spherical dielectric shell changes with the angle θ with the horizontal as $\sigma = \sigma_0\cos\theta$, find the electric field at the centre of the sphere.

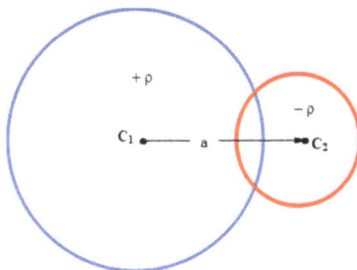

Solution

(a)

(i) At any point P in the common portion which is neutral in its volume, the net field is

$$\vec{E} = \vec{E_1} + \vec{E_2} \tag{1.136}$$

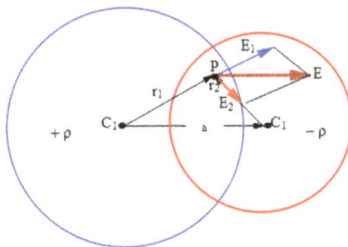

The field due to sphere 1 is

$$\overrightarrow{E_1} = \frac{\rho}{3\varepsilon_0}\overrightarrow{r_1} \tag{1.137}$$

The field due to sphere 2 is

$$\overrightarrow{E_2} = \frac{(-\rho)}{3\varepsilon_0}\overrightarrow{r_2} \tag{1.138}$$

Using the last three equations, we have

$$\overrightarrow{E} = \frac{\rho}{3\varepsilon_0}\overrightarrow{r_1} + \frac{(-\rho)}{3\varepsilon_0}\overrightarrow{r_2}$$

$$= \frac{\rho}{3\varepsilon_0}(\overrightarrow{r_1} - \overrightarrow{r_2})$$

Since $\overrightarrow{a} + \overrightarrow{r_2} = \overrightarrow{r_1}$, we can substitute $\overrightarrow{r_1} - \overrightarrow{r_2} = \overrightarrow{a}$-in the last expression obtain.

$$\overrightarrow{E} = \frac{\rho}{3\varepsilon_0}\overrightarrow{a}$$

This means that in the neutral position the electron field remains uniform having the magnitude.

$$E = \frac{\rho}{3\varepsilon_0}a$$

The field parts in the direction of $\overrightarrow{a}$ which represents a vector joining the centres of the sphere.

(ii) Since the common portion is chargeless (volume charge density = 0), this portion behaves as a cavity.

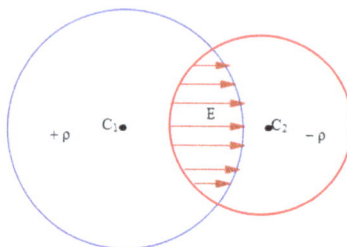

(b) So, when we assume a spherical portion inside any charged sphere of volume charge density ρ, we will have a uniform electric field inside the cavity which can be given by the above expression

$$\overrightarrow{E} = \frac{\rho}{3\varepsilon_0}\overrightarrow{a} \quad \text{Ans.}$$

(c) If we superimpose two identical spheres carrying equal and opposite charges with uniform volume charge densities $+\rho$ and $-\rho$ respectively, using the previous argument, we have a uniform electric field inside the common portion (or cavity) which is given as

$$\vec{E} = \frac{\rho}{3\varepsilon_0}\vec{a}$$

If the magnitude of $\vec{a}$ is very small, $\vec{E}$ will be very small. If the centres C_1 and C_2 coincide, $\vec{a} = \vec{O}$. Then $\vec{E} = 0$.

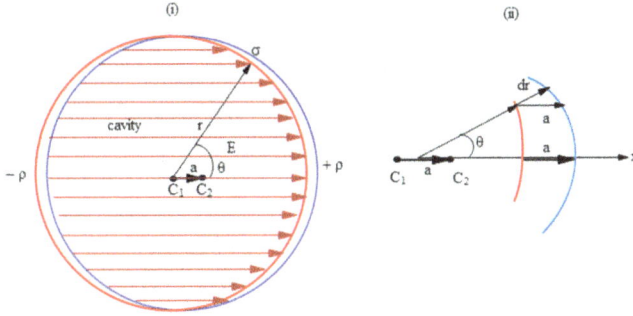

If we take an element at P of volume dV, it has a charge

$$dq = \rho dV = \rho(dA)(dr),$$

where $dr = a \cos\theta$

$$\Rightarrow dq = \rho(dA)(a \cos\theta)$$

$$\Rightarrow \frac{dq}{dA} = \rho a \cos\theta, \text{ where } \frac{dq}{dA} = \text{ surface charge density } = \sigma.$$

$$\Rightarrow \sigma = \rho a \cos\theta \text{ Ans.}$$

(d) Putting $\rho a = \sigma_{max} = \sigma_0$, we have $\sigma = \sigma_0 \cos\theta$

So, when the surface charge density of a thin spherical shell changes obeying this law we can understand that it is the combination two identical spheres of volume charge densities $+\rho$ and $-\rho$ whose centres are separated by a small distance a. Using the previous method we have electric field:

$$\vec{E} = \frac{\rho}{3\varepsilon_0}\vec{a}, \text{ where } \rho = \frac{\sigma_0}{a}$$

$$\Rightarrow \vec{E} = \left(\frac{\sigma_0}{a}\right)\frac{\vec{a}}{3\varepsilon_0}$$

$$\Rightarrow \vec{E} = \frac{\sigma_0}{3\varepsilon_0}\vec{a} \text{ Ans.}$$

Problem 17 A particle of mass m and charge q_0 is placed in between the fixed charges Q and q. (a) Find the value of x/l so that the net force acting on charge q_0 is zero. (b) what is the frequency of oscillation at the point if the charge is displaced slightly along the line joining the charges.

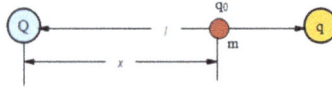

Solution

(a) For equilibrium $F = 0$ at a point where the repulsion of one cancels that of the other

$$\Rightarrow K\frac{Qq_0}{x^2} = \frac{Kqq_0}{(\ell - x)^2}$$

$$\Rightarrow \frac{x}{\ell - x} = \sqrt{\frac{Q}{q}}$$

$$\Rightarrow \frac{x}{\ell} = \frac{\sqrt{Q}}{\sqrt{Q} + \sqrt{q}} \quad \text{Ans.}$$

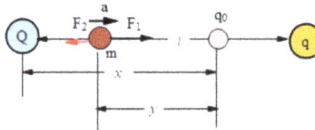

(b) If we disturb charge q_0 by a small displacement y along the x-axis, the understood force acting on the particle is

$$F_{\text{net}} = \frac{KQq_0}{(x - y)^2} - \frac{Kqq_0}{(\ell - x + y)^2}$$

$$= Kq_0\left[\frac{Q}{(x - y)^2} - \frac{q}{(\ell - x + y)^2}\right]$$

$$= Kq_0\left[\frac{Q(\ell - x + y)^2 - q(x - y)^2}{(x - y)^2(\ell - x + y)^2}\right]$$

$$\cong Kq_0\left[\frac{Q(\ell - x)^2 + Qy^2 + 2y(\ell - x)Q - q(x^2 + y^2 - xy)}{x^2(\ell - x)^2}\right]$$

Obtained after neglecting the value of y in comparison to x and l,
Putting $Q(\ell - x)^2 - q(x^2) = 0$ we have

$$F_{\text{net}} = Kq_0\left[\frac{(q + Q)y^2 + 2y(\ell - x)Q + 2xyq}{\{x(\ell - x)\}^2}\right]$$

Putting $x = \frac{\sqrt{Q}\ell}{\sqrt{Q} + \sqrt{q}}$ and $\ell - x = \frac{\sqrt{q}\ell}{\sqrt{Q} + \sqrt{q}}$ and $(q + Q)y^2 \simeq 0$ $(\because y < <)$

$$\Rightarrow \vec{F}_{\text{net}} = \frac{2Kq_0\{(\ell - x)Q + qx\}y}{\left\{\left(\frac{\sqrt{Q}\ell}{\sqrt{Q} + \sqrt{q}}\right)\left(\frac{\sqrt{q}\ell}{\sqrt{Q} + \sqrt{q}}\right)\right\}^2}$$

$$= \frac{q_0(\sqrt{Q} + \sqrt{q})^4}{2\pi\varepsilon_0 Qq\ell^4}\left\{\left(\frac{\sqrt{q}\ell}{\sqrt{Q} + \sqrt{q}}\right)Q + \frac{q\sqrt{Q}\ell}{\sqrt{Q} + \sqrt{q}}\right\}$$

$$= \frac{q_0(\sqrt{Q} + \sqrt{q})^4(\sqrt{Q} + \sqrt{q})}{2\pi\varepsilon_0\sqrt{Q}q\ell^3(\sqrt{Q} + \sqrt{q})}y$$

$$\Rightarrow F_{\text{net}} = \frac{(\sqrt{Q} + \sqrt{q})^4}{2\pi\varepsilon_0\sqrt{Qq}\ell^3}y = k_{\text{eff}}y \text{ (let)}$$

$$\Rightarrow k_{\text{eff}} = \frac{(\sqrt{Q} + \sqrt{q})^4}{2\pi\varepsilon_0\sqrt{Qq}}q_0$$

Then the angular frequency of oscillation is

$$\omega_{\text{osc}} = \sqrt{\frac{k_{eff}}{m}}$$

$$\Rightarrow \omega_{\text{osc}} = \sqrt{\frac{(\sqrt{Q} + \sqrt{q})^4 q_0}{2\pi\varepsilon_0\sqrt{Qq}\ell^3}} \text{ Ans.}$$

Problem 18 The surface charge density of a thin spherical dielectric shell is uniform, $\rho = \vec{a}.\vec{r}$. Find the electric field at the centre of the sphere is uniform due to its half portion (hemisphere).

Solution

The electric field of the ring at C is

$$E_{\text{ring}} = \frac{q_{\text{ring}}x}{4\pi\varepsilon_0 R^3} \tag{1.139}$$

Then the total field at C due to the sphere is

$$E = \int E_{\text{ring}} \tag{1.140}$$

Using these two equations,

$$E = \int \frac{q_{\text{ring}}x}{4\pi\varepsilon_0 R^3} = \frac{1}{4\pi\varepsilon_0 R^3}\int q_{\text{ring}}x,$$

where $q_{\text{ring}} = \sigma 2\pi r R d\theta$

$$\Rightarrow E = \frac{1}{4\pi\varepsilon_0 R^3} \int (\sigma 2\pi r R d\theta) x$$

$$\Rightarrow E = \frac{1}{2\varepsilon_0 R^2} \int \sigma x d\theta$$

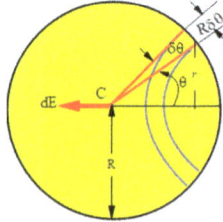

Putting the values of x, we have

$$E = \frac{\sigma_0}{2\varepsilon_0 R^2} \int_0^{\frac{\pi}{2}} r\, x d\theta, \quad \text{where } r = R\sin\theta \text{ and } x = R\cos\theta$$

$$\Rightarrow E = \frac{\sigma_0}{2\varepsilon_0 R^2} R^2 \int_0^{\frac{\pi}{2}} \sin\theta \cos\theta$$

$$\Rightarrow E = \frac{\sigma_0}{4\varepsilon_0} \int_0^{\frac{\pi}{2}} \sin 2\theta$$

$$= \frac{\sigma_0}{4\varepsilon_0}\left(-\frac{1}{2}\cos 2\theta\right)\Big|_0^{\frac{\pi}{2}}$$

$$= -\frac{\sigma_0}{8\varepsilon_0}[\cos\pi - \cos 0]$$

$$= \frac{\sigma_0}{4\varepsilon_0} \text{ Ans.}$$

Problem 19 Three identical charge particles each of mass m and charge Q are released from a given position. Find (a) the velocity of the particles (b) the tension in the strings (c) the acceleration of the particles when all particles lie on the same straight line.

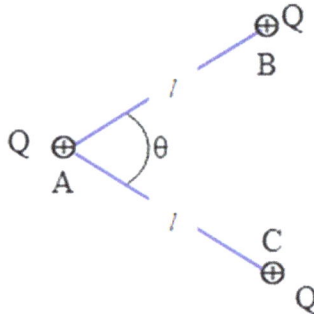

Solution

(a) If the system is released due to the effect of electrostatic repulsion between the balls B and C, both these balls move relative to the base A. Since no net force acts on the system of three balls, its momentum remains conserved, that is, equal to zero.

$$P = C$$

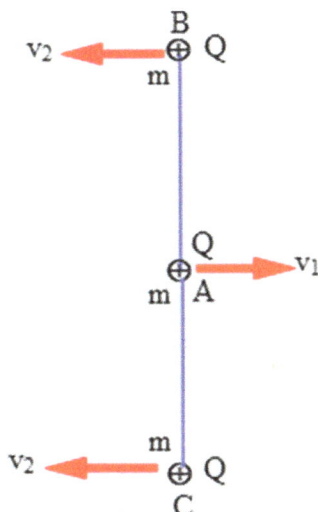

Let the velocities of A be v_1 and B and C be v_2. Then

$$P_{\text{system}} = mv_1 - 2mv_2 = 0$$

$$\Rightarrow v_1 = 2v_2 \tag{1.141}$$

Energy conservation:

$$K_i + U_i = K_f + U_f$$

$$\Rightarrow 0 + \frac{KQ^2}{\ell'} + \frac{KQ^2}{\ell^2} + \frac{KQ^2}{\ell}$$

$$= \frac{1}{2}mv_1^2 + \frac{1}{2}mv_2^2 + \frac{KQ^2}{(2\ell)} + \frac{KQ^2}{\ell} + \frac{KQ^2}{\ell}$$

$$\Rightarrow \frac{1}{2}m(v_1^2 + 2v_2^2) = KQ^2\left(\frac{1}{\ell'} - \frac{1}{2\ell}\right) \tag{1.142}$$

Using equations (1.141) and (1.142)

$$\frac{1}{2}m\{(2v_2)^2 + 2v_2^2\} = KQ^2\left(\frac{1}{\ell'} - \frac{1}{2\ell}\right)$$

Putting $\ell' = 2\ell \sin \frac{\theta}{2}$, we have

$$3mv_2^2 = \frac{KQ^2}{2\ell}\left(\operatorname{cosec} \frac{\theta}{2} - 1\right)$$

$$\Rightarrow v_2 = \sqrt{\frac{KQ^2}{6m\ell}\left(\operatorname{cosec} \frac{\theta}{2} - 1\right)}$$

$$v_1 = 2v_2 = \sqrt{\frac{2KQ^2}{6m\ell}\left(\operatorname{cosec} \frac{\theta}{2} - 1\right)}$$

where $K = \frac{1}{4\pi\varepsilon_0}$ Ans.

(b) The relative velocity between the balls A and C or (A and B) is

$$v_{\text{rel}} = v_1 + v_2 = 2v_2 + v_2 = 3v_2$$

$$= 3\sqrt{\frac{KQ^2}{6m\ell}\left(\operatorname{cosec} \frac{\theta}{2} - 1\right)}$$

$$\Rightarrow v_{\text{rel}} = \sqrt{\frac{3KQ^2}{2m\ell}\left(\operatorname{cosec} \frac{\theta}{2} - 1\right)} \qquad (1.143)$$

Then, the tension is given as

$$\overrightarrow{T} = m\overrightarrow{a_C} = m(\overrightarrow{a_{CA}} + \overrightarrow{a_A})$$

Since $a_A = \frac{F_A}{m} = 0$

$$(\because F_A = \text{net force acting on } A = 0)$$

we have, $T = ma_{C_A}$, where $a_{CA} = \frac{v_{CA}^2}{\ell} = \frac{v_{\text{rel}}^2}{\ell}$

$$\Rightarrow T = \frac{mv_{\text{rel}}^2}{\ell} \qquad (1.144)$$

Using equations (1.143) and (1.144)

$$T = \frac{m}{\ell}\left\{\frac{3KQ^2}{2m\ell}\left(\operatorname{cosec} \frac{\theta}{2} - 1\right)\right\}$$

$$T = \frac{3KQ^2}{2\ell^2}\left(\operatorname{cosec} \frac{\theta}{2} - 1\right) \text{ Ans.}$$

(c) Then $a_C = a_B = \frac{T}{m} = \frac{3KQ^2}{2m\ell^2}(\cos \frac{\theta}{2} - 1)$ and $a_A = \frac{(F_{\text{net}})_A}{m} = 0$ Ans.

N.B: Particle B will accelerate vertically down, and particle C will accelerate up vertically, whereas particle A does not accelerate.

Problem 20 Two infinite charged rods each of linear charge density λ are fixed on an insulated smooth horizontal surface. A particle of mass m and charge q is placed at the mid-point of the line joining the rods. If the particle is displaced to right or left slightly, find the frequency of oscillation.

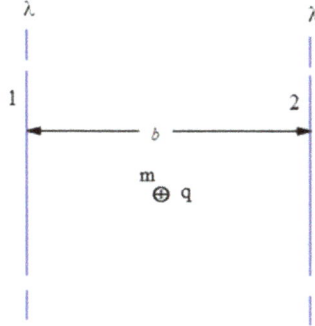

Solution

If we displace the charged particle by a small distance x, the unbalanced force acting on it is

$$F = q(E_1 - E_2)$$

$$= q\left[\frac{\lambda}{2\pi\varepsilon_0(b-x)} - \frac{\lambda}{2\pi\varepsilon_0(b+x)}\right]$$

$$= \frac{\lambda q}{2\pi\varepsilon_0}\left(\frac{1}{b-x} - \frac{1}{b+x}\right)$$

$$= \frac{\lambda q(2bx)}{2\pi\varepsilon_0\left(b^2 - x^2\right)}$$

$$\Rightarrow F \simeq \frac{\lambda q}{\pi\varepsilon_0 b^2}x$$

Comparing the last equation with $F = k_{\text{eff}}\,x$, we have $k_{\text{eff}} = \frac{\lambda q}{\pi\varepsilon_0 b^2}$

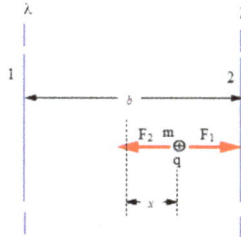

Then the particle will oscillate with an angular frequency

$$\omega = \sqrt{\frac{k_{\text{eff}}}{m}} = \sqrt{\frac{\lambda q}{\pi\varepsilon_0 m b^2}} \quad \text{Ans.}$$

Problem 21 Two fixed thin rigid long wires each of linear charge density λ are placed on the y-axis at $y = +b$ and $y = -b$ so that they are parallel to the z-axis . A particle of mass m and charge $-q$ is placed at the origin. If the particle is displaced along the x-axis slightly by a small distance x, (a) find the restoring force, (b) draw the graph of the variation of force or field versus x, (c) the frequency of oscillation.
  Solution

(a) At any small displacement x, the restoring (net) force is

$$F = 2F' \cos\theta, \quad \text{where } F' = Eq$$

$$\Rightarrow F = 2qE \cos\theta, \quad \text{where } E = \frac{\lambda}{2\pi\varepsilon_0 r} \quad \text{and} \quad \cos\theta = \frac{x}{r}$$

$$\Rightarrow F = \frac{\lambda q}{\pi\varepsilon_0 r^2} x$$

$$\Rightarrow F = \frac{\lambda q x}{\pi\varepsilon_0(b^2 + x^2)} \quad \text{Ans.}$$

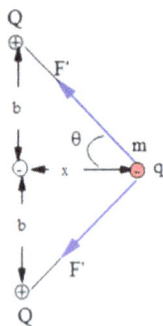

(b) If $x \to 0$, $F \to 0$

$$\text{If } x \to \infty, \quad F \to 0$$

Then the force will be maximum at certain value of x for which

$$\frac{dF}{dx} = 0$$

$$\Rightarrow \frac{d}{dx}\left(\frac{x}{b^2 + x^2}\right) = 0$$

$$\Rightarrow x = \pm b \quad \text{Ans.}$$

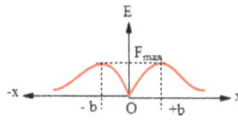

(c) For small value of x, $b^2 + x^2 \simeq b^2$

$$\Rightarrow F \simeq \frac{\lambda q x}{\pi \varepsilon_0 b^2} = k_{\text{eff}} x (\text{Let})$$

$$\Rightarrow k_{\text{eff}} = \frac{\lambda q}{\pi \varepsilon_0 b^2}$$

Then the particle oscillates with an angular frequency

$$\omega = \sqrt{\frac{k_{\text{eff}}}{m}} = \sqrt{\frac{\lambda q}{\pi m t_0 b^2}} \quad \text{Ans.}$$

Problem 22 Two fixed point charges each of charge Q are placed on an insulated smooth horizontal surface. A particle of mass m and charge $+q$ in figure (i) and $-q$ in figure (ii), is placed at the mid-point of the line joining the charges. If the particle is displaced slightly along the y-axis in figure (i) and x-axis in figure (ii) by a small distance, find (a) the frequency of oscillation in (a) figure (i) (b) the frequency of oscillation in figure (ii) (c) the ratio of frequency of oscillation.

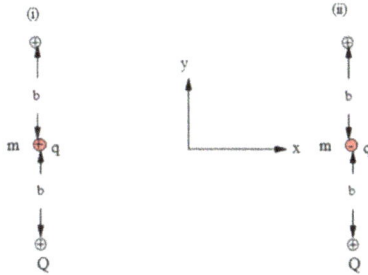

Solution

(a) In the left-hand side figure (i) the restoring force is

$$F = \frac{KQq}{(b-y)^2} - \frac{KQq}{(b+y)^2}$$

$$= KQq \frac{(b+y)^2 - (b-y)^2}{\left(b^2 - y^2\right)}$$

For small y, $b^2 - y^2 \simeq b^2$

$$\Rightarrow F = \frac{KQq}{b^4}(4by) = \frac{Qq}{4\pi\varepsilon_0 b^4}(4by)$$

$$\Rightarrow F = \frac{Qq}{\pi\varepsilon_0 b^3}y = k_{\text{eff}}\,y$$

$$\Rightarrow k_{\text{eff}} = \frac{Qq}{\pi\varepsilon_0 b^3}$$

Then, the angular frequency of oscillation of the particle is

$$\omega_{\text{osc}} = \sqrt{\frac{k_{\text{eff}}}{m}} = \sqrt{\frac{Qq}{\pi m\varepsilon_0 b^3}} \quad \text{Ans.}$$

(b) In figure (ii), the restoring force is

$$F = 2F' \cos\theta$$

$$= 2\frac{KQq}{r^2} \cdot \frac{x}{r} = \frac{2KQqx}{r^3}$$

$$\Rightarrow F = \frac{Qqx}{2\pi\varepsilon_0 b^3}(\because K = \frac{1}{4\pi\varepsilon_0} \text{ and } r \simeq b \text{ for small } x)$$

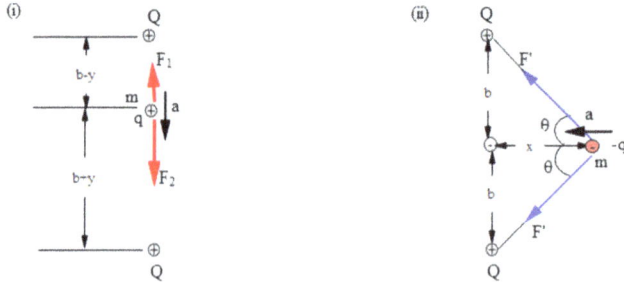

Comparing this equation with $F = k_{\text{eff}}\,x$, we have

$$k_{\text{eff}} = \frac{Qq}{2\pi\varepsilon_0 b^3}$$

Then,

$$\omega'_{\text{osc}} = \sqrt{\frac{k_{\text{eff}}}{m}} = \sqrt{\frac{Qq}{2\pi\varepsilon_0 m b^3}} \quad \text{Ans.}$$

(c) The ratio of angular frequencies is

$$\frac{\omega}{\omega'} = \sqrt{2} \text{ Ans.}$$

N.B: The frequency of oscillation along the y-axis is double that along the x-axis,

Problem 23 Two infinite uniform charged rods are placed at a distance of separation r on an insulated horizontal plane. The line charge densities of the rods are λ_1 and λ_2, respectively. Find (a) the point where the net electric field is zero, (b) the electric force experienced by each rod per unit length, (c) the work done by an external agent in slowly increasing the separation by η fold.

Solution

(a) At point 3 let the field be zero; then the field strengths are equal and opposite;

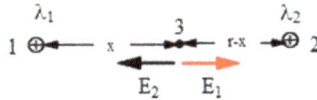

$$E_1 = E_2$$

$$\Rightarrow \frac{\lambda_1}{2\pi\varepsilon_0 x} = \frac{\lambda_2}{2\pi\varepsilon_0(\ell - x)}$$

$$\Rightarrow \frac{x}{\ell - x} = \frac{\lambda_1}{\lambda_2}$$

$$\Rightarrow \frac{x}{\ell} = \frac{\lambda_1}{\lambda_1 + \lambda_2}$$

$$\Rightarrow x = \frac{\lambda_1 \ell}{\lambda_1 + \lambda_2} \quad \text{Ans.}$$

(b) The force of interaction between the parallel rods can be found by taking an element $d\ell$ of one rod, and finding the force dF on it due to the other rod as:

$$\delta F = (E_{\text{others}})dq$$

$$\Rightarrow dF = E_{\text{others}}\lambda_1 d\ell$$

$$\Rightarrow \frac{dF}{d\ell} = \frac{\lambda_2}{2\pi\varepsilon_0 r}\lambda_1$$

$$\Rightarrow \frac{dF}{d\ell} = \frac{\lambda_1 \lambda_2}{2\pi\varepsilon_0 r} \quad \text{Ans.}$$

(c) External work done per unit length in separating the rods by an elementary distance dr is

$$\frac{dW_{ext}}{d\ell} = \frac{\delta F}{\delta \ell} dr = \frac{\lambda_1 \lambda_2}{2\pi\varepsilon_0 r} \cdot dr$$

$$\Rightarrow \frac{dW_{ext}}{d\ell} = \frac{\lambda_1 \lambda_2}{2\pi\varepsilon_0} \int_{r_0}^{\eta r_0} \frac{dr}{r}$$

$$\Rightarrow \frac{dW_{ext}}{d\ell} = \frac{\lambda_1 \lambda_2}{2\pi\varepsilon_0} \ln \eta \text{ Ans.}$$

Problem 24 Two uniformly charged rings A and B carrying charges Q and q, respectively, are separated by a distance l. (a) If at $x = R/2$ from point 1 the net electric field is zero, find Q/q. (b) If $Q = 2q$, find the potential difference between 1 and 2. Put $l = R$ and $r = R/2$.

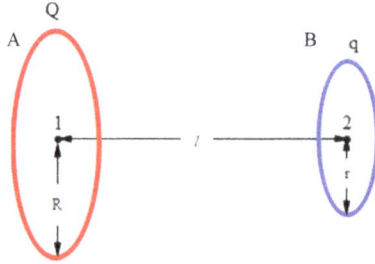

Solution

(a) The field at P is given as

$$E = E_1 - E_2 = 0$$

$$0 = \frac{Qx}{4\pi\varepsilon_0(R^2 + x^2)^{\frac{3}{2}}} - \frac{q(\ell - x)}{4\pi\varepsilon_0\{(\ell - x)^2 + r_2\}^{\frac{3}{2}}}$$

$$\Rightarrow \frac{Qx}{(R^2 + x^2)^{\frac{3}{2}}} = \frac{q(\ell - x)}{\{(\ell - x)^2 + r^2\}^{\frac{3}{2}}}$$

$$\Rightarrow \frac{Q}{q} = \frac{\ell - x}{x} \frac{(R^2 + x^2)^{\frac{3}{2}}}{\{(\ell - x)^2 + r^2\}^{\frac{3}{2}}}$$

$$\Rightarrow \frac{Q}{q} = \frac{R - \frac{R}{2}}{\frac{R}{2}} \left\{ \frac{R^2 + \frac{R^2}{4}}{(R - \frac{R}{2})^2 + \frac{R^2}{4}} \right\}^{\frac{3}{2}}$$

$$= \left\{ \frac{\frac{5R^2}{4}}{\frac{R^2}{2}} \right\}^{\frac{3}{2}} = \left(\frac{5}{2} \right)^{\frac{3}{2}} = \frac{5\sqrt{5}}{2\sqrt{2}} \text{ Ans.}$$

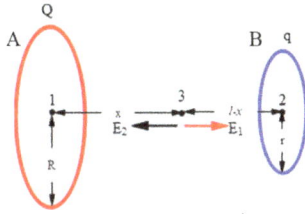

(b) The potential at A is given as

$$V_A = V_{1A} + V_{2A}$$

$$= \frac{KQ}{R} + \frac{Kq}{\sqrt{r^2 + \ell^2}}$$

$$= \frac{KQ}{R} + \frac{Kq}{\sqrt{\frac{R^2}{4} + R^2}}$$

$$\Rightarrow V_A = \frac{KQ}{R} + \frac{2Kq}{\sqrt{5}\,R} \tag{1.145}$$

The potential at B is given as

$$V_B = V_{1B} + V_{2B}$$

$$= \frac{KQ}{\sqrt{R^2 + \ell^2}} + \frac{Kq}{r}$$

$$= \frac{KQ}{\sqrt{R^2 + R^2}} + \frac{Kq}{\frac{R}{2}}$$

$$\Rightarrow V_B = \frac{KQ}{\sqrt{2}\,R} + \frac{2Kq}{R} \tag{1.146}$$

The potential difference between A and B is

$$\Delta V = V_B - V_A$$

$$= \frac{KQ}{R}\left(\frac{1}{\sqrt{2}} - 1\right) + \frac{Kq}{R}\left(2 - \frac{2}{\sqrt{5}}\right)$$

$$= \frac{KQ}{R}\left(\frac{1}{\sqrt{2}} - 1\right) + \frac{KQ}{R}\left(1 - \frac{1}{\sqrt{5}}\right)$$

$$= \frac{KQ}{R}\left[\frac{1}{\sqrt{2}} - \frac{1}{\sqrt{5}}\right] \text{ Ans.}$$

N.B: So, point B is at higher potential than point A.

Problem 25 Two concentric rings of uniform charge distribution of charge Q and $-q$ are placed in a vertical plane. If $(r/R)^2 = 1/8$, where R and r are the radii of the rings. Find the (a) the critical point on the axis where the electric field is zero, (b) the escape speed of a charged particle from the common centre of the rings. Put $Q = 8q$.

Solution

(a) Let the critical point be B where the electric field is zero. Between A and B the field is attractive and beyond B the field is repulsive.

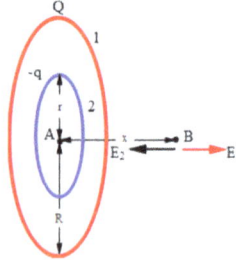

The field at B is given as

$$E_B = \frac{KQx}{(R^2 + x^2)^{\frac{3}{2}}} - \frac{Kqx}{(r^2 + x^2)^{\frac{3}{2}}} = 0$$

$$\Rightarrow \left(\frac{R^2 + x^2}{r^2 + x^2}\right)^{\frac{3}{2}} = \frac{Q}{q}$$

$$\Rightarrow \frac{R^2 + x^2}{r^2 + x^2} = \left(\frac{Q}{q}\right)^{\frac{2}{3}}$$

$$\Rightarrow \frac{R^2 + x^2}{r^2 + x^2} = 8^{\frac{2}{3}} = 4$$

$$\Rightarrow R^2 + x^2 = \frac{R^2}{2} + 4x^2 \left(\because x^2 = \frac{R^2}{8}\right)$$

$$\Rightarrow 3x^2 = \frac{R^2}{2}$$

$$\Rightarrow x = \frac{R}{\sqrt{6}} \text{ Ans.}$$

(b) If the charged particle crosses point B, it will never return to A. So we can conserve the energy between A and B. The potential at A is

$$V_A = \frac{KQ}{R} + \frac{K(-q)}{r}$$

$$= \frac{KQ}{R} + \frac{K}{\frac{R}{2\sqrt{2}}}\left(\frac{-Q}{8}\right)$$

$$= \frac{KQ}{R}\left(1 - \frac{1}{2\sqrt{2}}\right)$$

The potential at B is

$$V_B = \frac{KQ}{\sqrt{R^2 + x^2}} + \frac{K(-q)}{\sqrt{r^2 + x^2}}$$

$$= \frac{KQ}{\sqrt{R^2 + \frac{R^2}{6}}} - \frac{K\left(\frac{Q}{8}\right)}{\sqrt{\frac{R^2}{8} + \frac{R^2}{6}}}$$

$$= \frac{3}{4}\frac{KQ}{R}\left(\sqrt{\frac{6}{7}}\right)$$

Then, the potential difference is

$$V_B - V_A = \frac{KQ}{R}\left[\frac{3}{4}\sqrt{\frac{6}{7}} - 1 + \frac{1}{2\sqrt{2}}\right] \tag{1.147}$$

Energy conservation

$$\frac{1}{2}mv_e^2 + (+V_A) = (V_B)$$

$$\Rightarrow v_e = \sqrt{2\frac{(V_B - V_A)}{m}} \tag{1.148}$$

By using equations (1.147) and (1.148), we have

$$v_e = \sqrt{\frac{2}{m}\left[\frac{KQ}{R}\left\{\frac{3}{4}\sqrt{\frac{6}{7}} - 1 + \frac{1}{2\sqrt{2}}\right\}\right]}$$

$$= \sqrt{\frac{Q}{2\pi\varepsilon_0 mR}\left(\frac{3}{4}\sqrt{\frac{6}{7}} - 1 + \frac{1}{2\sqrt{2}}\right)} \text{ Ans.}$$

Problem 26 A square frame of sides $2a$ has a total charge Q. (a) Find the electric potential at the centre of the frame. (b) What will be the escape speed of a charged particle from the centre of the square?
  Solution

(a) The square has four identical rods. Let us find the potential due to each charged rod. Taking an element at a distance x, the potential is

$$dV = \frac{Kdq}{r} = \frac{K(\lambda\, dx)}{\sqrt{h^2 + x^2}}$$

Then the total potential due to the rod is

$$V' = \int dV'$$

$$= 2K\lambda \int_0^a \frac{dx}{\sqrt{h^2 + x^2}}$$

$$= 2K\lambda \ln\left(\sqrt{h^2 + x^2} + x\right)\Big|_0^a$$

$$= 2K\lambda \ln\left[\frac{\left(\sqrt{h^2 + a^2} + a\right)}{h}\right]$$

Since $h = a$, we have

$$V' = 2K\lambda \ln(\sqrt{2} + 1)$$

Then, the total potential due to the square is

$$V = 4V' = 4K\lambda \ln(\sqrt{2} + 1)$$

$$= \frac{\lambda}{\pi\varepsilon_0} \ln(\sqrt{2} + 1)$$

The total charge of the square frame is

$$Q = \lambda(8a)$$

$$\Rightarrow \lambda = \frac{Q}{8a}$$

Putting this value of λ in the last equation,

$$V = \frac{\left(\frac{Q}{8a}\right)}{\pi\varepsilon_0} \ln(\sqrt{2} + 1)$$

$$= \frac{Q}{8\pi\varepsilon_0 a} \ln(\sqrt{2} + 1) \text{ Ans.}$$

(b) The total energy of the particle at A is

$$E_A = \frac{1}{2}mv_e^2 + (-q)V$$

$$\Rightarrow E_A = \frac{1}{2}mv_e^2 - qV$$

Equating this with energy of the particle at infinity, that is, zero, we have

$$\frac{1}{2}mv_e^2 - qV = 0$$

$$\Rightarrow v_e = \sqrt{\frac{2qV}{m}}$$

$$\Rightarrow v_e = \sqrt{\frac{2q}{m}\left\{\frac{Q}{8\pi\varepsilon_0 a}\ln(\sqrt{2}+1)\right\}}$$

$$= \sqrt{\frac{Qq}{4\pi\varepsilon_0 ma}\ln(\sqrt{2}+1)} \text{ Ans.}$$

Problem 27 A disc and a ring of uniform charge Q and radii R are kept coaxially at a distance of separation R. If a point charge q and mass m is projected along the axis from point 1 with a velocity v_0, find the velocity of the particle at point 2.

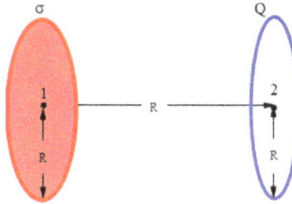

Solution

The potential at point 1 is given as

$$V_1 = V_{\text{disc}} + V_{\text{ring}} \tag{1.149}$$

The potential of the disc at 1 is

$$V_{\text{disc}} = \frac{\sigma R}{2\varepsilon_0} \tag{1.150}$$

The potential of the ring at A is

$$V_{\text{ring}} = \frac{Q}{4\pi\varepsilon_0 \sqrt{R^2 + R^2}} \tag{1.151}$$

Then using the last three equations the total potential at 1 is

$$V_1 = \frac{\sigma R}{2\varepsilon_0} + \frac{Q}{4\sqrt{2}\,\pi\varepsilon_0 R}, \quad \text{where } \sigma = \frac{Q}{\pi R^2}$$

$$\Rightarrow V_1 = \frac{QR}{2\pi\varepsilon_0 R^2} + \frac{Q}{4\sqrt{2}\,\pi\varepsilon_0 R}$$

$$\Rightarrow V_1 = \frac{Q}{2\pi\varepsilon_0 R}\left(1 + \frac{1}{2\sqrt{2}}\right) \tag{1.152}$$

Similarly, the total potential at 2 is

$$V_2 = V_{\text{disc}} + V_{\text{ring}}$$

$$= \frac{\sigma}{2\varepsilon_0}\left(\sqrt{R^2 + R^2} - R\right) + \frac{Q}{4\pi\varepsilon_0\sqrt{R}},$$

where $\sigma = \frac{Q}{\pi R^2}$

$$\Rightarrow V_2 = \frac{\sigma R}{2\varepsilon_0}(\sqrt{2} - 1) + \frac{Q}{4\pi\varepsilon_0 R}$$

$$= \left(\frac{Q}{\pi R^2}\right)\frac{R(\sqrt{2} - 1)}{2\varepsilon_0} + \frac{Q}{4\pi\varepsilon_0 R}$$

$$= \frac{Q}{2\pi\varepsilon_0 R}\left(\sqrt{2} - \frac{1}{2}\right)$$

Then, the potential difference between A and B is

$$V_2 - V_1 = \frac{Q}{2\pi\varepsilon_0 R}\left[\sqrt{2} - \frac{1}{2} - 1 - \frac{1}{2\sqrt{2}}\right]$$

$$\Rightarrow \Delta V = -\frac{3Q}{8\pi\varepsilon_0 R}[2 - \sqrt{2}] \tag{1.153}$$

Conserving energy we have,

$$\frac{1}{2}mv_0^2 - \frac{1}{2}mv^2 = q\Delta V$$

$$\Rightarrow v = \sqrt{v_0^2 - \frac{2q\Delta V}{m}} \tag{1.154}$$

By using equations (1.153) and (1.154)

$$v = \sqrt{v_0^2 + \frac{3Qq}{8\pi\varepsilon_0 Rm}(2 - \sqrt{2})} \text{ Ans.}$$

Problem 28 An annular disc of radius $2a$ has a uniform charge density σ. Find (a) the electric potential at the centre of the disc, (b) electric field of the disc at an axial distance x from the centre of the disc, (c) frequency of oscillation of a particle of mass m and negative charge $-q$ near the centre of the disc, (d) the distance at which electric field will be maximum and draw the E–x graph. Neglect gravity.
   Solution

   (a) The potential at A can be calculated by another technique called the principle of superposition. According to this principle the given figure (circular ring) is made equivalent to the superposition of a full disc of radius $R = 2a$ with surface charge density $+\sigma$ and another disc of radius $R' = a$ with surface charge density $-\sigma$. Then the potential at A is given as

$$V = V_{bigger} + V_{smaller}$$

$$= \frac{\sigma}{2\varepsilon_0}R + \frac{-\sigma}{2\varepsilon_0}R'$$

$$= \frac{\sigma}{2\varepsilon_0}(R - R') = \frac{\sigma}{2\varepsilon_0}(2a - a)$$

$$\Rightarrow V = \frac{\sigma a}{2\varepsilon_0} \text{ Ans.}$$

(b) Similarly, the electric field at any axial point located at a distance x from A can be given as the superposition of these two discs of uniform charge densities σ and $-\sigma$. The electric field at an axial point is

$$\overrightarrow{E} = \overrightarrow{E}_{bigger} + \overrightarrow{E}_{smaller}$$

$$= \frac{\sigma}{2\varepsilon_0}\left(1 - \frac{x}{\sqrt{R^2 + x^2}}\right)\hat{i} + \frac{(-\sigma)}{2\varepsilon_0}\left(1 - \frac{x}{\sqrt{R'^2 + x^2}}\right)\hat{i}$$

$$= \frac{\sigma x}{2\varepsilon_0}\left(\frac{1}{\sqrt{R'^2 + x^2}} - \frac{1}{\sqrt{R^2 + x^2}}\right)\hat{i}$$

Putting $R = 2a$ and $R' = a$, we have

$$\overrightarrow{E} = \frac{\sigma x}{2\varepsilon_0}\left(\frac{1}{\sqrt{a^2 + x^2}} - \frac{1}{\sqrt{4a^2 + x^2}}\right)\hat{i} \text{ Ans.}$$

(c) For small x, $a^2 + x^2 \simeq a^2$ and $4a^2 + x^2 \simeq 4a^2$

$$\Rightarrow \overrightarrow{E} \cong \frac{\sigma x}{2\varepsilon_0}\left(\frac{1}{a} - \frac{1}{2a}\right)$$

$$\Rightarrow \overrightarrow{E} = \frac{\sigma}{4\varepsilon_0 a}x\hat{i}$$

As we can see near the origin $E \propto x$, for a negative change q the force will be resting given as

$$F = -\frac{\sigma a}{4\varepsilon_0 a}x = -k_{eff}x$$

$$\Rightarrow k_{eff} = \frac{\sigma q}{4\varepsilon_0 a}$$

Then, the angular frequency of oscillation is

$$\omega = \sqrt{\frac{k_{eff}}{m}} = \frac{1}{2}\sqrt{\frac{\sigma a}{\varepsilon_0 a m}} \text{ Ans.}$$

(d) E is maximum when $\frac{dE}{dx} = 0$ and $\frac{d^2E}{dx^2} < 0$

Then, first we part $\frac{dE}{dx} = 0$

$$\Rightarrow \frac{d}{dx} \frac{\sigma}{2\varepsilon_0} \left\{ \frac{x}{\sqrt{a^2 + x^2}} - \frac{x}{\sqrt{4a^2 + x^2}} \right\} = 0$$

$$\Rightarrow d\left(\frac{x}{\sqrt{a^2 + x^2}} \right) = \frac{d}{dx}\left(\frac{x}{\sqrt{4a^2 + x^2}} \right)$$

$$\Rightarrow \frac{\sqrt{a^2 + x^2} - \frac{x}{2}(a^2 + x^2)^{-\frac{1}{2}}(2x)}{a^2 + x^2} = \frac{\sqrt{4a^2 + x^2} - \frac{x}{2}(4a^2 + x^2)^{-\frac{1}{2}}(2x)}{4a^2 + x^2}$$

$$\Rightarrow \frac{a^2 + x^2 - x^2}{(a^2 + x^2)^{\frac{3}{2}}} = \frac{4a^2 + x^2 - x^2}{(4a^2 + x^2)^{\frac{3}{2}}}$$

$$\Rightarrow (4a^2 + x^2)^{\frac{3}{2}} = 4(a^2 + x^2)^{\frac{3}{2}}$$

$$\Rightarrow 4a^2 + x^2 = 4^{\frac{2}{3}}(a^2 + x^2)$$

$$\Rightarrow x^2\left(2^{\frac{4}{3}} - 1 \right) = \left(4 - 2^{\frac{2}{3}} \right)a^2$$

$$\Rightarrow x = x_0 = \pm\sqrt{\frac{4 - 2^{\frac{2}{3}}}{2^{\frac{4}{3}} - 1}}\, a \text{ Ans.}$$

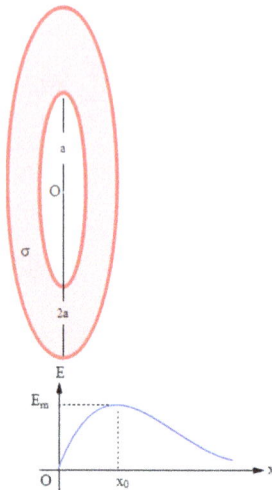

$$\text{As } E = \frac{\sigma}{2\varepsilon_0}\left(\frac{x}{\sqrt{a^2 + x^2}} - \frac{x}{\sqrt{4a^2 + x^2}}\right)$$

1. if we put $x = 0$, $E = 0$,
2. If we put $x \to \infty$, $E \to 0$.

This tells us that at $x = x_0$, E becomes maximum. So, the graph can be drawn as an asymptote. Ans.

Problem 29 The pattern of equipotential lines or surface (V-lines) is given in two dimensions. Find the electric field pattern.

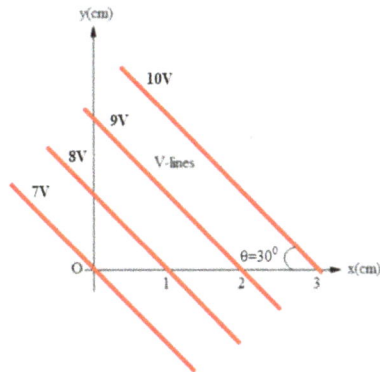

Solution

Since the flat equipotential lines are parallel, we can imagine the electric field perpendicular to these lines. This means that E-lines are perpendicular to V-lines. The E-lines point in the direction of decreasing potential with maximum value of the potential gradient. In other words, potential decreases with maximum gradient, in the direction of E-field.

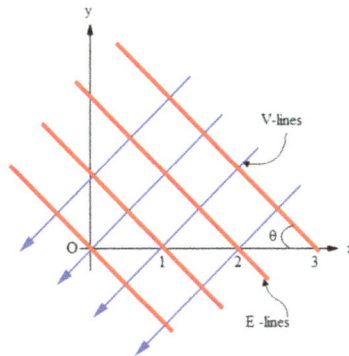

$$E = \frac{dV}{d\ell} = \frac{\Delta V}{\Delta \ell}$$

$$V_x = (V_x)_0 + \frac{\Delta V_x}{\Delta x} \cdot x, \quad \text{where } \Delta V = 1 \text{ volt and } \Delta x = 1 \text{ cm},$$

$$\Rightarrow V_x = 0 + \frac{(1) \text{ volt}}{1 \text{ cm}} x$$

$$V_x = x \tag{1.155}$$

Similarly, $V_y = (V_y)_0 + \frac{\Delta V_y}{\Delta y} y$, where $\Delta V_y = 1$ volt and $\Delta y = \frac{1}{\sqrt{3}}$ cm

$$\Rightarrow V_y = 0 + \frac{1}{\frac{1}{\sqrt{3}}} y$$

$$\Rightarrow V_y = \sqrt{3} y \tag{1.156}$$

Then, the electric field can be given as

$$\vec{E} = -\frac{\partial V_x}{\partial x}\hat{i} - \frac{\partial V_y}{\partial y}\hat{j}$$

$$= -\frac{\partial}{\partial x}x\hat{i} - \frac{\partial}{\partial y}(\sqrt{3}y)\hat{j}$$

$$\vec{E} = -(\hat{i} + \sqrt{3}\hat{j})N/C$$

The field pattern is shown in the diagram. Ans.

Problem 30 The pattern of equipotential lines or surface (V-lines) are given in two dimensions. Find the (a) $V = f(r)$, (b) $E = f(r)$, (c) ratio of the field intensity at $r = 10$ cm and $r = 29$ cm, (d) E and V pattern and the charge that causes the field and potential.

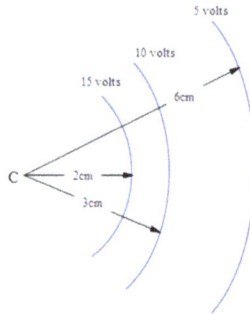

Solution

(a) From the given pattern of V-lines or equipotential lines we can see (by hit and trial method) that product of potential and radial distance is a constant quality. This means that at any radial distance r the potential V can be related as

$$V_r = C/r,$$

where the constant $C = (2\ \text{cm})(15\ \text{volt}) = \dfrac{2}{100} \times 15 = 0.3\ \text{Volt. m}$

$$\Rightarrow V = \frac{0.3}{r}\ \text{Volt Ans.}$$

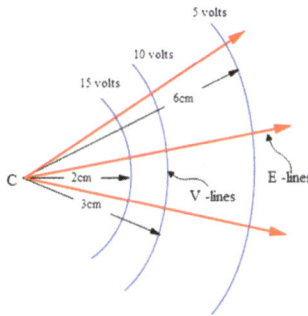

(b) The field at any radial distance is

$$E = -\frac{\partial V}{\partial r}\hat{r} = -\frac{\partial}{\partial r}\left(\frac{0.3}{r}\right)\hat{r}$$

$$\Rightarrow \overrightarrow{E} = \frac{0.3}{r^2}\hat{r}\ \text{Ans.}$$

This means that electric field obeys inverse square law possessing spherical symmetry. It points radially outward, as shown in the figure. We can see that the spacing of both E and V-lines increases with distance.

(c) The expression $E = \frac{0.3}{r^2}$ gives us the ratio,

$$\frac{E_1}{E_2} = \left(\frac{r_2}{r_1}\right)^2 = \left(\frac{20\ \text{cm}}{10\text{cm}}\right) = 4\ \text{Ans.}$$

(d) This E-pattern is given by a point change q because

$$E = \frac{Kq}{r^2},\quad \text{where } Kq = C = 0.3$$

$$\Rightarrow q = \frac{0.3}{K} = \frac{0.3}{9 \times 10^9} = \frac{1}{3} \times 10^{-10}C \text{ Ans.}$$

Problem 31 A spherical dielectric sphere of outer and inner radii a and b has a uniform volume charge density ρ. Find the (a) electric field as the function of radial distance r, (b) potential as the function of radial distance r, (c) ratio of potentials at $r = a$ and $r = b$.

Solution

(a) The charge enclosed by the dotted sphere is

$$q = \rho \frac{4}{3}\pi(r^3 - a^3) \tag{1.157}$$

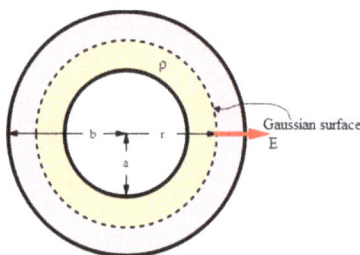

Applying Gauss' law, the electric field at a radial distance r inside the spherical shell is

$$E = \frac{q}{4\pi\varepsilon_0 r^2} \tag{1.158}$$

Using equations (1.157) and (1.158)

$$E = \frac{\rho \frac{4}{3}\pi(r^3 - a^3)}{4\pi\varepsilon_0 r^2}$$

$$\Rightarrow E = \frac{\rho}{3\varepsilon_0}\left(r - \frac{a^3}{r^2}\right); r < b$$

Furthermore, $E = 0$ for $r < a$ because it is a uniformly charged spherical shell.

For $r > b$, $E = \frac{Q}{4\pi\varepsilon_0 r^2}$, when Q = total change = $\frac{4}{3}\pi(b^3 - a^2)\rho$

$$\Rightarrow E = \frac{\rho}{3\varepsilon_0}\frac{(b^3 - a^3)}{r^2}$$

Putting all the results in one place, we have

$$E = \begin{cases} 0; \; r \leqslant a \\ \dfrac{\rho}{3\varepsilon_0}\left(r - \dfrac{a^3}{r^2}\right); \; a \leqslant r \leqslant b \\ \dfrac{\rho\left(b^3 - a^3\right)}{3\varepsilon_0 r^2}; \; r \geqslant b \end{cases}$$

We can see that E is continuous friction.

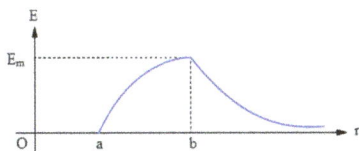

N.B: For $a \leqslant r \leqslant b$, we have

$$E = \frac{\rho}{3\varepsilon_0}\left(r - \frac{a^3}{r^2}\right)$$

So, $\frac{dE}{dr}$ (= gradient of the field) slope of E–r graph $= \frac{\rho}{3\varepsilon_0}(1 + \frac{2a^3}{r^3})$ which decreases as r increases. The slope of the graph decreases from $\frac{\rho}{3\varepsilon_0}(1 + \frac{2a^3}{a^3})$ $= \frac{3\rho}{3\varepsilon_0} = \frac{\rho}{\varepsilon_0}$ to $\frac{\rho}{3\varepsilon_0}(1 + \frac{2a^3}{b^3})$ from $r = a$ to $r = b$. So, the slope of E–r graph is not zero at $r = b$ at the left side of the point $x = b$. We can see that at the right side of $x = b$, the slope of the E–r graph is given as

$$\frac{dE}{dr} = \frac{d}{dr}\frac{\rho}{3\varepsilon_0}\frac{\left(b^3 - a^3\right)}{r^2}$$

$$= -\frac{2\rho\left(b^3 - a^3\right)}{3\varepsilon_0 r^3}|_{r=b} = -\frac{2\rho}{3\varepsilon_0}\left(1 - \frac{a^3}{b^3}\right) < 0$$

But the slope of E–r graph at $r = b$ is different on both sides of the point.
(b) The potential at any point at radial distance r ($a < r < b$) is given as

$$V(r) = -\int_\infty^r E \, dr$$

$$= -\int_\infty^b E \, dr - \int_b^r E \, dr$$

$$= -\int_\infty^b \frac{\rho}{3\varepsilon_0}\frac{\left(b^3 - a^3\right)}{r^2}dr - \int_b^r \frac{\rho}{3\varepsilon_0}\left(r - \frac{a^3}{r^2}\right)dr$$

$$= -\frac{\rho}{3\varepsilon_0}\left[-\frac{\left(b^3 - a^3\right)}{r}|_\infty^b + \frac{r^2}{2} + \frac{a^3}{r}|_b^r\right]$$

$$\Rightarrow V(r) = -\frac{\rho}{3\varepsilon_0}\left[-\frac{\left(b^3 - a^3\right)}{b} + \frac{r^2}{2} + \frac{a^3}{r} - \frac{b^2}{2} - \frac{a^3}{b}\right]$$

$$= -\frac{\rho}{3\varepsilon_0}\left[-\frac{3b^2}{2} + \frac{r^2}{2} + \frac{a^3}{r}\right]$$

$$V = \frac{\rho}{6\varepsilon_0}(3b^2 - r^2 - \frac{2a^3}{r}) \text{ Ans.}$$

V at $r \leqslant a$ is given by putting $r = a$ in the above expression.

$$V(r) = \frac{\rho}{6\varepsilon_0}(3b^2 - a^2 - 2a^2); \ r \leqslant a$$

$$= \frac{\rho}{2\varepsilon_0\left(b^2 - a^2\right)}; \ r \leqslant a$$

V at $r = b$ is given as

$$V|_{r=b} = \frac{\rho}{6\varepsilon_0}\left(3b^2 - b^2 - \frac{2a^3}{b}\right)$$

$$= \frac{\rho}{3\varepsilon_0}\left(\frac{b^3 - a^3}{b}\right)$$

V at $r > b$ is given as

$$V(r) = -\int_\infty^r E \, dr = -\int_\infty^r \frac{\rho}{3\varepsilon_0}\frac{(b^3 - a^3)}{r^2} \, dr$$

$$\Rightarrow V(r) = \frac{\rho(b^3 - a^3)}{3\varepsilon_0 r}; \ r \geqslant b$$

Recapitulating,

$$V = \begin{cases} \frac{\rho}{2\varepsilon_0}(b^2 - a^2) = V_0; \ r \leqslant a \\ \frac{\rho}{6\varepsilon_0}\left(3b^2 - r^2 - \frac{2a^3}{r}\right); \ a \leqslant r \leqslant b \\ \frac{\rho}{3\varepsilon_0}\frac{\left(b^3 - a^3\right)}{r}; \ r \geqslant b \end{cases}$$

For $a \leqslant r \leqslant b$, the slope of the V–r graph is given as

$$\frac{dV}{dr} = \frac{\rho}{6\varepsilon_0}\left(-2r + \frac{2a^3}{r^2}\right)$$

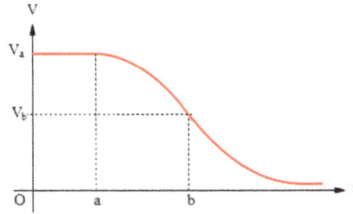

$$\text{At } r = a, \quad \frac{dV}{dr} = 0$$

For $r \geqslant b$, $\frac{dV}{dr} = -\frac{\rho}{3\varepsilon_0}\frac{(b^3 - a^3)}{r^2}$

Then,

$$\frac{dV}{dr}\Big|_{r=b} = -\frac{\rho}{3\varepsilon_0}\frac{b^3 - a^3}{b^2}$$

This means that the slope of V–r graph is continuous (same) at both sides of the point at $r = b$. From $r = a$ to $r = b$ the slope of V–r graph decreases from zero to a negative value of $-\frac{\rho}{3\varepsilon_0}\frac{b^3 - a^3}{b^2}$ as shown in the graph.

(c) The potential at $r = a$ is

$$V_1 = V_a = \frac{\rho}{2\varepsilon_0}(b^2 - a^2)$$

The potential at $r = b$ is

$$V_2 = V_b = \frac{\rho}{3\varepsilon_0}\frac{b^3 - a^3}{b}$$

Then

$$\frac{V_1}{V_2} = \frac{3b(b^2 - a^2)}{2(b^3 - a^3)}$$

$$= \frac{3b(a + b)}{2(a^2 + ab + b^2)} \quad \text{Ans.}$$

Problem 32 In the previous problem (problem 31), find the electric energy stored due to the shell. Assume that relative permittivity of the sphere is equal to one.

Solution

The energy stored due to the electric field of the sphere is given as

$$U = \frac{1}{2}\int V \, dq, \quad \text{where } dq = \rho dV = \rho(4\pi r^2 dr)$$

$$\Rightarrow U = \frac{4\pi\rho}{2}\int_a^b V(r)r^2 dr$$

$$\Rightarrow U = 2\pi\rho \int_a^b V(r)r^2 dr \tag{1.159}$$

In the last example we have

$$V(r) = \frac{\rho}{6\varepsilon_0}\left(3b^2 - r^2 - \frac{2a^3}{r}\right) \tag{1.160}$$

By using these two equations, we have

$$U = \frac{\rho}{6\varepsilon_0} \times 2\pi\rho \int_a^b (3b^2 - r^2 - \frac{2a^3}{r})r^2 dr$$

$$= \frac{\pi\rho^2}{3\varepsilon_0}\left[3b^2 \int_a^b r^2\, dr - \int_a^b r^4\, dr - 2a^3 \int_a^b r\, dr\right]$$

$$= \frac{\pi\rho^2}{3\varepsilon_0}\left[b^2(r^3)|_a^b - \frac{r^5}{5}|_a^b - 2a^3(\frac{r^2}{2})|_a^b\right]$$

$$= \frac{\pi\rho^2}{3\varepsilon_0}\left[b^5 - a^3b^2 - \frac{(b^5 - a^5)}{5} - (a^3b^2 - a^5)\right]$$

$$= \frac{\pi\rho^2}{3\varepsilon_0}\left[\frac{4b^5}{5} + \frac{7a^5}{5} - 2a^3b^2\right]$$

$$= \frac{\pi\rho^2}{15\varepsilon_0}(7a^5 + 4b^5 - 10a^3b^2) \text{ Ans.}$$

N.B: If asked to find the energy stored in a precise region i.e. $a \leqslant r \leqslant b$, we have to use the formula:

$$U = \frac{\varepsilon_0}{2} \int E^2\, dv$$

$$U = 2\pi\varepsilon_0 \int_a^b r^2 E(r)\, dr$$

Problem 33 A spherical dielectric sphere of radius R has a uniform volume charge density ρ. Find the (a) total electrostatic energy stored, (b) electrostatic energy stored inside the sphere, (c) electrostatic energy stored outside the sphere, (d) ratio of electrostatic energy stored inside and outside the sphere.
   Solution

   (a) For a uniform volume charge distribution

$$E = \frac{\rho}{3\varepsilon_0}r; \ r \leqslant R \tag{1.161}$$

For $r \geqslant R$,

$$E = \frac{u}{4\pi\varepsilon_0 r^2} = \frac{\left(\frac{4}{3}\pi R^3 \rho\right)}{4\pi\varepsilon_0 r^2}$$

$$\Rightarrow E = \frac{\rho R^3}{3\varepsilon_0 r^2} \; ; \; r \geqslant R \tag{1.162}$$

So, we have the energy stored

$$U = \frac{E_0}{2}\int E^2 dV$$

$$\Rightarrow U = \frac{t_0}{2}\int_0^R E^2 dV + \int_R^\infty E^2 dV \tag{1.163}$$

Using the function of E from equations (1.161) and (1.162) in equation (1.163), we have

$$U = \frac{t_0}{2}\left[\int_0^R \left(\frac{\rho}{3\varepsilon_0}\right)^2 dV + \int_0^\infty \left(\frac{\rho R^3}{3\varepsilon_0 r^2}\right)^2 dV\right]$$

where $dV = 4\pi r^2 \, dr$

$$\Rightarrow U = \frac{\rho^2(4\pi)}{18\,\epsilon_0}\left[\int_0^R r^4 dr + R\int_R^\infty \frac{dr}{r^2}\right]$$

$$\Rightarrow U = \frac{2\pi\rho^2}{9\varepsilon_0}\left(\frac{r^5}{5}\Big|_0^R + R^6\left(-\frac{1}{r}\right)\Big|_R^\infty\right)$$

$$= \frac{2\pi\rho^2}{9\varepsilon_0}\left(\frac{R^5}{5} + R^5\right)$$

$$= \frac{2\pi\rho^2}{9\varepsilon_0}\left(\frac{6R^5}{5}\right)$$

$$= \frac{4\pi\rho^2 R^5}{15\varepsilon_0}$$

Putting $\rho = \frac{Q}{\frac{4}{3}\rho R^3} = \frac{3Q}{4\pi R^3}$, we have

$$U = \frac{4\pi R^5}{15\varepsilon_0}\left(\frac{3Q}{4\pi R^3}\right)^2$$

$$= \frac{4\times 9}{15\times 16}\frac{Q^2}{\pi\varepsilon_0 R}$$

$$= \frac{3Q^2}{20\pi\varepsilon_0 R} \; \text{Ans.}$$

(b) The energy stored inside the sphere is

$$U_{\text{inside}} = \frac{\varepsilon_0}{2} \int_0^R E^2 dV$$

$$= \frac{\varepsilon_0}{2} \int_0^R \left(\frac{Qr}{3\varepsilon_0}\right)^2 4\pi r^2 dr,$$

where

$$\rho = \frac{3Q}{4\pi R^3}$$

$$= \frac{\varepsilon_0}{2} \int_0^R \left(\frac{Qr}{4\pi\varepsilon_0 R^3}\right)^2 4\pi r^2 dr$$

$$= \frac{Q^2}{8\pi\varepsilon_0 R^6} \int_0^R r^4 \, dr$$

$$= \frac{Q^2}{40\pi\varepsilon_0 R} \text{ Ans.}$$

(c) The energy stored outside the sphere is

$$U_{\text{outside}} = \frac{\varepsilon_0}{2} \int_R^\infty E^2 \, dV$$

$$= \frac{\varepsilon_0}{2} \int_R^\infty \left(\frac{Q}{4\pi\varepsilon_0 r^2}\right)^2 4\pi r^2 dr$$

$$= \frac{Q^2}{8\pi\varepsilon_0 R} \text{ Ans.}$$

(d) Then the ratio of energy stored is

$$\frac{U_{\text{in}}}{U_{\text{out}}} = \frac{1}{5} \text{ Ans.}$$

Alternative method:
The potential friction inside the sphere is given as

$$V = -\int_\infty^r E \, dr$$

$$= -\int_\infty^R E \, dr - \int_R^r E \, dr$$

$$= -\int_\infty^R \frac{Q}{4\pi\varepsilon_0 r^2} dr - \int_R^r \left(\frac{Qr}{4\pi\varepsilon_0 R^3}\right) dr$$

$$= \frac{Q}{4\pi\varepsilon_0 R} - \frac{Q}{4\pi\varepsilon_0 R^3}\left(\frac{r^2 - R^2}{2}\right)$$

$$= \frac{Q}{4\pi\varepsilon_0 R}\left(1 + \frac{1}{2} - \frac{r^2}{2R^2}\right)$$

$$= \frac{q}{4\pi\varepsilon_0 R}\left(\frac{3}{2} - \frac{r^2}{2R^2}\right)$$

$$V = \frac{Q}{8\pi\varepsilon_0 R}\left(3 - \frac{r^2}{R^2}\right); \quad 0 \leqslant r \leqslant R$$

Then we substitute this potential function in the formula

$$U = \frac{1}{2}\int V \, dq$$

$$\Rightarrow U = \frac{1}{2}\int_0^R \left\{\frac{Q}{8\pi\varepsilon_0 R}\left(3 - \frac{r^2}{R^2}\right)\right\}(\rho 4\pi r^2 dr)$$

$$= \frac{Q\rho}{4\varepsilon_0 R}\int_0^R \left(3 - \frac{r^2}{R^2}\right)r^2 \, dr$$

$$= \frac{Q\left(\frac{Q}{\frac{4}{3}\pi R^3}\right)}{4\varepsilon_0 R}\left(\frac{3r^3}{3} - \frac{r^5}{5R^2}\right)\Big|_0^R$$

$$= \frac{3Q^2}{16\pi\varepsilon_0 R^4}\left(\frac{R^3}{3} - \frac{R^3}{5}\right)$$

$$= \frac{3Q^2}{20\pi\varepsilon_0 R} \quad \text{Ans.}$$

N.B: In this method we cannot calculate the energy in a specific region. This method gives 'total' potential energy stored in 'all' space.

Problem 34

(a) A soap bubble has radius R, thickness t and volume charge density ρ. find
   (i) the potential of the soap bubble;
   (ii) energy stored due to the soap bubble.
(b) If the bubble bursts, a drop is formed. Then, find
   (i) the potential of the drop at its surface and center;
   (ii) ratio of energy stored in the drop and bubble.

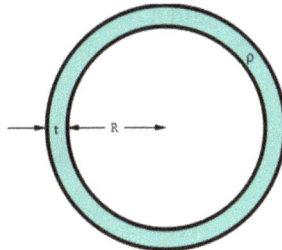

Solution

(a)   (i) The charge of the bubble is $q = \rho$ (volume), where 'volume' of the stuff of the bubble is

$$Vol = 4\pi R^2 t$$

$$\Rightarrow q = 4\pi R^2 \rho t \tag{1.164}$$

The potential of the bubble at its surface is

$$V_b = \frac{Q}{4\pi\varepsilon_0 R} \tag{1.165}$$

Using equations (1.164) and (1.165), we have

$$V_b = \frac{4\pi R^2 \rho t}{4\pi\varepsilon_0 R}$$

$$\Rightarrow V_b = \frac{R\rho t}{\varepsilon_0} \text{ Ans.}$$

(ii) The energy stored is

$$U = \frac{1}{2}\int V_b \, dq$$

$$= \frac{V}{2}\int_0^q dq (\because V_b = \text{Constant}) \tag{1.166}$$

$$= \frac{qV_b}{2}$$

Putting q from equation (1.164) and the obtained value of V_b in equation (1.166), we have

$$U = \tfrac{1}{2}(4\pi R^2 \rho t)\left(\frac{R\rho t}{\varepsilon_0}\right)$$

$$= \frac{2\pi R^3 \rho^2 t^2}{\varepsilon_0} \text{ Ans.}$$

(b)   (i) If the bubble is burst and a drop is formed, we can conserve the mass and charge. Conserving mass or volume, we have

$$m_{\text{bubble}} = m_{\text{drop}}$$

$$\Rightarrow Vol_{\text{bubble}} = Vol_{\text{drop}}$$

$$\Rightarrow 4\pi R^2 t = \frac{4}{3}\pi r^3$$

$$\Rightarrow r = (3R^2 t)^{\frac{1}{3}} \tag{1.167}$$

Then the potential of the drop (solid sphere) at its surface is

$$V_{\text{drop}} = \frac{q}{4\pi\varepsilon_0 r} \qquad (1.168)$$

Using equations (1.164), (1.167) and (1.168)

$$V_{\text{drop}} = \frac{4\pi R^2 \rho t}{4\pi\varepsilon_0 (3R^2 t)^{\frac{1}{3}}}$$

$$\Rightarrow V_{\text{drop}} = \frac{R^3 \rho t^{\frac{2}{3}}}{3^{\frac{1}{3}}\varepsilon_0} \text{ Ans.}$$

With reference to the previous problem, the potential at the centre of the drop is

$$V_{\text{center}} = \frac{3q}{8\pi\varepsilon_0 r}, \text{ where } q = 4\pi R^2 t\rho \text{ and } r = (3R^2 t)^{\frac{1}{3}}$$

$$\Rightarrow V_c = \frac{3(4\pi R^2 t\rho)}{8\pi\varepsilon_0 (3R^2 t)^{\frac{1}{3}}}$$

$$= \frac{R^{\frac{4}{3}}\rho t^{\frac{2}{3}}}{2 \times 3^{\frac{1}{3}}\varepsilon_0} \text{ Ans.}$$

(ii) The final and initial energy are

$$U_f = U' = \frac{3q^2}{20\pi\varepsilon_0 r}$$

$$U_i = U = \frac{q^2}{8\pi\varepsilon_0 R}$$

Then,

$$\frac{U'}{U} = \frac{3}{20} \times 8 \times \frac{R}{r} = \frac{6}{5}(\frac{R}{r}) \qquad (1.169)$$

By using equations (1.167) and (1.169), we have

$$\frac{U'}{U} = \frac{6}{5}\frac{R}{(3R^2 t)^{\frac{1}{3}}} = \frac{6}{5}(\frac{R}{3t})^{\frac{1}{3}} \text{ Ans.}$$

N.B: Since $t \ll R$, the final electrostatic energy is much greater than the initial energy.

Problem 35 A thin semi-circular frame is charged such that its linear charge density varies with the angle θ as $\lambda = \lambda_0 \cos\theta$. It is placed in a uniform electric field E. Find (a) the dipole moment of the frame, (b) energy possed by the dipole in the electric field.

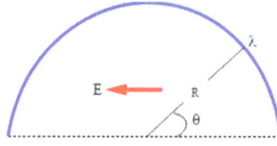

Solution

(a) Conserving two elementary charges dq and $-dq$, the dipole moment

$$dp = 2dq(R \cos \theta)$$

Then the net dipole moment due to the ring is given as

$$p = \int dp = 2R \int dq \cos \theta$$
$$= 2R \int (\lambda \, Rd\theta)\cos \theta$$
$$= 2R^2 \int \lambda \cos \theta d\theta$$
$$= 2R^2 \int_0^{\frac{\pi}{2}} (\lambda_0 \cos \theta)\cos \theta \, d\theta$$
$$= 2\lambda_0 R^2 \int_0^{\frac{\pi}{2}} \cos^2 \theta d\theta$$
$$= \frac{\lambda_0 R^2}{2} \int_0^{\frac{\pi}{2}} (1 + \cos 2\theta) \, d\theta$$
$$= \pi\lambda_0 R^2 \text{ Ans.}$$

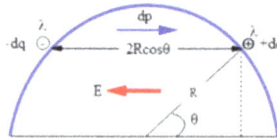

(b) The energy possessed by the dipole is

$$U = -pE \cos \theta$$
$$= -(\pi\lambda_0 R^2)(E)\cos 80°$$
$$= +\pi\lambda_0 R^2 E.$$

The dipole possesses a positive energy as its dipole moment points opposite to the electric field. Ans.

N.B: $q = \int \lambda Rd\theta = \lambda_0 R \int_0^{\pi} \cos \theta \, d\theta = 0$

Since the net charge is zero, the charge in right and left quadrant will be equal and opposite.

Problem 36 A thin semi-circular frame of radius R is comprised of two quadrants of charge $+Q$ and $-Q$. If the charge distribution is uniform in each quadrant, find the dipole moment of the frame.

Solution

The charges of elements $-dq$ and dq form a dipole of moment given as

$$dp = dq(2x) = 2dq \, x$$
$$= 2(\lambda \, Rd\theta)R \cos \theta$$

$$\Rightarrow dp = 2\lambda R^2 \cos \theta d\theta$$

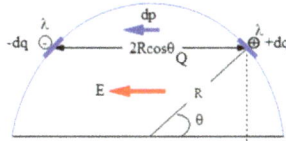

Then, the net dipole moment is

$$p = \int dp = 2\lambda R^2 \int_0^{\frac{\pi}{2}} \cos \theta \, d\theta$$

$$\Rightarrow p = 2\lambda R^2$$

Putting $\lambda = \frac{Q}{\pi R}$, we have $p = \frac{2QR}{\pi}$. Ans.

Problem 37 A thin straight composite rod is comprised of two identical rods of charge $+q$ and $-q$. If the charge distribution is uniform in each rod, find (a) the dipole moment of the frame, (b) the angular acceleration of the dipole in the electric field E at the given position, (c) maximum angular speed of the rod. Put $\theta = 60°$.

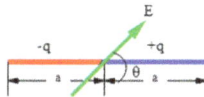

Solution

(a) Take two elementary charges $-dq$ and $+dq$ at a distance $2x$. The dipole moment of these two is

$$dp = dq(x), \quad \text{where } dq = \frac{q}{a}dx$$

$$\Rightarrow dp = \frac{q}{a}dx \cdot 2x = \frac{2q}{a}x \, dx$$

Then, the net dipole is

$$p = \int dp = \frac{2q}{a} \big|_0^a \, x dx$$

$$\Rightarrow p = \frac{2q}{a} \frac{a^2}{2} = qa \text{ Ans.}$$

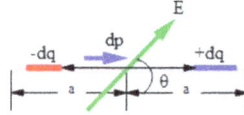

(b) The torque experienced by the dipole is

$$\tau = pE \sin \theta$$
$$= (qa)E \sin \theta$$

Since the rod is not pivoted, it is free to rotate about the centre of mass. The moment of inertia about the centre of mass is

$$I_c = \frac{ma^3}{3}$$

Then, the angular acceleration of the rod is

$$\alpha = \frac{\tau}{I_c} = \frac{qaE \sin \theta}{\frac{ma^2}{3}}$$

$$\Rightarrow \alpha = \frac{3qE \sin \theta}{ma} \text{ Ans.}$$

(c) Putting $\alpha = \frac{\omega d\omega}{d\theta}$, we have $\omega d\omega = -\alpha d\theta$ (because θ decreases)

$$\Rightarrow \omega d\omega = -\frac{3qE \sin \theta}{ma} d\theta$$

$$\Rightarrow \int_0^\omega \omega d\omega = -\frac{3qE}{ma} \int_0^\theta \sin \theta \, d\theta$$

$$\Rightarrow \frac{\omega^2}{2} = \frac{3qE}{ma}\left(1 - \frac{1}{2}\right)$$

$$\Rightarrow \omega = \sqrt{\frac{3qE}{ma}} \text{ Ans.}$$

Alternative method:
Conserving energy, we have

$$U_i + K_i = U_f + K_f$$

$$\Rightarrow (-pE \cos 60°) + 0 = (-pE \cos 0) + K_f$$

$$\Rightarrow K_f = \left(1 - \frac{1}{2}\right)pE$$

$$\Rightarrow \frac{1}{2}I_c\omega^2 = \frac{1}{2}pE$$

$$\Rightarrow \omega = \sqrt{\frac{pE}{I_c}}, \quad \text{where } p = qa \text{ and } I_c = \frac{ma^2}{3}$$

$$\Rightarrow \omega = \sqrt{\frac{(qa)E}{\frac{ma^2}{3}}} = \sqrt{\frac{3qE}{ma}} \text{ Ans.}$$

N.B: The angular speed will be maximum when angular acceleration is zero at the stable equilibrium position. At this position the dipole moment points in the direction of the electric field.

Problem 38 A thin semi-circular frame of mass m, radius R has charge $+q$ which is uniformly distributed throughout the length. It is placed in a uniform electric field E on an insulated smooth horizontal plane. A point charge $-q$ is fixed at the smooth pivot O. The pivot is attached with the ring by a thin light rigid insulated rod (not shown in the figure). Find (a) the dipole moment of the frame, (b) torque experienced by the dipole in the electric field, (c) angular acceleration of the frame, at the given position.

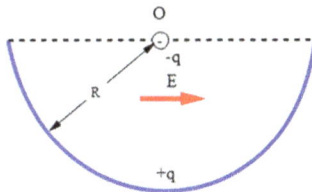

Solution

(a) The semicircle is symmetrical about the y-axis. From a symmetrical point of view, the centre of charge lying on the y-axis is given as

$$y_c = \frac{2 \int y \, dq}{\int dq} = \frac{2 \int_0^{\frac{\pi}{2}} (R \sin \theta)(\lambda R d\theta)}{Q}$$

$$= \frac{2R^{2\lambda} \int_0^{\frac{\pi}{2}} \sin \theta d\theta}{Q},$$

where $\lambda = \frac{\theta}{\pi R}$ and $\int_0^{\frac{\pi}{2}} \sin\theta d\theta = 1$

$$\Rightarrow y_c = \frac{2R^2\left(\frac{Q}{\pi R}\right)}{Q} = \frac{2R}{\pi}$$

Then the dipole moment is

$$p = qy_c = q\left(\frac{2R}{\pi}\right) \Rightarrow p = \frac{2Rq}{\pi} \text{ Ans.}$$

(b) The torque exerted by the electric field and the frame is

$$\overrightarrow{\tau} = \overrightarrow{p} \times \overrightarrow{E}$$

$$\Rightarrow \overrightarrow{\tau} = pE \sin 90° \hat{k}$$

$$\Rightarrow \overrightarrow{\tau} = pE\hat{k}$$

Putting the value of $p = \frac{2Rq}{\pi}$,

$$\overrightarrow{\tau} = \frac{2qER}{\pi}\hat{k} \text{ Ans.}$$

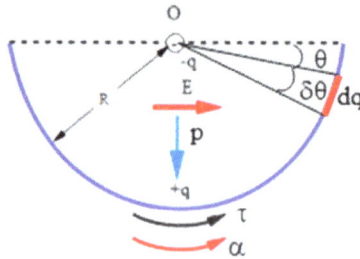

(c) The moment of inertia of the half ring about O is
$$I_O = mR^2$$

Then, $\overrightarrow{\tau_O} = I_O \overrightarrow{\alpha}$

$$\text{or, } \overrightarrow{\alpha} = \frac{\overrightarrow{\tau_O}}{I_O} = \frac{\frac{2qER\hat{k}}{\pi}}{mR^2}$$

$$\Rightarrow \overrightarrow{\alpha} = \frac{2qE}{\pi mR}\hat{k} \text{ Ans.}$$

N.B: If the system is free (not pivoted) and we are asked to find the angular acceleration α, we can apply the concept of centre of mass to find it. If the body is free to rotate about its centre of mass, first of all by using parallel and perpendicular axis theorem we find the moment of inertia I_c about the z-axis passing through its centre of mass, then, apply the equation

$$\vec{\tau} = I_c \vec{\alpha}$$

Please note that the torque $\vec{\tau} = \vec{p} \times \vec{E}$ is frame-independent (does not depend upon the reference frame).

Problem 39 At time $t = 0$, the particle of mass m and charge q is given a velocity v_0 in a uniform electric field inside a charged capacitor. The distance between the plates of the capacitor is equal to d. Find the (a) total time after which the particle hits the left-side plate of the capacitor, (b) average velocity of the charged particle. Assume that the particle cannot touch the right- side plate of the capacitor and the charge density as 'σ'.

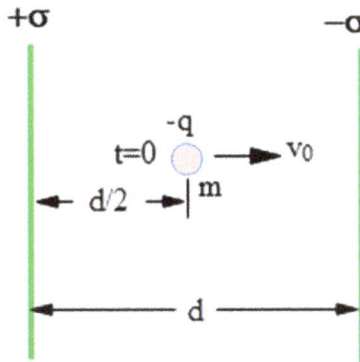

Solution

(a) The time taken by the charged particle q to reach point C is

$$t_1 = \frac{v_0}{a}, \quad \text{where } a = \frac{qE}{m}$$

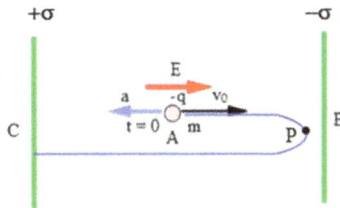

$$\Rightarrow t_1 = \frac{mv_0}{qE}, \quad \text{where } E = \frac{\sigma}{2\varepsilon_0} + \frac{a}{2\varepsilon_0} = \frac{\sigma}{\varepsilon_0}$$

$$\Rightarrow t_1 = \frac{mv_0\varepsilon_0}{q\sigma}$$

The time from P to C is found by the formula

$$PC = \frac{1}{2}at_2^2$$

$$\Rightarrow t_2 = \sqrt{\frac{2PC}{a}},$$

where

$$PC = AC + AP$$

$$= \frac{d}{2} + \frac{v_0^2}{2a} = \frac{1}{2}\left(d + \frac{v_0^2}{a}\right)$$

$$\Rightarrow t_2 = \sqrt{\frac{2}{a} \times \frac{1}{2}\left(d + \frac{v_0^2}{a}\right)}$$

$$= \sqrt{\frac{d}{a} + \frac{v_0^2}{a^2}} = \frac{v_0}{a}\sqrt{\frac{ad}{v_0^2} + 1},$$

where

$$a = \frac{qE}{m} = \frac{q}{m}\frac{\sigma}{\varepsilon_0}$$

$$\Rightarrow t_2 = \frac{mv_0\varepsilon_0}{q\sigma}\sqrt{\frac{q\sigma d}{m\varepsilon_0 v_0^2} + 1}$$

Then the total time is

$$T = t_1 + t_2$$

$$= \frac{mv_0\varepsilon_0}{q\sigma} + \frac{mv_0\varepsilon_0}{q\sigma}\sqrt{1 + \frac{q\sigma d}{m\varepsilon_0 v_0^2}}$$

$$T = t_{AB} + t_{BC}$$

$$= \frac{v_0}{a} + \sqrt{\frac{2BC}{a}}$$

$$= \frac{v_0}{a} + \sqrt{\frac{2\left(\frac{d}{2} + \frac{v_0^2}{2a}\right)}{2}}$$

$$= \frac{v_0}{a} + \sqrt{\frac{d}{a} + \frac{v_0^2}{a^2}}$$

$$T = \frac{v_0}{a}\left(1 + \sqrt{\frac{ad}{v_0^2} + 1}\right) \text{ Ans.}$$

(b) Then, the average velocity over time T is

$$\vec{v}_{av} = \frac{\vec{S}}{T}$$

$$= \frac{-\frac{d}{2}\hat{i}}{\frac{v_0}{a}\left(1 + \sqrt{1 + \frac{ad}{v_0^2}}\right)}$$

$$= \frac{-da}{\left(v_0^2 + \sqrt{v_0^2 + ad}\right)},$$

where

$$a = \frac{qE}{m} = \frac{q\sigma}{m\varepsilon_0}$$

$$\Rightarrow \vec{v}_{av} = -\frac{\frac{q\sigma}{m\varepsilon_0}\hat{i}}{v_0^2 + \sqrt{v_0^2 + \frac{q\sigma d}{m\varepsilon_0}}} \quad \text{Ans.}$$

(c) The average velocity is

$$v_{av} = \frac{s}{T} = \frac{\frac{d}{2}}{\frac{mv_0\varepsilon_0}{9\sigma}[1 + \sqrt{1 + \frac{q\sigma d}{mt_0 d^2}}]}$$

$$= \frac{q\sigma d}{2mv_0\varepsilon_0(1 + \sqrt{1 + \frac{q\sigma d}{mt_0 d^2}})} \quad \text{Ans.}$$

Alternative method:
It can also be found by the formula

$$V_{av} = \frac{\vec{u} + \vec{v}}{2}, \quad \text{where } \vec{u} = v_0\hat{i} \text{ and } \vec{v} = -\sqrt{v_0^2 + 2ab}\,\hat{i}$$

$$\Rightarrow \vec{v}_{av} = \frac{\sqrt{v_0^2 + 2as} - v_0}{2}\hat{i}$$

$$= \frac{\sqrt{v_0^2 + 2\left(\frac{q\sigma}{m\varepsilon_0}\right)\left(\frac{d}{2}\right)} - v_0}{2}\hat{i}$$

$$= \frac{\sqrt{v_0^2 + \frac{q\sigma d}{m\varepsilon_0}} - v_0}{2}\hat{i} \quad \text{Ans.}$$

N.B: Kinematically, you can verify that both expressions of average velocities are the same.

Problem 40 A dielectric sphere of radius R has a volume charge density ρ that varies with the radial distance r as $\rho = \rho_0 r/R$. Find (a) the electric field, (b) potential of the sphere as the function of radial distance r, (c) potential difference between the center and surface of the sphere.

   Solution

   (a) Applying Gauss' law,

$$\oint E \, dA = \frac{q_{en}}{\varepsilon_0}$$

$$\Rightarrow E 4\pi r^2 = \frac{q_{en}}{\varepsilon_0}$$

$$\Rightarrow E = \frac{q_{en}}{4\pi\varepsilon_0 r^2} \tag{1.173}$$

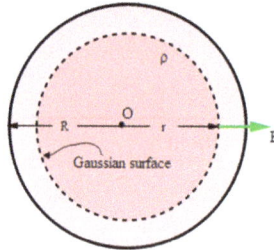

The charge enclosed is

$$q_{en} = \int \rho dv$$
$$= \int_0^r \rho \, 4\pi r^2 dr,$$

   where

$$\rho = \rho_0 \frac{r}{R}$$
$$= \int_0^r \rho_0 \frac{r}{R} 4\pi r^2 dr$$
$$= \frac{4\pi\rho_0}{R} \int_0^r r^3 dr$$

$$\Rightarrow q_{en} = \frac{4\pi\rho_0 r^4}{4R} = \frac{\pi\rho_0 r^4}{R} \tag{1.174}$$

The total charge enclosed or charge of the sphere is

$$Q = q_{en}|_{r=R} = \frac{\pi \rho_0 R^4}{R} = \pi \rho_0 R^3$$

$$\Rightarrow \rho_0 = \frac{Q}{\pi R^3} \tag{1.175}$$

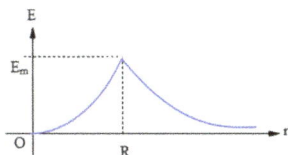

Using all three equations,

$$E = \frac{\pi r^4 \left(\frac{Q}{\pi R^3}\right)}{4\pi \varepsilon_0 r^2}$$

$$\Rightarrow E = \frac{Q r^2 R}{4\pi \varepsilon_0 R^4}$$

The sphere behaves as point charge for $r \geqslant R$.
Then $E(r) = \frac{Q}{4\pi \varepsilon_0 r^2}$; $r \geqslant R$ Ans.

(b) The potential at any point inside the sphere is

$$V = -\int_{\infty}^{r} E\, dr$$

$$= -\int_{\infty}^{R} E\, dr - \int_{R}^{r} E\, dr$$

$$= -\int_{\infty}^{R} \frac{Q}{4\pi \varepsilon_0 r^2} dr - \int_{R}^{r} \frac{Q r^2}{4\pi \varepsilon_0 R^4}\, dr$$

$$= -\frac{Q}{4\pi \varepsilon_0}\left[-\frac{1}{r}\Big|_{\infty}^{R} + \frac{r^3}{3R^4}\Big|_{R}^{r} \right]$$

$$= -\frac{Q}{4\pi \varepsilon_0}\left[-\frac{1}{R} + \frac{(r^3 - R^3)}{3R^4} \right]$$

$$= -\frac{Q}{4\pi \varepsilon_0}\left[-\frac{4}{3R} + \frac{r^3}{3R^4} \right]$$

$$\Rightarrow V = +\frac{Q}{12\pi \varepsilon_0 R}\left(4 - \frac{r^3}{R^3} \right); \quad r \leqslant R$$

The potential outside the sphere is

$$V = \frac{Q}{4\pi \varepsilon_0 x}; \quad r \geqslant R \text{ Ans.}$$

(c) Putting

$$r = 0, \quad V = V_C = \frac{Q}{3\pi\varepsilon_0 R}$$

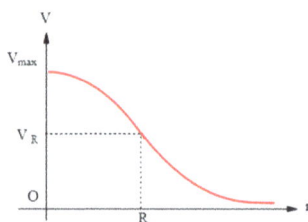

Putting

$$r = R, \quad V = V_D = \frac{Q}{4\pi\varepsilon_0 R}$$

Then,

$$V_C - V_D = \frac{Q}{12\pi\varepsilon_0 R} \quad \text{Ans.}$$

N.B: The slope of the V–r graph of $\frac{dV}{dr} = 0$ at $r = 0$, furthermore $\frac{dV}{dr} =$ same at both left- and right-hand side of the point at the surface of the sphere. This physically signifies that electric field $E = -\frac{dV}{dr}$ is continuous at the surface or boundary as shown in the E–r graph.

Problem 41 In Thomson's model of the hydrogen atom, calculate (a) the frequency of oscillation of the electron, (b) the potential difference between the centre and the surface of the atom.

Solution

(a) In Thomson's model of the atom, the radius B of a uniformly charged sphere of radius R having volume charge density

$$\rho = \frac{e}{\frac{4}{3}\pi R^3}$$

$$\Rightarrow \rho = \frac{3e}{4\pi R^3}$$

The electric field inside the nucleus at a radial distance r is

$$E = \frac{\rho}{3\varepsilon_0} r$$

Then, the electron will experience a resting force

$$F = -eE = -\frac{\rho e}{3\varepsilon_0} r$$

at any radial distance.

Since $F = -kx$, the electron will oscillate with an angular frequency

$$\omega = 2\pi f_{osc} = \sqrt{\frac{k}{m}}, \quad \text{where } k = \frac{\rho e}{3\varepsilon_0}$$

$$\Rightarrow f_{osc} = \frac{1}{2\pi}\sqrt{\frac{\rho e}{3m\varepsilon_0}}, \quad \text{where } \rho = \frac{3e}{4\pi R^3}$$

$$\Rightarrow f_{osc} = \frac{1}{2\pi}\sqrt{\frac{e^2}{4\pi\varepsilon_0 m R^3}}, \quad \text{where } \frac{1}{4\pi\varepsilon_0} = 9 \times 10^9 \text{ Nm}^2 \text{ C}^{-2}$$

$$\Rightarrow f_{osc} = \frac{1}{2 \times \frac{22}{7}}\sqrt{\frac{(1.6 \times 10^{-19}) \times 9 \times 10^9}{(9.1 \times 10^{-31})(10^{-10})}}$$

$$= 2.5 \times 10^{15} \text{ Hz Ans.}$$

(b) The potential function is given as

$$V = \frac{e}{8\pi\varepsilon_0 R}\left(3 - \frac{r^2}{R^2}\right)$$

At the centre, $r = 0$

$$\Rightarrow V_c = \frac{3e}{8\pi\varepsilon_0 R}$$

At the periphery $r = R$

$$\Rightarrow V_D = \frac{e}{4\pi\varepsilon_0 R}$$

Then the potential difference is

$$\Delta V = V_D - V_C = -\frac{3}{8\pi\varepsilon_0 R},$$

where $e = 1.6 \times 10^{-19} C, \frac{1}{4\pi\varepsilon_0} = 9 \times 10^9, R = 10^{-14}$ m

$$\Rightarrow \Delta V = -\frac{(1.6 \times 10^{-19}) \times 9 \times 10^9}{2 \times 10^{-14}}$$

$$= -7.2 \times 10^{-4} \text{ volt} \simeq 0.07 \text{ MVolt Ans.}$$

Problem 42 The potential function of an electric field varies with x and y as $V = K(y^2 - x^2)$, where K is a positive constant. (a) Find the field. (b) Draw the field and potential lines. potential pattern.
Solution

(a) It is given that
$$V = K(y^2 - x^2)$$

Then, the electric field is
$$\overrightarrow{E} = -\frac{\partial V}{\partial x}\hat{i} - \frac{\partial V}{\partial y}\hat{j}$$
$$= -K\left[\frac{\partial}{\partial x}(y^2 - x^2)\hat{i} \frac{\partial}{\partial y}(y^2 - x^2)\hat{j}\right]$$
$$= -K(-2x\hat{i} + 2y\hat{j})$$
$$\overrightarrow{E} = 2K(x\hat{i} - y\hat{j})\text{ Ans.}$$

(b) The components of E are,
$$E_x = 2Kx \text{ and } E_y = -2Ky$$

Then
$$\frac{E_y}{E_x} = \frac{dy}{dx} = -\frac{2Ky}{2Kx}$$
$$\Rightarrow \frac{dy}{dx} = -\frac{y}{x}$$

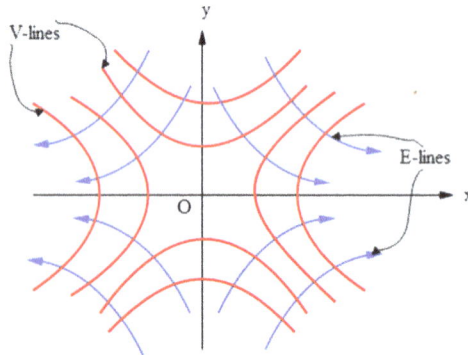

$$\Rightarrow x\,dy + y\,dx = 0$$
$$\Rightarrow d(xy) = 0$$
$$\Rightarrow xy = \text{Constant}$$

Then the equation of field pattern is rectangular hyperbolic. Ans.

N.B: The direction of $\overrightarrow{E}$ or $\overrightarrow{E}$ – lines are computed by a simple technique by looking at the expression.

$$\overrightarrow{E} = 2K(x\hat{i} - y\hat{j})$$

Let $x = 0$; then we have $\overrightarrow{E} = -2Ky\hat{j}$

In the first and second quadrants y is +ve. So, $\overrightarrow{E}$ points along the '$-\hat{j}$' direction. Similarly at the third and fourth quadrants, y is −ve.

So, $\overrightarrow{E}$ points along the $+\hat{j}$ direction.

You can also check the direction E along the x-axis by putting $y = 0$ in the above equation of $\overrightarrow{E}$.

Problem 43 The potential function of an electric field varies with x and y as, $V = axy$, where a is a positive constant. Find (a) the field, (b) field pattern.

Solution

(a) The given equation for potential is

$$V = axy$$

Then, the electric field is given as

$$\overrightarrow{E} = -\frac{\partial V}{\partial x}\hat{i} = \frac{\partial V}{\partial y}\hat{j}$$

$$= -\frac{\partial}{\partial x}(a\,xy)\hat{i} - \frac{\partial}{\partial x}(a\,xy)\hat{j}$$

$$\Rightarrow \overrightarrow{E} = -(y\hat{i} + x\hat{j}) \quad \text{Ans.}$$

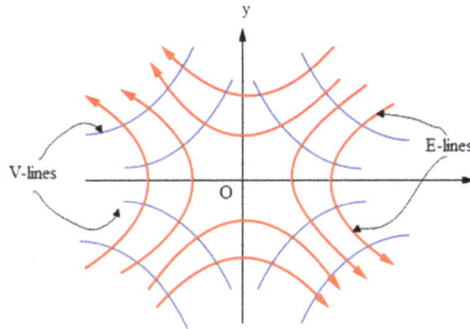

(b) Putting $\overrightarrow{E} = E_x\hat{i} + E_y\hat{j}$ and comparing this with the last equation, we have

$$E_x = y$$

$$E_y = x$$

$$\Rightarrow \frac{E_y}{E_x} = \frac{x}{y} = \frac{dy}{dx}$$

$$\Rightarrow x \, dx - y \, dy = 0$$

$$\Rightarrow x^2 - y^2 = \text{Constant}$$

This equation is hyperbolic. This means that field or E-lines are hyperbolic. Ans.

N.B: Since the slope of the curves is given by

$$m = \frac{dy}{dx} = \frac{x}{y}$$

The equation of the field is

$$\overrightarrow{E} = -a(y\hat{i} + x\hat{j})$$

In the first quadrant x and y are positive. So both $\overrightarrow{E_x}$ and $\overrightarrow{E_y}$ will be negative (point towards $-$ve axes).

Problem 44 The potential function of an electric field varies with x and y as, $V = K(x^2 + y^2)$ where K is a positive constant. Draw the field and potential pattern.
   Solution
   The equation equipotential or V-lines is

$$V = K(x^2 + y^2)$$

Then, the electric field is

$$\overrightarrow{E} = -\frac{\partial V}{\partial x}\hat{i} - \frac{\partial V}{\partial y}\hat{j}$$

$$= -2kx\hat{i} - 2ky\hat{j}$$

So we have the components

$$E_x = -2Kx$$

$$E_y = -2Ky$$

$$\tan \theta = \frac{E_y}{E_x} = \frac{dy}{dx}$$

$$\frac{dy}{dx} = \frac{-2Ky}{-2Kx} = \frac{y}{x}$$

$$\Rightarrow x \, dy = y \, dx$$

$$\Rightarrow \frac{dy}{y} = \frac{dx}{x}$$

Integrating both sides,

$$\int \frac{dy}{y} = \int \frac{dx}{x}$$

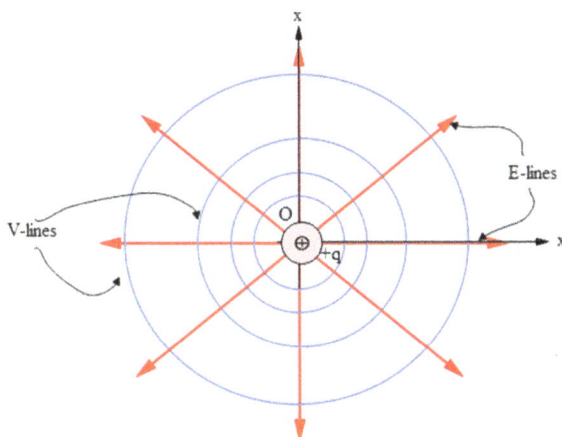

$$\Rightarrow \ln y - \ln x = C'$$

$$\Rightarrow \ln \frac{y}{x} = C'$$

$$\Rightarrow y = Cx$$

Then the electric field or E-lines are straight lines passing through the origin on centre of the circles, (centre of equipotential lines). Ans.

  N.B:
  1. It can be a point charge possessing spherical symmetry.
  2. It can be a line charge possessing cylindrical symmetry.

Problem 45 A dielectric sphere of radius R_1 having a uniform volume charge density ρ has a spherical cavity of radius R_2. Find the electric potential at the point P. Assume ρ = volume charge distribution of the sphere. Put $R_1 = 3a$, $R_2 = 2a$ and $PQ = a$. Assume that the relative permittivity of the dielectric is nearly one.

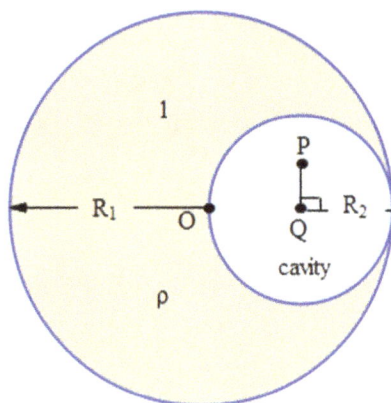

Solution

With reference to the problem 16, a uniform electric field inside the cavity is given as

$$\vec{E} = \frac{\rho}{3\varepsilon_0}\vec{a}$$

With reference to the example 22, a uniform electric field inside the cavity is given as

$$V = \frac{KQ}{2R^3}\left(3R^2 - r^2\right) = \frac{\rho}{6\varepsilon_0}\left(3R^2 - r^2\right); r \leqslant R, \tag{1.176}$$

where R = radius of the sphere and r = radial distance of the point P inside the sphere.

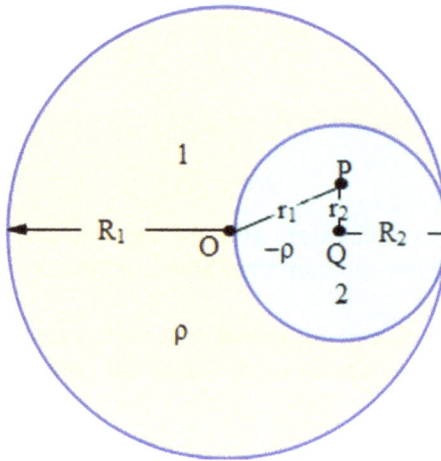

Using superposition principle, potential at the centre of the cavity = potential V_1 due to the sphere of radius R without the cavity (having a uniform charge density $+\rho$) + potential V_2 due to the sphere of radius b (having a uniform charge density $-\rho$ whose center coincides with the center of the cavity. So, we can write

$$V = V_1 + V_2 \tag{1.177}$$

Using equation (1.176), V_1 and V_2 are given as

$$V_1 = \frac{(+\rho)}{6\varepsilon_0}\left(3R_1^2 - r_1^2\right) \tag{1.178}$$

$$V_2 = \frac{\rho}{6\varepsilon_0}\left(3R_2^2 - r_2^2\right) \tag{1.179}$$

Using the last three equations, we have

$$V = \frac{\rho}{6\varepsilon_0}\left(3R_1^2 - r_1^2\right) + \frac{(-\rho)}{6\varepsilon_0}\left(3R_2^2 - r_2^2\right)$$

$$\Rightarrow V = \frac{\rho}{6\varepsilon_0}\left\{\left(3R_1^2 - r_1^2\right) - \left(3R_2^2 - r_2^2\right)\right\}$$

$$\Rightarrow V = \frac{\rho}{6\varepsilon_0}\left\{3\left(R_1^2 - R_2^2\right) - \left(r_1^2 - r_2^2\right)\right\}$$

Putting $R_1 = 3a$, $R_2 = 2a$, $r_1 = \sqrt{2}\,a$, $r_2 = a$, we have

$$V = \frac{\rho}{6\varepsilon_0}\left\{3\left(9a^2 - 4a^2\right) - \left(2a^2 - a^2\right)\right\} = \frac{7\rho a^2}{3\varepsilon_0} \quad \text{Ans.}$$

Problem 46

(a) Find the electric field due to a long dielectric cylinder of radius R having a uniform volume charge density ρ as the function of radial distance r. Draw the E–r graph.

(b) In part (a), if a long cylindrical portion of radius a is removed from the given cylinder, find the electric field inside the cylindrical cavity.

(c) If a particle of charge q and mass m is released from rest in the cavity, what is the maximum speed it can have? Neglect gravity on the charged particle. Assume that the relative permittivity of the dielectric is nearly 1.

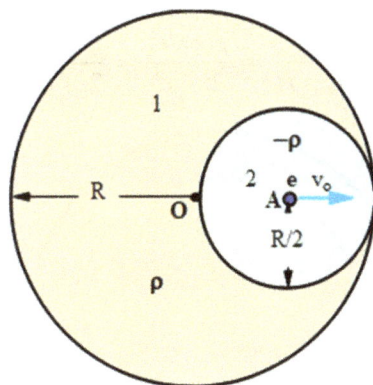

Solution

(a) Applying Gauss' law for cylindrical Gaussian surface,

For any inside point,

$$\oint E dA = \frac{q_{en}}{\varepsilon_0}$$

$$\Rightarrow E(2\pi r l) = \frac{q_{en}}{\varepsilon_0} \qquad (1.180)$$

$$\Rightarrow E = \frac{q_{en}}{2\pi\varepsilon_0 r l}$$

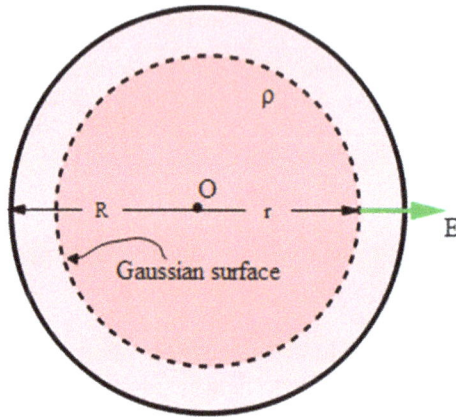

The charge enclosed is

$$q_{en} = \rho(\pi r^2 l) \qquad (1.181)$$

Using the last two equations,

$$E = \frac{q_{en}}{2\pi\varepsilon_0 r l} = \frac{\rho(\pi r^2 l)}{2\pi\varepsilon_0 r l} = \frac{\rho r}{2\varepsilon_0} \quad \text{Ans.}$$

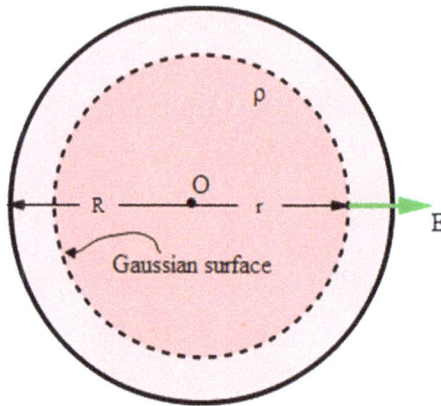

For any outside point, the charge enclosed is

$$q_{en} = \rho(\pi R^2 l) \tag{1.182}$$

Using the last two equations,

$$E = \frac{q_{en}}{2\pi\varepsilon_0 r l} = \frac{\rho(\pi R^2 l)}{2\pi\varepsilon_0 r l} = \frac{\rho R^2}{2\varepsilon_0 r} \quad \text{Ans.}$$

N.B: The cylinder behaves as a line charge for $r \geqslant R$.

Then $E(r) = \dfrac{\rho R^2}{2\varepsilon_0 r}; r \geqslant R$ and $E(r) = \dfrac{\rho r}{2\varepsilon_0}; r \leqslant R$.

Then, the field varies linearly inside and hyperbolically outside the charged cylinder with the radial distance. The E–r graph is given as follows:

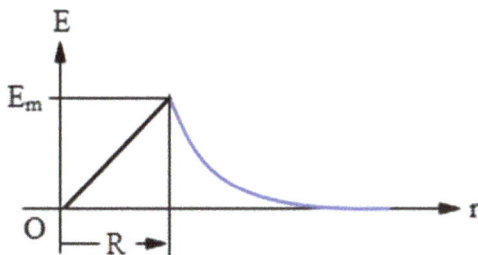

(b) Referring to problem 16, at any point P in the cavity, the net field is

$$\vec{E} = \vec{E}_1 + \vec{E}_2 \tag{1.183}$$

The field due to sphere 1 is

$$\vec{E}_1 = \frac{\rho}{2\varepsilon_0}\vec{r}_1 \tag{1.184}$$

The field due to sphere 2 is

$$\vec{E}_2 = \frac{(-\rho)}{2\varepsilon_0}\vec{r}_2 \tag{1.185}$$

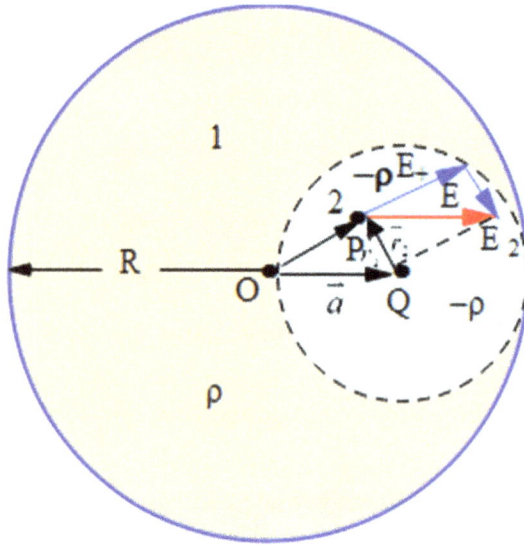

Using the last three equations, we have

$$\vec{E} = \frac{\rho}{2\varepsilon_0}\vec{r_1} + \frac{(-\rho)}{2\varepsilon_0}\vec{r_2}$$

$$= \frac{\rho}{2\varepsilon_0}\left(\vec{r_1} - \vec{r_2}\right)$$

Since $\vec{a} + \vec{r_2} = \vec{r_1}$, we can substitute $\vec{r_1} - \vec{r_2} = \vec{a}$ in the last expression to obtain.

$$\vec{E} = \frac{\rho}{2\varepsilon_0}\vec{a} \quad \text{Ans.}$$

(c) The work done by the electric field = change in KE of the charge particle

$$Eq = \frac{1}{2}mv^2$$

$$\Rightarrow v = \sqrt{2Eq/m}$$

Putting the value of E, we have

$$v = \sqrt{\frac{2\rho q}{2m\varepsilon_0}}a = \sqrt{\frac{\rho q a}{m\varepsilon_0}} \quad \text{Ans.}$$

1-150

Problem 47

(a) Find the electric field due to a dielectric sphere of radius R having a uniform volume charge density ρ as the function of radial distance r. Draw the E–r graph.

(b) In part (a), if a spherical portion of radius r is removed from the given sphere, find the electric field inside the cavity. Assume the distance between the centers of the sphere and cavity is equal to a.

(c) Find the electrostatic energy stored in the spherical cavity.

Solution

(a) Please refer to example 22.

(b) Referring to problem 16, at any point P in the cavity, the net field is

$$\vec{E} = \vec{E_1} + \vec{E_2} \tag{1.183}$$

The field due to the sphere 1 is

$$\vec{E_1} = \frac{\rho}{3\varepsilon_0}\vec{r_1} \tag{1.187}$$

The field due to the sphere 2 is

$$\vec{E_2} = \frac{(-\rho)}{3\varepsilon_0}\vec{r_2} \tag{1.188}$$

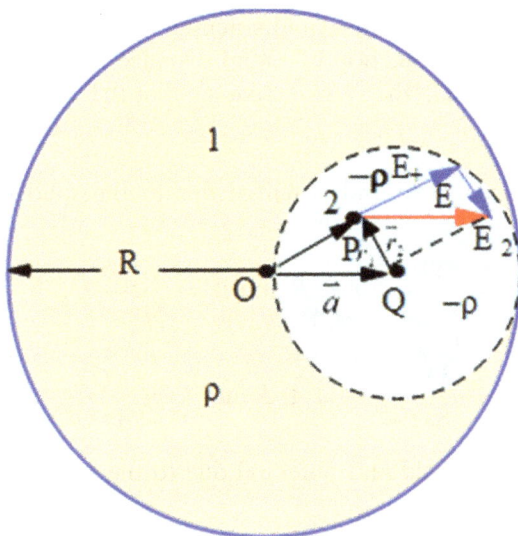

Using the last three equations, we have

$$\vec{E} = \frac{\rho}{3\varepsilon_0}\vec{r_1} + \frac{(-\rho)}{3\varepsilon_0}\vec{r_2}$$

$$= \frac{\rho}{2\varepsilon_0}(\vec{r_1} - \vec{r_2})$$

Since $\vec{a} + \vec{r_2} = \vec{r_1}$, we can substitute $\vec{r_1} - \vec{r_2} = \vec{a}$ in the last expression obtain.

$$\vec{E} = \frac{\rho}{3\varepsilon_0}\vec{a} \text{ Ans.}$$

(c) A uniform electric field exists inside the cavity parallel to the vector $\vec{a}$. So, the energy stored inside the cavity is

$$U = \frac{1}{2}\varepsilon_0 E^2(v_{\text{cavity}}) = \frac{1}{2}\varepsilon_0\left(\frac{\rho}{3\varepsilon_0}a\right)^2\left(\frac{4}{3}\pi r^2\right)$$

$$= \frac{1}{2}\varepsilon_0\left(\frac{Q/\frac{4}{3}\pi r^3}{3\varepsilon_0}a\right)^2\left(\frac{4}{3}\pi r^3\right)$$

$$= \frac{Q^2 a^2}{24\pi\varepsilon_0 R^3} \text{ Ans.}$$

Problem 48 A solid dielectric sphere of radius R having a uniform volume charge density carries a charge Q. It is coated with a thin layer of charge $-Q$ with a uniform distribution. Find the (a) electric field and potential of the system (dielectric sphere + thin charged layer) as the function of radial distance r. Draw the E–r graph, (b) electrostatic energy stored in the system. Assume that the relative permittivity of the dielectric is nearly 1.

Solution

(a) The electrostatic field and potential due to the solid sphere are given as follows:

$$\vec{E_1} = \frac{Qr}{4\pi\varepsilon_0 R^3}\hat{r}; r \leqslant R \text{ and } \vec{E_1} = \frac{Q}{4\pi\varepsilon_0 r^2}\hat{r}; r \geqslant R$$

$$V_1 = \frac{Q(3R^2 - r^2)}{8\pi\varepsilon_0 R^3}; r \leqslant R \text{ and } V_1 = \frac{Q}{4\pi\varepsilon_0 r}; r \geqslant R$$

The electrostatic field and potential due to the spherical shell are given as follows:

$$\vec{E_2} = 0; r < R \text{ and } \vec{E_1} = \frac{-Q}{4\pi\varepsilon_0 r^2}\hat{r}; r > R$$

$$V_2 = \frac{-Q}{4\pi\varepsilon_o R}; r \leqslant R \text{ and } V_1 = \frac{Q}{4\pi\varepsilon_o r}; r \geqslant R$$

By applying the principle of superposition, the electrostatic field and potential due to the spherical shell are given as follows:

$$\vec{E} = \vec{E}_1 + \vec{E}_2 = \frac{Qr}{4\pi\varepsilon_o R^3}\hat{r} + 0 = \frac{Qr}{4\pi\varepsilon_o R^3}\hat{r}; \; r < R$$

$$\vec{E} = \vec{E}_1 + \vec{E}_2 = \frac{Q}{4\pi\varepsilon_o r^2}\hat{r} - \frac{Q}{4\pi\varepsilon_o r^2}\hat{r} = \vec{0}; \; r > R$$

$$V = V_1 + V_2 = \frac{Q(3R^2 - r^2)}{8\pi\varepsilon_o R^3} + \frac{-Q}{4\pi\varepsilon_o R} = \frac{Q(R^2 - r^2)}{8\pi\varepsilon_o R^3}; r < R$$

$$V = V_1 + V_2 = \frac{Q}{8\pi\varepsilon_o R} + \frac{-Q}{8\pi\varepsilon_o R} = 0; \quad r > R$$

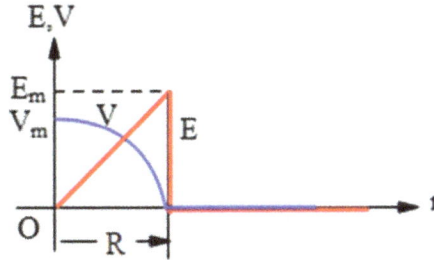

In the above graph, it is clear that the radially outward field increases from zero at $r = 0$ to $E_m = Q/4\pi\varepsilon_o R^2$ at $r = R$. Outside the sphere the net field is zero. Similarly, the potential decreases from $V_m = Q/8\pi\varepsilon_o R$ at $r = 0$ to zero at $r = R$. Outside the sphere the net potential is zero.

(b) The self-energy of the thin spherical shell is

$$U_2 = \frac{Q^2}{8\pi\varepsilon_o R}$$

The self-energy of the dielectric sphere is

$$U_2 = \frac{3Q^2}{20\pi\varepsilon_o R}$$

The mutual energy of the system is

$$U_{12} = \frac{-Q^2}{4\pi\varepsilon_o R}$$

The total electrostatic energy stored due to given system is

$$U = U_1 + U_2 + U_{12} = \frac{Q^2}{8\pi\varepsilon_o R} + \frac{3Q^2}{20\pi\varepsilon_o R} - \frac{Q^2}{4\pi\varepsilon_o R}$$

$$= \frac{-Q^2}{8\pi\varepsilon_o R} + \frac{3Q^2}{20\pi\varepsilon_o R} = \frac{Q^2}{40\pi\varepsilon_o R} \quad \text{Ans.}$$

N.B: Both field and potential are zero outside and non-zero just inside the system. So, they are discontinuous at the boundary (surface of the sphere), and the total electrostatic energy is stored inside the system.

Problem 49 Two large thin dielectric sheets each having a uniform surface charge density are placed perpendicular to each other. In figure (i) the sheets have $+\sigma_1$ and $+\sigma_2$ and in figure (ii) the sheets have surface charge densities $+\sigma_1$ and $-\sigma_2$ respectively. Find the electric field and draw the E- (or electric lines of force) and V-lines (or equipotential lines/surfaces).

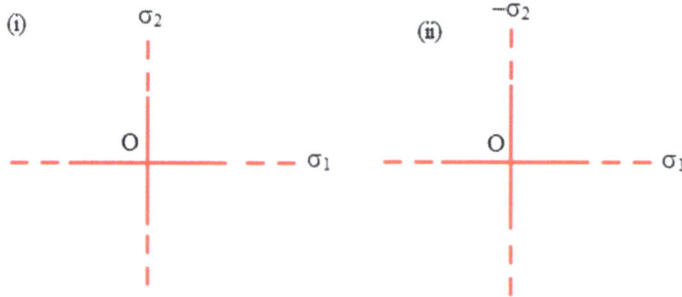

Solution
In figure (i), the electric field due to sheet 1 and 2 can be given as

$$\vec{E}_1 = \frac{\sigma_1}{2\varepsilon_0}\hat{j}, \ \vec{E}_2 = \frac{\sigma_2}{2\varepsilon_0}\hat{i}$$

Then, the electric field in the first quadrant can be given as

$$\vec{E} = \vec{E}_1 + \vec{E}_2 \tag{1.183}$$

$$\Rightarrow \vec{E} = \frac{\sigma_1}{2\varepsilon_0}\hat{j} + \frac{\sigma_2}{2\varepsilon_0}\hat{i} = \frac{1}{2\varepsilon_0}\left(\sigma_2\hat{i} + \sigma_1\hat{j}\right)$$

The slope of the straight line representing the E-field is

$$m = \tan\theta = \frac{E_1}{E_2} = \frac{\frac{\sigma_1}{2\varepsilon_0}}{\frac{\sigma_2}{2\varepsilon_0}} = \frac{\sigma_1}{\sigma_2}$$

The equation of the E-field lines is given as

$$y = mx + c = \frac{\sigma_1}{\sigma_2}x + c$$

where c is the y-intercept. Putting different values of 'c' we can get a series of parallel lines. As the electric field is uniform, the spacings of field lines must be uniform. Depending upon the quadrants, the signs of x and y will change, and the direction of electric field will change, keeping the magnitude constant. Once, E-lines are found, V-lines will be perpendicular to E-lines. Since E-lines are uniform, V-lines must also be uniform but without any directional properties because potential is a scalar quantity. Ans.

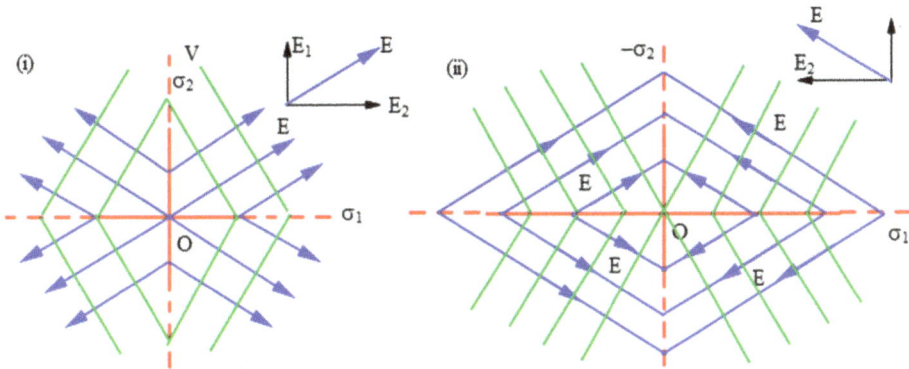

In figure (ii), the electric field due to sheet 1 and 2 can be given as

$$\vec{E}_1 = \frac{\sigma_1}{2\varepsilon_0}\hat{j}, \vec{E}_2 = -\frac{\sigma_2}{2\varepsilon_0}\hat{i}$$

Then, the electric field in the first quadrant can be given as

$$\vec{E} = \vec{E}_1 + \vec{E}_2 \qquad (1.183)$$

$$\Rightarrow \vec{E} = \frac{\sigma_1}{2\varepsilon_0}\hat{j} + \frac{-\sigma_2}{2\varepsilon_0}\hat{i} = \frac{1}{2\varepsilon_0}\left(-\sigma_2\hat{i} + \sigma_1\hat{j}\right)$$

The slope of the straight line representing the E-field is

$$m = \tan\theta = \frac{E_1}{E_2} = -\frac{\sigma_1}{\sigma_2}$$

The equation of the E-field lines is given as

$$y = mx + c = -\frac{\sigma_1}{\sigma_2}x + c$$

where c is the y-intercept. Putting different values of 'c' we can get a series of parallel lines. As the electric field is uniform, the spacings of field lines must be uniform. Depending upon the quadrants, the signs of x and y will change, and the direction of electric field will change, keeping the magnitude constant. Once, E-lines are found, V-lines will be perpendicular to E-lines. Since E-lines are uniform, V-lines must also be uniform but without any directional properties because potential is a scalar quantity. You can refer the problem 29 for depicting E- and V-lines. Ans.

Problem 50 A dielectric sphere of charge Q and radius R has a uniform volume charge density ρ, say. From the sphere a spherical portion of radius r is removed to form a cavity. The distance between the centers of the sphere and cavity is equal to a. (a) Find the electrostatic energy stored in the dielectric sphere. (b) If the cavity is concentric, find the total electrostatic energy stored. Assume that the relative permittivity of the dielectric is nearly 1.

Solution

(a) Referring to problem 32, the electrostatic self-energy energy stored due to the dielectric sphere of radius R is

$$U = \frac{3Q^2}{20\pi\varepsilon_o R}$$

The sphere with the cavity of radius r can be imagined as the superposition of a sphere of charge Q and another sphere of charge $-q$ of the same volume charge density $-\rho$. The electrostatic self-energy stored in the spherical dielectric sphere of charge $-q$ and radius r is

$$U_1 = \frac{3q^2}{20\pi\varepsilon_o R}$$

The interaction energy between these two spheres is

$$U_1 = -qV',$$

where V' is the potential of the bigger sphere at the center of the smaller sphere, given as

$$V' = \frac{Q\left(3R^2 - a^2\right)}{8\pi\varepsilon_o R^3}$$

Then, the interaction energy between these two spheres is

$$U\frac{Qq(3R^2 - a^2)}{8\pi\varepsilon_o R^3}_{\text{int}}$$

So, the net electrostatic field energy of the dielectric sphere with the cavity is

$$U_{\text{net}} = U_1 + U_2 + U\frac{3Q^2}{20\pi\varepsilon_o R}\frac{3q^2}{20\pi\varepsilon_o r}\frac{Qq(3R^2 - a^2)}{8\pi\varepsilon_o R^3}_{\text{int}}$$

For a uniform volume charge distribution, putting

$$q = \left(\frac{r}{R}\right)^3 Q$$

in the last expression, we have

$$U_{net} = \frac{3Q^2}{20\pi\varepsilon_o R} + \frac{3Q^2 r^5}{20\pi\varepsilon_o R^6} - \frac{Q^2(3R^2 - a^2)r^3}{8\pi\varepsilon_o R^6} \quad \text{Ans.}$$

(b) If the cavity is concentric, put $a = 0$ to obtain,

$$U_{net} = \frac{3Q^2}{20\pi\varepsilon_o R} + \frac{3Q^2 r^5}{20\pi\varepsilon_o R^6} - \frac{Q^2(3R^2 - 0)r^3}{8\pi\varepsilon_o R^6}$$

$$= \frac{3Q^2}{20\pi\varepsilon_o R} + \frac{3Q^2 r^5}{20\pi\varepsilon_o R^6} - \frac{3R^2 Q^2 r^3}{8\pi\varepsilon_o R^6}$$

$$= \frac{3Q^2}{20\pi\varepsilon_o R}\left(1 + \frac{r^5}{R^5} - \frac{5r^3}{2R^3}\right) \quad \text{Ans.}$$

Problem 51 A thin dielectric shell has a charge Q with a uniform distribution. It is filled with a liquid dielectric of uniform volume charge distribution of total charge Q. (a) Find the field and potential of the system (shell + liquid), (b) electrostatic energy stored in the system. Assume that the relative permittivity of the dielectric is nearly 1.

Solution

(a) The electrostatic field due to the liquid (dielectric) sphere is

$$E = 0; \ r < R \text{ and } \vec{E}_1 = \frac{Qr}{8\pi\varepsilon_o R^3}\hat{r}; r \leqslant R$$

$$\vec{E}_2 = \frac{Q}{4\pi\varepsilon_o r^2}\hat{r}; \ r > R$$

The electrostatic field due to the spherical shell is

$$\vec{E_2} = \frac{Q}{8\pi\varepsilon_o r^2}\hat{r};\, r > R \text{ and } E = 0;\, r \;\; < R$$

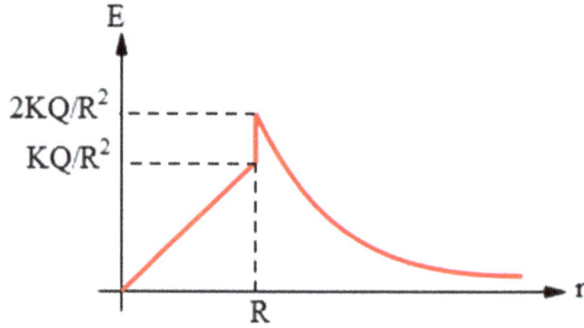

Then the net electric field is

$$\vec{E} = \vec{E_1} + \vec{E_2} = \frac{Qr}{4\pi\varepsilon_o R^3}\hat{r};\, r \;\; < R$$

$$\vec{E} = \vec{E_1} + \vec{E_2} = \frac{Q}{2\pi\varepsilon_o r^2}\hat{r};\, r > R$$

The electrostatic potential to the liquid (dielectric) sphere is

$$V_1 = \frac{Q(3R^2 - r^2)}{8\pi\varepsilon_o R^3};\, r \leqslant R$$

$$V_1 = \frac{Q}{8\pi\varepsilon_o r};\, r \geqslant R$$

The electrostatic potential to the spherical shell is

$$V_1 = \frac{Q}{4\pi\varepsilon_o R};\, r \leqslant R$$

$$V_1 = \frac{Q}{4\pi\varepsilon_o r};\, r \geqslant R$$

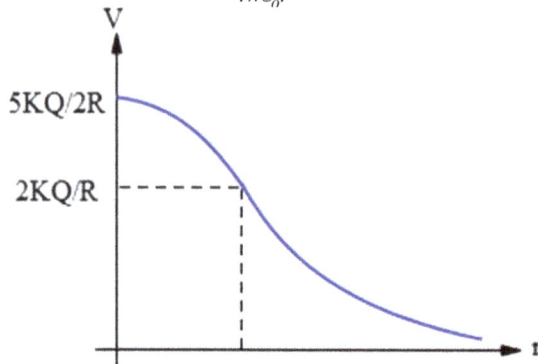

Then, the net potential is

$$V_1 = \frac{Q(3R^2 - r^2)}{8\pi\varepsilon_o R^3} + \frac{Q}{4\pi\varepsilon_o R}; r \leqslant R$$

$$\Rightarrow V_1 = \frac{Q(5R^2 - r^2)}{8\pi\varepsilon_o R^3}; r \leqslant R$$

$$V = \frac{Q}{2\pi\varepsilon_o R}; r \geqslant R \quad \text{Ans.}$$

(b) The electrostatic energy stored due to the thin spherical shell is

$$U_2 = \frac{Q^2}{8\pi\varepsilon_o R}$$

The total electrostatic energy stored due to the given system is

$$U = U_1 + U_2 + U_{12} = \frac{3Q^2}{20\pi\varepsilon_o R} + \frac{Q^2}{8\pi\varepsilon_o R} + \frac{Q^2}{4\pi\varepsilon_o R}$$

$$\Rightarrow U = \left(\frac{3}{5} + \frac{1}{2} + 1\right)\frac{Q^2}{4\pi\varepsilon_o R} = \frac{21Q^2}{40\pi\varepsilon_o R} \quad \text{Ans.}$$

IOP Publishing

Problems and Solutions in Electricity and Magnetism

Pradeep Kumar Sharma

Chapter 2

Properties of conductors and dielectrics

2.1 Introduction

Metals like aluminium, copper and their alloys are used for transmission of electrical power. Aluminium-conductor steel-reinforced (ACSR) is a suitable choice for overhead high-voltage electric lines due to the combination of aluminium's light weight, good conductivity and corrosion resistance of aluminium and mechanical strength of steel.

The coaxial cables made of conductors and insulators are used as waveguides in broadband Ethernet, cable television, commercial radio transmissions etc. The windings in electrical machines such as motors, generators and transformers are done by copper and aluminium wires and rods. In electrical circuits, conductors join the circuit elements such as capacitors, resistors, and inductors. Capacitors are made of conductors embedded with dielectrics. The materials that can be polarized in an electric field are called dielectrics. The examples of dielectrics are vacuum, air, distilled water etc. In order to avoid (or minimize) the electrical breakdown, the conductors are coated with a layer of insulators. In high-voltage engineering, the insulators such as mica, cotton etc, are used to withstand a high electric field. Insulators have very low dielectric constants compared with ideal dielectrics. Dielectrics can store a considerable amount of electrostatic energy in capacitors. In high-voltage lines, insulators such as ceramics like mica etc, are used to reduce the leakage current. In the cores of transformers, we use insulators to laminate the conducting sheets to reduce eddy current and hysteresis losses. So, the service of electromagnetism to humans is accomplished through the basic raw materials of conductors, dielectrics and insulators. Resistors are basically conductors of high resistivity. When a conducing wire is coiled it acts as a good inductor. We will talk about inductance in chapter 9. In this chapter, we will talk about the response of a conductor to an external electric field by carrying a current and inducing the charges. First of all, we need to understand the meaning of an electrical conductor.

doi:10.1088/978-0-7503-6477-5ch2

2.2 Definition of an electrical conductor

An object that conducts an electric current is known as an electrical conductor. Metals like copper, aluminium and their alloys etc, are good conductors of electricity at ordinary temperature and frequency of the supply voltage. Copper is commonly used as electrical wires in household wiring and electrical winding in motors, generators and transformers. A good conductor of electricity has the density of conduction electrons in the order of 10^{28} cm^{-3}. The cluster of this large number of conduction electrons is called 'electron gas'. In the absence of an electric field inside the conductors, the electron gas does not move. Even for a small electric field in the order of 10^{-3} V m^{-1}, say, the electron gas gains a net velocity or momentum in the opposite direction of the applied field constituting an electric current inside a conductor in the order of an ampere.

If you take a mica sheet, with a strong electric field in the order of 10^5 V m^{-1} it does not produce an electric current. So, it is called an insulator as it restricts the flow of electric current. However, at a voltage greater than the breakdown voltage of mica (0.5 KV to 1.25 KV, say), (minimum voltage to make it conducting by producing free electrons), it can be an electrical conductor. Furthermore, a conductor at low frequency of 50 Hz, say, behaves as a non-conductor at a higher frequency (in the order of megahertz, say) of the applied voltage (electric field). Furthermore, a metal can be a superconductor at very low temperature approaching 0 K. Conversely, at high temperature, a metal will behave as a poor conductor. Although air is a non-conductor of electricity at ordinary voltage, it can also carry electric current at very high voltage in the form of sparks or glow or discharge in the electric circuits, thunderbolt, corona discharge (glow near sharp points) in power stations. At low pressure the gases can also conduct electricity in the form of tube lights, neon and mercury vapour lamps etc. Thus, in principle, a substance (solid, liquid and gas) can be an electric conductor depending on the temperature, pressure, magnitude and frequency of the applied electric field. But we will restrict our discussion of the metallic conductors under a constant (zero frequency) voltage or slowly varying (quasi-static or a small frequency of 50 Hz) voltage (or electric field). The applied electric field must be very small compared to the breakdown strength of air.

2.3 Conductor in an electric field, dynamic condition ($J \neq 0$)

Let us connect the ends of two copper wires to the (+ve and −ve) poles of an electric bulb and the other ends of the wires to the terminals of a battery through a key. If we press the key, the bulb glows, indicating a flow of electric current, which is a flow of electrons. The conventional current is assumed as the flow of positive charge taken opposite to the flow of electrons. It is a 'convention but not a concept' as suggested by Benjamin Fraklin.

Why do the electrons flow in a metal? Let's see. After closing the key, an electric field inside the metal (wire) is established along the wire. This drives the electron gas in the opposite direction of the field with a constant speed called drift speed v_d. It is in the order of 10^{-3} m s^{-1} in a copper wire.

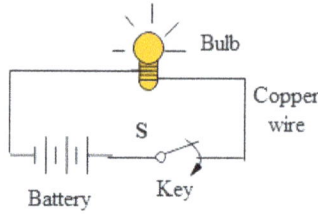

Let us assume that a charge (group of electrons) dQ passes through the cross-section of the conductor during time dt. Then, the rate of flow of charge, that is $\frac{dQ}{dt}$ is defined as an electric current. It is given as

$$i = \frac{dQ}{dt} \tag{2.1}$$

If n = free-electron density in metal, A = cross-sectional area of the wire and e = electronic charge, we can write

$$i = \frac{dQ}{dt} = n\,e\,v_d\,A \tag{2.2}$$

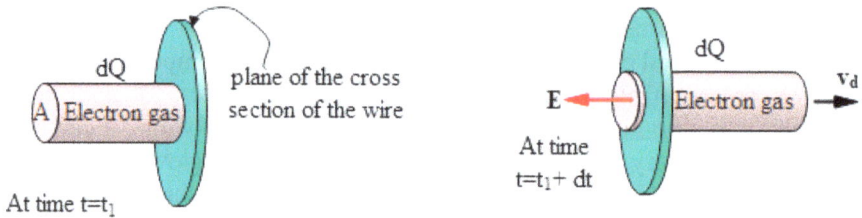

The electron cloud of charge dQ passes through the cross-section of the wire in time dt in the opposite direction of the applied electric field E in the wire causing an electric current i=dQ/dt

Then the ratio of current and cross-sectional area is defined as a vector known as 'current density' given as

$$J = \frac{i}{A} = n\,e\,v_d \tag{2.3}$$

Since, the drift velocity v_d is proportional to the applied field E (we will prove it later on); $v_d \propto E$, and the current density is proportional to the electric field, we can say that the current density will be proportional to the electric field. It is given as

$$\overrightarrow{J} = \sigma \overrightarrow{E}, \tag{2.4}$$

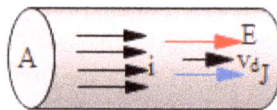

Current density, drift velocity
and electric field are parallel

2-3

where the constant of proportionality σ is called 'conductivity' of the metal (conductor). This is known as Ohm's law in point or vector form which will be discussed in detail in the next chapter. If you can establish a huge current by applying a small electric field, the ratio of J and E, that is, σ is high. It means that the specimen is a good conductor of electricity possessing a large conductivity. Practically, the noble metals such as copper, silver, gold etc, have very high conductivity.

Equation (2.4) tells us the fact that the presence of a non-zero current ($J \neq 0$) at any point in a conductor is caused by a non-zero electric field at that point. Due to the presence of a non-zero electric field inside the conductor, the potential inside the conductor varies from point to point. For instance, the potentials at different planes of the cross-sections of the current-carrying wire decrease in the direction of current (or electric field such as $V_1 > V_2 > V_3 > V_4$). However, all points of a cross-section (plane perpendicular to $\overrightarrow{E}$) of the conductor must have the same potential. We will discuss this in detail in later chapters.

So, in an 'electrodynamic (current-carrying) condition' the potential in a conductor cannot be equal at all points. It is contrary to the electrostatic case (when $J = 0$).

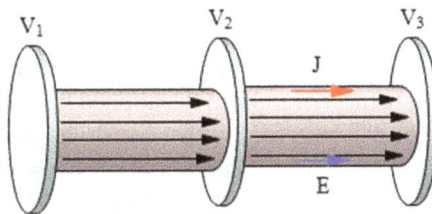

Indide a conductor electric field is non-zero and hence the conductor surface is not equipotential while it carries an electric current

2.4 Conductor in an electric field, static condition ($J = 0$)

Static condition: Let us imagine that, at time $t = 0$, a uniform electric field of intensity E suddenly appears in a region where a piece of conductor is kept. This electric field penetrates the conductor in the form of an electromagnetic wave. The electric field inside the conductor pushes the electron cloud (conduction electrons) opposite to its direction, and pushes the positive charges (nuclei + other bounded electrons) in the direction of the field. This causes an electric current inside the conductor. However, this current is temporary. It will be zero after a very short time called relaxation time. During this time, the electrons move opposite to the field and stop after reaching the boundary (surface) of the conductor. As the current ceases to exist, the electric field will be zero. This (zero electric field inside the conductor) is the 'electrostatic' situation. The transient current flows for extremely small time of 10^{-19} s. So, the process of rearrangement of the charge (free electrons) inside the conductor is practically instantaneous and known as 'electrostatic induction'. Needless to say, the metallic lattice does not move appreciably in response to the applied electric field. So, mostly the current is caused by the movement of electrons relative to the conductor or metal lattice.

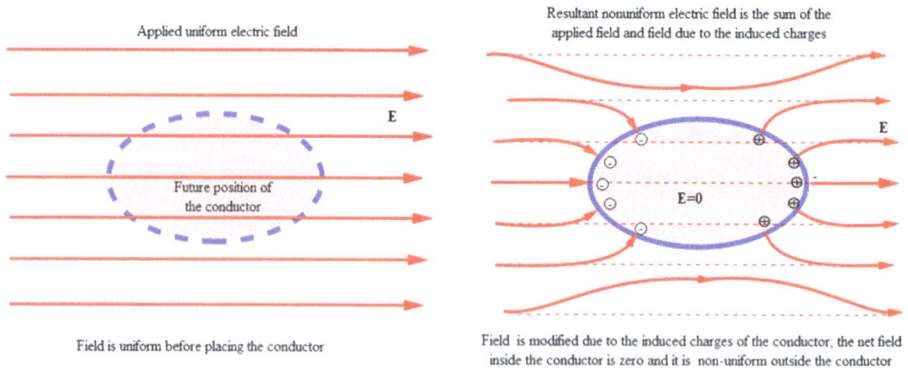

Applied uniform electric field

Resultant nonuniform electric field is the sum of the applied field and field due to the induced charges

E

E

Future position of the conductor

E=0

Field is uniform before placing the conductor

Field is modified due to the induced charges of the conductor; the net field inside the conductor is zero and it is non-uniform outside the conductor

2.4.1 Conductor properties

1. If a conductor is placed in an external electric field, electrostatic induction (charge separation) takes place. The induced charges will produce a field opposite to the applied field. This induced field will gradually increase, whereas the applied field remains constant. So, after the relaxation time, the included field will completely kill (cancel or nullify) the applied field; then the net electric field inside the conductor will be zero. In other words, at any point of the conductor, electric field is zero in electrostatic condition; since $E = dV/dr$, the potential at all points of the conductor is a constant quantity. *So, the electric field is zero and potential is uniform inside a conductor when it does not carry a current.*

$$\overrightarrow{E}_{\text{inside}} = 0 \ \text{ and } \ V = \text{constant} \tag{2.5}$$

2. Let us take a Gaussian surface just below the surface of the conductor. Since $E = 0$ inside the conductor, the flux of E passing through the Gaussian surface is

$$\phi_E = \oint \overrightarrow{E} \cdot \overrightarrow{dA} = 0 \tag{2.6}$$

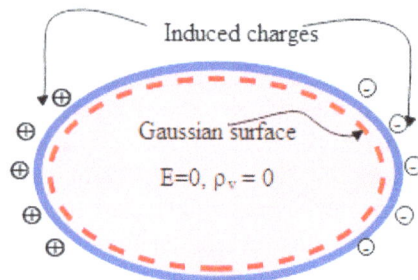

Induced charges

Gaussian surface

E=0, $\rho_v = 0$

The net electric field and volume charge density inside the conductor is zero

Gauss's law tells that the flux of electric field is

$$\phi_E = \frac{q_{\text{enclosed}}}{\varepsilon_0} \tag{2.7}$$

Using equations (2.6) and (2.7), the charge enclosed by the Gaussian surface is

$$q_{\text{encl}} = \int \rho dv = 0 \tag{2.8}$$

$$\Rightarrow \rho_V = 0, \tag{2.9}$$

where ρ_V = volume charge density.

This tells us that the excess (or induced) charge inside a conductor is zero in electrostatic condition. So, the excess (or induced) charge must reside (appear) at the surface (skin) of the conductor.

3. Each surface (or induced) charge is in equilibrium under the combined action of electrostatic forces due to other charges and mechanical forces offered by the wall of the conductor (as it does not allow the charges to fly apart (away) from the conductor). As the net force acting on each element of charge at the surface of the conductor is zero, the tangential acceleration of each surface charge is zero. Then, we can conclude that there is no tangential electric field $\overrightarrow{E_t}$ acting on the surface charges. In other words, *the tangential electric field along the surface of a conductor is zero.*

$$\Rightarrow \overrightarrow{E_t} = 0 \tag{2.10}$$

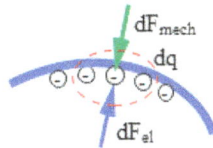

Each element of surface chage has zero
tangential acceleration; so,$E_{\text{tangential}}$ =0

Proof:

As the electrostatic field is conservative, we can write

$$\oint \overrightarrow{E} \cdot \overrightarrow{dl} = 0 \tag{2.11}$$

This means that the line integral of the electric field (work done per unit charge) taken along the closed loop 1 2 3 4 1 taken (inside and just outside the conductor) is zero. So, dividing this closed line integral into two parts, one along the path remaining inside, and the other lying outside the conductor, we can write

$$\int_{\text{inside}} \overrightarrow{E} \cdot dl + \int_{\text{outside}} \overrightarrow{E} \cdot \overrightarrow{dl} = 0. \tag{2.12}$$

Putting $E_{\text{inside}} = 0$ in equation (2.12), we have

$$\int_{\text{outside}} \vec{E} \cdot \vec{dl} = 0$$

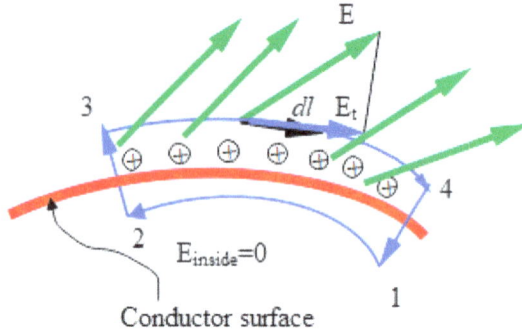

Conductor surface

Taking the component of the electric field along the path outside the conductor, that is the tangential component of the electric field, we have

$$\int E \, dl \, \cos \theta = \int E_t \, dl = 0 \text{ (where } E \cos \theta = E_t)$$

Or, $E_t = 0$ (Proved)

4. If the tangential electric field is absent just outside the conductor, *the net electric field just outside the conductor must be normal to the conductor surface.*

$$\vec{E}_{\text{outside}} = \frac{\sigma}{\varepsilon_0} \hat{n} \qquad (2.13)$$

where $\hat{n}$ = unit vector of the outward normal drawn at the surface of the conductor and σ = surface charge density at the point of consideration. If σ is positive, the field is normally outward; if σ is negative, the field is normally inward; So, we can conclude the following.

The lines of forces emanate from the induced positive charges or terminate on the induced negative charges normal to the surface of the conductor.

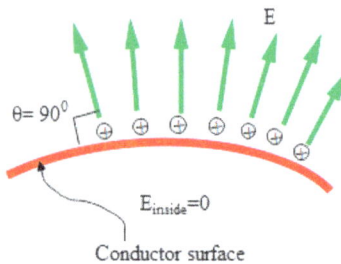

Conductor surface

Just outside of a charged conductor or conductor in an external electric field the net electrostatic field is normal to the conductor surface

Proof:

Let us take a pillbox shaped Gaussian surface enclosing a patch of area dA of the surface of the charged conductor. The flux out of the Gaussian surface is

$$\phi_E = E \cdot \Delta A \cos 0^0 \quad (\because E_{\text{inside}} = 0)$$

$$\Rightarrow \phi_E = E \, \Delta A \tag{2.14}$$

According to Gauss's law,

$$\phi_E = \frac{q_{\text{enclosed}}}{\varepsilon_0},$$

where q_{en} = the charge enclosed by the Gaussian surface = charge contained on the area ΔA of the conductor = $\sigma \, \Delta A$

$$\Rightarrow \phi_E = \frac{\sigma \, \Delta A}{\varepsilon_0} \tag{2.15}$$

Using equations (2.14) and (2.15), we have

$$\overrightarrow{E} = \frac{\sigma}{\varepsilon_0} \hat{n} \text{ (Proved)}$$

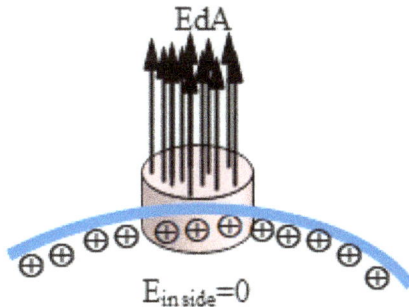

Example 1 Prove that the field due to a uniformly charged thin conducting plate is

$$\overrightarrow{E} = \frac{\sigma}{\varepsilon_0} \hat{n},$$

where σ = surface charge density of each face of the plate.

Method 1

Since there are two surfaces (1 and 2) of the thin conducting plate, the electric fields $\overrightarrow{E_1}$ and $\overrightarrow{E_2}$ will be added outside the plate to give a net field

$$E = E_1 + E_2,$$

where $E_1 = E_2 = \frac{\sigma}{2\varepsilon_0}$ (electric field due to a large charged sheet)

So, we have $E = \dfrac{\sigma}{\varepsilon_0}$ (Proved)

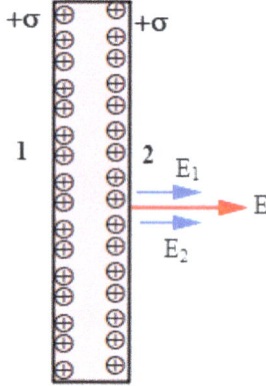

Method 2

Take a pillbox shaped Gaussian surface enveloping the area ΔA at both sides of the plate. The total flux out of both the flat faces of the pillbox is given as

$$\phi_E = \phi_1 + \phi_2$$
$$= E\,\Delta A + E\,\Delta A,$$

where

$$E = \frac{\sigma}{2\varepsilon_0}$$

$$\Rightarrow \phi_E = \frac{\sigma\,\Delta A}{\varepsilon_0} \tag{2.16}$$

According to Gauss's law,

$$\phi_E = \frac{q_{\text{em}}}{\varepsilon_0},$$

where $q_{\text{en}} = \sigma\,\Delta A + \sigma\,\Delta A$ (at both sides of the plate)

$$\Rightarrow \phi_E = \frac{2\sigma\,\Delta A}{\varepsilon_0} \tag{2.17}$$

Using equations (2.16) and (2.17), we obtain

$$E = \frac{\sigma}{\varepsilon_0} \text{ (proved)}$$

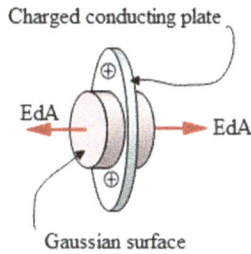

Charged conducting plate

EdA

EdA

Gaussian surface

Method 3

Taking a pillbox shaped Gaussian surface enveloping the inside and outside of surface 2 of the conductor, total flux out is

$$\phi_E = \sigma \, \Delta A \qquad (2.18)$$

(because $E = 0$ inside the conductor)

Since, $\phi_E = \frac{q_{en}}{\varepsilon_0}$ according to Gauss's law, where $q_{en} = \sigma \, \Delta A$ and $\phi_E = \sigma \, \Delta A$, we have

$$E = \frac{\sigma}{\varepsilon_0} \text{ (proved)}$$

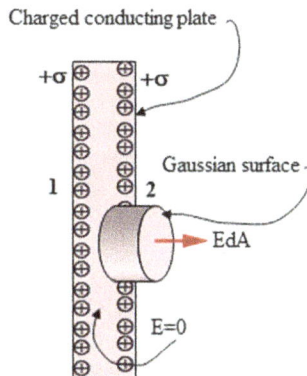

Charged conducting plate

$+\sigma$ $+\sigma$

1   2

Gaussian surface

EdA

E=0

N.B: $\vec{E} = \frac{\sigma}{\varepsilon_0}\hat{n}$ outside a conductor (or conducting plate), whereas $\vec{E} = \frac{\sigma}{2\varepsilon_0}\hat{n}$ due to a large sheet of charge (one side of the conducting plate). When we take the fields due to both sides, the net field outside the plate will be equal to $\frac{\sigma}{\varepsilon_0}\hat{n}$. Hence, there is no contradiction between these formulae.

2.5 Electrostatic induction

As discussed in the last section, when a conductor is placed in an external electrostatic field, equal and opposite charges are induced at the opposite surfaces of the conductor. The induced charges produce their own electric fields, which is known as 'induced electric field $(\vec{E}_{ind})$'. $\vec{E}_{ind}$ cannot be greater then $\vec{E}_{applied}$. However, it can exactly counteract the external (or applied) electric field inside the conductor to produce a zero electric field inside the conductor. So, we can write

$$\vec{E}_{int} = -\vec{E}_{app} \text{ (inside a conductor)} \tag{2.19}$$

However, these two electric fields cannot cancel each other outside the conductor. This means that the net field outside the conductor $(\vec{E}_{outside})$ is a vector sum of $\vec{E}_{ind}$ and $\vec{E}_{app}$.

$$\Rightarrow \vec{E}_{outsede} = \vec{E}_{ind} + \vec{E}_{app} = \frac{\sigma}{\varepsilon_0}\hat{n} \tag{2.20}$$

Hence, both the fields are responsible for the charge distribution at the surface of the conductor. As a result, the presence of a conductor modifies the field around it by inducing the charges at its surface.

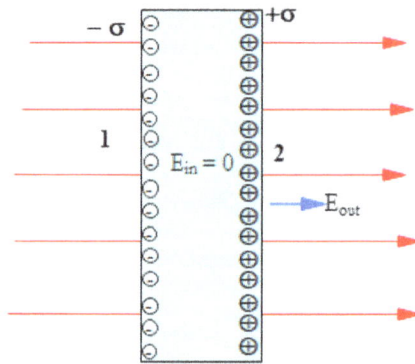

In electrostatic induction, the net field = resultant of induced field and the applied or external field;it is zero inside the the conductor as they are equal and opposite, but just outside the conductor it is non-zero and it is normal to the conductor surface

This instantaneous process of rearrangement of charges at the surface of a conductor when kept in an external static electric field is called 'electrostatic induction' and the charges appearing at the surface of the conductor are called 'induced charges'. Due to the vast reservoir of free electrons, a conductor can cancel any amount of applied electric field by inducing an appropriate amount of charge at its surface. In this process, no physical contact with a charged body (source of the applied electrostatic field) is required for charging the conductor.

2.6 Earthing

As a vast reservoir of water, a sea does not change its height or gravitational potential due to the inflow of water through different rivers and evaporation. Likewise, Earth is a vast reservoir of free electrons due to the presence of liquids, metals salts, minerals etc; so, it behaves as a good conductor of electricity and it does not change its electric potential when it receives the electrons from (or delivers the electrons to) any charged bodies.

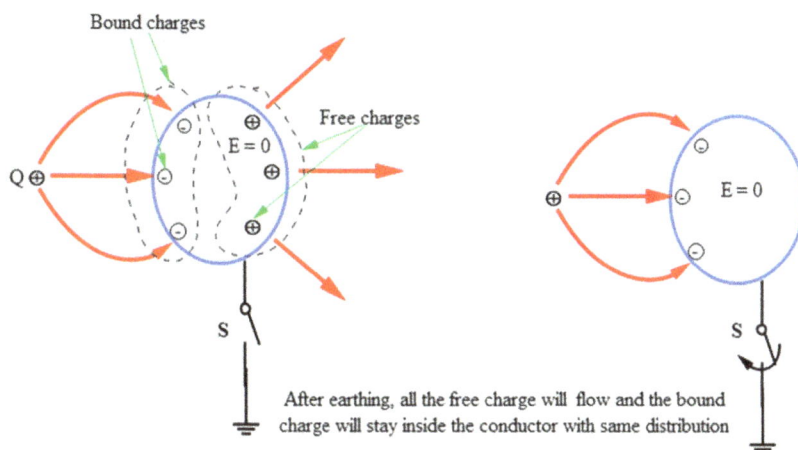

Bound charges

Free charges

$E = 0$

$Q \oplus$

$E = 0$

S

S

After earthing, all the free charge will flow and the bound charge will stay inside the conductor with same distribution

When we connect a charged conductor to Earth by a copper (or any conducting) wire, the excess charge (free electons) flows between earth and conductor in order to neutralize the excess free charges on the conductor. In this process, the potential of the conductor will be equal to the potential of Earth which is assumed to be zero even though Earth is at negative (electrical) potential. So, the process of connecting any electrical circuit or equipment to Earth to provide a safe path for the excess current to flow into the ground is called *earthing*. The flow of electrons between Earth and a charged conductor is called *earthing current*. By earthing a conductor, it will acquire a zero potential.

If the charged conductor is at positive potential, electrons flow from Earth to the conductor; if the conductor is at negative potential, electrons flow from the conductor to Earth. After earthing, the potential of the conductor will be 'zero'. During the earthing, the free (but not bound) charges will flow to Earth or get neutralized. The flow of total charge does not depend on the location of the point of the conductor connected to Earth.

Example 2 When we bring a charge $+q$ from infinity to the vicinity of a neutral spherical conductor of radius R, (a) define free and bound induced charge, (b) find the potential of the conductor.

Solution

(a) When we bring the charge q, let us assume that the charges $-q'$ and q' are induced on the surface of the conductor due to the electric field of q. As the charge q binds the charges $-q'$ it is called bound charge. The charge $+q'$ will appear at the opposite (right) surface of the conductor, which is called free charge; the lines of force emanating from the free charges terminate at infinity.

The charges q' and $-q'$ are distributed in such a way that their net field cancels the field due to the charge q inside the conductor. As a result, the field inside the conductor is zero to obey the property of a conductor.

$$\Rightarrow \vec{E}_{\text{net}} = \vec{E}_q + \vec{E}_{q'} + \vec{E}_{-q'} = 0 \text{ (inside the conductor)}$$

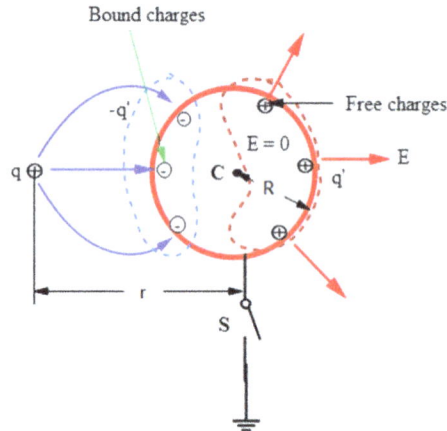

Before earthing, all the free charge q' will appear at the
right-side surface of the sphere and the bound charge -q'
will at the left-side surface of the conductor.

Let the potential at any point of the conductor be equal to V, say. By choosing the centre C of the sphere, its potential is given as the sum of the potentials of q, $+q'$ and $-q'$ at C. So, we can write

$$V_{\text{conductor}} = V_c = V_q + V_{-q'} + V_{q'}$$

$$= \frac{q}{4\pi\varepsilon_0 r} + \int \frac{-dq'}{4\pi\varepsilon_0 R} + \int \frac{dq'}{4\pi\varepsilon_0 R} = \frac{q}{4\pi\varepsilon_0 r} \tag{2.21}$$

N.B: As the positive charge q is brought from infinity, the potential of the conductor increases from zero (because q was at infinity) to $\frac{q}{4\pi\varepsilon_0 r}$. The potential due to the induced charges is equal to zero because the induced charges q' and $-q'$ produce equal and opposite potential at the center. This is because each element of the induced charge at the surface of the spherical conductor is equidistant from the center C of the sphere. So, the net potential of the conductor is equal to the potential of the positive point charge q at the center C.

2.7 Electrostatic shielding by a cavity in a conductor (conducting shell)

Since the electrostatic field inside a conductor is zero, the conductor acts as a *shield* to the external electric fields. As no excess or net charge can stay inside a conductor, the bulk (*meat*) of the conductor is useless. So, if we remove the total inner portion of the conductor, its charge carrying capacity still remains the same because the net charge always stays at the surface or skin of the conductor. After removing the 'meat of the conductor', we are left behind with the *skin* of the conductor that behaves as a thin conducting shell. However, removing some (not the entire meat) portion of the conductor, we will get a cavity in the conductor. Now a question arises: is a metallic shell (or cavity in a conductor) still able to shield or screen the external electric field? Let us see.

Let us take a Gaussian surface inside the conductor enclosing the cavity. Since $E = 0$ inside the conductor, the net flux passing through the Gaussian surface is zero.

$$\text{So, } \phi_E = \oint \vec{E}.d\,\vec{A} = 0 \qquad (2.22)$$

$$\text{According to Gauss's law, the flux out} = \phi_E = \frac{q_{en}}{\varepsilon_0} \qquad (2.23)$$

Using equations (2.23) and (2.24), we have

$$q_{en} = 0 \qquad (2.24)$$

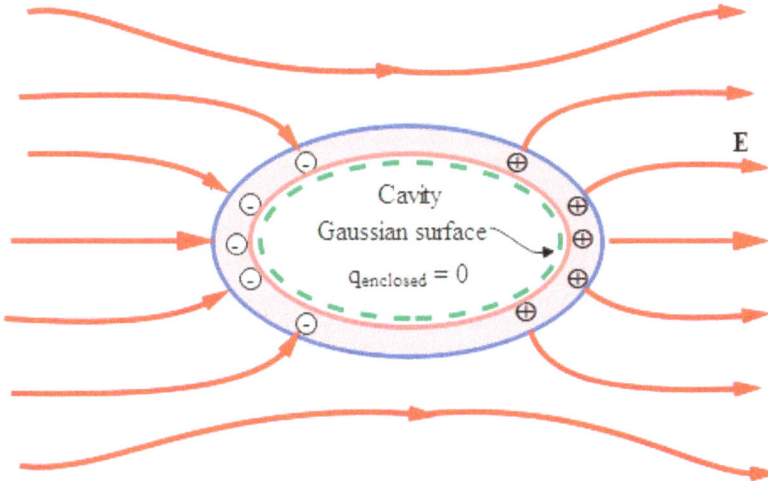

Metallic shell can shield or screen external electrostatic field

This tells us that no *net* charge can stay inside the conductor. As we have got an inner surface of the cavity we can expect that equal and opposite charge would be induced on it causing no net charge. As the induced charges produce their own electric field $\vec{E}$ inside the cavity (as a consequence of the applied field E_{ext}), we can argue that the cavity cannot screen the external electric field $\vec{E}_{ext}$, but this is not true. We will clear this fact by the following arguments.

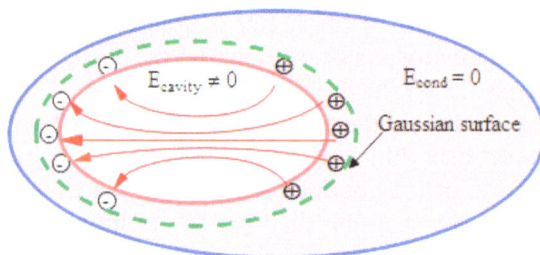

Let the positive and negative induced charges at the wall of
the cavity produce an electrostatic field inside the cavity

As static electric field is conservative,

$$\oint \vec{E} \cdot \vec{dl} = 0$$

in a closed loop which goes inside the cavity parallel to (along) $\vec{E}_{cav}$ and completes inside the conductor. So, we can write the closed line integral as a combination of that inside the cavity and conductor;

$$\oint \vec{E} \cdot \vec{dl} = \int \vec{E}_{cond} \cdot dl + \int \vec{E}_{cav} \cdot \vec{d}l = 0$$

Since $E_{cond} = 0$, the first integral will be zero; then, we have

$$\vec{E}_{cav} \cdot \vec{d}l = 0$$

Since dl is not zero, the electric field inside the cavity must be zero, which is given as

$$\vec{E}_{cav} = 0 \tag{2.25}$$

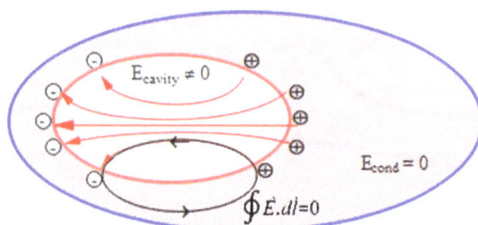

If the positive and negative induced charges on the wall of the cavity produce an
electrostatic field inside the cavity, due to its conservativeness its closed line integral
will be zero. As E=0, in the conductor, we can conclude that E=0 in the cavity

The last equation tells us that no static electric field can stay inside the cavity. Then how will it be possible for the assumed induced plus and minus charges to stay at the inner surface (wall) of the cavity? Then, we can argue that the plus and minus charges must meet to neutralize each other by sliding along the inner surface of the cavity. This eventually eliminates the presence of an electric field inside the cavity.

Recapitulating,

A *free cavity (a cavity that does not enclose a charge) can screen the external electric field. The external field cannot induce any charge on the inner surface of the cavity.* Now a question arises, can a conductor with a cavity screen its internal electric field? Let us see in the following example.

Example 3 A point charge q is placed in a cavity (we have shown it as a small sphere for the sake of simplicity) being suspended from the top of the cavity by a resin thread (the thread is not shown in the figure). Can the cavity screen or shield the internal electric field (produced by the induced charges and suspended charge) from external electric field (present outside the sphere)?

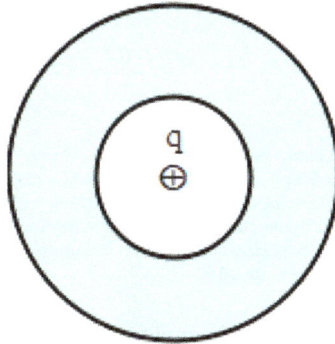

Solution

Let q' be the charge induced on the inner surface of the cavity. Then the net charge enclosed by the Gaussian surface drawn inside the conductor enclosing the inner surface of the cavity, is

$$q_{\text{enclosed}} = q + q' \tag{2.26}$$

According to Gauss's law,

$$\phi_E = \frac{q_{\text{en}}}{\varepsilon_0} \tag{2.27}$$

Using the last two equations, we have

$$\phi_E = \frac{q + q'}{\varepsilon_0} \tag{2.28}$$

Since the electric field in the conductor is zero, the net flux going out of the Gaussian surface is

$$\phi_E = \oint \vec{E} \cdot \vec{dA} = 0 \tag{2.29}$$

Using the last two equations, we have

$$q_{en} = q + q' = 0$$

$$\Rightarrow q' = -q \tag{2.30}$$

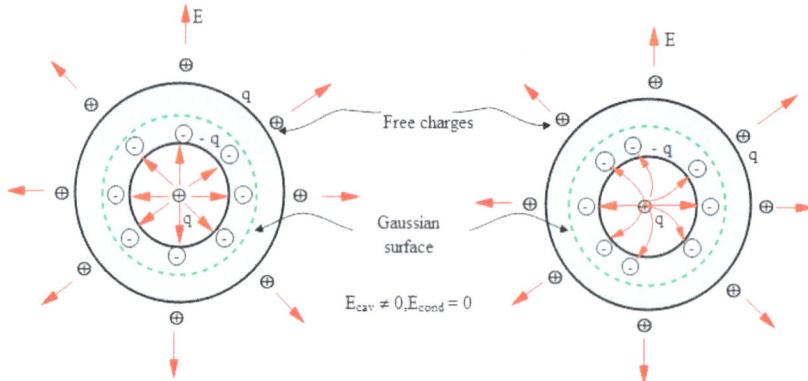

A free charge $+q$ appears at the outer surface of the conductor which is uniformly distributed irrespective of the position of the chage q in the cavity; a bound charge -q appears at the inner surface of the shell, In figure (a) the electric field inside the cavity is spherically symmetric, distribution of induced charge will be uniform, the net electrostatic force acting on the point charge $+q$ is zero; in figure (b) the chage $+q$ is displaced towards left from the centre that causes electric field inside the cavity unsymmetric and distribution of induced charge non-uniform (more charge on the near left-side face). The point charge $+q$ is attracted towards the left

This tells us that when a charge q is enclosed by a cavity, an equal and opposite charge $-q$ will be induced on the surface of the cavity, whereas $q_{ind}(=q') < q$ when the charge q is kept outside the conductor, as discussed in example 2.

A charge of $-q$ appears at the inner wall of the conductor. As the conductor is neutral, another charge of $+q$ must be induced at the outer surface of the conductor (to obey the conservation of charge). This charge $(+q)$ is free to interact with the charges at infinity if no other charges are placed near the conductor. However, the charge $-q'$ is a bound induced charge because it is directly linked with the charge q inside the cavity. Thus, the cavity is not a free cavity as it contains a charge $+q$.

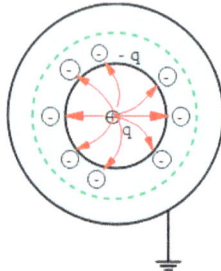

$E_{cav} \neq 0, E_{cond} = 0$; when grounded the free charge $+q$ will flow to earth and $E_{outside} = 0$

From the above discussion, we understand that, after placing a charge $+q$ inside the cavity, equal charge $+q$ will appear at the surface of the conductor. This charge will generate its electric field outside the conductor. In this way, a conducting shell (or a cavity in a conductor) cannot screen (shield) the effect of an internal electric field (field inside the cavity).

Thus, the charge kept inside the conducting shell communicates with the outer world (space) about its presence by inducing an equal charge at the surface of the conductor. Thus, a conducting shell cannot shield an internal electric field. However, after earthing the conductor, the outer free charge q will flow to the Earth. Then, the field outside the shell will be zero. Thus, a grounded or earthed conducting shell can screen the internal electric field $\overrightarrow{E}_{\text{int}}$.

Recapitulating, a free conducting shell can screen an external electric field. A grounded conducting shell can screen both internal and external electric field. If it is not grounded, it can shield an external electric field only.

2.8 Charge distribution on a conductor surface (uniqueness theorem)

Let us inject a charge $+q$ into a conductor by taking out electrons from it. The excess (injected) charge will be very quickly redistributed at the surface of the conductor such that the field inside the conductor will be zero. As $E = 0$ inside the conductor, the same potential V, say, will be felt at any point of the conductor.

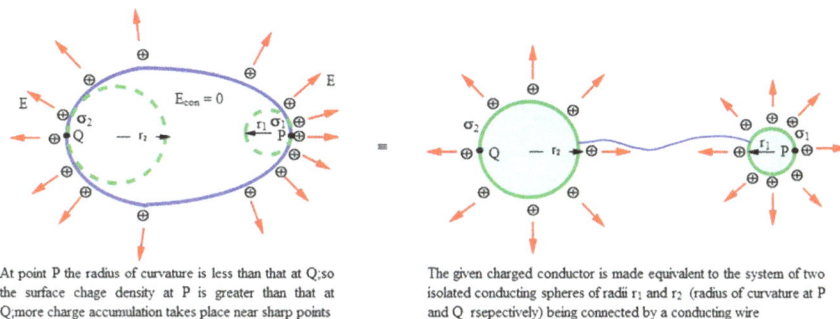

At point P the radius of curvature is less than that at Q; so the surface chage density at P is greater than that at Q; more charge accumulation takes place near sharp points

The given charged conductor is made equivalent to the system of two isolated conducting spheres of radii r_1 and r_2 (radius of curvature at P and Q rsepectively) being connected by a conducting wire

Let us take two points P and Q at the outer surface of the conductor. Let the radii of curvature at these points be r_1 and r_2, respectively. Since the potentials at P and Q are same, we can write

$$V_P = V_Q \tag{2.31}$$

To find these potentials let us imagine two spheres of radii be r_1 and r_2 passing through P and Q, respectively. Then mentally bring these two spheres out of the conductor and place them far apart to reduce their electrostatic induction to zero. Now, let us connect them by a long conducting wire. As the potentials of both spheres (or points P and Q of the spheres) will be equal, which is the necessary condition for the given conductor. So the system of these two spheres connected by a conducting wire is equivalent to the given system (conductor).

The potential of P is

$$V_P = \frac{Q_1}{4\pi\varepsilon_0 r_1} = \frac{\sigma_1 4\pi r_1^2}{4\pi\varepsilon_0 r_1} = \frac{\sigma_1 r_1}{\varepsilon_0} \tag{2.32}$$

The potential of Q is

$$V_Q = \frac{Q_2}{4\pi\varepsilon_0 r_2} = \frac{\sigma_2 4\pi r_2^2}{4\pi\varepsilon_0 r_2} = \frac{\sigma_2 r_2}{\varepsilon_0} \tag{2.33}$$

Using the last three equations, we have

$$\sigma_1 r_1 = \sigma_2 r_2 \tag{2.34}$$

This tells us the following.

The charge distribution takes place in a unique way at the surface of the conductor such that the product of surface charge density σ and radius of curvature r at any point of the conductor will be constant;

$$\sigma r = c \tag{2.35}$$

N.B: Since $E = \frac{\sigma}{\varepsilon_0}$ just outside the conductor and $\sigma r = c$, we can conclude that $E \propto \frac{1}{r}$. This means that the electric field near (very close to) any point on the conductor surface is inversely proportional to the radius of curvature of the conductor surface at that point. We must take convexity but not concavity of the surface into account to find the electric field; concavity refers to a *cavity* and field inside a free cavity is zero.

Example 4
 (a) What is corona discharge?
 (b) Find the minimum radius of curvature at a point of the conductor to ensure corona discharge.

Solution

 (a) Referring to the expression $E \propto \frac{1}{r}$ we understand that the electric field near the sharp points (very small radii of curvature) of a conductor would be very strong. When the electric field exceeds the dielectric strength (minimum electric field to ionize the air) of air which is equal to $(3 \times 10^6 \text{ V m}^{-1})$, the ionization of air takes place near these sharp points. In other words, when the free ions or charged particles undergo large acceleration due to strong electric fields near the sharp points of a charged conductor, their kinetic energy increases tremendously. These highly energetic particles further ionize air molecules forming a plasma (hot ionized gas) . Eventually, the air becomes conductive with a large number of positive and negative ions produced by the collision of energetic particles (ions etc). The ions of the same sign move away from the conductor with great acceleration and velocity and ions of opposite sign move towards the sharp points of the conductor. This phenomenon is followed by heat, sound and light due to the collision and successive ionization of the gases in air. Generally, we observe it as a spectacular faint bluish or violet glow with a hissing or cracking sound near the sharp points or edges of the conductors of the high voltage power stations in moist and foggy weather. This is known as 'corona discharge' which is an undesirable waste of electric energy due to the ionization of air near sharp charged conductors.

The consequences of corona discharge are many; leaking energy from an electrical system as light and heat to the surroundings, degrading the material by corrosion, disrupting the nearby electronics equipment by its radiation, disturbing the others by its hissing sounds etc. Ans.

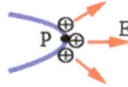

At the sharp point P of the conductor the electric field strength is so high that when it exceeds the breakdown strength of air corona discharge takes place

(b) Let r = radius of curvature of the conductor at a certain point. Since $\sigma r = C$ for a particular value of r, we have

$$\sigma_{max} = \varepsilon_0 E_{max} \left(\because E = \frac{\sigma}{\varepsilon_0} \right)$$

$$= 8.85 \times 10^{-12} \times 3 \times 10^6 = 26.55 \times 10^{-6} \, \text{C m}^{-2}$$

The critical radius of curvature of the conductor is given as

$$r_{critical}(=r_{min}) = \frac{C}{\sigma_{max}} = \frac{V}{E_{max}},$$

where C = constant (product of charge density and radius of curvature) and V = potential of the conductor. Ans.

2.9 Electrical force acting on the surface of a charged conductor

In order to find the electrostatic force on the charged conductor, let us take an elementary area dA of the conductor which has a charge $dq = \sigma dA$, where σ = surface charge density.

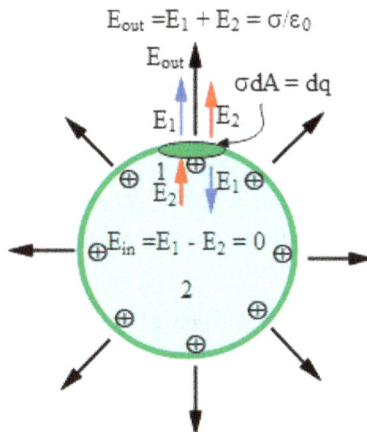

E_1 and E_2 are equal and opposite (negative vectors) inside the conductor just below the patch and just outside the patch they are parallel (equal vectors).

The elementary charge dq will experience a force dF due to the electric field E_2 of other charges (except dq) because a charge cannot experience a force by its own electric field. This is given as

$$dF = (dq)E_2 \tag{2.36}$$

Let E_1 = electric field of dq and E_2 = electric field of the others (except dq). We know that just below the elementary patch inside the conductor the net electric field is zero. This is possible only when $\vec{E_1}$ and $\vec{E_2}$ would cancel each other. Then, they are equal in magnitude and opposite in direction inside the conductor. So, we can write

$$E_1 = E_2 \tag{2.37}$$

On the other hand, $\vec{E_1}$ and $\vec{E_2}$ are parallel just outside the patch dA. These two fields combine to give a net field outside the conductor which is given as

$$E_{\text{outside}} = E_1 + E_2 = \frac{\sigma}{\varepsilon_0} \tag{2.38}$$

Using equations (2.37) and (2.38), we have

$$E_1 = E_2 = \frac{\sigma}{2\varepsilon_0} \tag{2.39}$$

Then, from equation (2.36) the force $dF = E_2 \, dq$, where $E_2 = \frac{\sigma}{2\varepsilon_0}$ and $dq = \sigma \, dA$.

$$\Rightarrow \frac{dF}{dA} = \frac{\sigma^2}{2\varepsilon_0} \tag{2.40}$$

Since, $E_{\text{net}}(=E) = \frac{\sigma}{\varepsilon_0}$ just outside the conductor, putting $\sigma = \varepsilon_0 E$ in equation (2.40), we have

$$\frac{dF}{dA} = p = \frac{\varepsilon_0 E^2}{2} \tag{2.41}$$

The last expression tells us the following.

The electrostatic stress $p \propto \sigma^2$ and $p \propto E^2$. This physically signifies the charges tend to fly apart but they are held by the inward mechanical force of the conductor.

Example 5 Find the pressure at the surface of a conductor at a point of surface charge density $\sigma = 2\mu \, \text{C m}^{-2}$.
   Solution
   The electrostatic pressure at the surface of the conductor is

$$P = \frac{\sigma^2}{2\varepsilon_0} = \frac{(2 \times 10^{-6})^2}{2 \times 8.85 \times 10^{-12}} \simeq 0.225 \, \text{N m}^{-2} \, \text{Ans.}$$

2.10 Dielectrics

The word 'dielectric' means 'electric field through (or in) matter'. It means that when we keep an object like wood or paper in an electric field, we can find an electric field inside that object. Hence wood, glass, rubber, plastics etc, are dielectrics. As discussed earlier, the electrons are loosely bound in a conductor. But in the dielectrics the electrons are bound to the nuclei of atoms and cannot move freely. So, the dielectrics do not conduct electricity under an electric field below a critical electric field which is known as *dielectric strength*. The dielectric strength of air is 3×10^6 V m^{-1} and that of ceramics and plastics varies between 17 and 40 million volts per meter. Beyond the critical electric field some electrons can be pulled out of their respective atoms and made available for conducting electricity. This is called '*breakdown of dielectrics*' which is followed by ionization of dielectric atoms and molecules by a large electric field. The maximum electric field that a dielectric can withstand is called 'dielectric strength which is in the order of 10^6 V m^{-1} as mentioned above. So, nothing is purely a non-conductor or insulator of electricity in a strict sense. Thus, everything can be treated as a dielectric. For instance, a conductor is a dielectric of infinite permittivity and a good insulator is a dielectric with a finite value of permittivity. Some good insulators such as ceramics have relative permittivity greater than 100.

2.11 Behaviour of a dielectric in an external electric field

It is interesting to see a rubbed comb attracting a piece of paper. When we rub a comb with hair, it gets certain negative charge and it creates an electric field which eventually polarizes a neutral piece of paper. The paper closer to the comb gets a positive charge and the other side will get a negative charge. The closer side of the paper experiences a greater attraction force than the repulsion force acting on the other side of the paper. So, the net force is an attraction. This tells us about the peculiar behaviour of a dielectric to an external non-uniform electric field. To understand the behaviours of a dielectric in an external electric field, we need to know the nature of its atoms and molecules. The molecules of a dielectric can be polar or non-polar.

Polar dielectrics in an external electric field: The polar dielectrics have polar molecules. In a polar molecule the centres of positive and negative charges are permanently shifted possessing a permanent dipole moment. For instance, $H_2^+O^-$, $H^+ Cl^-$ etc, are the polar molecules. Due to thermal agitation, all dipoles are randomly oriented in the absence of an external electric field. So, in the absence of an external electric field, the net dipole moment of the dielectric is zero (even though each molecule has a non-zero dipole moment). But, an external electric field applies a torque on each dipole. This torque tends to twist or rotate the dipoles so as to align them in the direction of the applied electric field. However, the complete alignment of the dipole may not be possible due to their thermal agitation. So, a polar dielectric acquires a net dipole moment under an external electric field following the process of twisting against the internal friction. If we remove the applied field, the dipoles cannot regain their mean positions completely due to the thermal agitation (internal friction). Eventually, some alignment will be retained even after the removal of the applied field. This property of a dielectric is called 'ferroelectrics' which is analogous to ferromagnetic in magnetism.

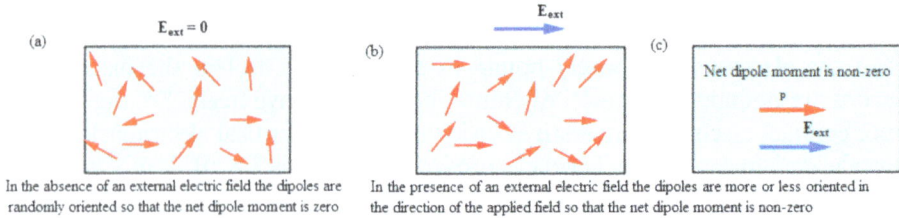

(a) $E_{ext} = 0$

(b) E_{ext}

(c) Net dipole moment is non-zero

p

E_{ext}

In the absence of an external electric field the dipoles are randomly oriented so that the net dipole moment is zero

In the presence of an external electric field the dipoles are more or less oriented in the direction of the applied field so that the net dipole moment is non-zero

Non-polar dielectrics in an external electric field: In a non-polar dielectric each atom is non-polar because the centre of positive charge (nucleus) and centre of negative charge (electron cloud) overlap. Hence, each atom or molecule has a zero dipole moment in the absence of an external electric field. Under the action of an applied (external) electric field, the centre of the positive charge will be pulled in the direction of the electric field and the centre of the negative charge will be pulled in the opposite direction to the electric field. In other words, the applied electric field stetches the molecule (or atom) against the electrostatic (Coulombic) force between the shifted centres of positive and negative charge. Due to the relative displacement of the positive and negative charge centres or stretching of the molecules or atoms by the external electric field, a permanent dipole moment is induced.

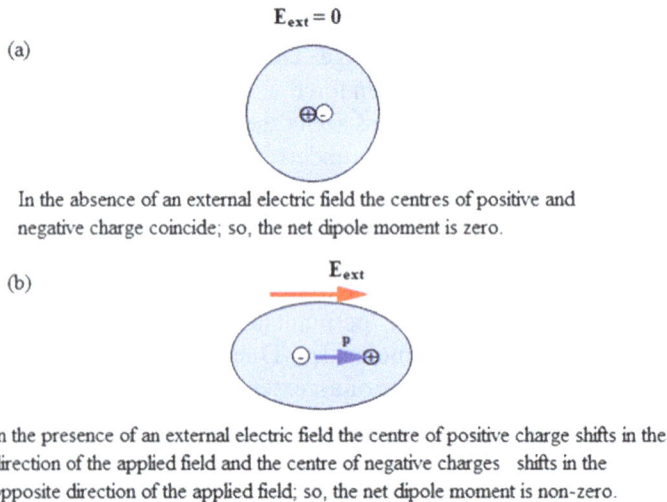

(a)

$E_{ext} = 0$

In the absence of an external electric field the centres of positive and negative charge coincide; so, the net dipole moment is zero.

(b)

E_{ext}

p

In the presence of an external electric field the centre of positive charge shifts in the direction of the applied field and the centre of negative charges shifts in the opposite direction of the applied field; so, the net dipole moment is non-zero.

However, if the applied field is too strong, some electrons will be removed permanently ionizing the atoms and molecules. This is called breakdown of the non-polar dielectrics. Hence, in the presence of an electric field (not too strong to breakdown the non-polar dielectrics), the induced dipole moment $\vec{p}$ of each atom is directly proportional to the applied field $\vec{E}$.

2-23

$$\vec{p} = \alpha \vec{E},$$

where $\alpha =$ atomic polarizability which depends on the atomic structure of the dielectric. Recapitulating, a non-polar dielectric induces a dipole moment under an external electric field. If the external field is withdrawn, the atoms get relaxed to yield zero dipole moment. You should note that beyond certain electric field the above linear relation will not hold good. *Each polar and nonpolar dielectric acquires a net electric dipole moment in the external electric field by the process of stretching and twisting of the atoms or molecules, respectively.*

Example 6 If a non-polar atom of radius R is placed in an external electric field E, find the induced dipole moment of the atom.

Solution

Let the applied electric field stretch the atom by a distance r. Assuming a uniform spherical distribution of the electron cloud, its electric field at a distance $r < R$ is

$$E' = \frac{Qr}{4\pi\varepsilon_0 R^3}$$

This electric field E' of the electron cloud pulls the nucleus of charge $+Q$ with a force $F' = QE'$ to the left. Another force $F = QE$ acts to the right on the nucleus $+Q$ due to the applied field. Since the charge $+Q$ is in equilibrium, $F = F'$. Using the last four expressions, we have

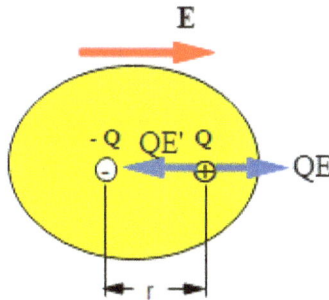

$$QE = \frac{Q^2 r}{4\pi\varepsilon_0 R^3}$$

So, the induced atomic dipole moment is

$$p = Qr = (4\pi\varepsilon_0 R^3)E$$

This tells us that the induced dipole moment p is directly proportional to the applied field E and the constant of proportionality is called atomic polarizability, given as

$$\alpha = 4\pi\varepsilon_0 R^3$$

This result holds good for many simple atoms. Ans.

2.12 Polarization of dielectrics

In last section we learnt that under the action of an electric field, a net dipole moment appears in the dielectrics. The principle behind the appearance of a net dipole moment in a polar dielectric is different from that in a non-polar dielectric. In polar dielectrics, it occurs due to twisting of the molecular dipoles, whereas in non-polar dielectrics it occurs in the process of stretching of molecules and atoms. However, in both cases, the centres of positive charges in each atom or molecule shift slightly in the direction of the applied or external electric field. Although this effect of charge separation is very small for an atom or molecule, its effect on the whole specimen is noticeable, inducing a considerable dipole moment because millions of dipoles orient in the same direction. This process of charge separation in a dielectric is called polarization. In a polarized dielectric, equal positive and negative bound charges appear at its opposite surfaces. This polarization of a dielectric is characterized by a vector called 'electric polarization' given as the ratio of net dipole of a dielectric under the application of an external electric field and the volume of the dielectric. If n = total number of dipoles and p = dipole moment of each atom, the polarization vector, that is, dipole moment per unit volume is given as

$$\vec{P} = \frac{\text{net dipole moment}}{\text{volume of the dielectric}} = \frac{n\vec{p}}{v}$$

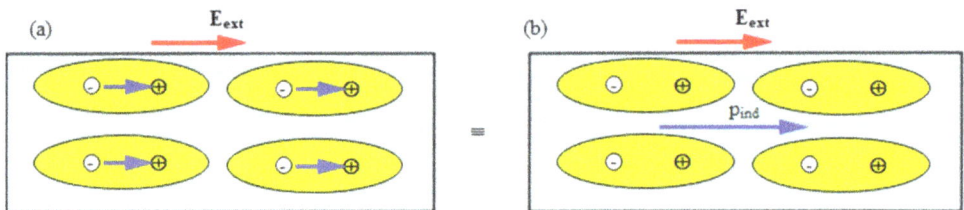

In the presence of an external electric field the tiny induced dipole moments (non-polar dielectric molecules) align in the direction of the applied field; so, the net induced dipole moment is non-zero.

Recapitulating, when a dielectric is placed in an electric field, it acquires a net dipole moment. Eventually, equal and opposite bound charges appear at the opposite surfaces of the dielectric which is called polarization of the dielectric. Polarization is defined as the dipole moment per unit volume.

2.13 Relation between polarization vector and surface charge density

If we place a rectangular dielectric slab in a uniform electric field, it becomes polarized. As a result, equal and opposite bound charges $-Q$ and Q appear on the left and right faces, respectively

The net dipole moment of the slab is given as $p_{net} = np$, where n = total number of dipoles and p = dipole moment of each dipole.

Substituting $p_{net} = Ql$, we have $Ql = np$.

Dividing both sides with the volume $v = Al$ of the specimen, we have

$$\frac{Ql}{Al} = \frac{np}{Al},$$

$$\Rightarrow \frac{Q}{A} = \left(\frac{np}{Al}\right),$$

where $\frac{np}{Al} = \frac{\text{total dipole moment}}{\text{total volume}} = P(\text{polarization})$ and $\frac{Q}{A} = \sigma_b$ (surface bound charge density). Then the relationship between the bound charge and polarization is given as

$$P = \sigma_b$$

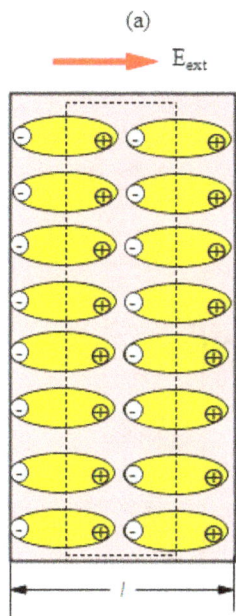

Polarised dielectric in the presence of an electric field

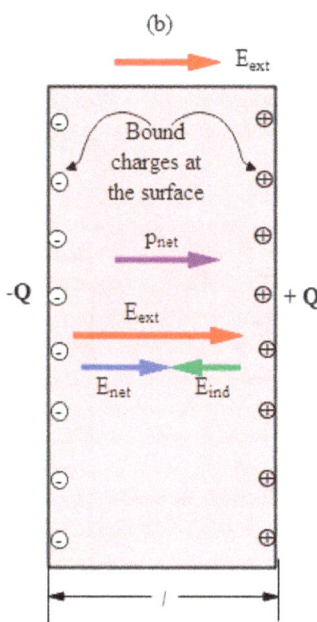

Removing the bulk zero net charges mentally in the polarised dielectric, we are left behind with the surface bound charges

This tells us that the polarization of a dielectric is numerically equal to the surface charge density of the bound charges.

2.14 Relative permittivity

As the dielectric slab gets polarized under an external (or applied) electric field $\overrightarrow{E}_{ext}$, the positive and negative induced (bound) charges at the opposite faces of the slab produce an internal electric field in the dielectric. Let us call it *induced* electric field denoted as $\vec{E}_{ind}$. As this induced field opposes the applied or external field, the net field inside the dielectric is given as the vector sum of these two electric fields. So,

$$\vec{E}_{net} = \vec{E}_{ext} + \vec{E}_{ind}$$

As the induced field $\vec{E}_{ind}$ is caused by the external field $\vec{E}_{ext}$, it cannot overcome the external field. So, the net field inside the dielectric is

$$E_{net} = E_{ext} - E_{ind}$$

For large plates, the net field due to the induced charges outside the plate is zero. So, the field outside the dielectric is equal to the applied or external field. The last expression tells us that the external field is stronger than the net electric field inside the dielectric by a factor $\frac{E_{ext}}{E_{net}}(>1)$, which is called 'relative permittivity or dielectric constant' given as

$$\varepsilon_r = \frac{E_{ext}}{E_{net}},$$

where $E_{net} = E_{ext} - E_{ind}$

$$\Rightarrow \varepsilon_r = \frac{E_{ext}}{E_{ext} - E_{ind}}$$

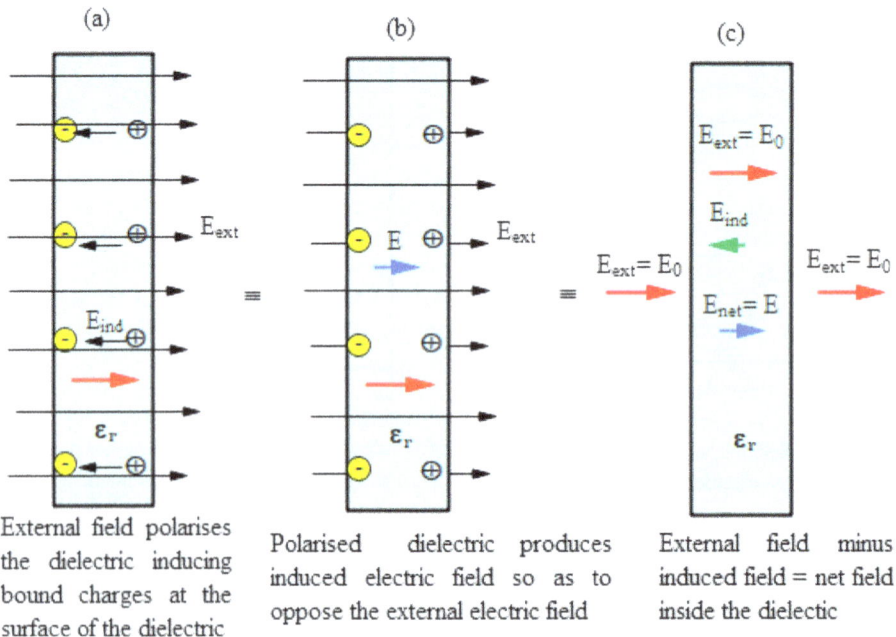

(a) External field polarises the dielectric inducing bound charges at the surface of the dielectric

(b) Polarised dielectric produces induced electric field so as to oppose the external electric field

(c) External field minus induced field = net field inside the dielectic

Hence, the dielectric constant is a measure of the extent to which a given material is polarized by an external electric field. If ε_r is greater, the induced electric field will be greater and vice versa. The dielectric constant of ceramic ranges from 9000 to

11 000; the dielectric constant of water is 82. Air or vacuum has the lowest dielectric constant of 1. The dielectric constant can also be termed as relative permittivity which is defined as the ratio of absolute permittivity ε of a material and the absolute permittivity ε_0 of air or vacuum. So, we can write

$$\varepsilon_r = \frac{\varepsilon}{\varepsilon_0}$$

So, the induced field in a dielectric is given as

$$E_{ind} = E_{ext}\left(1 - \frac{1}{\varepsilon_r}\right)$$

N.B: Putting $E_{net} = 0$ inside the conductor; $\vec{E}_{ind}$ is equal and opposite to the applied field $\vec{E}_{ext}$. Then, the relative permittivity of a conductor is

$$\varepsilon_r = E_{ext}/E_{net} = E_{ext}/0 = \infty$$

The relative permittivity of a conductor is infinite under constant or slowly varying electric fields.

Example 7 In a piece of mica, the electric field is four times weaker than the applied field $E = 2 \times 10^6$ V m^{-1}. What is the (a) dielectric constant of mica, (b) induced electric field?
   Solution

   (a) The dielectric constant is

$$\varepsilon_r = \frac{E_{ext}}{E_{inside}}, \quad \text{where } E_{ext} = E \text{ and } E_{inside} = \frac{E}{4}$$

$$\text{or } \varepsilon_r = \frac{E}{(E/4)} = 4 \text{ Ans.}$$

   (b) $E_{ind} = E_{ext} - E_{inside}$

$$= 2 \times 10^6 - \frac{2 \times 10^6}{4} = 1.5 \times 10^6 \text{ V m}^{-1} \text{ Ans.}$$

Alternative method: The induced field in a dielectric is given as

$$E_{ind} = E_{ext}\left(1 - \frac{1}{\varepsilon_r}\right) = E_{ext}\left(1 - \frac{1}{4}\right) = 3E_{ext}/4$$

Putting $E_{ext} = 2 \times 10^6$ V m^{-1}, we have $E_{ind} = 1.5 \times 10^6$ V m^{-1} Ans.

Problem 1
   (a) If a large thin conducting plate is kept in an external uniform electric field E_0, find the surface charge density of the induced charges on each surface of the plate.

(b) If the breakdown voltage (maximum potential gradient of air to become conductive) is $10^6 \, \text{V m}^{-1}$, find the maximum surface charge density of the induced charges on the plate.

Solution

(a) Since $\vec{E}_{net}$ inside the conductor is zero, the induced electric field $\vec{E}_{ind}$ exactly nullifies the external electric field; $\vec{E}_{ext} = \vec{E}_0$. Outside the plate, the equal and opposite surface charges contribute no net field. Hence, outside the plate, the net field is equal to the applied (external) field E_0.

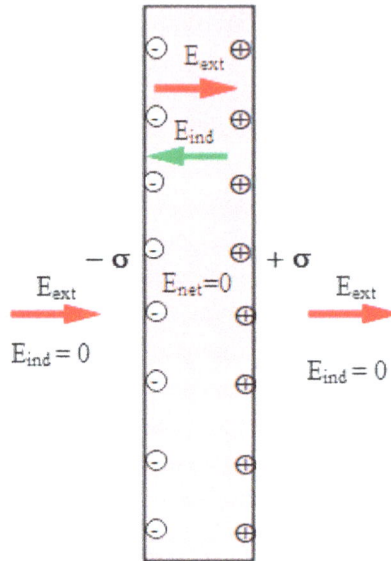

Since, $\vec{E}_{outside} = \frac{\sigma}{\varepsilon_0}\hat{n}$ and $\vec{E}_{outside} = E_0 \, \text{V m}^{-1}$, we can write $\sigma = \varepsilon_0 E_0$ Ans.

(b) Putting the breakdown strength of air $= E_{max} = 3 \times 10^6 \, \text{V m}^{-1}$, we have

$$\sigma_{max} = \varepsilon_0 E_{max} = 8.85 \times 10^{-12} \times 3 \times 10^6 = 26.55 \text{ micro Coulomb m}^{-2} \text{ Ans.}$$

N.B: The net field due to the induced charges is zero outside the conductor only when it is a large thin plate. However, in general, net electric field outside a conductor is not equal to the external (applied) electric field.

Problem 2

(a) When we bring a charge $+q$ from infinity and place it at a distance r from the centre of a neutral spherical conductor of radius R. If we ground the sphere, find
  (i) the amount of charge that flows to Earth;
  (ii) the induced charges q' of the sphere,

(b) If the initial charge of the sphere is Q, find the
   (i) charge flown after grounding;
   (ii) free charge;
   (iii) bound charge.

Solution

(a)

(i) Let $x =$ the charge flown to Earth. Then, the net charge of the conductor is $-x$ which generates a potential at C given as

$$V_{-x} = -\frac{x}{4 \pi \varepsilon_0 R}$$

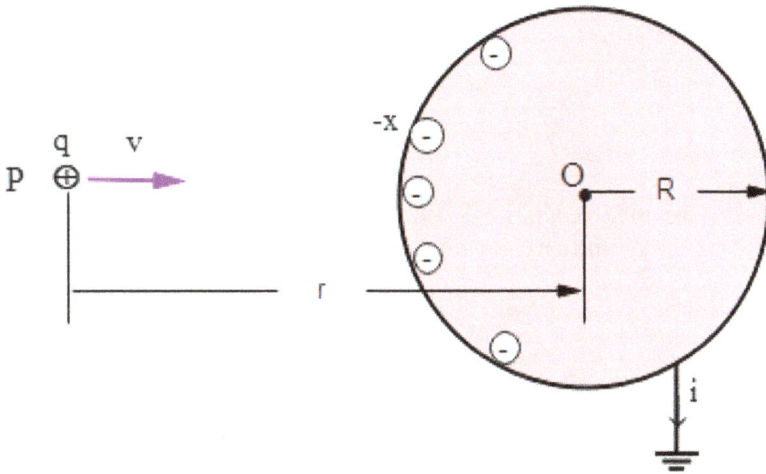

The potential due to q at the centre O of the sphere is

$$V_q = \frac{q}{4 \pi \varepsilon_0 r}$$

Then, the net potential of the conductor is

$$V = V_{-x} + V_q = \frac{1}{4 \pi \varepsilon_0}\left(\frac{-x}{R} + \frac{q}{r}\right)$$

Since, the conductor is grounded (or earthed), putting $V = 0$ in the last equation, we have

$$x = \frac{Rq}{r} \text{ Ans.}$$

(ii) Since, all free charge will flow to Earth, the charge flown to Earth must be equal to the free charge $+q'$ which is given as

$$q' = \frac{R}{r}q \text{ Ans.}$$

(b)

(i) Let $x =$ charge flowing to Earth. Then, the net charge on the sphere is $Q - x$. Then, the net potential of the conductor is

$$V = V_{Q-x} + V_q = \frac{1}{4 \pi \varepsilon_0} \left(\frac{Q - x}{R} + \frac{q}{r} \right)$$

Since the conductor is grounded (or earthed), putting $V = 0$ in the last equation, we have

$$\frac{1}{4 \pi \varepsilon_0} \left(\frac{Q - x}{R} + \frac{q}{r} \right) = 0$$

$$\Rightarrow x/R = \frac{Q}{R} + \frac{q}{r}$$

$$\Rightarrow x = Q + \frac{qR}{r} \text{ Ans.}$$

(ii) As the free charge flows when a charged conductor is earthed leaving the bound charges unaffected, the free charge is given as

$$q_{\text{free}} = x = Q + \frac{qR}{r} \text{ Ans.}$$

(iii) The bound charge is given as

$$q_{\text{bound}} = Q - x = -\frac{qR}{r} \text{ Ans.}$$

Problem 3 Find the (a) field, (b) potential due to an isolated spherical conductor of chare q and radius R.
Solution

(a) According to the first conductor property

$$E_{\text{inside}} = 0 \tag{2.42}$$

So, the net charge q is distributed on the surface of the conductor because the charge cannot stay inside the conductor. Due to the uniform radius of curvature of the surface of a sphere, the charge distribution will be uniform; so, the electric field $\overrightarrow{E}$ is spherically symmetrical. Applying Gauss's law at any radial distance r,

$$E(4\,\pi r^2) = \frac{q_{en}}{\varepsilon_0}, \quad \text{where } q_{en} = q$$

$$\Rightarrow E = \frac{q}{4\pi\varepsilon_0 r^2}; \, r > R \tag{2.43}$$

Since the radius of curvature is constant, the charge density will remain the same on the surface of the sphere. So, just outside the sphere the electric field is

$$\vec{E}_{\text{outside}} = \frac{\sigma}{\varepsilon_0}\hat{n} \tag{2.44}$$

The surface charge density of the sphere is

$$\sigma = \frac{q}{4\pi R^2} \tag{2.45}$$

Using the last two equations, we have

$$E_{\text{outside}} = \frac{q}{4\pi\varepsilon_0 R^2} \tag{2.46}$$

Equations (2.42) and (2.46) tell that E is discontinuous at the surface or boundary; but in reality, E does not vary abruptly at the skin of the conductor, rather E increases from zero to $\frac{q}{4\pi\varepsilon_0 R^2}$ smoothly or continuously within the skin (several micro meters of the conductor). If we assume an ideal skin as a *zero-thickness*, then the discontinuity of E-field will arise. Ans.

(b) The potential of the conductor is given as follows:

$$V = \frac{q}{4\pi\varepsilon_0 R}; \, r \leqslant R \tag{2.47}$$

$$V = \frac{q}{4\pi\varepsilon_0 r}; \, r \geqslant R \tag{2.48}$$

We can see that potential is continuous at the boundary (surface). An isolated charged spherical conductor possesses spherical symmetry of its surface charge density, electric field $\vec{E}$ and potential V. Hence, if we substitute the charged conductor by a point charge q placed at the centre of the sphere, the field and potential function remains unchanged outside the conductor.

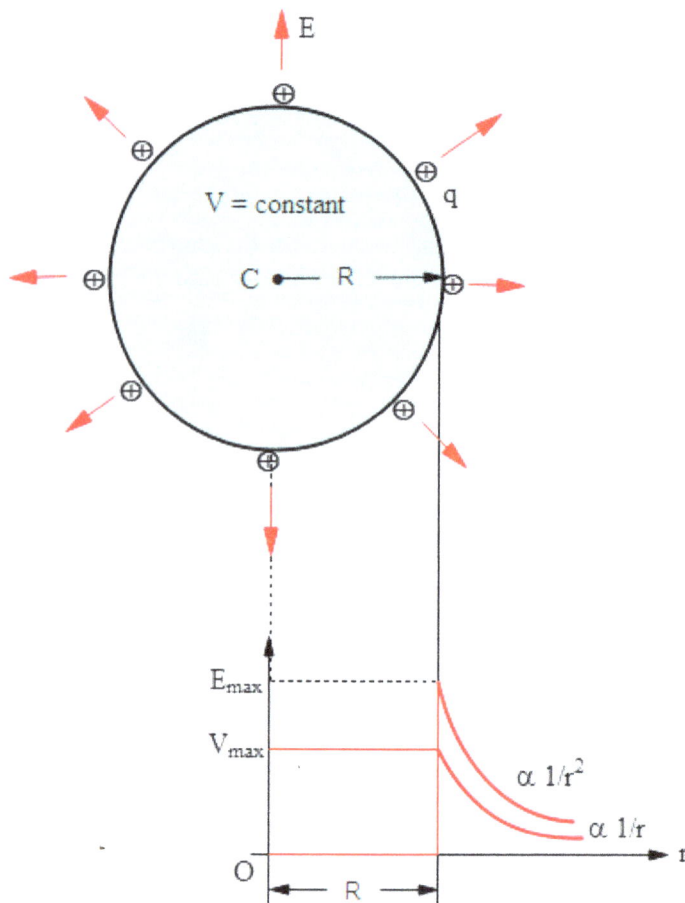

N.B: *Electric potential is continuous, whereas electric field is discontinuous across the boundary of a conductor. Potential varies with the 1/r rule and electric field varies with inverse square law of the radial distance outside the conducting sphere.*

Problem 4 Find the (a) induced charges, (b) field, (c) potential due to the system of two concentric conducting spherical shells of radii a and b carrying charges q_1 and q_2, respectively. Put $q_1 = -2q$, $q_2 = 3q$ and $b = 2a$.
  Solution

(a) Any charge q_1 placed inside a conducting shell, it induces an equal and opposite charge '$-q_1$' on the inner surface of the shell as discussed in example 3. Then the surplus charge $= q_2 - (-q_1) = q_1 + q_2$ will be distributed uniformly on the outer surface of the outer conductor as a free charge. Ans.

(b) At any point P inside the inner shell, both the shells contribute zero fields because the point P is lying inside both the shells.

   Since the individual fields $E_1 = 0$ and $E_2 = 0$

   the net field inside the inner shell is $E = \left| \vec{E_1} + \vec{E_2} \right| = 0;\ r < a$

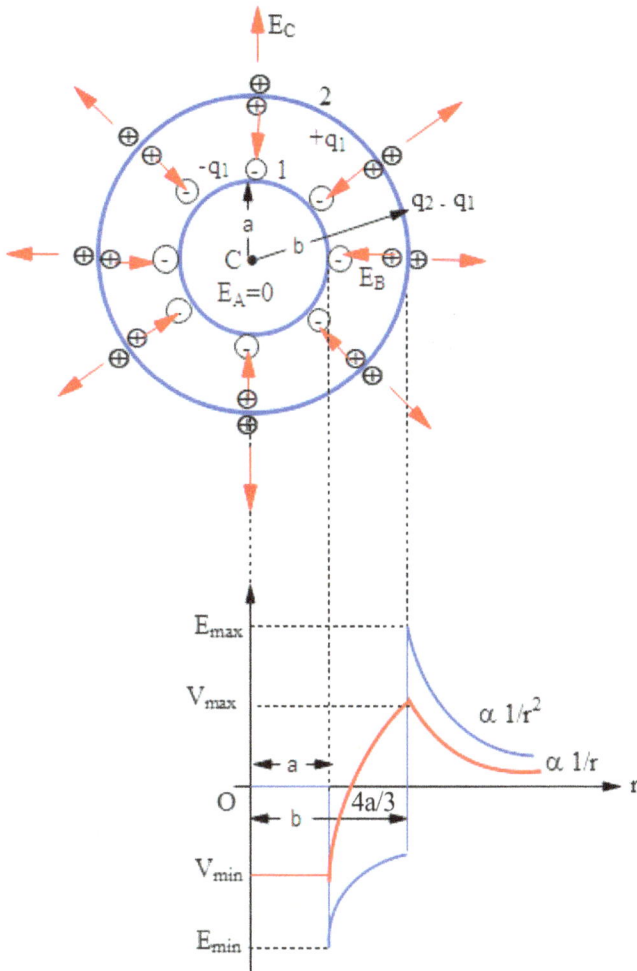

At any point Q between the shells, E is contributed due to the inner shell because the outer shell cannot contribute any field as point Q is lying inside the outer shell. Then putting

$$\overrightarrow{E_1} = \frac{q_1}{4\pi\varepsilon_0 r^2}\hat{r} \quad \text{and} \quad \overrightarrow{E_2} = 0,$$

the net field at Q is

$$\overrightarrow{E} = \overrightarrow{E_1} + \overrightarrow{E_2} = \frac{q_1}{4\pi\varepsilon_0 r^2}\hat{r}; \quad a < r < b$$

$$= \frac{-2q}{4\pi\varepsilon_0 r^2}\hat{r} = \frac{-q}{2\pi\varepsilon_0 r^2}\hat{r}; \quad a < r < b$$

Putting $r = a$, the minimum value of E which is radially inward is

$$\overrightarrow{E}_{\min} = \frac{-q}{2\pi\varepsilon_0 a^2}\hat{r}$$

If we take any point R outside the outer shell, both shells will contribute the fields which will reinforce. So, the net field at R is,

$$\vec{E} = \vec{E_1} + \vec{E_2}$$

$$= \frac{q_1 \hat{r}}{4\pi\varepsilon_0 r^2} + \frac{q_2 \hat{r}}{4\pi\varepsilon_0 r^2}$$

$$= \frac{q_1 + q_2}{4\pi\varepsilon_0 r^2}\hat{r}; \quad r > b$$

$$= \frac{-2q + 3q}{4\pi\varepsilon_0 r^2}\hat{r} = \frac{q}{4\pi\varepsilon_0 r^2}; \quad r > b$$

Putting $r = b$, the maximum value of E which is radially outward is

$$\vec{E}_{max} = \frac{q}{4\pi\varepsilon_0 (2a)^2}\hat{r} = \frac{q}{16\pi\varepsilon_0 a^2}\hat{r} \text{ Ans.}$$

Alternative method: We can also find $\vec{E}$ by Gauss's law as follows:

Let us take a spherical Gaussian surface of radius r whose centre lies at the centre of the spherical shells.

For any point inside the inner sphere, the charge enclosed by the Gaussian surface is zero; so, we can write

$$\phi_E = \oint \vec{E} \cdot \vec{dA} = 0$$

$$\Rightarrow E = 0; \quad r < a$$

This means that the electric field will be zero inside the inner shell.

For any point in between the shells, the charge enclosed is zero; so, we can write

$$\phi_E = \oint \vec{E} \cdot \vec{dA} = \frac{q_1}{\varepsilon_0}$$

$$\Rightarrow E(4\pi r^2) = \frac{q_1}{\varepsilon_0}$$

$$\Rightarrow E = \frac{q_1}{4\pi\varepsilon_0 r^2}; \quad a < r < b$$

For any point outside the outer shell, the charge enclosed is equal to the sum of the charges of both shells; so, we can write

$$\phi_E = \oint \vec{E} \cdot \vec{dA} = \frac{q_1 + q_2}{\varepsilon_0}$$

$$\Rightarrow E(4\pi r^2) = \frac{q_1 + q_2}{\varepsilon_0}$$

$$\Rightarrow E = \frac{q_1 + q_2}{4\pi\varepsilon_0 r^2}; \quad r > b \text{ Ans.}$$

(c) At any point, the net potential V is equal to algebraic sum of the potentials V_1 and V_2 due to each shell 1 and 2, respectively.

$$\Rightarrow V = V_1 + V_2$$

If point P lies inside both the shells,

$$V_1 = \frac{q_1}{4\pi\varepsilon_0 a} \quad \text{and} \quad V_2 = \frac{q_2}{4\pi\varepsilon_0 b}$$

Then, $V = V_1 + V_2 = \frac{1}{4\pi\varepsilon_0}\left(\frac{q_1}{a} + \frac{q_2}{b}\right); \ r \leqslant a$

$$\Rightarrow V = \frac{1}{4\pi\varepsilon_0}\left(\frac{-2q}{a} + \frac{3q}{2a}\right) = -\frac{q}{8\pi\varepsilon_0 a}; \ r \leqslant a$$

The minimum value of V is

$$V_{min} = -\frac{q}{8\pi\varepsilon_0 a}; \ r \leqslant a$$

For point Q that lies between the shells,

$$V_1 = \frac{q_1}{4\pi\varepsilon_0 r} \quad \text{and} \quad V_2 = \frac{q_2}{4\pi\varepsilon_0 b}$$

Then, $V = V_1 + V_2 = \frac{1}{4\pi\varepsilon_0}\left(\frac{q_1}{r} + \frac{q_2}{b}\right); \ a \leqslant r \leqslant b$

$$V = \frac{1}{4\pi\varepsilon_0}\left(\frac{-2q}{r} + \frac{3q}{2a}\right); \ a \leqslant r \leqslant b$$

For point R, as it lies outside of both shells,

$$V_1 = \frac{q_1}{4\pi\varepsilon_0 r} \quad \text{and} \quad V_2 = \frac{q_2}{4\pi\varepsilon_0 r}$$

$$\Rightarrow V = V_1 + V_2 = \frac{q_1 + q_2}{4\pi\varepsilon_0 r}; \ r \geqslant b$$

$$\Rightarrow V = \frac{-2q + 3q}{4\pi\varepsilon_0 r} = \frac{q}{4\pi\varepsilon_0 r}; \ r \geqslant b$$

Putting $r = 2a$, the maximum value of V is

$$V_{min} = \frac{q}{8\pi\varepsilon_0 a}; \ r \geqslant b$$

Putting $V = \frac{1}{4\pi\varepsilon_0}\left(\frac{-2q}{r} + \frac{3q}{2a}\right) = 0$, we have $r = 4a/3$.

We can note that the potential changes from negative to positive via zero at a point $r = 4a/3$ as shown in the figure. Ans.

Problem 5 Find the (a) induced charges at each face of plates, and (b) field distribution in the regions 1, 2 and 3 due to the parallel large conducting plates carrying charges Q_1 and Q_2.

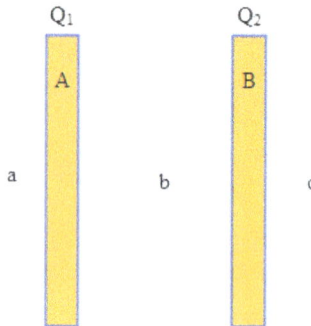

Solution

(a) Let $x =$ charge induced on the right-side face of plate A. Then equal and opposite bound charge $-x$ will appear on the left-side surface of plate B. Following the conservation of charge, the induced charges appearing at the outer surfaces of the plates are $(Q_1 - x)$ and $(Q_2 + x)$, respectively.

Let us now consider each face of the plates as infinite (very large) uniformly charged thin sheet each will contribute their fields at any point of the space. Let E_1, E_2, E_3 and E_4 be the fields due to the faces 1, 2, 3 and 4, respectively.

Consider a point P inside the conductor that holds the charge Q_1. The fields contributed by the charges are,

$$\overrightarrow{E_1} = \frac{\sigma_1}{2\varepsilon_0}\hat{i} = \left(\frac{Q_1 - x}{A}\right)\hat{i}/2\varepsilon_0 = \frac{Q_1 - x}{2\varepsilon_0 A}\hat{i}$$

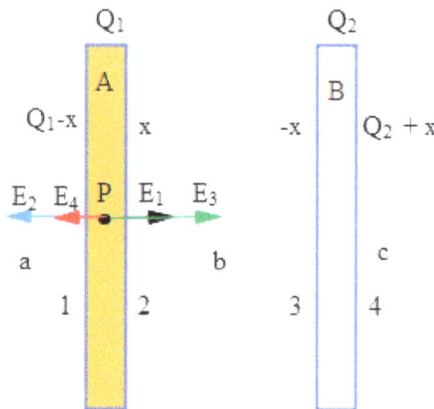

$$\vec{E_2} = \frac{\sigma_2}{2\varepsilon_0}(-\hat{i}) = -\frac{x\hat{i}}{2\varepsilon_0 A}$$

$$\vec{E_3} = \frac{\sigma_3}{2\varepsilon_0}\hat{i} = \frac{x\hat{i}}{2\varepsilon_0 A}\hat{i}$$

$$\vec{E_4} = \frac{\sigma_4}{2\varepsilon_0}(-\hat{i}) = -\frac{(Q_2 + x)}{2\varepsilon_0 A}\hat{i}$$

Then the net field at P is $\vec{E_P} = \vec{E_1} + \vec{E_2} + \vec{E_3} + \vec{E_4}$.

Substituting the values of the fields $\vec{E_1}$, $\vec{E_2}$, $\vec{E_3}$ and $\vec{E_4}$ and putting $\vec{E} = 0$ inside the conductor, we have

$$x = \frac{Q_1 - Q_2}{2} \text{ Ans.}$$

Then, the charges induced in the region (b) is $\left(\frac{Q_1 - Q_2}{2}\right)$ and $-\left(\frac{Q_1 - Q_2}{2}\right)$ due to the plate A and B respectively.

The charge induced at each outer surface of the plates is $\left(\frac{Q_1 + Q_2}{2}\right)$

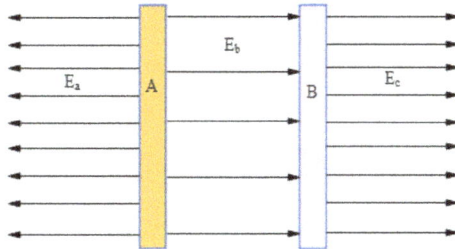

(b) The field at any point in region (a) is

$$\vec{E} = \vec{E_1} + \vec{E_2}$$
$$= \frac{Q_1(-\hat{i})}{2\varepsilon_0 A} + \frac{Q_2(-\hat{i})}{2\varepsilon_0 A} = -\frac{(Q_1 + Q_2)\hat{i}}{2\varepsilon_0 A} \text{ Ans.}$$

The field at any point in region (b) is

$$\vec{E} = \vec{E_1} + \vec{E_2} = \frac{Q_1\hat{i}}{2\varepsilon_0 A} + \frac{Q_2(-\hat{i})}{2\varepsilon_0 A} = -\frac{Q_1 - Q_2}{2\varepsilon_0 A}\hat{i} \text{ Ans.}$$

The field at any point in region (c) is

$$\vec{E} = \vec{E_1} + \vec{E_2} = \frac{Q_1}{2\varepsilon_0 A}\hat{i} + \frac{Q_2}{2\varepsilon_0 A}\hat{i} = \frac{Q_1 + Q_2}{2\varepsilon_0 A}\hat{i} \text{ Ans.}$$

N.B: Each space has uniform electric field if we neglect the fringing effects of the lines of forces.

Problem 6 *Charge distribution on a conductor*: Draw the electric field pattern and equipotential lines around a charged conductor.

Solution

As discussed earlier, $\overrightarrow{E} = \frac{\sigma}{\varepsilon_0}\hat{n}$ and $\sigma \propto \frac{1}{r}$; so, we have $E \propto \frac{1}{r}$.

This gives us three ideas about (i) E-pattern, (ii) distribution of charge σ, (iii) V-pattern.

The surface charge density σ is large where the radius of curvature r is small (at sharp point P); so, more charge disposition takes place near the pointed or sharp regions 1 and 2 than the blunt regions. This means that the equipotential lines are more closely spaced near the sharp points. At the region or point 3, the radius of curvature is also small (sharp point), but it is concave, the centre of curvature lies outside the conductor. So, individually, E- and V-lines shy away from each other.

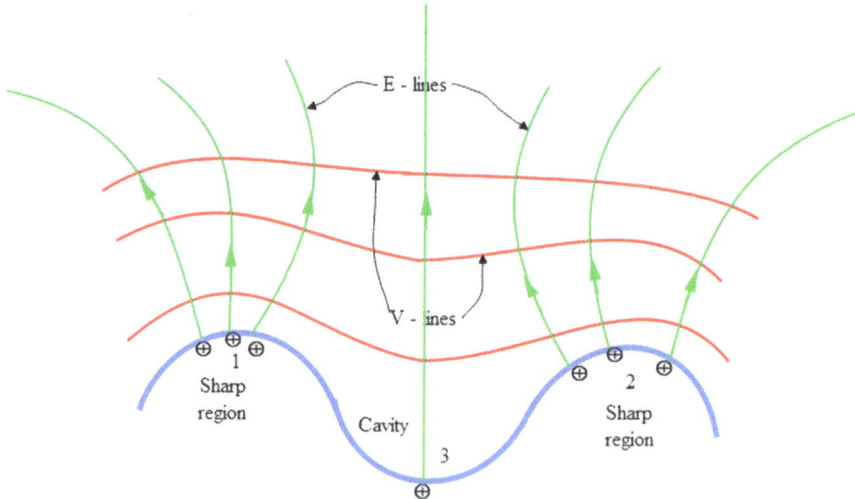

E - lines and V- lines always cross perpencular to each other;these lines are more crowded near the sharp points 1 and 2 and sighing away from the cavity 3;chrge density is maximum at the reigons 1 and 2 and no or feeble charge stays in a cavity

As we approach the conductor, the equipotential surfaces assume the shape of the surface of the conductor. Near the projection (pointed or sharp region) 1 and 2 equipotential lines are crowded ensuring more field strength $\overrightarrow{E}$ followed by crowding of more E-lines. In effect, the charge density σ is greatest at these points.

On the other hand, near the depression, (or cavity) 3, the densities of V (equipotential) and E-lines are lowest which signifies a minimum charge density in the cavity. The depressions behave as points in a cavity having a minimum or nearly zero field strength $\overrightarrow{E}$, minimum or nearly zero charge density σ and least crowding of V-lines because the minimum amount of charge stays in the cavity due to their mutual repulsion.

The density of charges with a given potential of a conductor is determined by the curvature of the surfaces. It grows with an increase in positive curvature (convexity) and diminishes with the negative curvature (concavity). Hence, only one (unique) way of charge distribution is possible in a conductor for a given electric field. This theorem is called uniqueness theorem. *This states that, for a given charge on a conductor, the charge distribution is uniquely defined; so, the electric field and potential in the region is uniquely determined.*

Problem 7 Find the electrostatic force of interaction between two halves of a spherical conductor of radius R carrying a charge Q.

Solution

Take an elementary area dA containing a charge dq on the elementary ring of radius r. The force acting on dq is

$$dF = \frac{\sigma^2}{2\varepsilon_0} dA \text{ (as derived earlier)}.$$

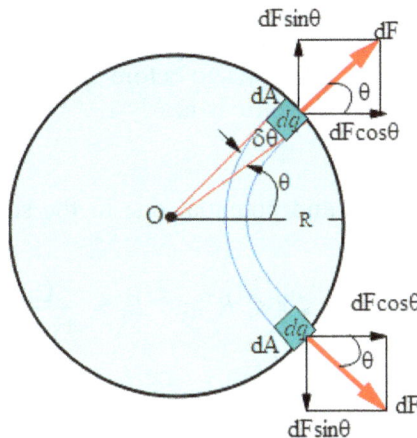

The net force on the ring is

$$F_{\text{ring}} = \int dF \sin\theta$$

$$= \int \frac{\sigma^2}{2\varepsilon_0} dA \sin\theta$$

$$= \frac{\sigma^2}{2\varepsilon_0} \sin\theta \int dA = \frac{\sigma^2}{2\varepsilon_0} \sin\theta (A_{\text{ring}}),$$

where, $A_{\text{ring}} = 2\pi r\, R\, d\theta = 2\pi (R\cos\theta) R\, d\theta = 2\pi R^2 \cos\theta\, d\theta$

$$\Rightarrow F_{\text{ring}} = \frac{\sigma^2 \pi R^2}{\varepsilon_0} \sin\theta \cos\theta\, d\theta \qquad (2.49)$$

The force acting on the hemisphere is the summation of force acting on all elementary rings.

$$\Rightarrow F_{\text{hemisphere}} = \int F_{\text{ring}} \tag{2.50}$$

Substituting F_{ring} from equation (2.49) in equation (2.50), we get

$$F_{\text{hemisphere}} = \frac{\sigma^2 \pi R^2}{\varepsilon_0} \int_{-\pi/2}^{+\pi/2} \sin\theta \cos\theta, \quad \text{where } \sigma = \frac{\theta}{4\pi R^2}$$

$$F_{\text{hemisphere}} = \frac{Q^2}{32\pi\varepsilon_0 R} \quad \text{Ans.}$$

Problem 8 A dielectric sphere of radius R has a charge Q distributed uniformly throughout its volume. The dielectric is surrounded by a neutral conducting shell of inner radius R and outer radius $2R$. (a) Find the electric field and potential due the system (sphere + shell) as the function of radial distance r. Draw the E–r graph. (b) Find the electric field and potential due the system (sphere + shell) as the function of radial distance r and draw the E–r graph after the conductor is earthed. (c) What is the electrostatic energy stored in the system before and after earthing. Assume that the relative permittivity of the dielectric is nearly 1.
   Solution

   (a) The electrostatic field and potential due to the solid sphere are given as follows:

$$\vec{E}_1 = \frac{Qr}{8\pi\varepsilon_o R^3}\hat{r}; \ 0 \leqslant r \leqslant R \ \text{ and } \ \vec{E}_1 = \frac{Q}{4\pi\varepsilon_o r^2}\hat{r}; r \geqslant R$$

$$V_1 = \frac{Q(3R^2 - r^2)}{8\pi\varepsilon_o R^3}; \ r \leqslant R \ \text{ and } \ V_1 = \frac{Q}{4\pi\varepsilon_o r}; r \geqslant R$$

   Due to the electrostatic induction, the charge induced at the inner surface of the conducting sphere is $-Q$, and $+Q$ charge will be induced at the outer surface of the conducting shell. The electrostatic field and potential due to the conducting shell are given as follows:

$$\vec{E}_2 = 0; \text{ in all regions.}$$

$$V_2 = \frac{-Q}{4\pi\varepsilon_o R} + \frac{+Q}{4\pi\varepsilon_o(2R)} = \frac{-Q}{8\pi\varepsilon_o R}; 0 \leqslant r \leqslant R$$

$$V_2 = \frac{-Q}{4\pi\varepsilon_o r} + \frac{Q}{4\pi\varepsilon_o(2R)} = \frac{-Q}{4\pi\varepsilon_o r} + \frac{Q}{8\pi\varepsilon_o R}; R \leqslant r \leqslant 2R$$

$$V_2 = \frac{-Q}{4\pi\varepsilon_o r} + \frac{Q}{4\pi\varepsilon_o r} = 0; r \geqslant 2R$$

By applying the principle of superposition, the electrostatic field and potential due to the entire charged system are given as follows:

$$\vec{E} = \vec{E}_1 + \vec{E}_2 = \frac{Qr}{8\pi\varepsilon_o R^3}\hat{r} + 0 = \frac{Qr}{8\pi\varepsilon_o R^3}\hat{r}; \ r < R$$

$E = 0$ (inside the conductor); $R < r < 2R$

$$\vec{E} = \vec{E}_1 + \vec{E}_2 = \frac{Q}{4\pi\varepsilon_o r^2}\hat{r} + \frac{-Q}{4\pi\varepsilon_o r^2}\hat{r} + \frac{Q}{4\pi\varepsilon_o r^2}\hat{r} = \frac{Q}{4\pi\varepsilon_o r^2}\hat{r}; \ r > 2R$$

$$V = V_1 + V_2 = \frac{Q(3R^2 - r^2)}{8\pi\varepsilon_o R^3} + \frac{-Q}{8\pi\varepsilon_o R} = \frac{Q(2R^2 - r^2)}{8\pi\varepsilon_o R^3}; r \leqslant R$$

$$V = V_1 + V_2 = \frac{Q}{8\pi\varepsilon_o R} + \frac{-Q}{8\pi\varepsilon_o R} = 0; r \geqslant R$$

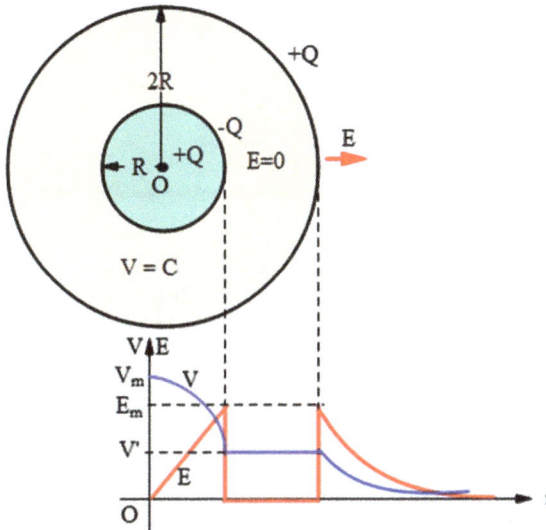

In the above graph, it is clear that the radially outward field increases from zero at $r = 0$ to Eo^2_{max} at $r = R$. Then it drops to zero abruptly inside the conductor. Outside the conducting sphere the net field drops from $E_m = Q/4\pi\varepsilon_o R^2$ to zero. Similarly, the potential decreases from its maximum value $V_m = Q/4\pi\varepsilon_o R$ to $Q/8\pi\varepsilon_o R$ at $r = R$. It remains constant inside the conductor. Outside the sphere the net potential drops from $Q/8\pi\varepsilon_o R$ to zero at infinity.

(b) After grounding, the free charge $+Q$ will flow from the outer surface of the conductor to earth. The electrostatic field and potential due to the solid sphere are given as follows:

$$\vec{E}_1 = \frac{Qr}{8\pi\varepsilon_o R^3}\hat{r}; 0 \leqslant r \leqslant R \ \text{ and } \ \vec{E}_1 = \frac{Q}{4\pi\varepsilon_o r^2}\hat{r}; r \geqslant R$$

$$V_1 = \frac{Q(3R^2 - r^2)}{8\pi\varepsilon_o R^3}; r \leqslant R \ \text{ and } \ V_1 = \frac{Q}{4\pi\varepsilon_o r}; r \geqslant R$$

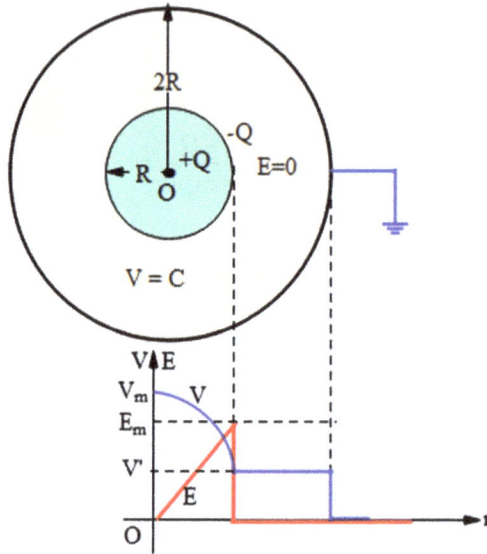

Due to the electrostatic induction, the charge induced at the inner surface of the conducting spherical shell is $-Q$ and no charge will be there on the outer surface of the shell due to the earthing. The electrostatic field due to the conducting shell are given as follows:

$$\vec{E}_2 = 0; r < 2R \ \text{ and } \ \vec{E}_2 = \frac{-Q}{4\pi\varepsilon_o r^2}\hat{r}; \ r > 2R$$

The electrostatic potentials due to the conducting shell are given as follows:

$$V_2 = \frac{-Q}{4\pi\varepsilon_o R} + 0 = \frac{-Q}{4\pi\varepsilon_o R}; 0 \leqslant r \leqslant R$$

$$V_2 = \frac{-Q}{4\pi\varepsilon_o r} + 0 = \frac{-Q}{4\pi\varepsilon_o r}; R \leqslant r \leqslant 2R$$

$$V_2 = \frac{-Q}{4\pi\varepsilon_o r} = 0; r \geqslant 2R$$

By applying the principle of superposition, the electrostatic field and potential due to the entire charged system are given as follows:

$$\vec{E} = \vec{E}_1 + \vec{E}_2 = \frac{Qr}{8\pi\varepsilon_o R^3}\hat{r} + 0 = \frac{Qr}{8\pi\varepsilon_o R^3}\hat{r};\ r < R$$

$E = 0$ (inside the conductor); $R < r < 2R$ and $E = 0$ (outside the conductor); $r > 2R$

$$V = V_1 + V_2 = \frac{Q(3R^2 - r^2)}{8\pi\varepsilon_o R^3} + \frac{-Q}{4\pi\varepsilon_o R} = \frac{Q(R^2 - r^2)}{8\pi\varepsilon_o R^3};\ r \leqslant R$$

$$V = V_1 + V_2 = \frac{Q}{8\pi\varepsilon_o R} + \frac{-Q}{8\pi\varepsilon_o R} = 0;\ r \geqslant R$$

In the above graph, it is clear that the radially outward field increases from zero at $r = 0$ to $E_m = Q/4\pi\varepsilon_o R^2$ at $r = R$. Then it drops to zero abruptly inside the conductor. Outside the conducting sphere the net field drops from $E_m = Q/4\pi\varepsilon_o R^2$ to zero. Similarly, the potential decreases from its maximum value $V_m = Q/4\pi\varepsilon_o R$ to $Q/8\pi\varepsilon_o R$ at $r = R$. It remains constant inside the conductor. Outside the sphere the net potential drops from $Q/8\pi\varepsilon_o R$ to zero at infinity.

(c) The self-energy stored due to the dielectric sphere of charge $+Q$ and radius R is

$$U_1 = \frac{3Q^2}{20\pi\varepsilon_o R}$$

The self-energy stored due to the spherical shell of charge $-Q$ and radius R is

$$U_2 = \frac{Q^2}{8\pi\varepsilon_o R}$$

The self-energy stored due to the spherical shell of charge $+Q$ and radius $2R$ is

$$U_3 = \frac{Q^2}{8\pi\varepsilon_o(2R)} = \frac{Q^2}{16\pi\varepsilon_o R}$$

The mutual energy between the dielectric sphere of charge $+Q$ of radius R and conducting spherical shell of charge $-Q$ and radius R is

$$U_1' = \frac{(-Q)(+Q)}{4\pi\varepsilon_o R} = -\frac{Q^2}{4\pi\varepsilon_o R}$$

The mutual energy between the dielectric sphere of charge $+Q$ and conducting spherical shell of charge $+Q$ and radius $2R$ is

$$U_2' = \frac{(+Q)(+Q)}{4\pi\varepsilon_o(2R)} = \frac{Q^2}{8\pi\varepsilon_o R}$$

The mutual energy between the shell of charge $-Q$ and radius R and shell of charge $+Q$ and radius $2R$ is

$$U_3' = \frac{(-Q)(+Q)}{4\pi\varepsilon_o(2R)} = \frac{-Q^2}{8\pi\varepsilon_o R}$$

The total electrostatic energy stored due to given system is

$$
\begin{aligned}
U &= U_1 + U_2 + U_3 + U_1' + U_2' + U_3' \\
&= \frac{3Q^2}{20\pi\varepsilon_o R} + \frac{Q^2}{8\pi\varepsilon_o R} + \frac{Q^2}{16\pi\varepsilon_o R} + \frac{-Q^2}{4\pi\varepsilon_o R} + \frac{Q^2}{8\pi\varepsilon_o R} + \frac{-Q^2}{8\pi\varepsilon_o R} \\
&= \frac{7Q^2}{80\pi\varepsilon_o R} \quad \text{Ans.}
\end{aligned}
$$

(d) As the conductor is earthed, the outer surface will have zero charge; so, we have two spheres of non-zero charges $+Q$, $-Q$ having radii R and R, respectively. Let us find the self- and mutual energy of each charged sphere.

The self-energy stored due to the dielectric sphere of charge $+Q$ and radius R is

$$U_1 = \frac{3Q^2}{20\pi\varepsilon_o R}$$

The self-energy stored due to the spherical shell of charge $-Q$ and radius R is

$$U_2 = \frac{Q^2}{8\pi\varepsilon_o R}$$

The mutual energy between the dielectric sphere of charge $+Q$ of radius R and conducting spherical shell of charge $-Q$ and radius R is

$$U_1' = \frac{(-Q)(+Q)}{4\pi\varepsilon_o R} = -\frac{Q^2}{4\pi\varepsilon_o R}$$

The total electrostatic energy stored due to the given system is

$$
\begin{aligned}
U &= U_1 + U_2 + U_1' \\
&= \frac{3Q^2}{20\pi\varepsilon_o R} + \frac{Q^2}{8\pi\varepsilon_o R} - \frac{Q^2}{4\pi\varepsilon_o R} \\
&= \frac{Q^2}{40\pi\varepsilon_o R} \quad \text{Ans.}
\end{aligned}
$$

Alternative method:

$$V = \frac{Q(R^2 - r^2)}{8\pi\varepsilon_o R^3}; r \leqslant R$$

Since $E = 0$ for the entire space for which $r > R$, no field energy is present in that region. So, the entire energy is stored inside the dielectric, which is given as

$$U = \frac{1}{2} \int_0^R V dq,$$

where $dq = 4\pi \rho r^2 dr$. Then, using all the last three expressions, we have

$$
\begin{aligned}
U &= \frac{Q}{2} \int_0^R \frac{(R^2 - r^2)}{8\pi\varepsilon_o R^3} \left(4\pi \rho r^2 dr\right) \\
&= \frac{Q\rho}{2\varepsilon_o R^3} \int_0^R (R^2 - r^2) r^2 dr
\end{aligned}
$$

$$\Rightarrow U = \frac{Q\rho}{2\varepsilon_o R^3} \left(R^2 \frac{R^3}{3} - \frac{R^5}{5}\right) = \frac{Q\rho R^2}{30\varepsilon_0}$$

Putting $\rho = \frac{3Q}{4\pi R^3}$ in the last expression, finally, we have

$$U = \frac{Q\rho R^2}{30\varepsilon_0} = \frac{QR^2}{30\varepsilon_0} \left(\frac{3Q}{4\pi R^3}\right) = \frac{Q^2}{40\pi\varepsilon_0 R} \quad \text{Ans.}$$

N.B: Both field and potential are zero outside the system. So, the total electrostatic energy is stored inside the system.

Problem 9 The parallel conducting plates placed very close to each other have charges Q_1 and Q_2, respectively. After closing the key in each case as shown in (i) figure (i), (ii) figure (ii), find the (a) charge flown through the key, (b) charge in each surface of the plates, (c) electric field distribution.

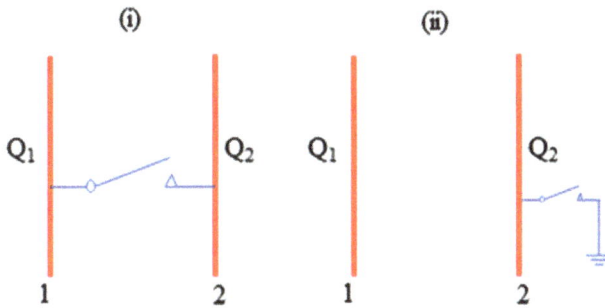

Solution

(i)

(a) In figure (ii), let $x =$ charge flown from plate 1 to plate 2. Then, plates 1 and 2 have charge $Q_1 - x$ and $Q_2 + x$, respectively. Their electric fields at point P are

$$\vec{E}_1 = \frac{(Q_1 - x)}{2\varepsilon_0 A}\hat{i}, \ \vec{E}_2 = \frac{(Q_2 - x)}{2\varepsilon_0 A}\hat{i}$$

respectively. Then, the net field at point P is

$$\vec{E} = \vec{E}_1 + \vec{E}_2$$

$$\Rightarrow \vec{E} = \frac{(Q_1 - x)}{2\varepsilon_0 A}\hat{i} - \frac{(Q_2 + x)}{2\varepsilon_0 A}\hat{i}$$

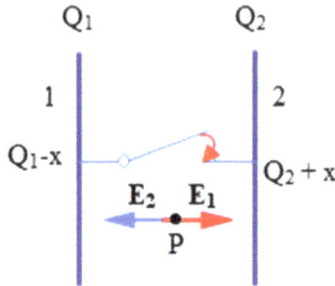

In steady state, the potential difference between the plates is zero after the key is closed; so, the electric field between the plates is zero. Then, we can write

$$\vec{E} = \frac{(Q_1 - x)}{2\varepsilon_0 A}\hat{i} - \frac{(Q_2 + x)}{2\varepsilon_0 A}\hat{i} = 0$$

$$x = \frac{Q_1 - Q_2}{2} \ \text{Ans.}$$

(b) This means that the facing surfaces of the plates do not have any charge and their outer surfaces will have equal charge given as

$$Q_1 - x = Q_1 - \frac{Q_1 - Q_2}{2} = \frac{Q_1 + Q_2}{2} \ \text{Ans.}$$

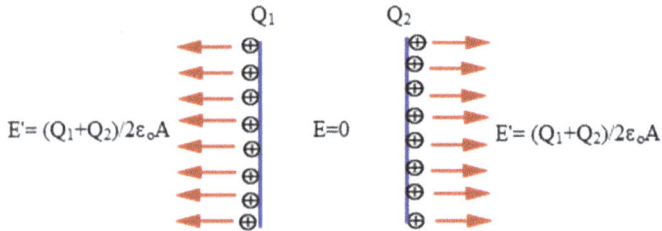

(c) So, the magnitudes of electric field at the outer surfaces will be

$$E = \frac{(Q_1 + Q_2)}{2\varepsilon_0 A}$$

The electric field points away from the outer surface of each plate assuming the charges as positive.

(ii)

(a) In figure (ii), let x = charge flown from plate 1 to plate 2. Then, plates 1 and 2 have charge $Q_1 - x$ and $Q_2 + x$, respectively. Their electric fields at the point P are

$$\vec{E}_1 = -\frac{(Q_1 - x)}{2\varepsilon_0 A}\hat{i}, \ \vec{E}_2 = -\frac{Q_2}{2\varepsilon_0 A}\hat{i}$$

respectively. Then, the net field at point P is

$$\vec{E} = \vec{E}_1 + \vec{E}_2$$

$$\Rightarrow \vec{E} = -\frac{(Q_1 - x)}{2\varepsilon_0 A}\hat{i} - \frac{Q_2}{2\varepsilon_0 A}\hat{i}$$

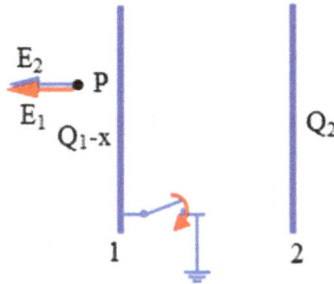

In steady state, the potential difference between the earth and plate 1 is zero after the key is closed; so, the electric field in between earth and plate 1 is zero. Then, we can write

$$\Rightarrow \vec{E} = -\frac{(Q_1 - x)}{2\varepsilon_0 A}\hat{i} - \frac{Q_2}{2\varepsilon_0 A}\hat{i}$$

$$x = Q_1 + Q_2 \ \text{Ans.}$$

(b) This means that the facing surfaces of the plates have equal charge given as

$$Q_1 - x = Q_1 - (Q_1 + Q_2) = -Q_2$$

So, their outer surfaces will have zero charge. Ans.

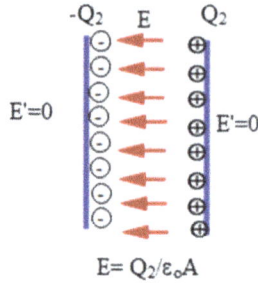

$$E = Q_2/\varepsilon_0 A$$

(c) So, the magnitudes of electric field at the outer surfaces will be zero and that in between the plates is

$$E = \frac{(Q_2 + Q_2)}{2\varepsilon_0 A} = \frac{Q_2}{\varepsilon_0 A} \text{ Ans.}$$

as shown in the following figure. The field points to the left assuming the charges as positive.

Problem 10 Three parallel conducting plates 1, 2 and 3 are placed very close to each other having charges Q_1, Q_2 and Q_3, respectively. The distance between the plates 1 and 2 is d_1 and that between 2 and 3 is d_2. If plate 1 is earthed, find the (a) charge flown to earth, (b) charge in each surface of the plates, (c) electric field distribution, (d) potentials of plates 2 and 3.

Solution

(a) Let x = charge flown from plate 1 to earth. Then, plate 1 will have charge $Q_1 - x$ and the charges on the other plates 2 and 3 remain unchanged. Their electric fields at point P are

$$\vec{E}_1 = -\frac{(Q_1 - x)}{2\varepsilon_0 A}\hat{i}, \ \vec{E}_2 = -\frac{Q_2}{2\varepsilon_0 A}\hat{i}, \ \vec{E}_2 = -\frac{Q_3}{2\varepsilon_0 A}\hat{i}$$

respectively. Then, the net field at point P is

$$\vec{E} = \vec{E}_1 + \vec{E}_2 + \vec{E}_3$$

$$\Rightarrow \vec{E} = -\frac{(Q_1 - x)}{2\varepsilon_0 A}\hat{i} - \frac{Q_2}{2\varepsilon_0 A}\hat{i} - \frac{Q_3}{2\varepsilon_0 A}\hat{i}$$

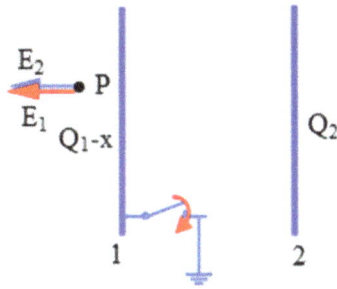

In steady state, the potential difference between the earth and plate 1 is zero after the earthing; so, the electric field in between earth and plate 1 is zero. Then, we can write

$$\vec{E} = -\frac{(Q_1 - x)}{2\varepsilon_0 A}\hat{i} - \frac{Q_2}{2\varepsilon_0 A}\hat{i} - \frac{Q_3}{2\varepsilon_0 A}\hat{i}$$

$$x = Q_1 + Q_2 + Q_3 \text{ Ans.}$$

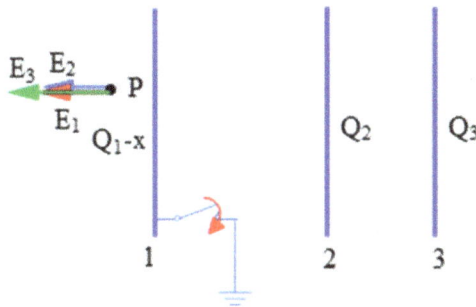

(b) The charge on plate 1 is given as

$$Q_1 - x = Q_1 - (Q_1 + Q_2 + Q_3) = -(Q_2 + Q_3)$$

So, their outer surfaces will have zero charge. Ans.

(c) With reference to the last problem, the magnitudes of electric field at the outermost surfaces (the outer surfaces of plates 1 and 3) will be zero; so, no charge will be induced on those faces. In other words, the entire charge of the outermost plates will reside on their inner surfaces, as shown in the figure. If the inner surface of plates 1 and 3 have charge $-(Q_2 + Q_3)$ and $+Q_3$, respectively, equal and opposite charges will appear on the opposite faces as shown in the following figure. Following the charge distributions, the electric field in regions A, B and C are given as follows.

$$\vec{E}_B = -\frac{(Q_2 + Q_3)}{\varepsilon_0 A}\hat{i} \quad \text{(to left)}$$

2-50

$$\vec{E}_C = -\frac{Q_3}{\varepsilon_0 A}\hat{i} \quad \text{(to left)}$$

The fields in each of the regions A and D are zero. The field points to the left assuming the charges as positive. Ans.

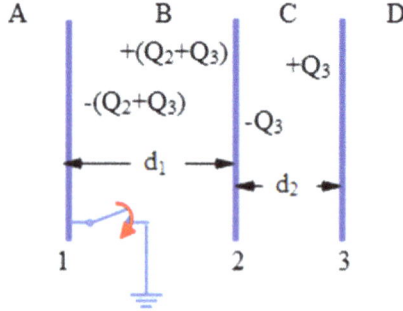

(d) The potential of plate 1 is zero because it is earthed. So, the potential of plate 2 is

$$V_2 = V_1 + E_B d_1 = 0 + \frac{(Q_2 + Q_3)}{\varepsilon_0 A}d_1$$

$$= \frac{(Q_2 + Q_3)}{\varepsilon_0 A}d_1 \ \text{Ans.}$$

The potential of plate 3 is

$$V_3 = V_1 + E_B d_1 + E_B d_1$$

$$= 0 + \frac{(Q_2 + Q_3)}{\varepsilon_0 A}d_1 + \frac{Q_2}{\varepsilon_0 A}d_2$$

$$= \frac{(Q_2 + Q_3)d_1 + Q_2 d_2}{\varepsilon_0 A} \ \text{Ans.}$$

Problem 11 In the last problem, if plate 2 is earthed instead of plate 1, after earthing the plate, find the (a) charge in each surface of the plates and charge flown through the key, (b) potentials of plates 2 and 3.
Solution

(a) By using the concept of inductive effect of an earthed conductor, we can directly attack this problem. Since Earth is a reservoir of a vast number of free electrons, it can supply or withdraw any number of electrons to or from a given conductor when earthed. As a result, the potential of the conductor will be equal to that of Earth which is taken as zero. So, the earthed conductor does not allow any charges to stay on the outermost surface of the extreme left and right conductors. Then, $+Q_1$ of plate 3 and $+Q_2$ of plate 1 will appear on their internal surfaces as shown in the figure. Thus, equal and opposite charges $-Q_1$ and $-Q_2$ will be induced on the left and right side (surface) of the plate 2. So, the total charge of plate 2 after earthing is equal to $-(Q_1 + Q_2)$. If x = charge flown from plate 2 to earth, the net charge remaining on the plate is equal to $Q_1 - x$. If we equate this charge with $-(Q_1 + Q_2)$, we will get $x = (Q_1 + Q_2 + Q_3)$.

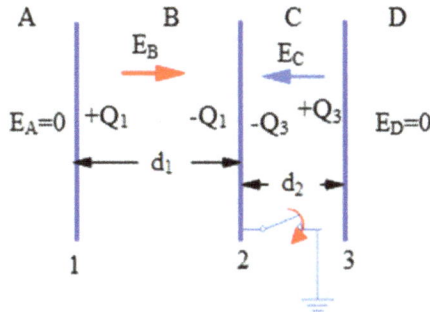

N.B: This means that if we earth any plate in a parallel grid of charged plates, the same charge will flow to earth, which is equal to the algebraic sum of charges on the system of the plates.

(b) Since the charge in the outermost surfaces is zero; electric field in regions A and D will be zero. Following the charge distributions, the electric field in regions B and C are given as follows.

$$\vec{E}_B = \frac{Q_1}{2\varepsilon_0 A}\hat{i} + \frac{Q_1}{2\varepsilon_0 A}\hat{i} = \frac{Q_1}{\varepsilon_0 A}\hat{i} \quad \text{(to right)}$$

$$\vec{E}_C = -\frac{Q_3}{\varepsilon_0 A}\hat{i} \quad \text{(to left) Ans.}$$

(c) The potential of plate 2 is zero because it is earthed. So, the potential of plate 1 is

$$V_1 = V_2 + E_B d_1 = 0 + \frac{Q_1}{\varepsilon_0 A}d_1 = +\frac{Q_1}{\varepsilon_0 A}d_1 \quad \text{Ans.}$$

The potential of plate 3 is

$$V_3 = V_2 + E_C d_2 = 0 + \frac{Q_3}{\varepsilon_0 A}d_2 = +\frac{Q_3}{\varepsilon_0 A}d_2 \quad \text{Ans.}$$

Problem 12 In the last problem, if the key is closed between plates 1 and 3, find the (a) charge flown through the key, (b) charge in each surface of the plates, (c) electric field in each region of space. Put $Q_1 = q$, $Q_2 = -2q$, $Q_3 = 4q$, $d_1 = d$, $d_2 = 2d$.

Solution

(a) If x = charge flown from plate 1 to plate 3, the net charge remaining on these plates will be equal to $Q_1 - x$ and $Q_2 + x$, respectively. The charge on plate 2 will still be Q_2. Following the charge distributions among the plates, the electric field in regions B and C are given as follows.

$$\vec{E}_B = \frac{Q_1 - x}{2\varepsilon_0 A}\hat{i} - \frac{Q_2}{2\varepsilon_0 A}\hat{i} - \frac{Q_3 + x}{2\varepsilon_0 A}\hat{i} = \frac{Q_1 - Q_2 - Q_3 - 2x}{2\varepsilon_0 A}\hat{i} \quad \text{(to right)} \quad (2.51)$$

$$\vec{E}_B = \frac{Q_1 - x}{2\varepsilon_0 A}\hat{i} + \frac{Q_2}{2\varepsilon_0 A}\hat{i} - \frac{Q_3 + x}{2\varepsilon_0 A}\hat{i} = \frac{Q_1 + Q_2 - Q_3 - 2x}{2\varepsilon_0 A}\hat{i} \quad \text{(to right)} \quad (2.52)$$

The potential of plate 3 is

$$V_3 = V_1 - E_B d_1 - E_C d_2$$

After the closing the key, the potential difference between plate 1 and 3 is zero. So, putting $V_1 = V_3$ in the last equations, we have

$$E_B d_1 + E_C d_2 = 0 \quad \text{Ans.} \quad (2.53)$$

Using the last three equations,

$$\Rightarrow \frac{(Q_1 - Q_2 - Q_3 - 2x)d_1}{\varepsilon_0 A} + \frac{(Q_1 + Q_2 - Q_3 - 2x)d_2}{\varepsilon_0 A} = 0$$

$$\Rightarrow (Q_1 - Q_2 - Q_3 - 2x)d_1 + (Q_1 + Q_2 - Q_3 - 2x)d_2 = 0$$

$$\Rightarrow (Q_1 - Q_2 - Q_3)d_1 + (Q_1 + Q_2 - Q_3)d_2 = 2x(d_1 + d_2)$$

$$\Rightarrow x = \frac{(Q_1 - Q_2 - Q_3)d_1 + (Q_1 + Q_2 - Q_3)d_2}{2(d_1 + d_2)} \quad \text{Ans.}$$

(b) Putting the given values, $Q_1 = q$, $Q_2 = -2q$, $Q_3 = 4q$, $d_1 = d$, $d_2 = 2d$ we have

$$x = \frac{\{(q) - (-2q) - (4q)\}d + \{(q) + (-2q) - (4q)\}2d}{2(d + 2d)}$$

$$\Rightarrow x = -11q/6 \quad \text{Ans.}$$

N.B: When the intermediate plates are interconnected, the charge flown depends upon the given charge in each conductor and the distance of separation between the conductors. So, in this case, generally, the charge on the outermost surfaces cannot be zero.

(c) Then, the charge on conductors 1 and 3 are given as
$Q_1' = Q_1 - x = q - (-11q/6) = 17q/6$ and
$Q_2' = Q_2 + x = 4q + (-11q/6) = 13q/6$, respectively. However, the charge on conductor 2 remains the same as $-2q$. Ans.

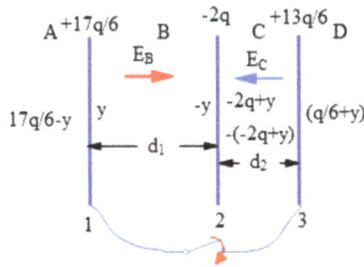

Let $y =$ charge induced on the inner surface of plate 1. Then, $-y$ charge will be there on the left surface of plate 2. The surplus charge $-2q + y$ will appear on the right side of plate 2. It induces an opposite charge of $-(-2q + y)$ on the right side of plate 3. The surplus charge of $13q/6 + (-2q + y) = q/6 + y$ will appear at the right side of plate 3. With this charge distribution, we can find the electric field inside conductor 1, say.

Since the equal and opposite charges cannot produce any electric field, the net field inside plate 1 is contributed by the charges on the outermost surfaces. Then, the net field inside plate 1 can be equated as zero (because the net field in a conductor is zero in the electrostatic case) which is given as

$$\vec{E} = \frac{17q/6 - y}{2\varepsilon_0 A}\hat{i} - \frac{q/6 + y}{2\varepsilon_0 A}\hat{i} = 0$$

This gives us $y = 8q/6$. Then, the charge at the outermost surfaces will be equal to

$$q/6 + y = q/6 + 8q/6 = 3q/2 \text{ Ans.}$$

(d) So, the electric fields in regions A and D can be given as

$$\vec{E}_A = -\frac{3q/2}{2\varepsilon_0 A}\hat{i} = -\frac{3q}{4\varepsilon_0 A}\hat{i}, \ \vec{E}_D = \frac{3q}{4\varepsilon_0 A}\hat{i}$$

respectively. In region B, electric field is

$$\vec{E}_B = \frac{y}{\varepsilon_0 A}\hat{i} = \frac{(8q/6)}{\varepsilon_0 A}\hat{i} = \frac{4q}{3\varepsilon_0 A}\hat{i}$$

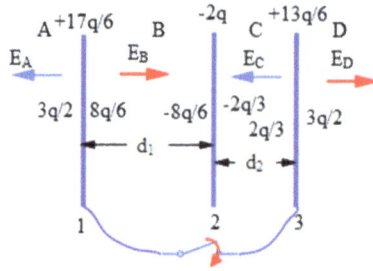

In region C, electric field is

$$\vec{E}_C = \frac{-2q + y}{\varepsilon_0 A}\hat{i} = \frac{(-2q + 8q/6)}{\varepsilon_0 A}\hat{i} = \frac{-4q}{3\varepsilon_0 A}\hat{i}$$

The electric field and charge distribution are given in the above figure.

Problem 13 Three conducting spherical shells 1, 2 and 3 are charged to Q_1, Q_2 and Q_3, respectively. A point charge $+q$ is placed inside shell 1 by hanging it by a resin thread that passes through the holes (not shown in the figure) made on the shells. (a) Find the charge distribution. (b) If the spherical shells (i) 1 (ii) 2 (iii) 3 are earthed separately, find the charge flown to earth. (c) If the spherical shells (i) 1 and 2 (ii) 2 and 3 (iii) 1 and 3 are connected separately by a conducting wire, find the charge flown to earth. Put $Q_1 = 2q$, $Q_2 = -3q$, $Q_3 = 4q$ and $a = b/2 = c/3$.

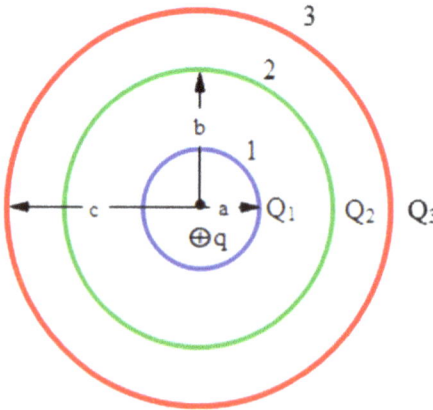

Solution

(a) Referring to example 3, the charge induced on the inner surface of a shell is equal to the total charge enclosed by the shell. Based upon this idea, the charge distribution is shown in the following figure.

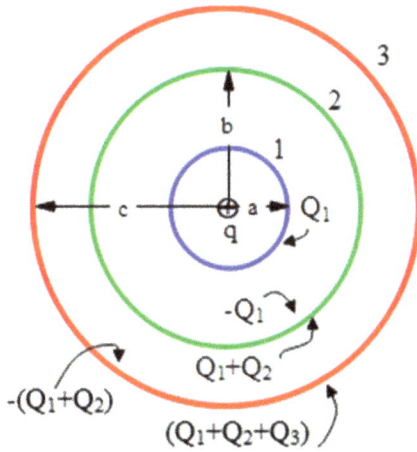

(b)

(i) If $x =$ charge flown from shell 1 to earth, the net charge remaining on the shell will be equal to $Q_1 - x$. The charge on the other shells will not change. Following the charge distributions among the shells, the potential of shell 1 is equal to the potentials contributed by all three shells and the point charge, which can be given as

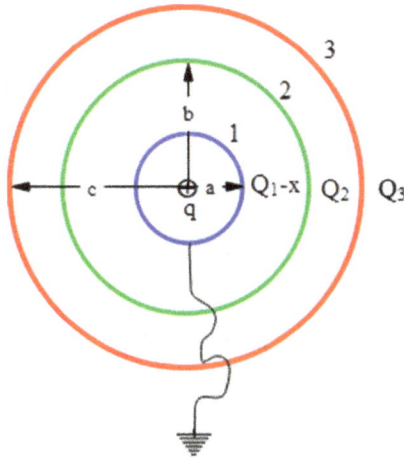

$$V_1 = V_q + V_{11} + V_{21} + V_{31}$$

$$= \frac{q}{4\pi\varepsilon_0 a} + \frac{Q_1 - x}{4\pi\varepsilon_0 a} + \frac{Q_2}{4\pi\varepsilon_0 b} + \frac{Q_3}{4\pi\varepsilon_0 c}$$

$$\Rightarrow V_1 = \frac{q}{4\pi\varepsilon_0 a} + \frac{Q_1 - x}{4\pi\varepsilon_0 a} + \frac{Q_2}{4\pi\varepsilon_0 b} + \frac{Q_3}{4\pi\varepsilon_0 c}$$

If we equate with the zero potential of earth, we have

$$V_1 = \frac{q}{4\pi\varepsilon_0 a} + \frac{Q_1 - x}{4\pi\varepsilon_0 a} + \frac{Q_2}{4\pi\varepsilon_0 b} + \frac{Q_3}{4\pi\varepsilon_0 c} = 0$$

$$\Rightarrow x = q + Q_1 + \frac{a}{b}Q_2 + \frac{a}{c}Q_3$$

$$\Rightarrow x = q + 2q + \frac{1}{2}(-3q) + \frac{1}{3}(4q)$$

$$\Rightarrow x = 17q/6 \ \ \text{Ans.}$$

(ii) If x = charge flown from shell 2 to earth, the net charge remaining on the shell will be equal to $Q_2 - x$. The charge on the other shells will not change. Following the charge distributions among the shells, the potential of shell 2 is equal to the potentials contributed by all three shells and the point charge, which can be given as

$$V_2 = V_q + V_{12} + V_{22} + V_{32}$$

$$= \frac{q}{4\pi\varepsilon_0 b} + \frac{Q_1}{4\pi\varepsilon_0 b} + \frac{Q_2 - x}{4\pi\varepsilon_0 b} + \frac{Q_3}{4\pi\varepsilon_0 c}$$

$$\Rightarrow V_2 = \frac{q}{4\pi\varepsilon_0 b} + \frac{Q_1}{4\pi\varepsilon_0 b} + \frac{Q_2 - x}{4\pi\varepsilon_0 b} + \frac{Q_3}{4\pi\varepsilon_0 c}$$

If we equate with the zero potential of earth, we have

$$\Rightarrow V_2 = \frac{q}{4\pi\varepsilon_0 b} + \frac{Q_1}{4\pi\varepsilon_0 b} + \frac{Q_2 - x}{4\pi\varepsilon_0 b} + \frac{Q_3}{4\pi\varepsilon_0 c} = 0$$

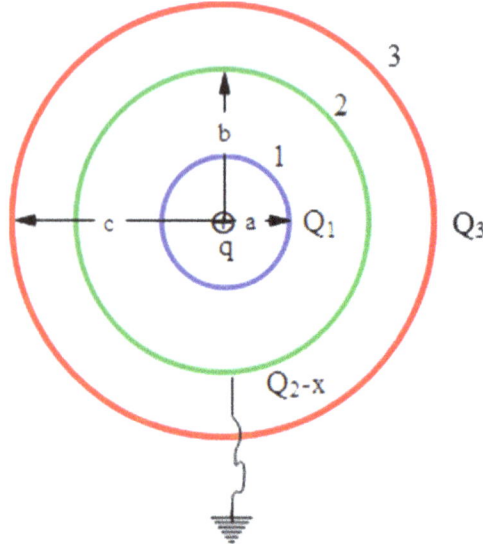

$$\Rightarrow x = q + Q_1 + Q_2 + \frac{b}{c}Q_3 = 0$$

$$\Rightarrow x = q + 2q + (-3q) + \frac{2}{3}(4q) = 8q/3 \ \text{Ans.}$$

(iii) If $x =$ charge flown from shell 3 to earth, the net charge remaining on the shell will be equal to $Q_3 - x$. The charge on the other shells will not change. Following the charge distributions among the shells, the potential of shell 3 is equal to the potentials contributed by all three shells and the point charge, which can be given as

$$V_3 = V_q + V_{13} + V_{23} + V_{33}$$

$$\Rightarrow V_3 = \frac{q}{4\pi\varepsilon_0 b} + \frac{Q_1}{4\pi\varepsilon_0 b} + \frac{Q_2}{4\pi\varepsilon_0 b} + \frac{Q_3 - x}{4\pi\varepsilon_0 c}$$

If we equate with the zero potential of earth, we have

$$\Rightarrow V_3 = \frac{q}{4\pi\varepsilon_0 b} + \frac{Q_1}{4\pi\varepsilon_0 b} + \frac{Q_2}{4\pi\varepsilon_0 b} + \frac{Q_3 - x}{4\pi\varepsilon_0 c} = 0$$

$$\Rightarrow x = q + Q_1 + Q_2 + Q_3$$

$$\Rightarrow x = q + 2q + (-3q) + 4q = 4q \ \text{Ans.}$$

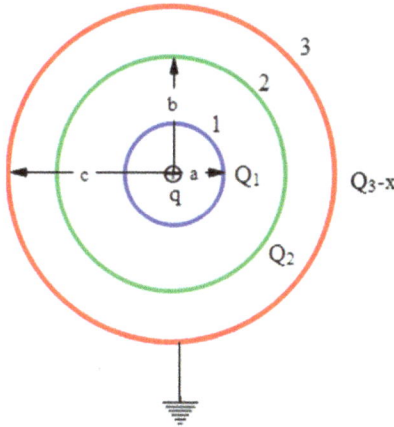

(c)

(i) If $x =$ charge flown from shell 1 to shell 2, the final charge remaining on the shells will be equal to $Q_1 - x$ and $Q_2 + x$, respectively. The charge on shell 3 will not change. Following the charge distributions among the shells, let us find the potentials of shells 1 and 2.

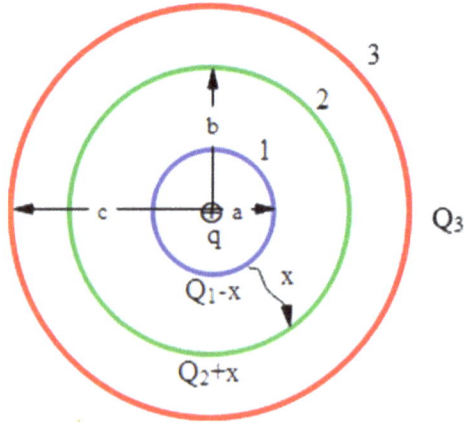

The potential of shell 2 is

$$V_1 = V_{q_1} + V_{11} + V_{21} + V_{31}$$

$$\Rightarrow V_1 = \frac{q}{4\pi\varepsilon_0 a} + \frac{Q_1 - x}{4\pi\varepsilon_0 a} + \frac{Q_2 + x}{4\pi\varepsilon_0 b} + \frac{Q_3}{4\pi\varepsilon_0 c} \qquad (2.54)$$

The potential of shell 2 is

$$V_2 = V_{q_2} + V_{12} + V_{22} + V_{32}$$

$$\Rightarrow V_2 = \frac{q}{4\pi\varepsilon_0 b} + \frac{Q_1 - x}{4\pi\varepsilon_0 b} + \frac{Q_2 + x}{4\pi\varepsilon_0 b} + \frac{Q_3}{4\pi\varepsilon_0 c} \qquad (2.55)$$

Since shells 1 and 2 are connected by a conducting wire, the potentials are equal. So, we can write

$$V_1 = V_2 \qquad (2.56)$$

Using the last three equations, we have

$$\frac{q}{4\pi\varepsilon_0 a} + \frac{Q_1 - x}{4\pi\varepsilon_0 a} + \frac{Q_2 + x}{4\pi\varepsilon_0 b} + \frac{Q_3}{4\pi\varepsilon_0 c} = \frac{q}{4\pi\varepsilon_0 b} + \frac{Q_1 - x}{4\pi\varepsilon_0 b} + \frac{Q_2 + x}{4\pi\varepsilon_0 b} + \frac{Q_3}{4\pi\varepsilon_0 c}$$

$$\Rightarrow \frac{q}{4\pi\varepsilon_0 a} + \frac{Q_1 - x}{4\pi\varepsilon_0 a} = \frac{q}{4\pi\varepsilon_0 b} + \frac{Q_1 - x}{4\pi\varepsilon_0 b}$$

$$\Rightarrow \left(\frac{1}{a} - \frac{1}{b}\right)Q_1 + \left(\frac{1}{a} - \frac{1}{b}\right)q = x\left(\frac{1}{a} - \frac{1}{b}\right)$$

$$\Rightarrow x = Q_1 + q = 2q + q = 3q \quad \text{Ans.}$$

(ii) If x = charge flown from shell 2 to shell 3, the final charge remaining on the shells will be equal to $Q_2 - x$ and $Q_3 + x$, respectively. The charge on shell 1 will not change. Following the charge distributions among the shells, let us find the potentials of shells 2 and 3.

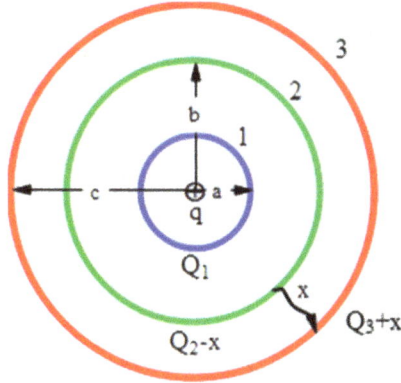

The potential of shell 2 is

$$V_2 = V_{q_2} + V_{12} + V_{22} + V_{32}$$

$$\Rightarrow V_2 = \frac{q}{4\pi\varepsilon_0 b} + \frac{Q_1}{4\pi\varepsilon_0 b} + \frac{Q_2 - x}{4\pi\varepsilon_0 b} + \frac{Q_3 + x}{4\pi\varepsilon_0 c}$$

The potential of shell 3 is

$$V_3 = V_{q_3} + V_{13} + V_{23} + V_{33}$$

$$\Rightarrow V_3 = \frac{q}{4\pi\varepsilon_0 c} + \frac{Q_1}{4\pi\varepsilon_0 c} + \frac{Q_2 - x}{4\pi\varepsilon_0 c} + \frac{Q_3 + x}{4\pi\varepsilon_0 c}$$

Equating the potentials of shells 2 and 3

$$\frac{q}{4\pi\varepsilon_0 b} + \frac{Q_1}{4\pi\varepsilon_0 b} + \frac{Q_2 - x}{4\pi\varepsilon_0 b} + \frac{Q_3 + x}{4\pi\varepsilon_0 c} = \frac{q}{4\pi\varepsilon_0 c} + \frac{Q_1}{4\pi\varepsilon_0 c} + \frac{Q_2 - x}{4\pi\varepsilon_0 c} + \frac{Q_3 + x}{4\pi\varepsilon_0 c}$$

$$\Rightarrow x = q + Q_1 + Q_2 = q + 2q - 3q = 0 \text{ Ans.}$$

(iii) If x = charge flown from shell 1 to shell 3, the final charge remaining on the shells will be equal to $Q_1 - x$ and $Q_3 + x$, respectively. The charge on shell 2 will not change. Following the charge distributions among the shells, let us find the potentials of shells 1 and 3.

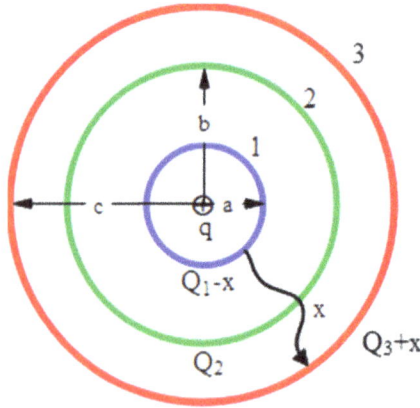

The potential of shell 1 is

$$V_1 = V_{q_1} + V_{11} + V_{21} + V_{31}$$

$$\Rightarrow V_1 = \frac{q}{4\pi\varepsilon_0 a} + \frac{Q_1 - x}{4\pi\varepsilon_0 a} + \frac{Q_2}{4\pi\varepsilon_0 b} + \frac{Q_3 + x}{4\pi\varepsilon_0 c}$$

The potential of shell 3 is

$$V_3 = V_{q_3} + V_{13} + V_{23} + V_{33}$$

$$\Rightarrow V_3 = \frac{q}{4\pi\varepsilon_0 c} + \frac{Q_1 - x}{4\pi\varepsilon_0 c} + \frac{Q_2}{4\pi\varepsilon_0 c} + \frac{Q_3 + x}{4\pi\varepsilon_0 c}$$

Equating the potentials of shells 1 and 3

$$\frac{q}{4\pi\varepsilon_0 a} + \frac{Q_1 - x}{4\pi\varepsilon_0 a} + \frac{Q_2}{4\pi\varepsilon_0 b} + \frac{Q_3 + x}{4\pi\varepsilon_0 c} = \frac{q}{4\pi\varepsilon_0 c} + \frac{Q_1 - x}{4\pi\varepsilon_0 c} + \frac{Q_2}{4\pi\varepsilon_0 c} + \frac{Q_3 + x}{4\pi\varepsilon_0 c}$$

$$\Rightarrow \left(\frac{1}{a} - \frac{1}{c}\right)x = \left(\frac{1}{a} - \frac{1}{c}\right)q + \left(\frac{1}{a} - \frac{1}{c}\right)Q_1 + \left(\frac{1}{b} - \frac{1}{c}\right)Q_2$$

$$\Rightarrow x = q + Q_1 + \frac{a(c - b)}{b(c - a)}Q_2$$

$$= q + 2q + (-3q)\frac{a(3a - 2a)}{2a(3a - a)} = 9q/4 \ \text{Ans.}$$

Problem 14 A soap bubble of charge Q expands form radius a to radius b under the electrostatic force due to its own surface charge and a central charge q. Find the work done by the electric force in expanding the bubble.

Solution

Initially, there are two objects; one is the central charge q of radius R, say; and the other is the soap bubble of charge Q and radius a. The energy of the system is the sum of their self- and mutual energies, given as

$$U_1 = U_{\text{self1}} + U_{\text{self2}} + U_{\text{mutual}}$$

$$\Rightarrow U_1 = \frac{q^2}{8\pi\varepsilon_0 R} + \frac{Q^2}{8\pi\varepsilon_0 a} + \frac{Qq}{4\pi\varepsilon_0 a} \tag{2.57}$$

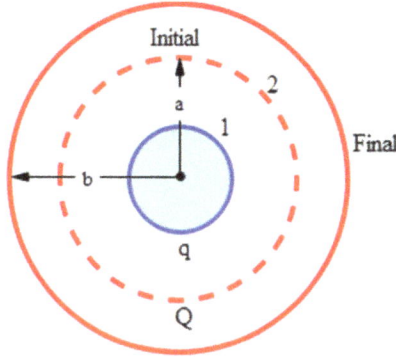

After expanding to the radius b, the new potential energy of the system is the sum of their self- and mutual energies, given as

$$U_2 = U_{\text{self1}} + U_{\text{self2}} + U_{\text{mutual}}$$

$$\Rightarrow U_1 = \frac{q^2}{8\pi\varepsilon_0 R} + \frac{Q^2}{8\pi\varepsilon_0 b} + \frac{Qq}{4\pi\varepsilon_0 b} \tag{2.58}$$

Using work–potential energy theorem, the work done by electrostatic forces is

$$W_{el} = -(U_2 - U_1) \tag{2.59}$$

Using the last three equations, we have

$$W_{el} = -\left\{\left(\frac{q^2}{8\pi\varepsilon_0 R} + \frac{Q^2}{8\pi\varepsilon_0 b} + \frac{Qq}{4\pi\varepsilon_0 b}\right) - \left(\frac{q^2}{8\pi\varepsilon_0 R} + \frac{Q^2}{8\pi\varepsilon_0 a} + \frac{Qq}{4\pi\varepsilon_0 a}\right)\right\}$$

$$W_{el} = \left(\frac{1}{a} - \frac{1}{b}\right)\left(\frac{Q^2}{8\pi\varepsilon_0} + \frac{Qq}{4\pi\varepsilon_0}\right)$$

$$\Rightarrow W_{el} = \left(\frac{1}{a} - \frac{1}{b}\right)\left(1 + \frac{2q}{Q}\right)\frac{Q^2}{8\pi\varepsilon_0} \quad \text{Ans.}$$

Alternative method:

By applying Gauss' law, the electric field at a radial distance r outside the soap bubble shell is given as

$$E 4\pi r^2 = \frac{Q + q}{\varepsilon_0}$$

Then, we have

$$E = \frac{Q + q}{4\pi\varepsilon_0 r^2} \tag{2.60}$$

The energy stored in the spherical region of radius from $r = a$ to $r = b$ is

$$U = \int_a^b \frac{\varepsilon_0}{2} E^2 dv = \frac{\varepsilon_0}{2} \int_a^b E^2 dv, \tag{2.61}$$

Where $dv = 4\pi r^2 dr$.

Using the last two equations,

$$U = \frac{\varepsilon_0}{2} \int_a^b \left(\frac{Q + q}{4\pi\varepsilon_0 r^2} \right)^2 4\pi r^2 dr$$

Evaluating the integral, we have

$$U = \frac{(Q + q)^2}{8\pi\varepsilon} \left(\frac{1}{a} - \frac{1}{b} \right)$$

We must deduct the self-energy of the charged sphere in the aforementioned region as

$$U' = \frac{q_0^2}{8\pi\varepsilon} \left(\frac{1}{a} - \frac{1}{b} \right)$$

Then the required energy (change in the self- plus interaction energy of the system $(q + q_0)$ is

$$U - U' = \frac{(Q + q)^2}{8\pi\varepsilon} \left(\frac{1}{a} - \frac{1}{b} \right) - \frac{q^2}{8\pi\varepsilon} \left(\frac{1}{a} - \frac{1}{b} \right)$$

$$\Rightarrow U - U' = \frac{(Q^2 + 2Qq)}{8\pi\varepsilon} \left(\frac{1}{a} - \frac{1}{b} \right)$$

$$= \frac{Q^2}{8\pi\varepsilon} \left(1 + \frac{2q}{Q} \right) \left(\frac{1}{a} - \frac{1}{b} \right)$$

So, $\quad W_{el} = -(U - U') = -\frac{Q^2}{8\pi\varepsilon} \left(1 + \frac{2q}{Q} \right) \left(\frac{1}{a} - \frac{1}{b} \right)$ Ans.

N.B: The work done can be +ve, zero and −ve depending upon the magnitude and nature of the charges.

Problem 15 A small conducting ball of radius a and charge q is hanging from a silk thread that passes through a small hole made on an insulated dielectric spherical shell. The ball is located at the center of a spherical shell of relative permittivity ε_r. The shell is kept on insulated ground. The inner and outer radii of the shell are b and c, respectively. (a) Find the field in each region and draw the graph. (b) Find the potential in each region and draw the graph.

   Solution

(a) By applying Gauss' law, the electric field inside a dielectric is given as

$$E = \frac{q_{enclosed}}{4\pi\varepsilon r^2} \qquad (2.62)$$

Using equation (2.62), the electric field in different regions are given as follows. By the property of conductors, inside the conducting ball $E = 0 \; r < a$; in between the solid sphere and dielectric shell, the electric field is

$$E = \frac{q}{4\pi\varepsilon_0 r^2}; \; a < r < b$$

Inside the dielectric the electric field decreases by a factor $1/\varepsilon_r$, which is given as

$$E = \frac{q}{4\pi\varepsilon_0 \varepsilon_r r^2}; \; b < r < c$$

Outside the dielectric shell, the electric field is given as

$$E = \frac{q}{4\pi\varepsilon_0 r^2}; \; r > c$$

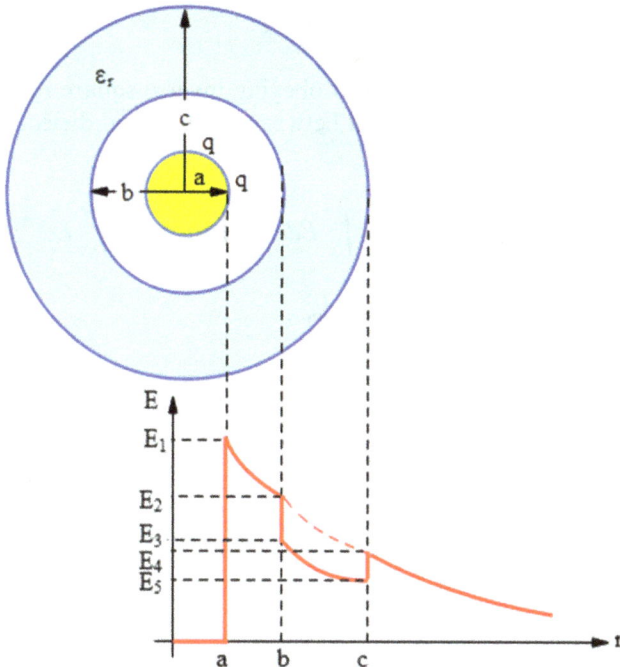

The values of electric field are shown in the graph; this tells us that the electric field is discontinuous over the boundary of the dielectric. First of all, electric field inside the conductor is zero. It increases from zero inside the conducting sphere to

$$E_1 = \frac{q}{4\pi\varepsilon_0 a^2}$$

just at the surface of the sphere. Then the field decreases to

$$E_2 = \frac{q}{4\pi\varepsilon_0 b^2}$$

just at the inner surface of the dielectric sphere. This field abruptly decreases from E_1 to

$$E_3 = \frac{q}{4\pi\varepsilon_0\varepsilon_r b^2}$$

Inside the dielectric sphere the field decreases from E_3 to

$$E_4 = \frac{q}{4\pi\varepsilon_0\varepsilon_r c^2}$$

Again, the electric field abruptly decreases from E_4 to

$$E_5 = \frac{q}{4\pi\varepsilon_0 c^2}$$

After that the field will decrease to zero obeying inverse square rule. Ans.

    (b) The potential at the point P in between the ball and dielectric shell is given as

$$V = -\int_\infty^r E dr = -\int_\infty^c E dr - \int_c^b E dr - \int_b^r E dr$$

Putting the value of electric fields,

$$V = -\int_\infty^c \frac{q}{4\pi\varepsilon_0 r^2} dr - \int_c^b \frac{q}{4\pi\varepsilon_0\varepsilon_r r^2} dr - \int_b^r \frac{q}{4\pi\varepsilon_0 r^2} dr$$

$$\Rightarrow V = -\frac{q}{4\pi\varepsilon_0}\left(\int_\infty^c \frac{dr}{r^2} + \frac{1}{\varepsilon_r}\int_c^b \frac{dr}{r^2} dr + \int_b^r \frac{dr}{r^2} dr\right)$$

$$\Rightarrow V = -\frac{q}{4\pi\varepsilon_0}\left\{-\frac{1}{c} - \frac{1}{\varepsilon_r}\left(\frac{1}{b} - \frac{1}{c}\right) - \left(\frac{1}{r} - \frac{1}{b}\right)\right\}$$

$$\Rightarrow V = \frac{q}{4\pi\varepsilon_0}\left\{-\frac{1}{r} + \left(1 - \frac{1}{\varepsilon_r}\right)\left(\frac{1}{b} - \frac{1}{c}\right)\right\}; a \leqslant r \leqslant b$$

So, putting $r = a$, the potential of the sphere(ball) is

$$V_1 = \frac{q}{4\pi\varepsilon_0}\left\{-\frac{1}{a} + \left(1 - \frac{1}{\varepsilon_r}\right)\left(\frac{1}{b} - \frac{1}{c}\right)\right\}; 0 \leqslant r \leqslant a$$

As the electric field inside the ball is zero, its potential is uniform inside it which is given by the last expression. Putting $r = b$, the potential of the sphere(ball) is

$$V_2 = \frac{q}{4\pi\varepsilon_0}\left\{-\frac{1}{b} + \left(1 - \frac{1}{\varepsilon_r}\right)\left(\frac{1}{b} - \frac{1}{c}\right)\right\}; r = b \text{ Ans.}$$

(c) The potential at the point P inside the dielectric is given as

$$V = -\int_\infty^r E dr = -\int_\infty^b E dr - \int_b^r E dr$$

Putting the value of electric fields,

$$V = -\int_\infty^c \frac{q}{4\pi\varepsilon_0 r^2} dr - \int_c^r \frac{q}{4\pi\varepsilon_0\varepsilon_r r^2} dr$$

$$\Rightarrow V = -\frac{q}{4\pi\varepsilon_0}\left(\int_\infty^c \frac{dr}{r^2} - \frac{1}{\varepsilon_r}\int_c^r \frac{dr}{r^2} dr\right)$$

$$\Rightarrow V = -\frac{q}{4\pi\varepsilon_0}\left\{-\frac{1}{c} - \frac{1}{\varepsilon_r}\left(\frac{1}{r} - \frac{1}{c}\right)\right\}$$

$$\Rightarrow V = \frac{q}{4\pi\varepsilon_0}\left\{\frac{1}{c} + \frac{1}{\varepsilon_r}\left(\frac{1}{r} - \frac{1}{c}\right)\right\}$$

$$\Rightarrow V = \frac{q}{4\pi\varepsilon_0}\left\{\frac{1}{c}\left(1 - \frac{1}{\varepsilon_r}\right) + \frac{1}{\varepsilon_r r}\right\}; b \leqslant r \leqslant c$$

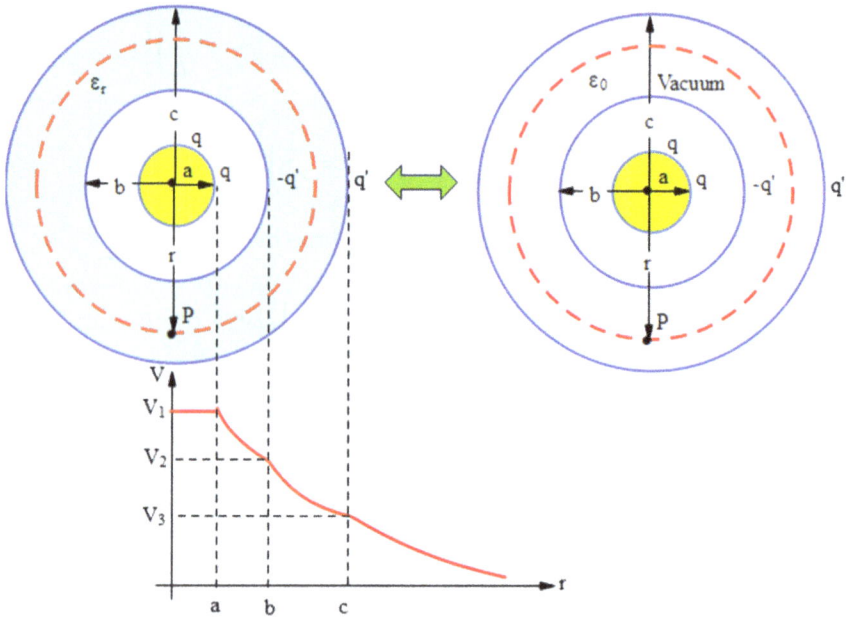

(d) The potential at the point P outside the dielectric shell is given as

$$V = -\int_{\infty}^{r} E dr = -\int_{\infty}^{r} E dr$$

Putting the value of electric fields,

$$V = -\int_{\infty}^{r} \frac{q}{4\pi\varepsilon_0 r^2} dr$$

$$\Rightarrow V = -\frac{q}{4\pi\varepsilon_0} \int_{\infty}^{r} \frac{dr}{r^2}$$

$$\Rightarrow V = -\frac{q}{4\pi\varepsilon_0}\left\{-\frac{1}{c} - \frac{1}{\varepsilon_r}\left(\frac{1}{r} - \frac{1}{c}\right)\right\}$$

$$\Rightarrow V = \frac{q}{4\pi\varepsilon_0 r}; r \geqslant c$$

So, putting $r = c$, we have

$$V_3 = \frac{q}{4\pi\varepsilon_0 c}; r = c.$$

Alternative method to find the potential in the dielectric:

Let q' and $-q'$ be the polarization charges appearing (induced) at the inner and outer surface of the dielectric, respectively, as shown in the last figure. These charges are distributed uniformly over the surfaces of the dielectric sphere. So, we can replace the dielectric sphere with two thin concentric spherical shells of radii b and c placed in a vacuum. Now we have a system of three charged bodies; the first one is the conducting sphere of charge q and radius a, the second one is the thin spherical shell of charge q' and radius b and the third is the is the thin spherical shell of charge q' and radius c as shown in the following figure. So, the potential of the system at point A is the sum of potential contributed by all the three charged bodies at P.

The potential of the first body (conducting sphere of radius a) at P is

$$V_{1st} = \frac{q}{4\pi\varepsilon_0 r}(\because r \geqslant a)$$

The potential of the second body (thin shell of radius c) at P is

$$V_{2nd} = \frac{-q'}{4\pi\varepsilon_0 r}(\because r \geqslant b)$$

The potential of the third body (thin shell of radius c) at P is

$$V_{2nd} = \frac{q'}{4\pi\varepsilon_0 c}(\because r \leqslant c)$$

Then, the total potential at P is

$$V = V_{1st} + V_{2nd} + V_{3rd} = \frac{q}{4\pi\varepsilon_0 r} + \frac{-q'}{4\pi\varepsilon_0 r} + \frac{+q'}{4\pi\varepsilon_0 c}$$

$$\Rightarrow V = \frac{1}{4\pi\varepsilon_0}\left(\frac{q}{r} - \frac{q'}{r} + \frac{q'}{c}\right)$$

$$\Rightarrow V = \frac{1}{4\pi\varepsilon_0}\left\{\frac{q}{r} - \left(\frac{1}{r} - \frac{1}{c}\right)q'\right\}$$

Putting $q' = \left(1 - \frac{1}{\varepsilon_r}\right)q$ as found earlier, we have

$$V = \frac{1}{4\pi\varepsilon_0}\left\{\frac{q}{r} - \left(\frac{1}{r} - \frac{1}{c}\right)\left(1 - \frac{1}{\varepsilon_r}\right)q\right\}$$

$$V = \frac{q}{4\pi\varepsilon_0}\left\{\frac{1}{\varepsilon_r r} - \frac{1}{c}\left(1 - \frac{1}{\varepsilon_r}\right)\right\}; b \leqslant r \leqslant c \text{ Ans.}$$

Using this method, we can also find the potential at other regions.

Problem 16 In the last problem, (a) what is the total electrostatic energy of the system (ball + shell)? (b) If ball is lifted slowly to a height h (so that the sphere does not interact with the shell) by pulling the string through the hole made on the shell, find the (i) final electrostatic energy stored in the system (ball − shell), (ii) change in electrostatic energy of the system. (c) What is the work done by the external (pulling) agent in doing so by (i) assuming m = mass of the ball, (ii) neglecting the mass of the ball, (iii) assuming that the shell is conducting and the ball is light?

Solution

(a) The total electrostatic energy of the system is given as

$$U = \int_0^\infty \frac{\varepsilon}{2} E^2 dv = \frac{\varepsilon_0}{2} \int_0^a E^2 dv + \frac{\varepsilon_0}{2} \int_a^b E^2 dv + \frac{\varepsilon}{2} \int_b^c E^2 dv + \frac{\varepsilon_0}{2} \int_c^\infty E^2 dv,$$

Where $dv = 4\pi r^2 dr$. Then, we can write

$$U = \frac{4\pi\varepsilon_0}{2}\left(\int_0^a E^2 dr + \int_a^b E^2 dr + \varepsilon_r \int_b^c E^2 dr + \int_c^\infty E^2 dr \right)$$

Since $E = 0$ (inside the conducting ball) $r < a$, the total energy stored in the system lies in between $r = a$ to $r = \infty$, which is given as

$$U = \frac{4\pi\varepsilon_0}{2}\left(\int_a^b E^2 dr + \varepsilon_r \int_b^c E^2 dr + \int_c^\infty E^2 dr \right)$$

Since the ball is conducting, the electric field inside it is zero. Putting the respective electric fields, we have energy stored in the system

$$U = \frac{4\pi\varepsilon_0}{2}\left\{ \int_a^b \left(\frac{q}{4\pi\varepsilon_0 r^2} \right)^2 r^2 dr + \frac{\varepsilon_r}{2} \int_b^c \left(\frac{q}{4\pi\varepsilon_0\varepsilon_r r^2} \right)^2 dr + \int_c^\infty \left(\frac{q}{4\pi\varepsilon_0 r^2} \right)^2 r^2 dr \right\}$$

$$\Rightarrow U = \frac{q^2}{8\pi\varepsilon_0}\left\{ \int_a^b \frac{dr}{r^2} + \frac{1}{\varepsilon_r}\int_b^c \frac{dr}{r^2} dr + \int_c^\infty \frac{dr}{r^2} \right\}$$

$$\Rightarrow U = \frac{q^2}{8\pi\varepsilon_0}\left\{ -\left(\frac{1}{b} - \frac{1}{a} \right) - \frac{1}{\varepsilon_r}\left(\frac{1}{c} - \frac{1}{b} \right) - \left(\frac{1}{\infty} - \frac{1}{c} \right) \right\}$$

$$\Rightarrow U = \frac{q^2}{8\pi\varepsilon_0}\left\{ \left(\frac{1}{a} - \frac{1}{b} \right) + \frac{1}{\varepsilon_r}\left(\frac{1}{b} - \frac{1}{c} \right) + \frac{1}{c} \right\} \quad \text{Ans.}$$

(b)

(i) After separating the ball from the shell, the polarization will be absent in the shell. As the shell is neutral, it does not store any electrical energy. However, the conducting sphere will possess a self-energy given as

$$U' = \frac{q^2}{8\pi\varepsilon_0 a}$$

(ii) So, the change in electrical potential energy of the system is

$$U' - U = \frac{q^2}{8\pi\varepsilon_0 a} - \frac{q^2}{8\pi\varepsilon_0}\left\{\left(\frac{1}{a} - \frac{1}{b}\right) + \frac{1}{\varepsilon_r}\left(\frac{1}{b} - \frac{1}{c}\right) + \frac{1}{c}\right\}$$

$$\Rightarrow U' - U = \frac{q^2}{8\pi\varepsilon_0}\left\{\left(-\frac{1}{b}\right) + \frac{1}{\varepsilon_r}\left(\frac{1}{b} - \frac{1}{c}\right) + \frac{1}{c}\right\}$$

$$\Rightarrow \Delta U_{el} = -\frac{q^2}{8\pi\varepsilon_0}\left(1 - \frac{1}{\varepsilon_r}\right)\left(\frac{1}{b} - \frac{1}{c}\right) \quad \text{Ans.}$$

(c)

(i) The work done by the external agent is equal to sum of change in electrical plus gravitational energy, given as

$$W_{ext} = \Delta U_{el} + \Delta U_{gr} = \frac{q^2}{8\pi\varepsilon_0}\left(1 - \frac{1}{\varepsilon_r}\right)\left(\frac{1}{b} - \frac{1}{c}\right) + mg(h - c)$$

(ii) If we ignore the mass of the sphere, we have

$$W_{ext} = \frac{q^2}{8\pi\varepsilon_0}\left(1 - \frac{1}{\varepsilon_r}\right)\left(\frac{1}{b} - \frac{1}{c}\right) \quad \text{Ans.}$$

(iii) If the spherical shell is also conducting, putting $\varepsilon_r = \infty$, we have

$$W_{ext} = \frac{q^2}{8\pi\varepsilon_0}\left(\frac{1}{b} - \frac{1}{c}\right) \quad \text{Ans.}$$

Problem 17 A dielectric solid ball of radius a, relative permittivity ε_r and charge q uniformly distributed in its volume is kept (by mechanical means) at the centre of a spherical conducting shell. The shell is kept on insulated ground and it has inner and outer radii of the shell b and c, respectively. A charge Q is given to the shell. (a) Find the field in each region and draw the graph, (b) find the potential in each region and draw the graph.

Solution

(a)

    (i) By referring to example 10, (chapter 1), the electric field inside the dielectric sphere is given as

$$E = \frac{qr}{4\pi\varepsilon_0 a^3}; 0 \leqslant r \leqslant a$$

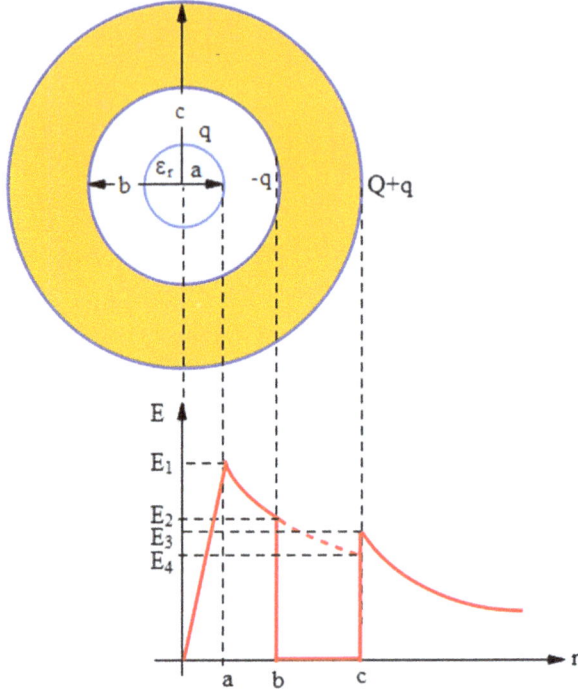

    (ii) By applying Gauss' law, the electric field between the sphere and shell is given as

$$E = \frac{q}{4\pi\varepsilon r^2}; \ a \leqslant r < b$$

    (iii) By the property of conductors, inside the conducting shell

$$E = 0; \ b < r < c$$

The charge induced uniformly on the inner surface of the shell is equal to $-q$; so, the surplus charge $Q + q$ will be uniformly distributed over the outer surface of the shell.

    (iv) By applying Gauss' law, the electric field outside the shell is given as

$$E = \frac{q_{\text{enclosed}}}{4\pi\varepsilon r^2},$$

Putting $q_{\text{enclosed}} = Q + q$ in the last expression, we have

$$E = \frac{Q + q}{4\pi\varepsilon_0 r^2}; \ r > c$$

Field distribution: The values of electric field are shown in the graph; this tells us that the electric field is continuous over the boundary of the dielectric. First of all, electric field inside the conductor is zero. Inside the conducting sphere, electric field increases linearly from zero to

$$E_1 = \frac{q}{4\pi\varepsilon_0 a^2}$$

just at the surface of the sphere. Then by obeying the inverse square rule, the field decreases to

$$E_2 = \frac{q}{4\pi\varepsilon_0 b^2}$$

just outside the inner surface of the conducting sphere. This field abruptly decreases from E_2 to zero inside the conducting shell. Again, it abruptly increases from zero to

$$E_4 = \frac{Q + q}{4\pi\varepsilon_0 c^2}$$

just outside the shell. If the shell had zero net charge, the field just outside the outer surface of the shell would have been

$$E_3 = \frac{q}{4\pi\varepsilon_0 c^2}$$

But due to the extra initial charge on the shell, the electric field is greater than this value. Then the field decreases to zero obeying the inverse square law.

(b)

(i) The potential at point P inside the conducting shell is given as

$$V = -\int_{\infty}^{r} E dr = -\int_{\infty}^{c} E dr - \int_{b}^{c} E dr - \int_{b}^{a} E dr - \int_{a}^{r} E dr$$

Putting the value of electric fields,

$$V = -\int_{\infty}^{c} \frac{Q + q}{4\pi\varepsilon_0 r^2} dr - \int_{b}^{c} (0) dr - \int_{b}^{a} \frac{q}{4\pi\varepsilon_0 r^2} dr - \int_{a}^{r} \frac{qr}{4\pi\varepsilon_0 a^3} dr$$

$$\Rightarrow V = -\frac{1}{4\pi\varepsilon_0}\left(\int_{\infty}^{c} \frac{Q + q}{r^2} dr + \int_{b}^{a} \frac{q}{r^2} dr + \int_{a}^{r} \frac{qr}{a^3} dr\right)$$

$$\Rightarrow V = -\frac{1}{4\pi\varepsilon_0}\left\{-\frac{Q+q}{c} - \left(\frac{q}{a} - \frac{q}{b}\right) + \frac{q(r^2 - a^2)}{2a^3}\right\}$$

$$\Rightarrow V = -\frac{1}{4\pi\varepsilon_0}\left\{-\frac{Q+q}{c} + \frac{q}{b} + \frac{q(r^2 - 3a^2)}{2a^3}\right\}$$

$$\Rightarrow V = \frac{1}{4\pi\varepsilon_0}\left\{\frac{Q+q}{c} - \frac{q}{b} + \frac{q(3a^2 - r^2)}{2a^3}\right\}; 0 \leqslant r \leqslant a$$

So, putting $r = 0$, the potential at the centre of the ball is

$$\Rightarrow V_1 = \frac{1}{4\pi\varepsilon_0}\left\{\frac{Q+q}{c} - \frac{q}{b} + \frac{3q}{2a}\right\}; r = 0$$

So, putting $r = a$, the potential at the surface of the ball is

$$\Rightarrow V_2 = \frac{1}{4\pi\varepsilon_0}\left\{\frac{Q+q}{c} - \frac{q}{b} + \frac{q}{2a}\right\}; r = a$$

(ii) The potential at point P in between the conducting ball and conducting shell is given as

$$V = -\int_\infty^r E dr = -\int_\infty^c E dr - \int_b^c E dr - \int_b^r E dr$$

Putting the value of electric fields,

$$V = -\int_\infty^c \frac{Q+q}{4\pi\varepsilon_0 r^2} dr - \int_b^c (0) dr - \int_b^r \frac{q}{4\pi\varepsilon_0 r^2} dr$$

$$\Rightarrow V = -\frac{1}{4\pi\varepsilon_0}\left(\int_\infty^c \frac{Q+q}{r^2} dr + \int_b^r \frac{q}{r^2} dr\right)$$

$$\Rightarrow V = -\frac{1}{4\pi\varepsilon_0}\left\{-\frac{Q+q}{c} - \left(\frac{q}{r} - \frac{q}{b}\right)\right\}$$

$$\Rightarrow V = \frac{1}{4\pi\varepsilon_0}\left(\frac{Q+q}{c} - \frac{q}{b} + \frac{q}{r}\right); a \leqslant r \leqslant b$$

So, putting $r = b$, the potential at the surface of the shell is

$$V = \frac{Q+q}{4\pi\varepsilon_0 c}; r = b$$

(iii) Since the electric field is zero inside the conducting shell, potential remains uniform inside it.

So, at $r = c$, the potential at the surface of the ball is

$$V = \frac{Q + q}{4\pi\varepsilon_0 c}; b \leqslant r \leqslant c$$

(iv) The potential at the point P outside the conducting shell is given as

$$V = -\int_{\infty}^{r} E dr.$$

Putting the value of electric fields,

$$V = -\int_{\infty}^{r} \frac{Q + q}{4\pi\varepsilon_0 r^2} dr$$

$$\Rightarrow V = -\frac{Q + q}{4\pi\varepsilon_0} \int_{\infty}^{r} \frac{dr}{r^2}$$

$$\Rightarrow V = \frac{Q + q}{4\pi\varepsilon_0 r}; r \geqslant c$$

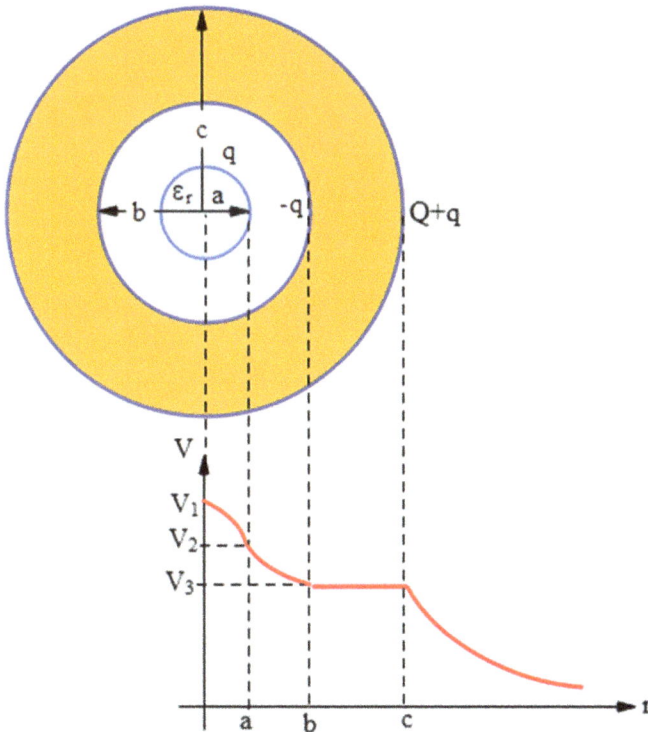

Potential distribution: The values of electric potential are shown in the last graph; this tells us that the electric potential is continuous over the boundary of the dielectric and conductor. First of all, potential inside the dielectric sphere decreases from V_1 to V_2 obeying parabolic law. Then it decreases from V_2 to V_3 hyperbolically ($1/r$ rule). Inside the conducting sphere, potential remains constant, given as

$$V_3 = \frac{Q+q}{4\pi\varepsilon_0 c}$$

Then it decreases hyperbolically obeying the *1/r* rule from $r = c$ to infinity.

Alternative method to find the potential in the dielectric:

The charge $-q$ and $-(Q + q)$ appear at the inner and outer surface of the conducting shell, respectively, as shown in the last figure. These charges are distributed uniformly over the surfaces of the dielectric sphere. So, we can replace the conducting shell with two thin concentric spherical shells of radii b and c placed in a vacuum having charges $-q$ and $-(Q + q)$, respectively. Now we have a system of three charged bodies; the first one is the dielectric sphere of charge q and radius a, the second one is the thin spherical shell of charge $-q$ and radius b and the third is the thin spherical shell of charge $Q + q$ and radius c, as shown in the following figure. So, the potential of the system at point A is the sum of potential contributed by all the three charged bodies at P.

(i) At any point P inside the dielectric sphere, the potential of the first body (dielectric sphere of radius a) is

$$V_{1st} = \frac{3a^2 - r^2}{8\pi\varepsilon_0 a^3}(\because r \leqslant a)$$

The potential of the second body (thin shell of radius b) at P is

$$V_{2nd} = \frac{-q}{4\pi\varepsilon_0 b}(\because r \leqslant b)$$

The potential of the third body (thin shell of radius c) at P is

$$V_{2nd} = \frac{Q+q}{4\pi\varepsilon_0 c}(\because r \leqslant c)$$

Then, the total potential at P is

$$V = V_{1st} + V_{2nd} + V_{3rd} = \frac{3a^2 - r^2}{8\pi\varepsilon_0 a^3} + \frac{-q}{4\pi\varepsilon_0 b} + \frac{Q+q}{4\pi\varepsilon_0 c}$$

$$\Rightarrow V = \frac{1}{4\pi\varepsilon_0}\left\{\left(\frac{3a^2 - r^2}{2a^3}\right)q - \frac{q}{b} + \frac{Q+q}{c}\right\} \quad \text{Ans.}$$

(ii) Between the dielectric sphere and conducting shell, **the** potential of the first body (dielectric sphere of radius a) at P is

$$V_{1st} = \frac{q}{4\pi\varepsilon_0 r}(\because r \geqslant a)$$

The potential of the second body (thin shell of radius b) at P is

$$V_{2nd} = \frac{-q}{4\pi\varepsilon_0 b}(\because r \leqslant b)$$

The potential of the third body (thin shell of radius c) at P is

$$V_{2nd} = \frac{Q+q}{4\pi\varepsilon_0 c}(\because r \leqslant c)$$

Then, the total potential at P is

$$V = V_{1st} + V_{2nd} + V_{3rd} = \frac{q}{4\pi\varepsilon_0 r} + \frac{-q}{4\pi\varepsilon_0 b} + \frac{Q+q}{4\pi\varepsilon_0 c}$$

$$\Rightarrow V = \frac{1}{4\pi\varepsilon_0}\left(\frac{q}{r} - \frac{q}{b} + \frac{Q+q}{c}\right) \text{ Ans.}$$

(iii) Inside the dielectric sphere, **the** potential of the first body (dielectric sphere of radius a) at P is

$$V_{1st} = \frac{q}{4\pi\varepsilon_0 r}(\because r \geqslant a)$$

The potential of the second body (thin shell of radius b) at P is

$$V_{2nd} = \frac{-q}{4\pi\varepsilon_0 r}(\because r \geqslant b)$$

The potential of the third body (thin shell of radius c) at P is

$$V_{2nd} = \frac{Q+q}{4\pi\varepsilon_0 c}(\because r \leqslant c)$$

Then, the total potential at P is

$$V = V_{1st} + V_{2nd} + V_{3rd} = \frac{q}{4\pi\varepsilon_0 r} + \frac{-q}{4\pi\varepsilon_0 r} + \frac{Q+q}{4\pi\varepsilon_0 c}$$

$$\Rightarrow V = \frac{1}{4\pi\varepsilon_0}\left(\frac{q}{r} - \frac{q}{r} + \frac{Q+q}{c}\right)$$

$$\Rightarrow V = \frac{Q+q}{4\pi\varepsilon_0 c} \text{ Ans.}$$

(iv) The potential of the first body (dielectric sphere of radius a) at P is

$$V_{1st} = \frac{q}{4\pi\varepsilon_0 r}(\because r \geqslant a)$$

The potential of the second body (thin shell of radius b) at P is

$$V_{2nd} = \frac{-q}{4\pi\varepsilon_0 r}(\because r \geqslant b)$$

The potential of the third body (thin shell of radius c) at P is

$$V_{2nd} = \frac{Q+q}{4\pi\varepsilon_0 r}(\because r \geqslant c)$$

Then, the total potential at P is

$$V = V_{1st} + V_{2nd} + V_{3rd} = \frac{q}{4\pi\varepsilon_0 r} + \frac{-q}{4\pi\varepsilon_0 r} + \frac{Q+q}{4\pi\varepsilon_0 c}$$

$$\Rightarrow V = \frac{1}{4\pi\varepsilon_0}\left(\frac{q}{r} - \frac{q}{r} + \frac{Q+q}{c}\right)$$

$$\Rightarrow V = \frac{Q+q}{4\pi\varepsilon_0 c} \text{ Ans.}$$

Problem 18 In the last problem, if we remove the ball out of the spherical shell to infinity through the hole made on it, what will be work done by the electric force?
Solution

(a) The total electrostatic energy of the system is given as

$$U = \int_0^\infty \frac{\varepsilon}{2}E^2 dv = \frac{\varepsilon}{2}\int_0^a E^2 dv + \frac{\varepsilon_0}{2}\int_a^b E^2 dv + \frac{\varepsilon_0}{2}\int_b^c E^2 dv + \frac{\varepsilon_0}{2}\int_c^\infty E^2 dv,$$

Where $dv = 4\pi r^2 dr$. Then, we can write

$$U = \frac{4\pi\varepsilon_0}{2}\left(\varepsilon_r \int_0^a E^2 dr + \int_a^b E^2 dr + \int_b^c E^2 dr + \int_c^\infty E^2 dr\right)$$

Since $E = 0$ (inside the conducting ball) $r < a$, the total energy stored in the system lies in between $r = a$ to $r = \infty$, which is given as

$$U = \frac{4\pi\varepsilon_0}{2}\left(\varepsilon_r \int_a^b E^2 dr + \int_b^c E^2 dr + \int_c^\infty E^2 dr\right)$$

Since the ball is conducting, the electric field inside it is zero. So, energy stored in it is zero.

After putting the values of the electric fields, we have

$$U = \frac{4\pi}{2}\left(\varepsilon \int_0^a \left(\frac{qr}{4\pi\varepsilon a^3} \right)^2 dr + \varepsilon_0 \int_a^b \left(\frac{q}{4\pi\varepsilon_0 r^2} \right)^2 dr + \varepsilon_0 \int_c^\infty \left(\frac{Q+q}{4\pi\varepsilon_0 r^2} \right)^2 dr \right)$$

$$\Rightarrow U = \frac{q^2}{40\pi\varepsilon_0\varepsilon_r a} + \frac{q^2}{8\pi\varepsilon_0}\left(\frac{1}{a} - \frac{1}{b} \right) + \frac{(Q+q)^2}{8\pi\varepsilon_0 c} \qquad (2.63)$$

After shifting the ball to infinity, we have two isolated charged bodies, namely the dielectric ball of charge q and radius a and the other is the conducting shell of charge Q, which entirely stays on the outer surface of the shell with uniform surface distribution.

Let us now find the final energy of the system (shell +sphere).

The energy due to the dielectric is

$$U = \int_0^a \frac{\varepsilon}{2} E^2 dv + \int_a^\infty \frac{\varepsilon_0}{2} E^2 dv$$

$$U_1 = \frac{4\pi}{2}\left\{ \varepsilon \int_0^a \left(\frac{qr}{4\pi\varepsilon a^3} \right)^2 dr + \varepsilon_0 \int_a^\infty \left(\frac{q}{4\pi\varepsilon_0 r^2} \right)^2 dr \right\}$$

$$\Rightarrow U_1 = \frac{q^2}{40\pi\varepsilon_0\varepsilon_r a} + \frac{q^2}{8\pi\varepsilon_0 a}$$

The energy of the charged shell is

$$U_2 = \int_c^\infty \frac{\varepsilon_0}{2} E^2 dv$$

$$U_2 = \frac{4\pi}{2}\left\{ \varepsilon_0 \int_c^\infty \left(\frac{Q}{4\pi\varepsilon_0 r^2} \right)^2 dr \right\}$$

$$\Rightarrow U_2 = \frac{Q^2}{8\pi\varepsilon_0 c}$$

Then the total energy of the system is

$$U' = U_1 + U_2 = \frac{q^2}{40\pi\varepsilon_0\varepsilon_r a} + \frac{q^2}{8\pi\varepsilon_0 a} + \frac{Q^2}{8\pi\varepsilon_0 c} \qquad (2.64)$$

Now, the difference in the energy before and after removing the dielectric sphere is

$$U' - U = \left(\frac{q^2}{40\pi\varepsilon_0\varepsilon_r a} + \frac{q^2}{8\pi\varepsilon_0 a} + \frac{Q^2}{8\pi\varepsilon_0 c} \right)$$
$$- \left\{ \frac{q^2}{40\pi\varepsilon_0\varepsilon_r a} + \frac{q^2}{8\pi\varepsilon_0}\left(\frac{1}{a} - \frac{1}{b}\right) + \frac{(Q+q)^2}{8\pi\varepsilon_0 c} \right\}$$

$$\Rightarrow \Delta U_{el} = \left\{ \frac{Q^2}{8\pi\varepsilon_0 c} + \frac{q^2}{8\pi\varepsilon_0 b} - \frac{(Q+q)^2}{8\pi\varepsilon_0 c} \right\}$$

$$\Rightarrow \Delta U_{el} = \left\{ \frac{q^2}{8\pi\varepsilon_0}\left(\frac{1}{b} - \frac{1}{c}\right) - \frac{Qq}{4\pi\varepsilon_0 c} \right\} \quad \text{Ans.}$$

N.B:

1. When the conducting shell is uncharged, putting $Q = 0$, we have our previous result given as

$$\Delta U_{el} = \frac{q^2}{8\pi\varepsilon_0}\left(\frac{1}{b} - \frac{1}{c}\right)$$

2. When the conducting shell is very thin, putting $b = c$, we have

$$\Delta U_{el} = \frac{q^2}{8\pi\varepsilon_0}\left(\frac{1}{b} - \frac{1}{c}\right)$$

Problem 19 A point charge q is slowly brought from infinity to the center of a shell relative through a diametrical chute made inside the shell. The inner and outer radii of the shell are b and c, respectively. If the shell is (a) dielectric and neutral having relative permittivity ε_r, (b) conducting and having a charge q, find the external work done.

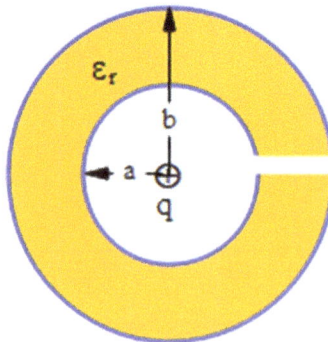

Solution

(a) We do not know whether the charged ball is a dielectric or conductor. Although it is a point charge, for mathematical simplicity, let us assume that it has a small but finite size possessing certain self-energy U_0, say. At infinity, the ball has only self-energy. Since the dielectric shell has zero net charge, it will not possess electrical energy. So, the total energy of the system after separation is equal to U_0 the (self-energy of the ball only) which is given as.

$$U' = \frac{q^2}{8\pi\varepsilon_0 a} = U_0$$

When the point charge is brought from infinity, referring to problem 16, the total energy of the system is

$$U = \frac{q^2}{8\pi\varepsilon_0}\left\{\left(\frac{1}{a} - \frac{1}{b}\right) + \frac{1}{\varepsilon_r}\left(\frac{1}{b} - \frac{1}{c}\right) + \frac{1}{c}\right\}$$

Then, the difference in energy is

$$\Delta U_{el} = U - U' = \frac{q^2}{8\pi\varepsilon_0}\left(1 - \frac{1}{\varepsilon_r}\right)\left(\frac{1}{b} - \frac{1}{c}\right) \text{ Ans.}$$

The energy increases and positive work will be done in bringing the charge from infinity to the center of the shell. If the shell is thin, no change of energy will take place.

(b) Initially, the total energy is the sum of their self-energy because the interaction energy is zero as they were electrically isolated. Since there is no charge staying inside the cavity (shell), no charge will be induced at the inner surface of the shell; so, the entire charge Q will reside on the outer surface of the shell with a uniform distribution. So, the self-energy of the shell is

$$U_{s1} = \frac{Q^2}{8\pi\varepsilon_0 c}$$

The self-energy of the point charge (small ball) is

$$U_{s2} = \frac{q^2}{8\pi\varepsilon_0 a}$$

Then, the initial energy of the system is

$$U' = U_{1s} + U_{2s} = \frac{Q^2}{8\pi\varepsilon_0 c} + \frac{q^2}{8\pi\varepsilon_0 a} \qquad (2.65)$$

After placing the ball (point charge $+q$ at the center of the shell), the charge $-q$ will appear uniformly at the inner surface of the shell and the surplus charge $Q + q$ will be uniformly distributed over the outer surface of the shell.

The self-energy of the small ball is

$$U_1 = \frac{q^2}{8\pi\varepsilon_0 a}$$

The self-energy of the inner surface of the conducting shell is

$$U_1 = \frac{q^2}{8\pi\varepsilon_0 b}$$

The self-energy of the outer surface of the conducting shell is

$$U_1 = \frac{(Q + q)^2}{8\pi\varepsilon_0 c}$$

The interaction (mutual) energy between the ball and inner surface of the shell is

$$U_{m1} = \frac{(q)(-q)}{4\pi\varepsilon_0 b} = -\frac{q^2}{4\pi\varepsilon_0 b}$$

The interaction (mutual) energy between the ball and outer surface of the shell is

$$U_{m2} = \frac{(q)(Q + q)}{4\pi\varepsilon_0 a}$$

The interaction (mutual) energy between the inner and outer surface of the shell is

$$U_{m3} = \frac{(-q)(Q + q)}{4\pi\varepsilon_0 a}$$

By summing up self-energy and interaction energy, the total energy is

$$U = \frac{q^2}{8\pi\varepsilon_0}\left(\frac{1}{a} - \frac{1}{b}\right) + \frac{(Q + q)^2}{8\pi\varepsilon_0 c} \tag{2.66}$$

From equations (i) and (ii), the difference in electrical energy is

$$\Delta U_{el} = U - U' = \left\{\frac{q^2}{8\pi\varepsilon_0}\left(\frac{1}{b} - \frac{1}{c}\right) - \frac{Qq}{4\pi\varepsilon_0 c}\right\} \quad \text{Ans.}$$

N.B: In this problem, for the sake of simplicity we have assumed the point charge as a small conducting sphere of radius a possessing a self-energy U_0. However, the answer is independent of the size of the point charge. We have to do it like this because a 'theoretical point charge' will have 'infinite-energy', which is physically impossible; so when we subtract initial from final energy, it will come as 'infinite minus infinite' which is an indeterminant form. So, to avoid these two mathematical fallacies, we can assume a point charge as a small (tiny) sphere.

Problem 20 A point charge q is placed at a distance a from the center of a conducting spherical shell of charge Q. The inner and outer radii of the shell are b and c, respectively. (a) If another point charge q' is kept outside the shell at a radial distance r, as shown in the figure, find the (i) charge distribution of the shell, (ii) potential at the centre of the shell, (iii) value of q' such that the potential at the center will be zero. (b) If the outer point charge is absent, find the (i) potential of the conductor (shell), (ii) potential at the centre of the conductor, (iii) difference between these two potentials. Interpret the result physically.

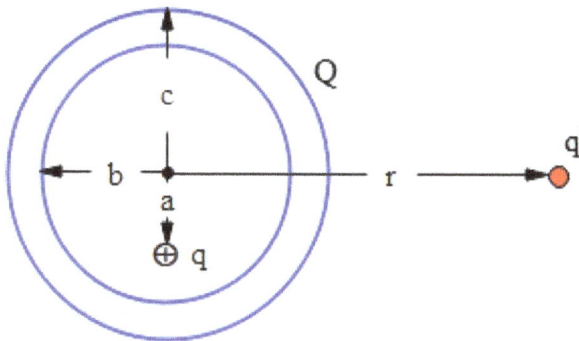

Solution

(a)

(i) Referring to example 3, the charge induced on the inner surface of the shell is equal to $-q$ which will be nonuniformly distributed over the inner surface of the shell. It is because the point charge is not placed at the center violating the geometrical symmetry. The field lines emanating from the point charge $+q$ and terminating on the distributed negative charges on the inner surface of the shell, will be always meeting the surface normally obeying the conductor property. On the other hand, the surplus charge $Q + q$ will be distributed nonuniformly over the outer surface of the shell. As we discussed in examples 2 and 3, this is a free charge. So, the field distribution will be radially outward, possessing spherical symmetry outside the shell.

(ii) The potential due to the point charge at the centre C is

$$V_1 = \frac{q}{4\pi\varepsilon_0 a} \tag{2.67}$$

Since each elementary charge dq is at same distance b from the centre C, the integration of its potential gives us the net potential V_2 due to the inner surface charge.

$$V_2 = \int_0^{-q} \frac{dq}{4\pi\varepsilon_0 b} = -\frac{q}{4\pi\varepsilon_0 b} \tag{2.68}$$

Following the same logic, the net potential V_3 due to the outer surface charge is

$$V_3 = \int_0^{Q+q} \frac{dq}{4\pi\varepsilon_0 c} = \frac{Q+q}{4\pi\varepsilon_0 c} \tag{2.69}$$

The potential contributed by the point charge q' is

$$V_4 = \frac{q'}{4\pi\varepsilon_0 r} \tag{2.70}$$

The potential at the center is given by

$$V = V_1 + V_2 + V_3 \tag{2.71}$$

From the equations (2.67)–(2.71), we have

$$V = \frac{q}{4\pi\varepsilon_0 a} - \frac{q}{4\pi\varepsilon_0 b} + \frac{Q+q}{4\pi\varepsilon_0 c} + \frac{q'}{4\pi\varepsilon_0 r}$$

$$\Rightarrow V = \frac{1}{4\pi\varepsilon_0}\left(\frac{q}{a} - \frac{q}{b} + \frac{Q+q}{c} + \frac{q'}{r}\right)$$

(iii) Putting, $V = 0$, we have

$$V = \frac{1}{4\pi\varepsilon_0}\left(\frac{q}{a} - \frac{q}{b} + \frac{Q+q}{c} + \frac{q'}{r}\right) = 0$$

$$\Rightarrow \left(\frac{q}{a} - \frac{q}{b} + \frac{Q+q}{c} + \frac{q'}{r}\right) = 0$$

$$\Rightarrow q' = -\left\{\left(\frac{1}{a} - \frac{1}{b} + \frac{1}{c}\right)qr + \frac{rQ}{c}\right\} \quad \text{Ans.}$$

(b)

(i) In the absence of outer point charge, the charge distribution at the outer surface of the shell will be uniform. Then, the electric field due to the conductor is contributed due to the outer charge because the field due to the inner charge (point charge q plus the induced charge $-q$) will be zero for any outside point. Thus, the internal electric field is screened by the conductor. Then, the potential of the conductor at a radial distance r can be given as follows:

$$V(r) = \frac{Q+q}{4\pi\varepsilon_0 r}; r \geqslant R$$

This tells us that the potential of the conductor is

$$V' = \frac{Q+q}{4\pi\varepsilon_0 c}$$

(ii) In the absence of outer point charge, the potential at the center of the conductor is

$$V = \frac{1}{4\pi\varepsilon_0}\left(\frac{q}{a} - \frac{q}{b} + \frac{Q+q}{c}\right)$$

(iii) So, the potential difference between the conductor and its center is

$$V - V' = \frac{1}{4\pi\varepsilon_0}\left(\frac{q}{a} - \frac{q}{b} + \frac{Q+q}{c}\right) - \frac{Q+q}{4\pi\varepsilon_0 c}$$

$$V - V' = \frac{q}{4\pi\varepsilon_0}\left(\frac{1}{a} - \frac{1}{b}\right) \text{ Ans.}$$

N.B:

1. Potential of the conductor and potential at any point inside the cavity of the conductor need not be equal.
2. The potential remains uniform inside the conductor (but not necessarily inside the cavity made in the conductor).

$$E(r) = \frac{Q+q}{4\pi\varepsilon_0 r^2}; r > R$$

Problem 21 There are two conducting thin spherical shells. The inner shell has charge Q and the outer shell has charge q. The inner and outer radii of the shell are a and b, respectively. (a) Find the field and potential functions and draw their graphs.

(b) Find the ratio of the charges so the potential of (i) inner and (ii) outer shell will be zero, and draw the field and potential.

Solution

(a) *Electric field equations:* Since no charge is present inside the inner shell, the charge q will entirely lie on the outer surface of the inner sphere. The charge induced on the inner surface of the outer shell is equal to $-q$ and the surplus charge $Q + q$ will be distributed uniformly over the outer surface of the outer shell. All charges are distributed uniformly over the surfaces due to the spherical symmetry. So, the electric field distribution can be given by the following equations.

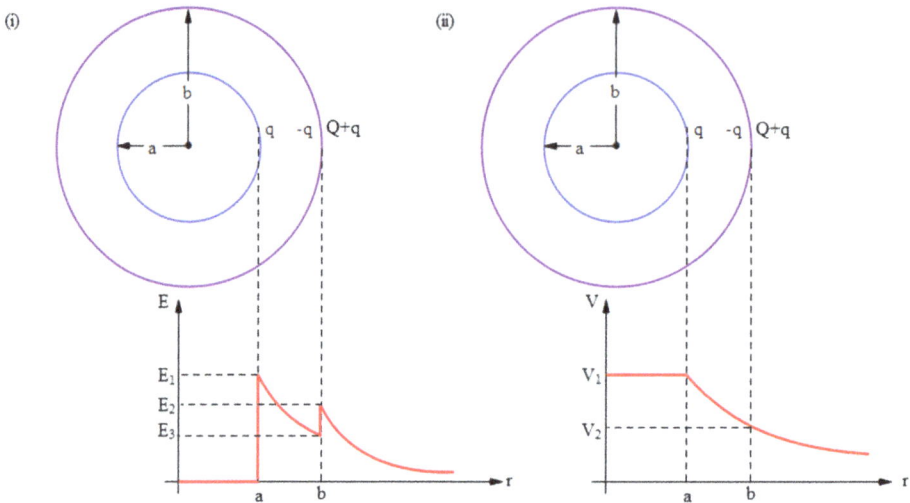

By applying Gauss' law, and the property of conductors, inside the conducting shell

$$E = 0; \ 0 < r < a$$

The electric field between the shells and shell is given as

$$E = \frac{q}{4\pi\varepsilon r^2}; \ a \leqslant r < b$$

The charge induced uniformly on the inner surface of the shell is equal to $-q$; so, the surplus charge $Q + q$ will be uniformly distributed over the outer surface of the shell. By applying Gauss' law, the electric field outside the shell is given as

$$E = \frac{q_{\text{enclosed}}}{4\pi\varepsilon r^2},$$

where $q_{enclosed} = Q + q$. Then, we have

$$E = \frac{Q + q}{4\pi\varepsilon_0 r^2}; \ r > c$$

Field distribution: The values of electric field are shown in the graph; this tells us that the electric field is discontinuous over the boundary of the conductors. First of all, electric field inside the inner shell is zero. It increases abruptly from zero to

$$E_1 = \frac{q}{4\pi\varepsilon_0 a^2}$$

just outside the surface of the inner shell. Then by obeying the inverse square rule, the field decreases to

$$E_2 = \frac{q}{4\pi\varepsilon_0 b^2}$$

just outside the inner surface of the outer shell. Again, it abruptly increases from zero to

$$E_3 = \frac{Q + q}{4\pi\varepsilon_0 c^2}$$

just outside the shell. Then the field decreases to zero obeying the inverse square law.

$$E(r) = \frac{Q + q}{4\pi\varepsilon_0 r^2}; \ r > R$$

Electric potential equations: Referring to the method described in problem 16, at any point P inside the inner shell, the potential of the first body (inner shell of radius a) is

$$V_{1st} = \frac{q}{4\pi\varepsilon_0 a}\left(\because r \leqslant a\right)$$

The potential of the second body (thin shell of radius b) at P is

$$V_{2nd} = \frac{Q}{4\pi\varepsilon_0 b}\left(\because r \leqslant b\right)$$

Then, the total potential at P is

$$V = V_{1st} + V_{2nd} = \frac{q}{4\pi\varepsilon_0 a} + \frac{Q}{4\pi\varepsilon_0 b}$$

At any point P between the inner and outer shell and conducting shell, the potential of the first body (inner shell of radius a) is

$$V_{1st} = \frac{q}{4\pi\varepsilon_0 r} \left(\because r \geqslant a \right)$$

The potential of the second body (outer shell of radius b) at P is

$$V_{2nd} = \frac{Q}{4\pi\varepsilon_0 b} \left(\because r \leqslant b \right)$$

Then, the total potential at P is

$$V = V_{1st} + V_{2nd} = \frac{q}{4\pi\varepsilon_0 r} + \frac{Q}{4\pi\varepsilon_0 b}$$

At any point P outside the outer shell, the dielectric sphere, the potential of the first body (inner shell of radius a) is

$$V_{1st} = \frac{q}{4\pi\varepsilon_0 r} \left(\because r \geqslant a \right)$$

The potential of the second body (thin shell of radius b) at P is

$$V_{2nd} = \frac{Q}{4\pi\varepsilon_0 r} \left(\because r \geqslant b \right)$$

Then, the total potential at P is

$$V = V_{1st} + V_{2nd} = \frac{q}{4\pi\varepsilon_0 r} + \frac{Q}{4\pi\varepsilon_0 r}$$

$$\Rightarrow V = \frac{Q + q}{4\pi\varepsilon_0 r} \quad \text{Ans.}$$

Electric potential distribution: The potential inside the inner shell is uniform, given as

$$V_{inner} = \frac{q}{4\pi\varepsilon_0 a} + \frac{Q}{4\pi\varepsilon_0 b} = V_1$$

Then the potential decreases obeying the $1/r$ rule from V_1 to

$$V_{outer} = \frac{Q + q}{4\pi\varepsilon_0 b} = V_2$$

Then the potential decreases obeying the $1/r$ rule from V_2 to zero outside the outer shell, as shown in the last figure.

(b)

(i) *Condition:* If the inner sphere acquires zero potential, putting $V_{inner} = 0$, we have

$$V_{inner} = \frac{q}{4\pi\varepsilon_0 a} + \frac{Q}{4\pi\varepsilon_0 b} = V_1 = 0$$

$$\Rightarrow Q = \frac{-qb}{a} \quad \text{Ans.}$$

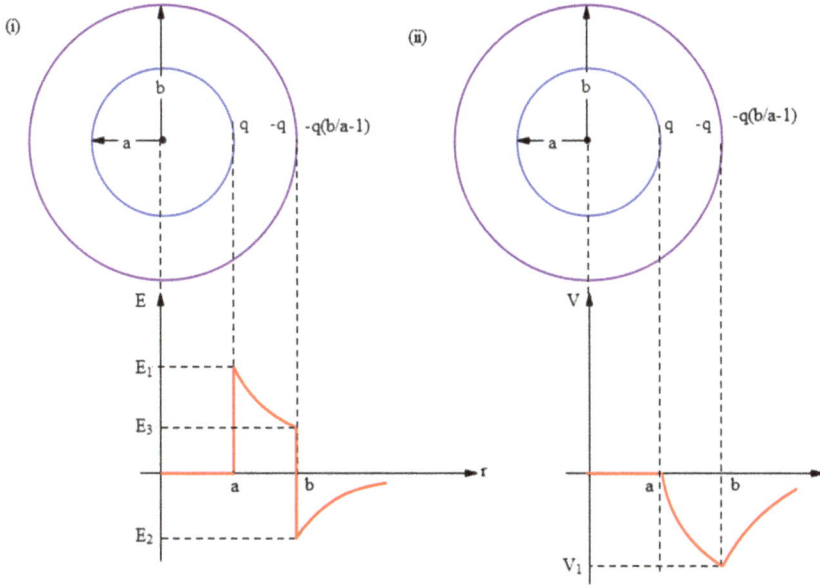

(i) (ii)

E

E_1

E_3

E_2

V

V_1

Potential equations: We can see that the outer shell will have to acquire a negative charge $-q(b/a - 1)$. This negative charge will generate negative potential and radially inward electric field; so, both E and V are negative; for a vector, negative means negative direction (radially outward refers to the +ve and radially inward refers to the −ve direction). However, in the graphs, we have used the magnitudes of E and V. The equations of E and V are given as follows:

The potential is zero inside the inner shell, given as

$$V = 0; r \leqslant a$$

The potential is negative in between the shells that varies as

$$V = \frac{q}{4\pi\varepsilon_0 r} + \frac{Q}{4\pi\varepsilon_0 b} = -\frac{q}{4\pi\varepsilon_0}\left(\frac{1}{a} - \frac{1}{r}\right)\left(\because Q = \frac{-qb}{a}\right); a \leqslant r \leqslant b$$

The potential is negative in between the outer shell that varies as

$$V = \frac{Q + q}{4\pi\varepsilon_0 r} = -\frac{q(b/a - 1)}{4\pi\varepsilon_0 r}\left(\because Q = \frac{-qb}{a}\right); r \geqslant b$$

So potential decreases from zero from the inner shell to a minimum value of

$$V_1 = -\frac{q(b-a)}{4\pi\varepsilon_0 ab}$$

Then it increases to zero at infinity obeying the $1/r$ rule.

Electric field equations: The electric field is zero inside the inner shell.

$$\vec{E} = 0; r \leqslant a$$

Then the field intensity abruptly increases from zero (inside the inner shell) to

$$\vec{E} = \frac{q}{4\pi\varepsilon_0 a^2}\hat{r}$$

just outside the inner shell, which is radially outward.

Then its magnitude decreases obeying inverse law in between the shells, given as

$$\vec{E} = \frac{q}{4\pi\varepsilon_0 r^2}\hat{r} = \frac{q(b/a-1)}{4\pi\varepsilon_0 r^2}\hat{r}\left(\because Q = \frac{-qb}{a}\right); 0 \leqslant r \leqslant b \quad \text{(radially outward)}$$

$$\vec{E} = \frac{Q+q}{4\pi\varepsilon_0 r^2}\hat{r} = -\frac{q(b/a-1)}{4\pi\varepsilon_0 r^2}\hat{r}\left(\because Q = \frac{-qb}{a}\right); r \geqslant b \quad \text{(radially inward)}$$

We can see that at $r = b$, the electric field changes from

$$\vec{E} = \frac{q(b-a)}{4\pi\varepsilon_0 ab^2}\hat{r}$$

at just inside of the outer shell to

$$\vec{E} = -\frac{q(b-a)}{4\pi\varepsilon_0 ab^2}\hat{r}$$

just outside of the outer shell. The magnitude remains constant, but the direction changes from radially outward to radially inward. This results in flipping from the $+y$ axis to the $-y$ axis that represents the electric field in the graph. So, the graph takes care of both sign and direction of E and sign of V. Ans.

(ii) *Electric potential equations:* If the outer sphere acquires zero potential, putting $V_{outer} = 0$, we have

$$V_{outer} = \frac{q}{4\pi\varepsilon_0 b} + \frac{Q}{4\pi\varepsilon_0 b} = V_2 = 0$$

$$\Rightarrow Q = -q \quad \text{Ans.}$$

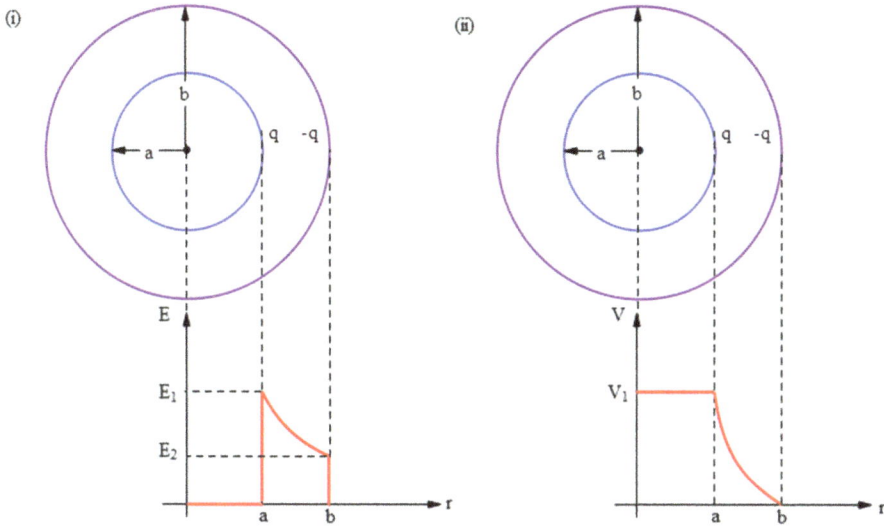

We can see that the outer shell will have to acquire a negative charge $-q(b/a - 1)$. This negative charge will generate negative potential and radially inward electric field; so, both E and V are negative; for a vector, negative means negative direction (radially outward refers to +ve and radially inward refers to −ve direction). However, in the graphs, we have used the magnitudes of E and V. The equations of E and V are given as follows:

The potential is zero inside the inner shell, given as

$$V = 0; r \leqslant a$$

The potential is negative in between the shells that varies as

$$V = \frac{q}{4\pi\varepsilon_0 r} + \frac{Q}{4\pi\varepsilon_0 b} = -\frac{q}{4\pi\varepsilon_0}\left(\frac{1}{a} - \frac{1}{r}\right)\left(\because Q = \frac{-qb}{a}\right); a \leqslant r \leqslant b$$

The potential is negative in between the outer shell that varies as

$$V = \frac{Q+q}{4\pi\varepsilon_0 r} = -\frac{q(b/a - 1)}{4\pi\varepsilon_0 r}\left(\because Q = \frac{-qb}{a}\right); r \geqslant b$$

So potential decreases from zero from the inner shell to a minimum value of

$$V_1 = -\frac{q(b - a)}{4\pi\varepsilon_0 ab}$$

Then it increases to zero at infinity obeying the $1/r$ rule.

Electric field equations: The electric field is zero inside the inner shell.

$$\vec{E} = 0; r \leqslant a$$

Then the field intensity abruptly increases from zero (inside the inner shell) to

$$\vec{E} = \frac{q}{4\pi\varepsilon_0 a^2}\hat{r}$$

just outside the inner shell which is radially outward.
Then its magnitude decreases obeying the inverse law in between the shells, given as

$$\vec{E} = \frac{q}{4\pi\varepsilon_0 r^2}\hat{r} = \frac{q(b/a - 1)}{4\pi\varepsilon_0 r^2}\hat{r}\left(\because Q = \frac{-qb}{a}\right); 0 \leqslant r \leqslant b \quad \text{(radially outward)}$$

$$\vec{E} = \frac{Q + q}{4\pi\varepsilon_0 r^2}\hat{r} = -\frac{q(b/a - 1)}{4\pi\varepsilon_0 r^2}\hat{r}\left(\because Q = \frac{-qb}{a}\right); r \geqslant b \quad \text{(radially inward)}$$

We can see that at $r = b$, the electric field changes from

$$\vec{E} = \frac{q(b - a)}{4\pi\varepsilon_0 ab^2}\hat{r}$$

at just inside of the outer shell to

$$\vec{E} = -\frac{q(b - a)}{4\pi\varepsilon_0 ab^2}\hat{r}$$

just outside of the outer shell. The magnitude remains constant, but the direction changes from radially outward to radially inward. This results in flipping from the $+y$ axis to the $-y$ axis that represents the electric field in the graph. So, the graph takes care of both sign and direction of E and sign of V. Ans.

IOP Publishing

Problems and Solutions in Electricity and Magnetism

Pradeep Kumar Sharma

Chapter 3

Capacitance

3.1 Introduction

A capacitor is a passive circuit element that stores energy in the form of an electrostatic field. Since a capacitor can store or condense a distributed electric field in it, as an energy storing device, it can also be named as a condenser. It has basically a system of conductors embedded in dielectrics. The most commonly used capacitors have two conductors sandwiching a piece of dielectric between them. We cannot think of any electrical and electronics circuits without capacitors. In addition to energy storage, capacitors are used in many practical applications such as tuning of resonant circuits at specific frequency (a tuning knob of a radio set is a variable capacitor), improving the power factor of alternating current (AC) machines, smoothening the voltage pulsation, filtering unwanted frequencies, blocking the direct current and allow the alternating current to pass, eliminating the sparking in the ignition system in automobiles etc.

3.2 Capacitance

Generally, a capacitor consists of two conductors and a dielectric. The presence of dielectric increases the electric energy stored in it. Two conductors must be identical having equal and opposite charges so that no electric field can stay outside the capacitor; so, we can confine the total electric field in between the plates. Due to this mutual electric field, we can define the capacitance of the system of two conductors, in the form of parallel-plate capacitor, cylindrical and spherical capacitors. Having said that, let us first explain how a single conductor can also have a capacitance called self-capacitance.

3.2.1 Isolated conductor (self-capacitance)

In last chapter we explained how a piece of conductor can hold its charge at its surface. Let us imagine an isolated conductor, a conducting sphere of radius R, say, and give a positive charge Q to the sphere. As a result, its electric potential rises from zero to

doi:10.1088/978-0-7503-6477-5ch3
3-1

$$V = \frac{KQ}{R}$$

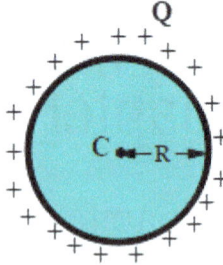

This tells us that potential of an isolated spherical conductor is directly proportional to its charge, which can be given as

$$V \propto Q$$

Although for the sake of mathematical simplicity we took a spherical conductor to prove this relation, it does not depend on the shape and size of the conductor. This means that the ratio of charge given to an isolated conductor and its potential rise, that is, $\frac{Q}{V}$, is a constant quantity. This is known as 'self-capacitance' of the isolated conductor denoted by the letter C. So, we can write the capacitance of an isolated conductor as

$$C = \frac{Q}{V}$$

If we put $V = 1$ volt, we have $C = Q$. This tells us that, the capacitance of an isolated conductor is numerically equal to the 'charge required to raise the potential of the conductor by 1 volt'. If more charge is required for a conductor to raise its potential by 1 volt, its capacitance is more and vice versa. The unit of capacitance is the *farad* (Coulomb volt^{-1}).

N.B: The definition of *electrical capacity* (capacitance) of an electrical conductor is analogous to heat capacity of a thermal conductor. Heat capacity of a conductor is defined as the ratio of heat given Q to increment of temperature ΔT of the conductor, given as

$$C = \frac{Q}{\Delta T}$$

If we put $\Delta T = 1$ K, we have $C = Q$. This tells us that, the heat capacity of an isolated conductor is numerically equal to the 'heat given to raise the temperature of the conductor by one Kelvin or one degree Celsius'. If more heat is required for a conductor to raise its temperature by 1 K, its heat capacity is more and vice versa. The unit of heat capacity is Joule Kelvin^{-1}.

Example 1 Find the capacitance of Earth assuming it as a perfect sphere and good conductor of electricity.

Solution

If we give a charge Q, the potential rise of the earth (sphere of radius R) is

$$V = \frac{Q}{4\pi\varepsilon_0 R}$$

Thus, the capacitance of Earth is

$$C = \frac{Q}{V} = 4\pi\varepsilon_0 R$$

Putting the value of $R = 6371$ km and $4\pi\varepsilon_0 = 10^{-9}/9$, we have $C = 709.5\ \mu\text{F}$ after evaluation. Ans.

3.2.2 System of conductors (mutual capacitance)

Let us consider two conducting spheres 1 and 2 having charge $+Q$ and $-Q$, respectively. If they are placed at a large distance, they are least affected by each other's electric fields. Thus, they are said to be electrically isolated experiencing zero force of attraction. Let us assume that the potential difference between them is V_0. If we bring them closer, they attract each other linking some of their E-lines between each other. This physically signifies that the electrostatic field is more confined in the space between the conductors. So, the more electrostatic field energy is stored in between these two conductors. This is one advantage of taking two conductors over a single (isolated) conductor.

When we bring the charged conductors closer, the net potential of conductor 1 decreases and the net potential of conductor 2 increases. This means that the new potential difference (V, say) between the conductors will increase. In other words, more plus and minus charge should be added to 1 and 2, respectively, to bring the potential difference V back to the initial value V_0. This tells us that the charge storing capacity of each conductor increases in the presence of the other. Thus, the capacitance of the two-conductor system increases. Now we can understand that, in a system of two conductors, each conductor can have (store) more charge when placed closer to each other than when they are isolated.

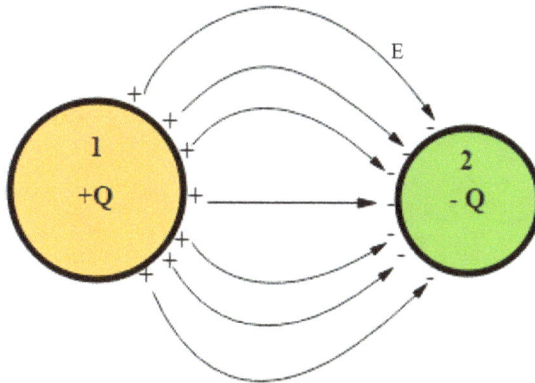

Two equal and oppositely charged conductors can store most of the electrostatic energy in the gap between them.

The potential difference between the conductors can be given as

$$V = V_+ - V_-,$$

where V_+ and V_- are the potentials of the conductors 1 and 2, respectively. Since, $|V_+|$ and $|V_-|$ vary linearly with the magnitude of charge Q, we have

$$V \propto Q$$

Then, the ratio of Q and V, that is, $\frac{Q}{V}$, is a constant quantity, which is defined as the capacitance of the system of two conductors or mutual capacitance.

So, the ratio of amount of charge supplied by the battery to each conductor (or the charge transferred between the plates or conductors by the battery) and the potential difference between the conductors is defined as the capacitance of the system of two conductors. It is given as

$$C = \frac{Q}{V} = \frac{\text{Charge flowing through the battery}}{\text{Rise in potential difference}} = \frac{Q}{-\int_1^2 E. \ dr}$$

Capacitance = charge on each plate/potential difference between the plates.

Example 2 (Spherical capacitor)

Find the capacitance of a system of two concentric metallic spheres of radii a and b.

Solution

Let us give a charge $+Q$ to the inner sphere and $-Q$ to the outer sphere. Then, the potential difference between the spheres is

$$V = -\int_+^- E \cdot dr,$$

where $E = E_+ + E_- = \frac{Q}{4\pi\varepsilon_0 r^2}$ (because $E_- = 0$ as the outer sphere cannot contribute any field inside it).

$$\Rightarrow V = -\int_a^b \frac{Q}{4\pi\varepsilon_0 r^2} \ dr$$

$$= \frac{Q}{4\pi\varepsilon_0}\left(\frac{1}{a} - \frac{1}{b}\right)$$

Then, the capacitance of the system is

$$C = \frac{Q}{V} = \frac{4\pi\varepsilon_0 ab}{b - a} \quad \text{Ans.}$$

N.B: If the outer sphere is very large, the capacitance of the system is given as $C = 4\pi\varepsilon_0 a$ which is the capacitance of the isolated sphere of radius a. So, in the presence of the outer conducting shell (sphere), the capacitance increases by a factor $b/(b-a)$

3.3 Parallel plate capacitor

Let us bring two identical conducting plates of area A and place them parallel to each other at a distance of separation much less than the length and breadth of the plates. Now we give a charge $+Q$ to one plate and $-Q$ to the other plate by connecting the plates to a battery. As the plates are separated by a small distance d, a uniform electric field is felt in the gap between the plates. However, there is a non-uniform electric field due to the fringing of lines of forces at the edge of the plates. But, for the sake of simplicity of calculation of capacitance, we ignore the fringing of lines of force so as to confine a uniform electric field in it. In other words, we can say that it can store electrostatic energy in the gap between the plates. So, we call it a parallel-plate capacitor. The net electric field in the space between the plates inside the capacitor is given as

$$\vec{E} = \vec{E}_+ + \vec{E}_- = \frac{Q\hat{i}}{2\varepsilon_0 A} + \frac{Q\hat{i}}{2\varepsilon_0 A}$$

$$\Rightarrow E = \frac{Q}{\varepsilon_0 A}$$

Then, the potential difference developed between the plates is

$$V = Ed = \frac{Qd}{\varepsilon_0 A}$$

Now, the capacitance of the parallel-plate capacitor is

$$C = \frac{Q}{V} = \frac{\varepsilon_0 A}{d}$$

N.B: The above formula is an approximation formula obtained after neglecting the fringing of lines of force. It can also be used for the parallel curved plates (cylindrical, spherical etc) placed very close to each other.

Fringing of lines of force

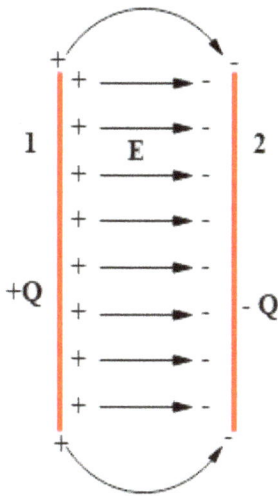

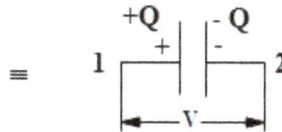

Fringing of lines of force

Representation of capacitor

Example 3 Find the capacitance of the system of two concentric conducting spherical shells of radii $r_1 = a$ and $r_2 = d + a$ where the distance of separation d between the shells is much less than their radii; $d \ll a$.

Solution

Since $d \ll a$, the system can be treated as a parallel-plate capacitor whose capacitance can be given as

$$C = \frac{\varepsilon_0 A}{d}, \quad \text{where } A = 4\pi a^2$$

$$\Rightarrow C = \frac{4\pi\varepsilon_0 a^2}{d} \text{ Ans.}$$

Alternative method:

Put $b \simeq a$ and $b - a = d$ in the formulae $C = \frac{4\pi\varepsilon_0 A ab}{b-a}$ derived in example 2 to obtain $C \simeq \frac{4\pi\varepsilon_0 a^2}{d}$. Ans.

N.B: Following the above example, if we substitute the spherical shells by two coaxial cylindrical conductors of the same radii, you can show that the capacitance per unit length can be given as $C = \frac{2\pi a\varepsilon_0}{d}$.

3.4 Energy stored in a capacitor

You have seen that the electric field is mostly confined in a practical capacitor (space between the conductors). Therefore, a charged capacitor can store electrostatic energy. No energy can stay outside of a capacitor due to the absence of electric field outside the capacitor. We can release the stored energy in the form of heat, light, sound etc, by connecting the plates of a charged capacitor by a thin conducting wire. This is known as discharging of the capacitor, which will be discussed in the next chapter. The stored energy in the capacitor can be used to supply energy to flash lamps on cameras, electric vehicles etc.

The reverse process of discharging a capacitor is known as charging. The process of charging is associated with a continuous pumping of charge from one conductor (plate in a parallel plate capacitor) to the other by the battery. If a charge dq is brought from plate 2 to plate 1 against their electric field E, the work done by the battery is given as

$$dW = (Ed)dq = Vdq, \quad \text{where } V = \frac{q}{C} = Vdq$$

$$\Rightarrow dW = \frac{q}{C}dq$$

Then, the total work done by the battery in transferring a charge Q between the plates is

$$W = \int dW = \int_0^Q \frac{q\, dq}{C} = \frac{Q^2}{2C}$$

This work is done by the battery at the expense of its chemical energy. The work done by the battery is stored the capacitor in the form of electrostatic field energy U, which is given as

$$U = \frac{Q^2}{2C}$$

Putting $Q = CV$, the other two expressions of energy can be written as

$$U = \frac{1}{2}QV = \frac{CV^2}{2}$$

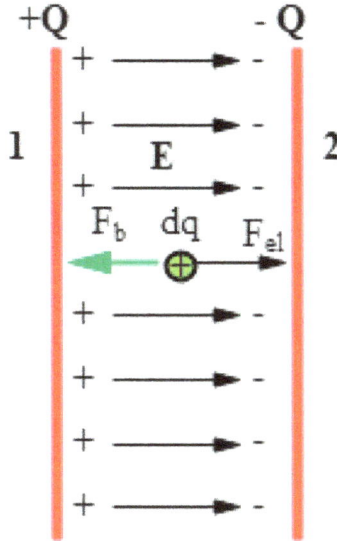

Alternative method:
We can say that a charged capacitor is a group of two conductors 1 and 2 of charge $+Q$ and $-Q$, respectively. With reference to the last chapter, the energy possessed by a charged conductor is

$$U = \frac{1}{2}q\,V,$$

where q and V are the charge and potential of the conductor, respectively.

The charge of one conductor is $Q_1 = +Q$ and potential $V_1 = V_+$ and the charge of other conductor is $Q_2 = -Q$ and potential $V_2 = V_-$. Then, the total energy possessed by the capacitor (two conducting plates) is

$$U = \frac{1}{2}Q_1 V_1 + \frac{1}{2}Q_2 V_2$$

$$= \frac{1}{2}(+Q)(V_+) + \frac{1}{2}(-Q)(V_-)$$

$$= \frac{Q}{2}(V_+ - V_-),$$

where $V_+ - V_- = V$ (potential difference between the conducting plates)

$$\Rightarrow U = \frac{1}{2}QV, \text{ as obtained earlier.}$$

The energy stored in a capacitor is numerically equal to the work done in charging the capacitor by bringing the charges from infinity to place them on the conductors of the capacitor.

3.5 Energy density in a parallel-plate capacitor

As the field is uniform in a parallel-plate capacitor, the distribution of electric field energy is uniform inside the capacitor. To understand the distribution of energy in an electric field, we need to find the density of energy, that is, energy stored per unit volume.

As derived earlier, the electrostatic energy stored in the parallel-plate capacitor is

$$U = \frac{1}{2}CV^2, \text{ where } C = \frac{\varepsilon_0 A}{d} \text{ and } V = Ed$$

$$\Rightarrow U = \frac{1}{2}\left(\frac{\varepsilon_0 A}{d}\right)(Ed)^2 = \frac{1}{2}\varepsilon_0 E^2(Ad),$$

where $Ad = v$ (volume of the capacitor).

$$\Rightarrow U = \frac{1}{2}\varepsilon_0 E^2 v$$

Then, the energy stored per unit volume can be given as

$$\frac{dU}{dv}(=u_v) = \frac{1}{2}\varepsilon_0 E^2$$

We can find the electric field energy in a region by using the last expression if the variation of field intensity is known in that region.

3.6 Capacitor with a dielectric

In a simple practical capacitor, a thin dielectric is placed in between the conducting plates so as to increase the capacitance and energy stored. Let us introduce a dielectric slab of area A and thickness x in between the plates of a parallel-plate air capacitor of plate area A and distance of separation $d(>x)$. Let the charge of the plates of the capacitor be $+Q$ and $-Q$. This charge is called free charge. The electric field due to this free charge is the electric field in the free space between the plates, which is given as

$$E_0 = \frac{\sigma}{\varepsilon_0} = \frac{Q}{\varepsilon_0 A} \tag{3.1}$$

The electric field due to the free charge, that is, $\vec{E_0}$, is also known as applied or external field to the dielectric slab introduced between the plates. In the last chapter you learnt that when a dielectric is placed in an external or applied field E_0, the electric field E inside the dielectric will decrease by a factor ε_r, which is given as

$$E = \frac{E_0}{\varepsilon_r} \qquad\qquad (3.2)$$

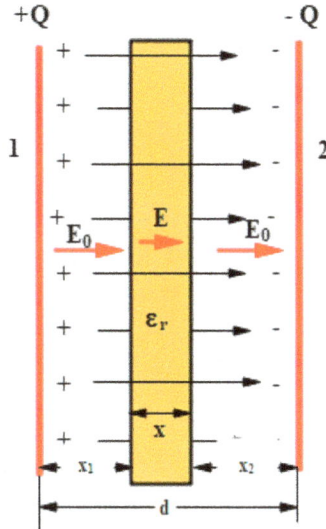

Putting $E_0 = \frac{\sigma}{\varepsilon_0} = \frac{Q}{A\varepsilon_0}$ from equation (3.1) in equation (3.2), we have

$$E = \frac{Q}{\varepsilon_0 \varepsilon_r A}$$

As the field is uniform, the potential difference between the plates is

$$V = E_0 x_1 + Ex + E_0 x_2 = E_0(x_1 + x_2) + Ex$$

$$= E_0(d - x) + \frac{E_0 x}{\varepsilon_r} \left(\because x_1 + x_2 = d - x \text{ and } E = \frac{E_0}{\varepsilon_r} \right)$$

$$\Rightarrow V = \frac{Q}{\varepsilon_0 A}\left(d - x + \frac{x}{\varepsilon_r} \right)\left(\because E_0 = \frac{Q}{\varepsilon_0 A} \right)$$

Then, the capacitance of the system is

$$C = \frac{Q}{V} = \frac{\varepsilon_0 A}{d - x + \frac{x}{\varepsilon_r}}$$

In the absence of dielectric, the capacitance is

$$C_0 = \frac{\varepsilon_0 A}{d}(<C)$$

This means that, the capacitance increases after inserting a dielectric slab as the potential difference decreases between the isolated charged plates. *The presence of a dielectric increases the capacitance.*

Example 4 A parallel-plate capacitor has capacitance C_0 in the absence of a dielectric. (a) If a dielectric slab of relative permittivity ε_r is completely filled inside the air capacitor, find the new capacitance. (b) If the plates are connected by a conducting wire, find the capacitance of the parallel-plate capacitor.

Solution

(a) If the dielectric is completely filled in between the capacitor plates, putting $x = d$ in the expression

$$C = \frac{\varepsilon_0 A}{\left(d - x + \frac{x}{\varepsilon_r}\right)},$$

we have

$$C = \frac{\varepsilon_0 \varepsilon_r A}{d}, \quad \text{where} \quad \frac{\varepsilon_0 A}{d} = C_0$$

$$\Rightarrow C = \varepsilon_r C_0 \text{ Ans.}$$

(b) Connecting the plates by a conducting wire is equivalent to fitting a conducting slab completely between the plates because it will not allow the potential difference between the plates to change in steady state or electrostatic condition. Then we can say that infinite charge is required to raise the potential difference between the plates by one volt. In other words, the capacitance becomes infinite.

After connecting the plates by a conducting wire, the effective medium between the plates is a conductor which always maintains a zero electric field (zero potential difference) between the plates. Then by putting $\varepsilon_r = \infty$ (for a conducting slab) in the last expression $C = \varepsilon_r C_0$, we get an infinite capacitance. Ans.

N.B: Capacitance increases ε_r times if a dielectric slab fills the space between the plates completely. It does not matter whether the capacitor is (i) parallel plate or not, (ii) connected or disconnected from the supply.

Example 5 (a) In a parallel-plate capacitor with a dielectric, find the induced charge at the surface of the dielectric slab of the parallel-plate capacitor. (b) If we replace the dielectric by a metallic slab, find the induced charge.

Solution

(a) Let the induced charge be Q'. The induced charges are distributed uniformly on the surfaces of the dielectric. So, we can think of the dielectric slab as the system of two thin plates of charge $-Q'$ and Q'; the net electric field produced by them is

$$E_{\text{ind}} = \frac{Q'}{2\varepsilon_0 A} + \frac{Q'}{2\varepsilon_0 A} = \frac{Q'}{\varepsilon_0 A} \tag{3.3}$$

The applied or external electric field is given as

$$E_{\text{ext}} = \frac{Q}{\varepsilon_0 A} \tag{3.4}$$

In last chapter, we have obtained the expression for the induced electric field as

$$E_{\text{ind}} = E_{\text{ext}}\left(1 - \frac{1}{\varepsilon_r}\right) \tag{3.5}$$

Using the last three equations, we have

$$Q' = Q\left(1 - \frac{1}{\varepsilon_r}\right) \text{ Ans.}$$

Alternative method:

(a) Draw a Gaussian surface (dotted line) that encloses one face of the dielectric slab and the nearest internal face of the conducting plate. Applying Gauss's law, we have the net flux out = charge enclosed/ε_0

$$\Rightarrow EA = \frac{Q - Q'}{\varepsilon_0}, \quad \text{where } E = \frac{Q}{\varepsilon_0 \varepsilon_r A}$$

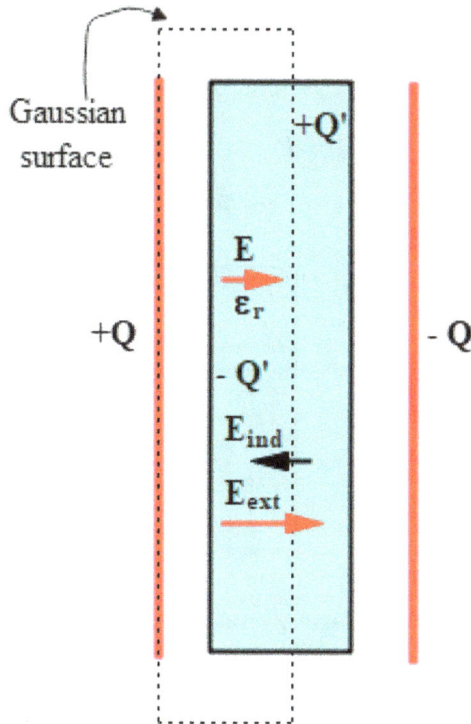

This gives us the same expression

$$Q' = Q\left(1 - \frac{1}{\varepsilon_r}\right)$$

which does not depend on the distance of separation between the plates of the capacitor. Ans.

(b) Putting $\varepsilon_r = \infty$ for the conducting slab, we have $Q' = Q$. This means that the face of the conducting slab facing the left side plate of charge $+Q$ will get $-Q$ and the other face will induce a charge $+Q$. Then the net field (due to all charges) in the conducting slab will be zero satisfying the first conductor property of zero electric field inside a conductor. Ans.

3.7 Relation between polarization $\overrightarrow{P}$ and applied field $\overrightarrow{E}$

Since the induced charge on the surface of a dielectric in a parallel-plate capacitor is

$$Q' = Q\left(1 - \frac{1}{\varepsilon_r}\right),$$

putting $Q' = \sigma'A$ and $Q = \sigma A$, we have

$$\sigma' = \sigma\left(1 - \frac{1}{\varepsilon_r}\right)$$

Then, putting $\sigma' = P$(polarization) and $\frac{\sigma}{\varepsilon_0 \varepsilon_r} = E_{net}$, we have

$$P = \varepsilon_0(\varepsilon_r - 1)E_{net}$$

$$\Rightarrow \overrightarrow{P} = \varepsilon_0 \chi \overrightarrow{E},$$

where E = field inside the dielectric and $\chi = (\varepsilon_r - 1)$, called susceptibility of the dielectric.

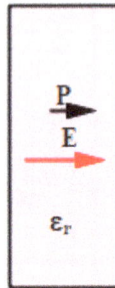

Polarization of a dielectric is directly proportional to (or linearly dependent on) the applied field so also the net field inside the dielectric. This is the condition of a linear dielectric. The dielectrics obeying the above relation are called 'linear dielectrics'. We will calculate the work, energy and force for the linear dielectrics in later sections.

3.8 Grouping of capacitors

For a given number of capacitors, we can connect them in different ways to draw different capacitance according to our requirement. So, it is important to know the types of connection of capacitors in electrical circuits.

3.8.1 Series combination

When we connect the capacitors end to end between two terminals A and B, say, it is known as series connection. If we connect the terminals to a battery, let it send a charge q through each capacitor. The voltage dropped across the combination is equal to the sum of voltage dropped across each capacitor. It is given as

$$V_{AB}(=V) = \sum V_i, \quad \text{where} \quad V_i = \frac{q}{C_i}$$

$$\Rightarrow V = q \sum \frac{1}{C_i} \tag{3.6}$$

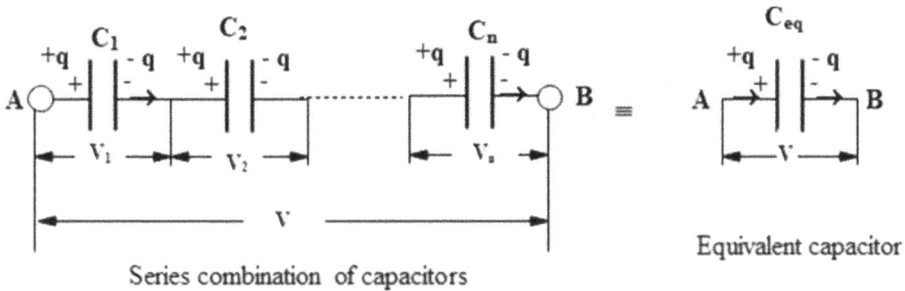

Series combination of capacitors   Equivalent capacitor

If you substitute the entire combination by a single capacitor and apply the same voltage V between its terminals, let the same charge q flow through it. Then, we can call it equivalent capacitance C_{eq}, given as

$$V = \frac{q}{C_{eq}} \tag{3.7}$$

Comparing the equations (3.6) and (3.7), we have

$$\frac{1}{C_{eq}} = \sum \frac{1}{C_i}.$$

3.8.2 Parallel combination

When one end of each capacitor is connected to the positive terminal of a battery and the other ends of the capacitors are connected to the negative terminals of the battery, this is known as parallel combination.

In parallel combination, the same voltage is dropped (same potential difference V is felt) across each capacitor.

So, the total charge flowing through the combination (between the terminals A and B) is the sum of charges flowing across each capacitor. It is given as

$$q = \sum q_i, \quad \text{where} \quad q_i = \text{charge of } i\text{th capacitor } C_i = C_i V$$

$$\Rightarrow q = \left(\sum C_i\right) V \tag{3.8}$$

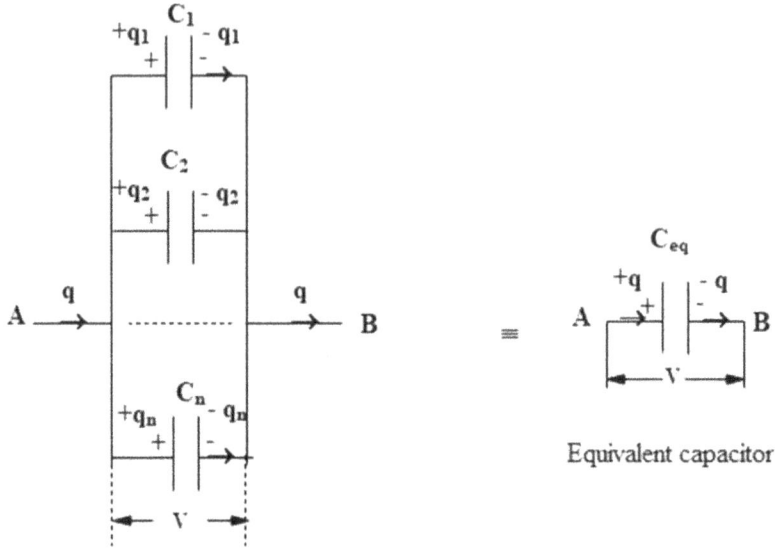

Parallel combination of capacitors

Equivalent capacitor

Let the combination be replaced by a single capacitor so that the same voltage V will cause the same charge q to flow through the capacitor, or draw same charge from the battery, we can call it the equivalent capacitor whose capacitance C_{eq} is given as

$$q = C_{eq} V \tag{3.9}$$

Comparing equations (3.8) and (3.9), we have

$$C_{eq} = \sum C_i$$

Example 6 Find the effective capacitance between A and B.

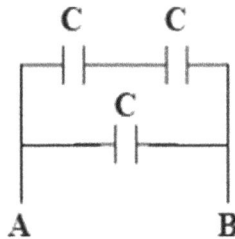

Solution

Since the upper capacitors 1 and 2 each of capacitance C are in series, their effective capacitance is

$$C' = \frac{C \times C}{C + C} = \frac{C}{2}$$

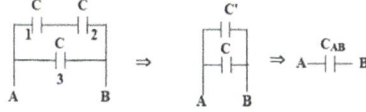

Since C' and C_3 are in parallel,

$$C_{AB} = C' + C_3 = \frac{C}{2} + C = \frac{3C}{2} \quad \text{Ans.}$$

3.9 Energy stored in a capacitor with a dielectric

Referring to section 3.4, the energy stored in a capacitor is

$$U = \frac{1}{2}QV$$

This formula can also be used for the capacitors with dielectrics.

Constant voltage: At constant voltage V, putting $Q = CV$, the energy stored in a capacitor is

$$U = \frac{1}{2}CV^2$$

Referring to example 4, putting $C = \frac{\varepsilon_0 \varepsilon_r A}{d}$ and $V = Ed$ in the last expression of energy, we have

$$U = \frac{1}{2}\frac{\varepsilon_0 \varepsilon_r A}{d}(E^2 d^2), \quad \text{where} \quad \varepsilon_0 \varepsilon_r = \varepsilon$$

$$\Rightarrow U = \frac{1}{2}\varepsilon E^2 (Ad) = \frac{1}{2}\varepsilon E^2 v,$$

where $v = Ad = $ volume of the capacitor.

Then, energy stored per unit volume of the capacitor is

$$\frac{dU}{dv} = u_v = \frac{1}{2}\varepsilon E^2$$

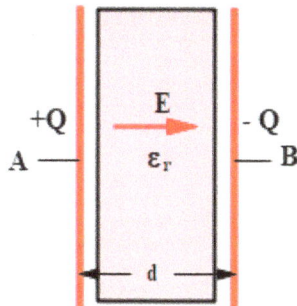

Constant charge: At constant charge Q, putting $Q = CV$, in the expression $U = \frac{1}{2}QV$, the energy stored in a capacitor is

$$U = \frac{Q^2}{2C}, \quad \text{where } Q = \varepsilon E A \text{ and } C = \frac{\varepsilon_0 \varepsilon_r A}{d}$$

$$\Rightarrow U = \frac{\varepsilon E^2 v}{2} \text{ (volume), where } v = Ad$$

Then, the volume density of electrostatic field is given as

$$u_V = \frac{\varepsilon E^2}{2},$$

where $\varepsilon =$ permittivity of the dielectric and $E =$ field inside the dielectric.

So, the density of the energy stored in a capacitor with a dielectric increases by the factor of ε_r. This does not depend upon the condition (constant charge or constant voltage).

3.10 Polarization energy

If you insert a dielectric that fits completely inside an isolated charged parallel-plate capacitor, its energy stored is

$$U = \frac{Q^2}{2C}$$

Before insertion of the dielectric, energy stored in the capacitor is

$$U_0 = \frac{Q^2}{2C}$$

Then, the difference in energy after inserting the dielectric is

$$U - U_0 = -\frac{Q}{2C_0}\left(1 - \frac{C_0}{C}\right), \quad \text{where } C = \varepsilon C_0$$

$$\Rightarrow U - U_0 = -\frac{Q^2}{2C_0}\left(1 - \frac{1}{\varepsilon_r}\right)$$

This surplus energy is called polarization energy denoted as U_{pol}. So, we can write

$$U_{\text{pol}} = -\frac{Q^2}{2C_0}\left(1 - \frac{1}{\varepsilon_r}\right),$$

where

$$\frac{Q^2}{2C_0} = \frac{1}{2}\varepsilon_0 E_0^2 v$$

$$\Rightarrow U_{\text{pol}} = -\left(1 - \frac{1}{\varepsilon_r}\right)\varepsilon_0 E_0^2 v$$

Hence, the polarization energy stored per unit volume is given as

$$\frac{dU_{\text{pol}}}{dv} = u_{\text{pol}} = -\frac{1}{2}\varepsilon_0 \, E_0^2 \frac{(\varepsilon_r - 1)}{\varepsilon_r}$$

$$= -\frac{1}{2}\varepsilon_0(\varepsilon_r - 1)\frac{E_0}{\varepsilon_r} \times E_0$$

$$\Rightarrow u_{\text{pol}} = -\frac{1}{2}\varepsilon_0(\varepsilon_r - 1)E \cdot E_0$$

Putting $P = \varepsilon_0(\varepsilon_r - 1)E$ in the last expression, we have

$$\Rightarrow u_{\text{pol}} = -\frac{1}{2}P \cdot E_0,$$

where E_0 = external (or applied) electric field or field due to the free charges.

Hence, the energy possessed by a dielectric of polarization $\overrightarrow{P}$ kept in an applied (external) field intensity E_0 can be given as

$$U_{\text{pol}} = -\frac{1}{2}\int \overrightarrow{P} \cdot \overrightarrow{E_0} dV$$

U_{pol} is the excess energy required to polarize the dielectric in the process of stretching and rotating the molecules and atoms of the dielectric inducing equal and opposite bound charges. Then, we can write

$$U_{\text{pol}} = U_{\text{bound}} + U_{\text{stretching}} + U_{\text{twisting}}$$

If the external applied field $E_{\text{ext}}(=E_0)$ is removed, we will get all energy back in the ideal case by neglecting the residual polarization inside the dielectric.

Alternative method: In the process of polarization, let the elementary charge $+\delta q$ moves in the direction of $\overrightarrow{E}$ by a distance dl_+ and $-\delta q$ moves in the opposite direction of $\overrightarrow{E}$ by a distance dl_-.

The work performed by the field during this process is

$$dW = (E \, \delta q_+)dl_+ + (E \, \delta q_-)dl_-$$
$$= E \, \delta q \cdot (dl_+ + dl_-)$$
$$= E \, dq \, dl, \quad \text{where } \delta q \cdot dl = dp \text{ (elementary change in the dipole moment)}$$
$$= E \, dp$$

Since $dW = -dU_{\text{pol}}$ (polarization energy stored in the dielectric), we have

$$dU_{\text{pol}} = -E \, dp$$

$$= -\left(E \frac{dP}{V}\right) \cdot (V)$$

$$dU_{\text{pol}} = -(EdP)V \quad \left(\because dP = \frac{dp}{V}\right)$$

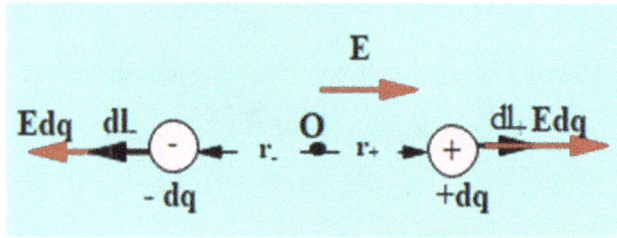

The elementary charges $+dq$ and $-dq$ shift in opposite directions of the field E. Then the total energy density is

$$\frac{U_{\text{pol}}}{V} = -\int_0^{E_0} E \, dP, \quad \text{where} \quad P = \varepsilon_0 \times \frac{E_0}{\varepsilon_r}$$

$$\Rightarrow \frac{U_{\text{pol}}}{V} = -\int_0^{E_0} E \frac{\varepsilon_0}{\varepsilon_r} \times dE$$

$$u_v = -\frac{1}{2} \times \frac{\varepsilon_0}{\varepsilon_r} E^2 = -\frac{1}{2}\left(x\varepsilon_0 \frac{E_0}{\varepsilon_r}\right). \quad E_0 = -\frac{1}{2}PE_0$$

Polarization energy is equal to the negative work done by an external agent in slowly inserting the dielectric against the electrostatic attraction due to the fringing of lines of forces near the edge of the plates. Please refer to problem 11.

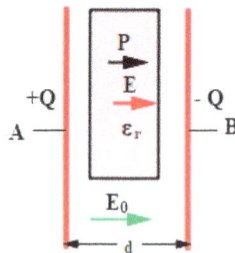

Polarization of a dielectric in a parallel plate capacitor reduces the electric field inside the dielectric: so, in an isolated capacitor(Q = Constant), the energy decreases which will be released after withdrawing the dielectric. This means that the polarization energy is negative.

3.11 Finding the field energy of different charged objects

The electric energy stored in a capacitor is numerically equal to energy spent by a battery in supplying free charges $+q$ and $-q$ to the capacitor plates. During the process of charging, the dielectric is polarized inducing bound charges by the method of twisting and stretching of the dipoles in response to the applied field. Thus, the increase in elastic and electrostatic energy stored in the dielectric and free space of the capacitor is termed as energy stored in the capacitor.

To find the energy stored due to a charged body, first of all we take a very small volume dv and find the energy density in that volume.

Then, find the energy confined in the elementary volume as

$$dU = U_v \, dv$$

Adding all elementary energy, the total energy stored is

$$U = \int dU$$

$$= \int u_v dv = \int \tfrac{1}{2}\varepsilon_0 E^2 dv$$

$$\Rightarrow U = \frac{\varepsilon_0}{2} \int E^2 \, dv$$

To use the last expression, the variation of E as the function of x (in Cartesian coordinate system) or r (in cylindrical or spherical coordinate system) must be given. So, we write $E = f(x)$ or $f(r)$. Then write $dv = A dx$ in Cartesian coordinate system, $dv = 2\pi r dr$ in cylindrical coordinate system and $dv = 4\pi r^2 dr$ in spherical coordinate system. Finally evaluate the integral for all regions, where E-field is present (inside or outside the matter).

Example 7 Find the electrostatic energy stored between two large conducting sheets of area A carrying charges q and $2q$ separated by a small distance d.

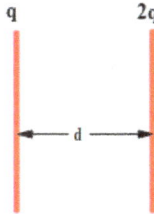

Solution
The field in the space between the plates is

$$\vec{E} = \vec{E_1} + \vec{E_2}$$

$$= \frac{q}{2\varepsilon_0 A}\hat{i} + \frac{2q}{2\varepsilon_0 A}(-\hat{i}) = -\frac{q}{2\varepsilon_0 A}\hat{i}$$

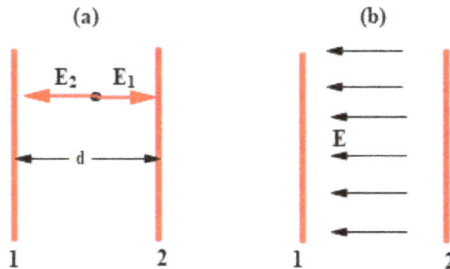

(a)　　　　　　(b)

The total energy stored in the space between the plates is

$$U = \frac{\varepsilon_0}{2} \int E^2 dV = \frac{\varepsilon_0}{2} E^2 \int_0^{Ad} dV = \frac{\varepsilon_0 E^2}{2} Ad$$

$$= \frac{\varepsilon_0}{2} Ad \left| \frac{-q}{2\varepsilon_0 A} \right|^2 = \frac{q^2 d}{8\varepsilon_0 A} \text{ Ans.}$$

3.12 Total, self- and mutual energy

If we consider two charged bodies 1 and 2 generate fields $\vec{E_1}$ and $\vec{E_2}$, respectively. The net field due to system of two bodies $(1 + 2)$ is

$$E = \left| \vec{E_1} + \vec{E_2} \right|$$

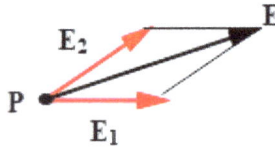

3.12.1 Total energy

Then, the total energy of the system can be given as

$$U = \frac{\varepsilon_0}{2} \int E^2 \, dV = \frac{\varepsilon_0}{2} \int \left| \vec{E_1} + \vec{E_2} \right|^2 dV$$

$$\Rightarrow U = \frac{\varepsilon_0}{2} \int E_1^2 dV + \frac{\varepsilon_0}{2} \int E_2^2 dV + \varepsilon_0 \int \vec{E_1} \cdot \vec{E_2} dV.$$

3.12.2 Self-energy

The first two terms $\frac{\varepsilon_0}{2} \int E_1^2 dV$ and $\frac{\varepsilon_0}{2} \int E_2^2 dV$ are called self-energies of the bodies 1 and 2 (fields $\vec{E_1}$ and $\vec{E_2}$), respectively. The self-energy is numerically equal to the minimum work done by an external agent to assemble the charges in the bodies.

3.12.3 Mutual energy

The third term $\varepsilon_0 \int \vec{E_1} \cdot \vec{E_2} dV$ is called mutual energy because both the fields are present in this expression. Since U_{mutual} has a dot (scalar) product of $\vec{E_1}$ and $\vec{E_2}$ it can be +ve, 0 and −ve. It is numerically equal to the work done by an external agent in bringing the charged bodies from infinity to the given position (without disturbing their charge distribution). *Total energy stored = self-energy + mutual energy; mutual energy is algebraic.*

Example 8 Find the self-energy of non-conducting thin spherical shells of radii r_1 and r_2 having charges Q_1 and Q_2, respectively and their mutual energy of the system of two spheres separated by a distance r.

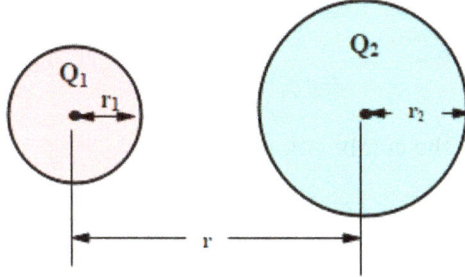

Solution

The self-energy Q_1 is

$$U_1 = \frac{\varepsilon_0}{2} \int E_1^2 \, dv = \frac{\varepsilon_0}{2} \int_0^{r_1} \left(\frac{Q_1}{4\pi\varepsilon_0 r^2} \right)^2 4\pi r^2 dr = \frac{Q_1^2}{8\pi\varepsilon_0 r_1}$$

Similarly, the self-energy of Q_2 is

$$U_2 = \frac{Q_2^2}{8\pi\varepsilon_0 r_2} \text{ Ans.}$$

The mutual energy between Q_1 and Q_2 is equal to the work done by an external agent in bringing them closer

$$U_{\text{mutual}} = U_{\text{ext}} = \int F \, dr = \int_\infty^r \frac{Q_1 Q_2}{4\pi\varepsilon_0 r^2} \times dr = \frac{Q_1 Q_2}{4\pi\varepsilon_0 r} \text{ Ans.}$$

N.B: Since $E = 0$ inside the spherical shells, electrostatic energy inside the shells is zero. So, the energy is stored in the space outside the spheres.

3.13 Multiple dielectric capacitors

3.13.1 Series combination of dielectrics

Let us join two dielectrics of thickness x_1 and x_2 and relative permittivity ε_{r_1} and ε_{r_2} end to end such that their field intensities become perpendicular to the interface of each dielectric. If the capacitor plates have charges Q and $-Q$, the applied field is

$$E_0 = \frac{Q}{\varepsilon_0 A}$$

Then, the field inside the dielectrics are

$$E_1 = \frac{E_0}{\varepsilon_{r_1}} = \frac{Q}{\varepsilon_0 \varepsilon_{r_1} A} \text{ and } E_2 = \frac{E_0}{\varepsilon_{r_2}} = \frac{Q}{\varepsilon_0 \varepsilon_{r_2} A}$$

Hence, the potential difference between the plates is

$$
\begin{aligned}
V &= V_1 + V_2 \\
&= E_1 x_1 + E_2 x_2 \\
&= \frac{Q}{\varepsilon_0 \varepsilon_{r_1} A} x_1 + \frac{Q}{\varepsilon_0 \varepsilon_{r_2} A} x_2 \\
&= \frac{Q}{\varepsilon_0 A}\left(\frac{x_1}{\varepsilon_{r_1}} + \frac{x_2}{\varepsilon_{r_2}}\right)
\end{aligned}
$$

Then, the capacitor of the combination is,

$$
C = \frac{Q}{V} = \frac{\varepsilon_0 A}{\frac{x_1}{\varepsilon_{r_1}} + \frac{x_2}{\varepsilon_{r_2}}}
$$

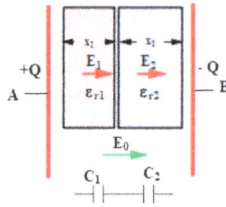

If there are n dielectrics connected side by side, the capacitance of the combination can be given as

$$
C = \frac{\varepsilon_0 A}{\sum\limits_{i=1}^{i=n} \frac{x_i}{\varepsilon_{r_i}}}
$$

The above expression can be written as

$$
\frac{1}{C} = \frac{1}{\varepsilon_0 A}\sum \frac{x_i}{\varepsilon_{r_i}} = \sum \frac{1}{\varepsilon_0 \varepsilon_r A / x_i},
$$

where $\frac{\varepsilon_0 \varepsilon_{r_i} A}{x_i} = C_i = $ capacitance of a parallel-plate capacitor of plate area A completely filled with a dielectric of permittivity ε_{r_i} and thickness x_i.

$$
\Rightarrow \frac{1}{C} = \sum \frac{1}{C_i}, \quad \text{where } C_i = \frac{\varepsilon_0 \varepsilon_{r_i} A}{x_i}
$$

When the dielectrics are placed side by side such that the field intensity is perpendicular to their interfaces, you can imagine each dielectric slab with two thin conducting plates touching both sides of the dielectric forming a parallel-plate capacitor. As a result, the given combination can be reduced to a series combination of n number of capacitors having the effective capacitance C given as $\frac{1}{C} = \sum \frac{1}{C_i}$, where $C_i = \frac{\varepsilon_0 \varepsilon_{r_i} A}{x_i}$.

3.13.2 Parallel combination of dielectrics

Let us connect two dielectrics each of length equal to the distance d between the parallel plates such that their interfaces are parallel to the field strengths $\overrightarrow{E_1}$ and $\overrightarrow{E_2}$. You can call it parallel combination of dielectrics because the potential difference across each dielectric is equal to the potential difference V between the plates of the capacitor. If the areas of the cross-section of the dielectrics touching the conducting plates of the capacitor are A_1 and A_2, respectively, let the free charges Q of the conducting plates be divided into two parts Q_1 and Q_2 in the areas A_1 and A_2, respectively. It means that the free charge densities of the conducting plates in the areas A_1 and A_2 are different, which can be given as

$$\sigma_1 = \frac{Q_1}{A_1} \text{ and } \sigma_2 = \frac{Q_2}{A_2}$$

Hence, the applied fields for the dielectrics are

$$E_{0_1} = \frac{\sigma_1}{\varepsilon_0} = \frac{Q_1}{\varepsilon_0 A_1} \text{ and } E_{0_2} = \frac{\sigma_2}{\varepsilon_0} = \frac{Q_2}{\varepsilon_0 A_2}$$

Then, the field inside the dielectrics are

$$E_1 = \frac{E_{0_1}}{\varepsilon_{r_1}} = \frac{Q_1}{\varepsilon_0 \varepsilon_{r_1} A_1} \text{ and } E_2 = \frac{E_{0_2}}{\varepsilon_{r_2}} = \frac{Q_2}{\varepsilon_0 \varepsilon_{r_2} A_2}$$

Since the potential difference V across each dielectric is equal, we can write

$$V = E_1 d = E_2 d$$

$$\Rightarrow E_1 = E_2 = \frac{V}{d}, \text{ where } E_1 = \frac{Q_1}{\varepsilon_0 \varepsilon_{r_1} A_1} \text{ and } E_2 = \frac{Q_2}{\varepsilon_0 \varepsilon_{r_2} A_2}$$

$$\Rightarrow Q_1 = \varepsilon_0 \varepsilon_{r_1} A_1 \frac{V}{d} \text{ and } Q_2 = \varepsilon_0 \varepsilon_{r_2} A_2 \frac{V}{d}$$

Hence, the total charge of each conducting plate is

$$Q = Q_1 + Q_2 = \frac{\varepsilon_0 V}{d} (\varepsilon_{r_1} A_1 + \varepsilon_{r_2} A_2)$$

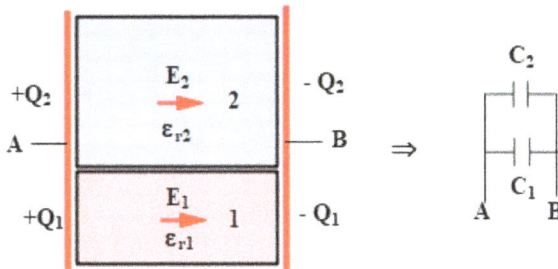

Then, the capacitance of the combination is

$$C = \frac{Q}{V} = \frac{\varepsilon_0}{d}(\varepsilon_{r_1} A_1 + \varepsilon_{r_2} A_2)$$

For n number of dielectrics connected in parallel

$$C = \frac{\varepsilon_0}{d}\sum \varepsilon_{r_i} A_i$$

The above expression can be written as

$$C = \sum \frac{\varepsilon_0 \varepsilon_{r_i} A_i}{d},$$

where $\frac{\varepsilon_0 \varepsilon_{r_i} A_i}{d}$ = capacitance of a parallel-plate capacitor with the dielectric of relative permittivity ε_{r_i} and plate area $A_i = C_i$, say.

Then, we have $C = \sum C_i$, where $C_i = \frac{\varepsilon_0 \varepsilon_{r_i} A_i}{d}$

When the dielectrics are placed one above the other such that the field intensity is parallel to their interfaces, you can imagine each dielectric slab with the given conducting plates touching both sides of the dielectric forming a parallel-plate capacitor. As a result, the given combination can be reduced to a parallel combination of n number of capacitors having the effective capacitance C given as, $C = \sum C_i$, where $C_i = \frac{\varepsilon_0 \varepsilon_{r_i} A_i}{d}$.

Example 9 Find the equivalent capacitance of an air capacitor of capacitance C_0 when the two dielectric slabs of permittivity $\varepsilon_{r_1} = 2$ and $\varepsilon_{r_2} = 1$ are introduced as shown in the below figures (a) and (b).

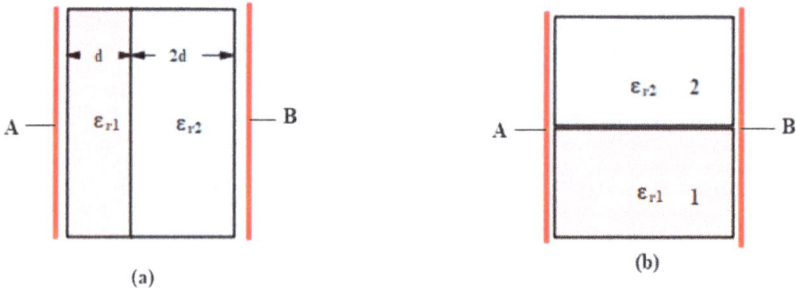

(a)

(b)

Solution

In figure (a) $C = \frac{\varepsilon_0 A}{\frac{x_1}{\varepsilon_{r_1}} + \frac{x_2}{\varepsilon_{r_2}}}$, where $x_1 = d$, $x_2 = 2d$, $\varepsilon_{r_1} = 2$ and $\varepsilon_{r_2} = 1$

$$\Rightarrow C = \frac{\varepsilon_0 A}{\frac{d}{2} + \frac{2d}{1}} = \frac{\varepsilon_0 A}{5d/2}, \quad \text{where } \frac{\varepsilon_0 A}{d} = C_0$$

$$\Rightarrow C = \frac{2C_0}{5} \text{ Ans.}$$

In figure (b), $C = \frac{\varepsilon_0}{d}(\varepsilon_{r_1} A_1 + \varepsilon_{r_2} A_2)$, where $\varepsilon_{r_1} = 2$, $\varepsilon_{r_2} = 1$, $A_1 = A_2 = \frac{A}{2}$

$$\Rightarrow C = \frac{\varepsilon_0}{d}\left(2 \times \frac{A}{2} + 1 \times \frac{A}{2}\right) = \frac{3\varepsilon_0 A}{2d}, \quad \text{where} \quad \frac{\varepsilon_0 A}{d} = C_0$$

$$\Rightarrow C = \frac{3}{2}C_0 \text{ Ans.}$$

3.14 Forces acting on conductors and dielectrics

Let us assume that two charged conductors 1 and 2 are placed in a dielectric. We want to find the force on the conductor 2. First of all, assume the position x of conductor 2 relative to a fixed point (conductor 1, say).

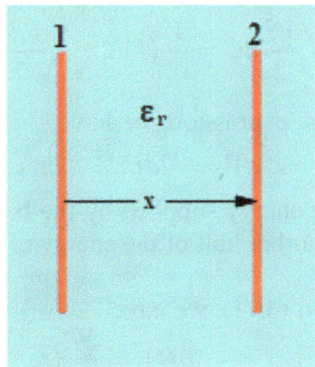

If you mentally pull conductor 2 by a small distance dx towards conductor 1, let the capacitance of the system increase by dC. Hence, an excess charge $dq = VdC$ flows from the supply as the capacitor is connected to the supply with a constant voltage V. In this process the battery (or supply or source) does positive work dW in sending a charge dq against the voltage V (or electromotive force ε of the battery). This is numerically equal to the energy supplied by the battery (sources). According to the conservation of energy, a fraction of the energy supplied by the source is spent in increasing the electrostatic potential energy of the system by dU and remaining amount of the supplied energy is utilized in doing mechanical work dW_{mech} on the system (conductor or dielectric on which you are intending to find the work), and energy loss dQ' in the form of heat, light and sound, etc. Balancing the energy, we can write

$$\begin{bmatrix} \text{Energy supplied} \\ \text{by the source} \\ \text{(or battery) } (dW_b) \end{bmatrix} = \begin{bmatrix} \text{Mechanical work done} \\ (dW_{mech.}) = Fdx \end{bmatrix} + \begin{bmatrix} \text{Change in electrostatic} \\ \text{potential energy } (dU) \end{bmatrix} + [dQ'],$$

where $dW_{\text{mech}} = F\,dx$; $F =$ desired force on conductor 2 (or dielectric) and the loss of energy dQ' can be neglected (in the battery and conductor).

For the sake of simplicity, we have taken all changes of energy as positive and ignored the energy loss in the conductors and dielectrics to derive a general expression of the force F acting on the conductor, which is given as

$$F = \frac{dW_b}{dx} - \frac{dU}{dx} \tag{3.10}$$

The above expression is valid for all processes such as constant charge and constant voltage. Let us discuss one by one.

Voltage V = Constant

If you want to find the force by connecting the capacitor with the supply of constant voltage V, the battery does work $dW_b(=Vdq)$ in sending a charge dq through a potential difference V. At the same time, the change in potential energy of the system can be given as

$$dU = d\left(\frac{1}{2}CV^2\right) = \frac{V^2}{2}dC = \frac{V}{2}(VdC) = \frac{Vdq}{2}, \quad (\because VdC = dq).$$

Putting $Vdq = dW_b$ in the last expression, we have

$$dW_b = 2dU \tag{3.11}$$

This tells us that, half of the energy supplied by the battery increases the potential energy of the system and the other half of the energy of the battery is spent in doing mechanical work.

Using equations (3.10) and (3.11), we have

$$F = +\frac{\partial U}{\partial x}\bigg|_{V=C} \tag{3.12}$$

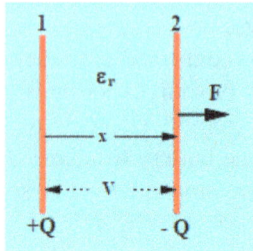

It does not matter which process you follow because the force F acting on the conductors (or dielectrics) does not depend on the types of processes. It only depends on the distribution of bound charges on the dielectric and the free charges on the conductor. However, if you follow the constant-charge method by mentally isolating (removing) the capacitor from the supply (even through the system (capacitor) really remains connected with the supply), you need not worry about the work done by the battery and the calculation will be easier. Let us use the above ideas in the following examples.

Charge $Q = $ Constant

If you want to find the force while disconnecting the capacitor with the supply, constant charge is maintained in each plate of the capacitor. As the supply (or battery) is disconnected, the work done by the battery will be zero.

$$dW_b = 0 \qquad (3.13)$$

Using the equations (3.10) and (3.13), we have

$$F = -\left.\frac{\partial U}{\partial x}\right|_{q=C} \qquad (3.14)$$

Both methods can be applied to find the force acting on the conductor or dielectric.

Example 10 (Forces on the conductor)

A constant supply voltage V is maintained between the plates of a parallel-plate air capacitor of plate area A. Let us immerse the capacitor completely in a liquid dielectric of relative permittivity ε_r. Find the force acting on the plates of the capacitor.

Solution

Using the constant voltage method, we have

$$F = \left.\frac{dU}{dx}\right|_{V=C}, \quad \text{where } U = \frac{1}{2}CV^2$$

$$\Rightarrow F = \frac{V^2}{2}\frac{dC}{dx}, \quad \text{where } C = \frac{\varepsilon_0\varepsilon_r A}{x}$$

$$\Rightarrow F = -\frac{\varepsilon_0\varepsilon_r A V^2}{2x^2} \text{ Ans.}$$

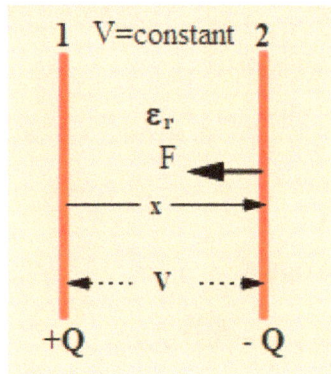

The negative sign signifies that the force F is attraction pointing to left. Ans.
Alternative method: By constant charge method we can write

$$F = -\frac{dU}{dx}, \quad \text{where} \quad U = \frac{Q^2}{2C};$$

$$\Rightarrow F = -\frac{q^2 d(1/C)}{2dx}, \quad \text{where} \quad C = \frac{\varepsilon_0 \varepsilon_r A}{x}$$

$$\Rightarrow F = -\frac{\varepsilon_0 \varepsilon_r A V^2}{2x^2} \quad \text{Ans.}$$

N.B: The force of attraction between the plates (conductors) decreases by a factor ε_r at constant charge Q and increases by a factor ε_r at $V =$ constant, after inserting the dielectric. Both methods are equally valid in all cases.

Problem 1 (Cylindrical capacitor)
Find the capacitance per unit length of a coaxial cable (a system of two thin long coaxial conducting cylinders with dielectric of absolute permittivity ε completely filling the gap between the cylinders). The radius of the core (inner cylinder) is a and the radius the outer cylinder is b.
Solution
Let us consider a length l of the coaxial cable. If the inner and outer cylinders are given charges $+Q$ and $-Q$, respectively, the potential difference between them is

$$V = \int_a^b E \cdot dr, \quad \text{where} \quad E = \frac{\lambda}{2\pi\varepsilon_0\varepsilon_r r}$$

$$\Rightarrow V = \int_a^b \frac{\lambda}{2\pi\varepsilon r} dr, \quad \text{where} \quad \lambda = Q/l$$

$$\Rightarrow V = \frac{Q}{2\pi\varepsilon l} \ln\frac{b}{a}$$

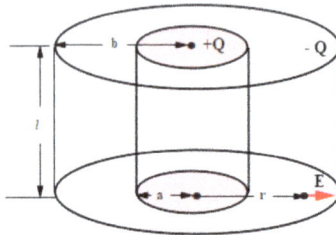

So, the capacitance of the system is

$$C = \frac{Q}{V} = \frac{2\pi\varepsilon l}{\ln\frac{b}{a}}$$

For the coaxial cable, the capacitance per unit length is given as

$$\frac{dC}{dl} = \frac{2\pi\varepsilon}{\ln\frac{b}{a}} \quad \text{Ans.}$$

Problem 2 A metallic slab of thickness x is filled in the parallel-plate capacitor of plate area A and distance of separation d.
   (a) Find the capacitance of the system.
   (b) By what distance should the plates be shifted to regain the original capacitance?

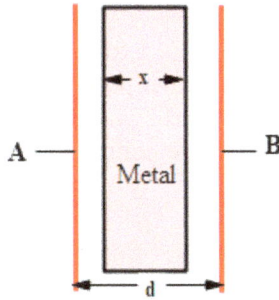

Solution

(a) The permittivity of the conducting slab is $\varepsilon_r = \infty$
The capacitance with a dielectric slab is given as

$$C = \frac{\varepsilon_0 A}{d - x + \frac{x}{\varepsilon_r}}$$

Putting $\varepsilon_r = \infty$, we have

$$C = \frac{\varepsilon_0 A}{d - x} \quad \text{Ans.}$$

(b) Then, the plates will have to be separated (pulled away) through an additional distance of x to get the original capacitance $C = \frac{\varepsilon_0 A}{d}$ Ans.

N.B: Since the field inside the metallic slab is zero. Electrostatic energy in the region occupied by the slab is zero. If the slab completely fills the space between the capacitor plates, put $x = d$ to obtain $C = \infty$. It is equivalent to connecting the plates by a conducting wire. The physical significance of infinite capacitance is that it requires an infinite charge to change the potential difference between the plates and hence it is impossible to create a potential difference between the plates when connected by a conducting wire or slab.

Problem 3 Three capacitors are connected between A and B as shown in the figure below. Find the effective capacitance C_{AB} between A and B. Put $C_1 = C$, $C_2 = 2C$ and $C_3 = C$.

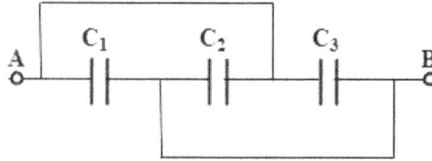

Solution

First of all, number the junctions as 1 and 2. Points A and 2 are at the same potential; points 1 and B are at the same potential. Then, you can see that C_1 lies between 1 and A, C_2 lies between 1 and 2 and C_3 lies between 2 and B. This means that these capacitors C_1, C_2 and C_3 are connected in parallel between A and B.

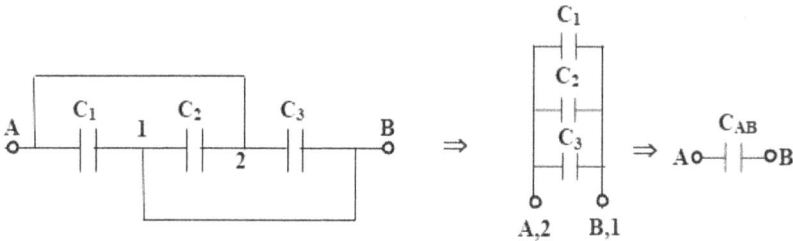

Hence, the effective capacitance between A and B is

$$C_{AB} = C_1 + C_2 + C_3$$
$$= C + 2C + C = 4C \text{ Ans.}$$

Problem 4 (a) Find the electrostatic energy stored due to a conducting sphere of radius R carrying a charge Q. (b) Find the radius of a sphere in which nth fraction of the total energy is stored.

Solution

Method 1:

Since $E = 0$ inside the conducting sphere, energy is stored in the surrounding space from $r_1 = R$ to $r_2 = \infty$. As the E-field due to a conducting sphere possesses spherical symmetry, total energy stored can be given as

$$U = \frac{\varepsilon_0}{2} \int_R^\infty E^2 dV,$$

where $E = \frac{Q}{4\pi\varepsilon_0 r^2}$; $r > R$ and $dV = 4\pi r^2 dr$

$$\Rightarrow U = \frac{\varepsilon_0}{2} \int_R^\infty (\frac{Q}{4\pi\varepsilon_0 r})^2 4\pi r^2 dr$$

$$= \frac{Q^2}{8\pi\varepsilon_0 R} \text{ Ans.}$$

(3.15)

Method 2:

Directly we can use the expression

$$U = \frac{Q^2}{2C},$$

where $C = 4\pi\varepsilon_0 R$ for spherical capacitor

$$\Rightarrow U = \frac{Q^2}{8\pi\varepsilon_0 R} \text{ Ans.}$$

Method 3:

(a) The energy stored by an isolated conductor of charge Q and potential V is

$$U = \frac{1}{2} QV$$

(3.16)

Since the conductor is spherical its potential is given as

$$V = \frac{Q}{4\pi\varepsilon_0 R}$$

(3.17)

Putting V from equation (3.17) in equation (3.16), we have

$$U = \frac{Q^2}{8\pi\varepsilon_0 R} \text{ Ans.}$$

(b) Using equation (3.15) the energy stored between $r = R$ and $r = r$ is given as

$$U' = \frac{\varepsilon_0}{2} \int_R^r \left(\frac{Q}{4\pi\varepsilon_0 r}\right)^2 4\pi r^2 dr$$

$$\Rightarrow U' = \frac{Q^2}{8\pi\varepsilon_0}\left(\frac{1}{R} - \frac{1}{r}\right)$$

Putting $U' = U/n$, we have

$$U' = \frac{Q^2}{8\pi\varepsilon_0}\left(\frac{1}{R} - \frac{1}{r}\right) = \frac{1}{n}\frac{Q^2}{8\pi\varepsilon_0 R}$$

$$\Rightarrow \left(\frac{1}{R} - \frac{1}{r}\right) = \frac{1}{nR}$$

$$\Rightarrow \frac{1}{r} = \frac{1}{R}\left(1 - \frac{1}{n}\right)$$

$$\Rightarrow r = \frac{nR}{n-1} \quad \text{Ans.}$$

Problem 5 Find the total electrostatic potential energy possessed by a sphere of radius R and charge Q having uniform volume charge distribution.

Solution

The total energy stored inside the sphere is

$$U_1 = \frac{\varepsilon_0}{2}\int E_1^2 dV,$$

where $E_1 = \frac{Q}{4\pi\varepsilon_0 R^3}r$; $r \leqslant R$ and $dV = 4\pi r^2 dr$

$$\Rightarrow U_1 = \frac{\varepsilon_0}{2}\int_0^R \left(\frac{Q}{4\pi\varepsilon_0 R^3}\right)^2 4\pi r^2 dr = \frac{Q^2}{40\pi\varepsilon_0 R}$$

The total energy stored outside the sphere is

$$U_2 = \frac{\varepsilon_0}{2}\int E_2^2 dV,$$

where $E_2 = \frac{Q}{4\pi\varepsilon_0 r}$ and $dV = 4\pi r^2\, dr$

$$U_2 = \frac{\varepsilon_0}{2}\int_R^\infty \left(\frac{Q}{4\pi\varepsilon_0 r}\right)^2 4\pi r^2\, dr = \frac{Q^2}{8\pi\varepsilon_0 R}$$

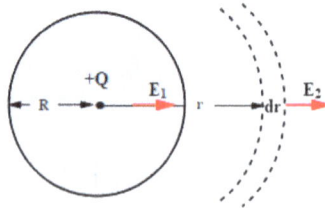

Then the total energy stored is

$$U = U_1 + U_2$$

$$= \frac{Q^2}{40\pi\varepsilon_0 R} + \frac{Q^2}{8\pi\varepsilon_0 R} = \frac{3Q^2}{20\pi\varepsilon_0 R} \quad \text{Ans.}$$

N.B: 1. We can solve this problem by using the formula $U = \frac{1}{2}\int V\, dq$.

2. If we put $R = 0$, the electrostatic energy stored due to a point charge Q will be infinite, which limits the classical mechanics and then quantum mechanics will come in to play to solve this problem of infinite energy.

Problem 6

(a) Find the electrostatic energy stored in the cylindrical region between $r_1 = a$ to $r_2 = b$ due to a long straight-line charge of linear charge density λ. (b) Referring to (a), find the radius r which divides the space into two regions where equal energy is stored.

Solution

For long straight-line charge the electric field at a radial distance r is

$$E = \frac{\lambda}{2\pi\varepsilon_0 r} \tag{3.18}$$

Take a thin cylindrical shell of space between r and $r + dr$. The electrostatic energy stored in this region is

$$dU = \frac{1}{2}\varepsilon_0 E^2 dv, \quad \text{where} \quad dv = 2\pi l dr \tag{3.19}$$

Using equations (3.18) and (3.19),

$$dU = \frac{1}{2}\varepsilon_0 \left(\frac{\lambda}{2\pi\varepsilon_0 r}\right)^2 2\pi r l dr = \frac{\lambda^2 l}{4\pi\varepsilon_0}\frac{dr}{r}$$

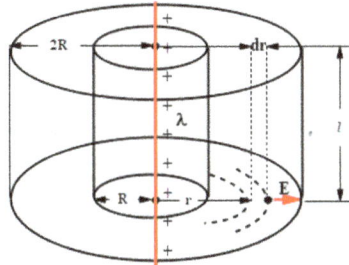

Then, the total energy stored in the cylindrical region of radii lying between $r_1 = R$ and $r_2 = 2R$ is

$$U = \int dU = \frac{\lambda^2 l}{4\pi\varepsilon_0}\int_a^b \frac{dr}{r} = \frac{\lambda^2 l}{4\pi\varepsilon_0}\ln(b/a) \text{ Ans.}$$

(b) The total energy stored in the cylindrical region of radii lying between $r_1 = a$ and $r_2 = r$ is

$$U' = \int dU = \frac{\lambda^2 l}{4\pi\varepsilon_0}\int_a^r \frac{dr}{r} = \frac{\lambda^2 l}{4\pi\varepsilon_0}\ln(r/a)$$

Putting $U' = U/2$, we have

$$U' = \frac{\lambda^2 l}{4\pi\varepsilon_0} \ln(r/a) = \frac{1}{2}\frac{\lambda^2 l}{4\pi\varepsilon_0} \ln(b/a)$$

$$\Rightarrow \ln(r/a) = \frac{1}{2}\ln(b/a)$$

$$\Rightarrow r = \sqrt{ab} \quad \text{Ans.}$$

Problem 7 Two parallel conducting plates have charge Q_1 and Q_2. The distance of separation d between the plates is much smaller than the linear dimensions of the plates. (a) Find the charge distribution at each face of the conducting plates having charged Q_1 and Q_2 such that the system of the charged plates possesses minimum electrostatic potential energy. (b) What is minimum electrostatic energy stored in the space between the plates.

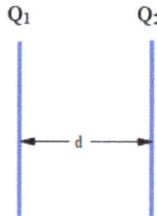

Solution

(a) Let $x =$ charge induced at the inner surface of the plate 1. Using the property of conductor and conservation of charge, the charge distribution is shown for all surfaces. The fields in the regions 1, 2 and 3 can be given as

$$E_1 = \frac{Q_1 - x}{\varepsilon_0 A}, \quad E_2 = \frac{x}{\varepsilon_0 A} \quad \text{and} \quad E_3 = \frac{Q_2 + x}{\varepsilon_0 A} \quad \text{respectively.}$$

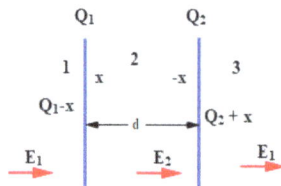

Let $v =$ volume (very large but finite) occupied in regions 1 and 3 individually. The field energy stored in the entire space is

$$U = \tfrac{1}{2}\varepsilon_0 E_1^2 v + \tfrac{1}{2}\varepsilon_0 E_2^2 Ad + \tfrac{1}{2}\varepsilon_0 E_3^2 v$$

$$= \tfrac{\varepsilon_0 v}{2}(E_1^2 + E_3^2) + \tfrac{\varepsilon_0 Ad}{2} E_2^2$$

$$= \tfrac{\varepsilon_0 v}{2}\left\{\left(\tfrac{Q_1 - x}{\varepsilon_0 A}\right)^2 + \left(\tfrac{Q_2 + x}{\varepsilon_0 A}\right)^2\right\} + \tfrac{\varepsilon_0 Ad}{2}\left(\tfrac{x}{\varepsilon_0 A}\right)^2$$

$$\Rightarrow U = \frac{v}{2\varepsilon_0}\left\{(Q_1 - x)^2 + (Q_2 + x)^2\right\} + \frac{Ad.\, x^2}{2\varepsilon_0}$$

For minimum electrostatic energy, $\frac{dU}{dx} = 0$ (for stable configuration)

$$\Rightarrow \frac{v}{2\varepsilon_0}\left\{2(Q_1 - x)(-1) + 2(Q_2 + x)\right\} + \frac{2Adx}{2\varepsilon_0} = 0$$

$$\Rightarrow x = \frac{Q_1 - Q_2}{2 + \frac{Ad}{V}}$$

Since V is very large, we have

$$x \simeq \frac{Q_1 - Q_2}{2} \text{ Ans.}$$

(b) The electric field between the plates is

$$E_2 = \frac{x}{2\varepsilon_0 A} + \frac{x}{2\varepsilon_0 A} = \frac{x}{\varepsilon_0 A} = \frac{Q_1 - Q_2}{2\varepsilon_0 A}$$

Then the electrostatic energy stored in between the plates is

$$U = \frac{1}{2}\varepsilon_0 E_2^2 Ad = \frac{1}{2}\varepsilon_0\left\{\frac{(Q_1 - Q_2)}{2\varepsilon_0 A}\right\}^2 Ad = \frac{(Q_1 - Q_2)^2 d}{8\varepsilon_0 A} \text{ Ans.}$$

N.B: So, the above results are not exact but approximated after taking an infinite value of the volume V. The equilibrium (electrostatic) charge distribution follows minimum potential energy condition.

Problem 8 Find the self, mutual and total energy of interaction of two concentric spherical conductors of charges Q_1 and Q_2 and radii r_1 and r_2 ($r_1 < r_2$).
  Solution
  Method 1:
  The self-energy of the inner shell is

$$U_1 = \tfrac{\varepsilon_0}{2}\int E_1^2 dV$$

$$= \tfrac{\varepsilon_0}{2}\int_{r_1}^{\infty}\left(\tfrac{Q_1}{4\pi\varepsilon_0 r^2}\right)^2 4\pi r^2 dr$$

$$= \tfrac{Q^2}{8\pi\varepsilon_0 r_1} \text{ Ans.}$$

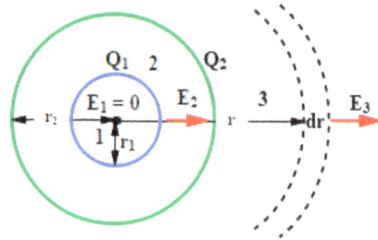

The self-energy of the outer shell is

$$U_2 = \frac{\varepsilon_0}{2} \int E_2^2 \, dV$$

$$= \frac{\varepsilon_0}{2} \int_{r_2}^{\infty} \left(\frac{Q_2}{4\pi\varepsilon_0 r^2} \right)^2 4\pi r^2 dr = \frac{Q_2^2}{8\pi\varepsilon_0 r_2} \text{ Ans.}$$

The mutual energy between the shells is

$$U_{12} = \frac{\varepsilon_0}{2} \int \vec{E_1} \cdot \vec{E_2} \, dV$$

$$= \frac{\varepsilon_0}{2} \int_{r_2}^{\infty} \left(\frac{Q_1}{4\pi\varepsilon_0 r^2} \hat{r} \right) \cdot \left(\frac{Q_2}{4\pi\varepsilon_0 r^2} \right) 4\pi r^2 dr$$

$$= \frac{Q_1 Q_2}{4\pi\varepsilon_0 r_2} \text{ Ans.}$$

Then, the total energy of the system is

$$U = U_1 + U_2 + U_{12}$$

$$= \frac{1}{8\pi\varepsilon_0} \left(\frac{Q_1^2}{r_1} + \frac{Q_2^2}{r_2} + \frac{2Q_1 Q_2}{r_2} \right) \text{ Ans.}$$

Method 2:

The total energy can also be found as

$$U = \frac{\varepsilon_0}{2} \left\{ \int_0^{r_1} E^2 dV + \int_{r_1}^{r_2} E^2 dV + \int_{r_2}^{\infty} E^2 dV \right\},$$

where

$$E = 0; \quad r < r_1$$

$$= \frac{Q_1}{4\pi\varepsilon_0 r^2}; \quad r_1 < r < r_2$$

$$= \frac{Q_1 + Q_2}{4\pi\varepsilon_0 r^2}; \quad r > r_2$$

N.B: You should not be tempted to write

$$U_{\text{mutual}} = \frac{Q_1^2}{8\pi\varepsilon_0} \left(\frac{1}{r_1} - \frac{1}{r_2} \right)$$

because it is the energy stored in between the shells but not the mutual energy because $\vec{E_1} \cdot \vec{E_2} = 0$ in region 2 between the shells.

Problem 9 Three dielectrics of relative permittivity $\varepsilon_{r_1} = 1$, $\varepsilon_{r_2} = 2$ and $\varepsilon_{r_3} = 3$ are introduced in a parallel-plate capacitor of plate area A and separation d. Find the effective capacitance between A and B. Assume that the height of each dielectric is the same.

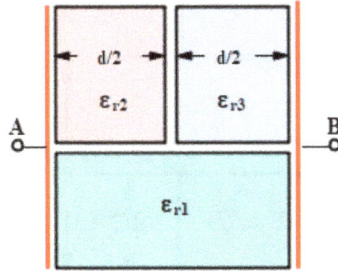

Solution

The given system can be made equivalent to two capacitors C_1 and C_2 in parallel.

For the lower part (area $= A/2$) the capacitance is comprised of one dielectric, which is given as

$$C_1 = \frac{\varepsilon_0 \varepsilon_{r_1} \left(\frac{A}{2}\right)}{d} = \frac{\varepsilon_0 A}{2d} \tag{3.20}$$

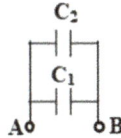

For the upper part (area $= A/2$) the capacitance is comprised of two dielectrics which is given as

$$C_2 = \frac{\varepsilon_0 \left(\frac{A}{2}\right)}{\frac{d/2}{\varepsilon_{r_2}} + \frac{d/2}{\varepsilon_{r_3}}} = \frac{\varepsilon_0 A/2}{\frac{d}{2}\left(\frac{1}{2} + \frac{1}{3}\right)} = \frac{6\varepsilon_0 A}{5d} \tag{3.21}$$

Then the effective capacitance between A and B is the parallel combination of C_1 and C_2.

$$\Rightarrow C_{AB} = C_1 + C_2 \tag{3.22}$$

Using equations (3.20), (3.21) and (3.22),

$$C_{AB} = \frac{\varepsilon_0 A}{2d} + \frac{6\varepsilon_0 A}{5d} = \frac{17\varepsilon_0 A}{10d} \text{ Ans.}$$

Problem 10 A dielectric slab thickness $d/2$ of relative permittivity ε_r is introduced in a parallel-plate capacitor of plate area A and separation d. Find the capacitance between the terminals A and B, if $\varepsilon_r = 2$.

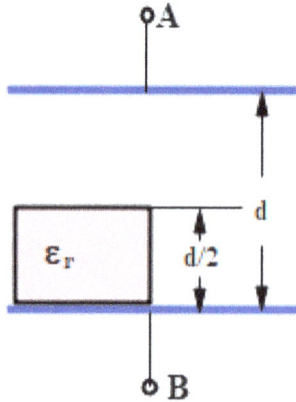

Solution

The given system is broken into three capacitors C_1, C_2 and C_3 as shown in the figure,

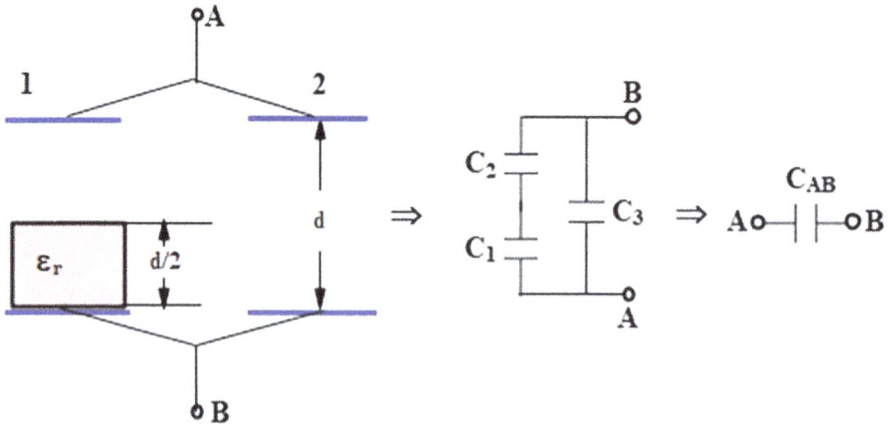

where $C_1 = \frac{\varepsilon_0 A/2}{d/2}$, $C_2 = \frac{\varepsilon_0 \varepsilon_r A/2}{d/2}$ and $C_3 = \frac{\varepsilon_0 A/2}{d}$.

The capacitors C_1 and C_2 are placed in series and this series combination is in parallel with C_3. Then, the equivalent capacitance between the terminals A and B is

$$C_{AB} = \frac{C_1 C_2}{C_1 + C_2} + C_3$$

Putting the values of the capacitors C_1, C_2 and C_3 in the last equation and simplifying the factors, we have

$$C_{AB} = \frac{\varepsilon_0 A}{d}\left(\frac{\varepsilon_r}{\varepsilon_r + 1} + \frac{1}{2}\right), \quad \text{where } \varepsilon_r = 2$$

$$\Rightarrow C_{AB} = \frac{7\varepsilon_0 A}{6d} \text{ Ans.}$$

N.B: It does not make any difference if the dielectric slab is displaced inside the capacitor.

Problem 11 (Force on the solid dielectric)

A dielectric slab of relative permittivity ε_r and thickness d is introduced in a parallel-plate capacitor of plate area A and separation d. Find the force acting on the smooth dielectric slab when the voltage between the plates is V and the length of the square plates is l.

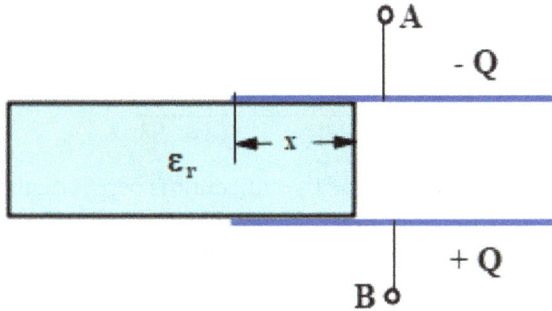

Solution

Method 1 (constant charge method):

Force acting on the dielectric slab can be given as

$$F = -\frac{dU}{dx}, \quad \text{where } U = \frac{Q^2}{2C}$$

$$\Rightarrow F = \frac{Q^2}{2C^2}\frac{dC}{dx}, \quad \text{where } C = \frac{\varepsilon_0 l}{d}(\varepsilon_r x + d - x)$$

$$= \frac{Q^2}{2C^2}(\varepsilon_r - 1)\frac{\varepsilon_0 l}{d}$$

If the capacitor is connected to a constant voltage, putting $\frac{Q}{C} = V$ in the last expression, we have

$$F = (\varepsilon_r - 1)\frac{\varepsilon_0 l V^2}{2d}$$

The positive sign signifies that the force points to right and the magnitude of the force on the dielectric is constant. Ans.

Method 2 (constant voltage method):
In this method the force is given as

$$F = +\frac{dU}{dx}, \quad \text{where } U = \frac{1}{2}CV^2$$

$$\Rightarrow F = \frac{V^2}{2}\frac{dC}{dx},$$

where

$$C = \frac{\varepsilon_0 l}{d}(\varepsilon_r x + d - x)$$

$$\Rightarrow F = (\varepsilon_r - 1)\frac{\varepsilon_0 l V^2}{2d} \quad \text{Ans.}$$

N.B: If the capacitor remains disconnected from the supply after acquiring a charge Q, the force acting on the dielectric slab will no more be a constant; it varies with x. In other words, $F = f(x)$, if $Q = C$. We can show that

$$F = \frac{Q^2(\varepsilon_r - 1)d}{2\varepsilon_0 l\{(\varepsilon_r - 1)x + l\}^2}$$

We can see that the electric field inside the capacitor is perpendicular to the dielectric slab. So, it cannot pull the slab into the capacitor. Then, what is responsible for accelerating the slab in to the capacitor? You can notice that the electric field near the edge of the plates of the capacitor is 'non-uniform field' which is represented by a curved lines of force. This is known as 'fringing of lines of force'. So, an element of the bound charge at the surface of the dielectric experiences an elementary force δF by the free charges of the capacitor plates deposited near its edge. The resultant of the component of $d\vec{F}$ parallel to the plates, that is, $\sum \delta \vec{F}_{\parallel}$, pulls the plate into the capacitor; the normal components of $\delta \vec{F}$, that is, $\sum \delta \vec{F}_n$, stretches the slab.

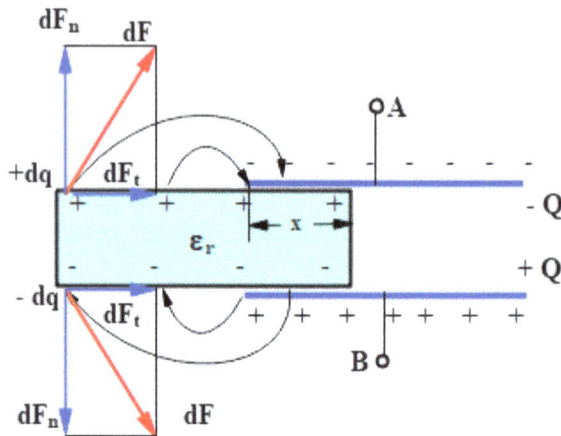

Please refer to problem 13 to find the stretching force of the dielectric. Remember that we have neglected the fringing electric field (E-lines) to simplify the calculation of capacitance and found an approximate value. It is because the exact variation of E is not known near the edge of the plates. But we should not misinterpret that the field outside the capacitor drops suddenly to zero, for which $\oint \vec{E} \cdot \vec{dl} \neq 0$, which is contrary to the conservativeness of the static electric field. Thus, the cause of pushing (but not pulling) force on the dielectric is the 'fringing field' which is generally neglected for the sake of mathematical simplicity of the calculation of capacitance, field, energy, etc. It should be noted that all calculations of electrostatics are based on the conservative property of the electrostatic field $\vec{E}$ which can be given as $\oint \vec{E} \cdot \vec{dl} = 0$.

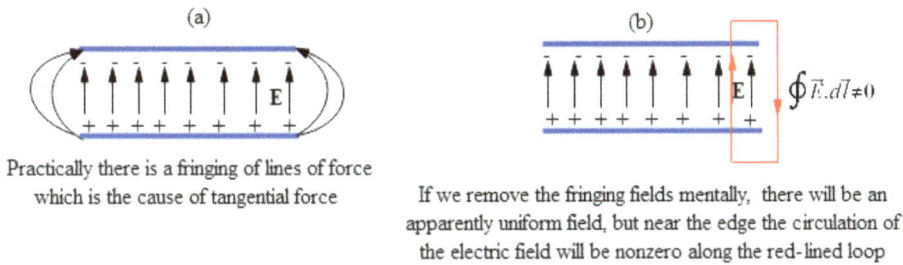

(a)

(b)

Practically there is a fringing of lines of force which is the cause of tangential force

If we remove the fringing fields mentally, there will be an apparently uniform field, but near the edge the circulation of the electric field will be nonzero along the red-lined loop

Problem 12 A dielectric slab of relative permittivity ε_r is placed on the bottom plate of the capacitor. Find the stress at the surface the dielectric slab. Assume σ = magnitude of surface charge density of the plates.

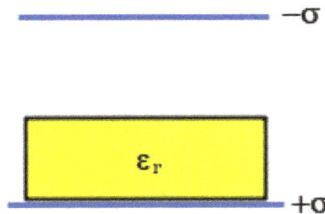

$-\sigma$

ε_r

$+\sigma$

Solution
The force acting on the free surface of the dielectric can be given as

$$F = -\frac{dU}{dx}\bigg|_{Q=C} \left(\text{or} = +\frac{dU}{dx}\bigg|_{V=C} \right),$$

where

$$U = \frac{1}{2}\frac{Q^2}{2C}$$

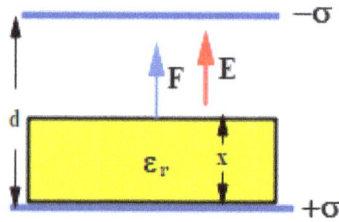

Then, the force acting on the dielectric is

$$F = \frac{Q^2}{2C^2}\frac{dC}{dx},$$

where $C = \frac{\varepsilon_0 A}{d - x + \frac{x}{\varepsilon_r}}$

$$\Rightarrow F = \frac{Q^2}{2C^2}\frac{\varepsilon_0 A}{\left(d - x + \frac{x}{\varepsilon_r}\right)^2}\left(\frac{1}{\varepsilon_r} - 1\right),$$

where $d - x + \frac{x}{\varepsilon_r} = \frac{\varepsilon_0 A}{C}$

$$\Rightarrow F = \frac{Q^2}{2A\varepsilon_0}\left(1 - \frac{1}{\varepsilon_r}\right)$$

Then, the pressure due to the perpendicular electric field is,

$$p = \frac{F}{A} = \frac{\sigma^2}{2\varepsilon_0}\left(1 - \frac{1}{\varepsilon_r}\right),$$

where $\frac{\sigma}{\varepsilon_0} = E$ (applied field or field outside the dielectric). Ans.

N.B: If the dielectric is a liquid, the height of the excess liquid column of the liquid due to the electrostatic pressure can be given as

$$h = \frac{\sigma^2}{2\varepsilon_0 \rho g}\left(1 - \frac{1}{\varepsilon_r}\right).$$

Problem 13 (Stretching force on a dielectric)

A dielectric slab of relative permittivity ε_r is placed in a parallel-plate capacitor. Find the pulling force acting on the surface of the dielectric due to the electric field of free and bound charges. Assume A = plate area and σ = surface charge density of the plates.

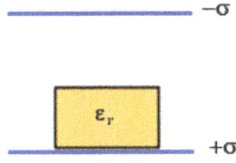

Solution

The upper surface of the dielectric slab experiences a force due to the net electric field E of the charges of the capacitor plates (free charges) and the charges induced at the lower surface of the dielectric. The upper surface cannot be influenced by the electric field due to the charges induced on it.

The electric field at any point on surface 3 due to surfaces 1, 2 and 4 is

$$\vec{E} = \vec{E_1} + \vec{E_2} + \vec{E_4}$$
$$= \frac{\sigma}{2\varepsilon_0}\hat{j} - \frac{\sigma'}{2\varepsilon_0}\hat{j} + \frac{\sigma}{2\varepsilon_0}\hat{j}$$
$$= \left(\frac{\sigma}{\varepsilon_0} - \frac{\sigma'}{2\varepsilon_0}\right)\hat{j},$$

where $\sigma' = \sigma\left(1 - \frac{1}{\varepsilon_r}\right)$

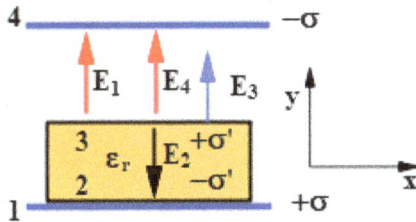

$$\Rightarrow \vec{E} = \frac{\sigma}{2\varepsilon_0}\left(1 + \frac{1}{\varepsilon_r}\right)\hat{j}$$

The force acting on surface 3 of the dielectric is

$$\vec{F} = q'\vec{E} = (\sigma'A)\vec{E}$$
$$= \sigma'A\frac{\sigma}{2\varepsilon_0}\left(1 + \frac{1}{\varepsilon_r}\right)\hat{j},$$

where $\sigma' = \sigma\left(1 - \frac{1}{\varepsilon_r}\right)$

$$\Rightarrow F = \frac{\sigma^2}{2\varepsilon_0}\left(1 - \frac{1}{\varepsilon_r^2}\right)A \text{ Ans.}$$

N.B: The above expression should not be misinterpreted as the net force acting on the surface. It is the force due to the fields of $+\sigma$, $-\sigma$ and $-\sigma'$. If you consider the downward elastic force acting on surface 3 during the polarization of the dielectric, you will get the net upward force $F = \frac{\sigma^2}{2\varepsilon_0}\left(1 - \frac{1}{\varepsilon_r}\right)$.

Problem 14 (Rise of the liquid dielectric)

A vertical charged parallel-plate capacitor is slowly introduced in a liquid dielectric of relative permittivity of ε_r. Find the equilibrium height of the liquid inside the capacitor. Assume that the plates are square plates of length l, $V =$ potential difference between the plates and $d =$ distance between the plates. Neglect the surface tension effect of the liquid.

Solution

The liquid dielectric gets polarized with the electric field near the edge of the capacitor plates. As a result, the induced or permanent dipoles of the liquid moves towards the region of stronger field inside the capacitor from the weaker field at the edge. Hence the liquid just outside the capacitor gets a push due to the non-uniform electric field. Eventually, the liquid rises till the push is balanced by the weight of the liquid column inside the capacitor.

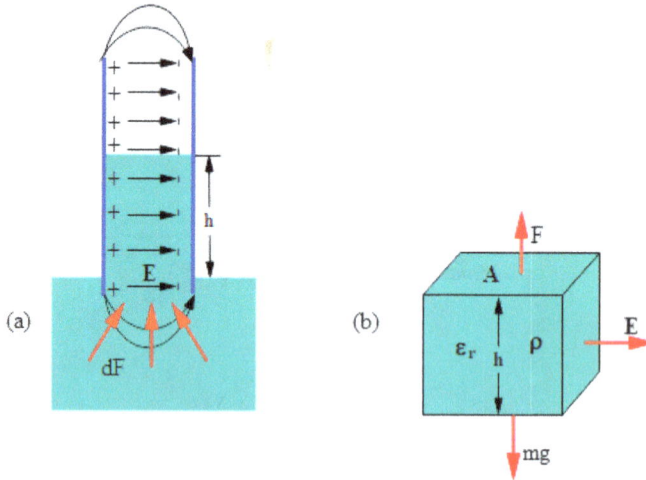

The push on the liquid dielectric is exactly equal to the push on the solid dielectric in problem 11, which can be given as

$$F = \frac{1}{2}(\varepsilon_r - 1)\varepsilon_0 l\frac{V^2}{d}$$

$$= \frac{1}{2}(\varepsilon_r - 1)\varepsilon_0 ld\frac{V^2}{d^2}$$

The liquid will be stationary when the vertical (tangential force) acting on the liquid is equal to weight of the liquid.

$$F = \frac{1}{2}(\varepsilon_r - 1)\varepsilon_0 dl\frac{V^2}{d^2} = mg$$

$$\Rightarrow \frac{1}{2}(\varepsilon_r - 1)\varepsilon_0 dl\frac{V^2}{d^2} = dlh\rho g$$

$$\Rightarrow h = \frac{(\varepsilon_r - 1)\varepsilon_0 E^2}{2\rho g} \quad \text{Ans.}$$

Problem 15 (Stress in a liquid dielectric)

Referring to the last problem, find the total stress of the liquid of dielectric constant ε_r at the edge of the capacitor if the tangential and normal components of the electric field inside the dielectric are E_t and E_n, respectively.

Solution

The rise of liquid is due to the push on the liquid at the bottom of the capacitor but not pull at the surface of liquid inside the capacitor. In the last problem the pressure rises due to the component of the electric field tangential and normal to the dielectric. Putting $\frac{V}{d} = E$, $ld = A$ (area of the cross-section of the dielectric), in the expression

$$F = \frac{1}{2}(\varepsilon_r - 1)\varepsilon_0 ld\frac{V^2}{d^2},$$

obtained in the last problem, the stress on the dielectric due to the electric field which is tangential to the liquid surface is

$$p = \frac{F}{A} = \frac{\varepsilon_0}{2}(\varepsilon_r - 1)E^2$$

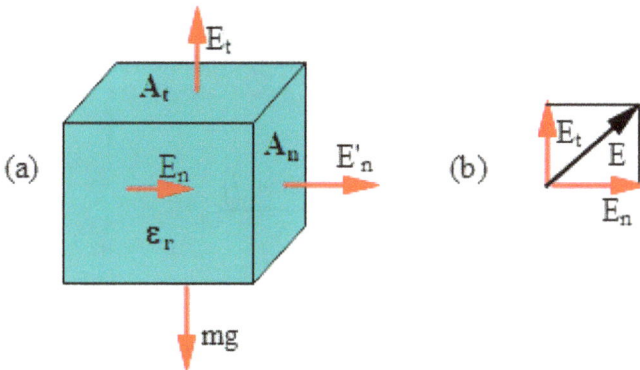

The rise in pressure due to the tangential component of $\vec{E}$, that is, E'_t, is given as

$$p_1 = \frac{\varepsilon_0(\varepsilon_r - 1)E'^2_t}{2} = \frac{\varepsilon_0(\varepsilon_r - 1)E^2_t}{2} (\because E_t = E'_t)$$

Referring to problem 12 the rise in pressure due to the normal component of E, that is E_n is

$$p_2 = \frac{\varepsilon_0 E'^2_n}{2}\left(1 - \frac{1}{\varepsilon_r}\right) = \frac{\varepsilon_0 \varepsilon_r E^2_n}{2}(\varepsilon_r - 1) \ (\because \varepsilon_r E_n = E'_n)$$

Then, the total pressure due to an electric field inside the dielectric is

$$p = p_1 + p_2$$
$$= \frac{\varepsilon_0(\varepsilon_r - 1)}{2}(E^2_t + \varepsilon_r E^2_n),$$

where E_t and E_n are the components of the electric field *inside* the dielectric. Ans.

Problem 16 A parallel plate air capacitor of square plate of length l is placed in a liquid dielectric of density ρ and relative permittivity ε_r. If the height of the liquid column in the capacitor is x when the capacitor just touches the liquid surface, find the charge of the capacitor. Assume that the capacitor is ideal and disconnected from the battery after its charging.

Solution
Let us use the expression

$$F = \frac{dU}{dx}\bigg|_{V=C}, \text{ where } U = \frac{1}{2}CV^2$$

$$\Rightarrow F = \frac{V^2}{2}\frac{dC}{dx} \tag{3.23}$$

After the length x of the liquid column enters the capacitor, the capacitance becomes

$$C = \frac{\varepsilon_0(\varepsilon_r A_1 + A_2)}{d}$$

Putting $A_1 = lx$, $A_2 = (l - x)l$, we have

$$C = \frac{\varepsilon_0 l\{(\varepsilon_r - 1)x + l\}}{d} \tag{3.24}$$

$$\text{Then, } \frac{dC}{dx} = \frac{\varepsilon_0 l(\varepsilon_r - 1)}{d} \tag{3.25}$$

The force acting on the free surface of the dielectric can be given as

$$F = -\frac{dU}{dx}\bigg|_{Q=C} \left(\text{or}= +\frac{dU}{dx}\bigg|_{V=C}\right),$$

Putting $U = \frac{1}{2}\frac{Q^2}{2C}$, we have

$$F = \frac{Q^2}{2C^2}\frac{dC}{dx} \tag{3.26}$$

Using the last two equations,

$$F = \frac{Q^2(\varepsilon_r - 1)\varepsilon_0 l}{2dC^2} \tag{3.27}$$

Since this force balances the weight of the liquid column, we can write

$$F = \frac{Q^2(\varepsilon_r - 1)\varepsilon_0 l}{2dC^2} = mg$$

$$Q = \sqrt{\frac{2mgd}{(\varepsilon_r - 1)\varepsilon_0 l}}\ C \tag{3.28}$$

Putting C from equation (3.23) in equation (3.28),

$$\Rightarrow Q = \frac{\varepsilon_0 l\{(\varepsilon_r - 1)x + l\}}{d}\sqrt{\frac{2mgd}{(\varepsilon_r - 1)\varepsilon_0 l}}$$

Putting $m = \rho l dx$ and simplifying the factors,

$$Q = l\{(\varepsilon_r - 1)x + l\}\sqrt{\frac{2\varepsilon_0 \rho g x}{(\varepsilon_r - 1)}} \quad \text{Ans.}$$

IOP Publishing

Problems and Solutions in Electricity and Magnetism

Pradeep Kumar Sharma

Chapter 4

Current, resistance and electromotive force

4.1 Introduction

This chapter starts with the definition of an electric current which is basically a flow of a charge. In this chapter we will discuss the concept of electromotive force (emf) which is responsible for producing an electric current in a conductor. How does a seat of emf produce an electric field in a metal? How do the electrons in a metal move by the action of this electric field? How does a current heat a piece of conductor? Will the electric field inside of a current-carrying conductor be zero? All these questions will be addressed in this chapter.So, this chapter will form the basis of understanding current electricity.

4.2 Electric current

4.2.1 Definition of current

Motion of a charge particle (an electron, proton or ions) constitutes an electric current. In hydrodynamics, liquid current is defined as the rate of flow of mass (or volume) of a fluid through a given cross-section or plane perpendicular to the motion of the fluid. Following the same idea, electric current can also be defined as the rate of flow of charge through a given area. If a charge ΔQ crosses a given plane perpendicularly during a time Δt, the average rate of flow of charge is defined as an average current.

$$i_{\mathrm{av}} = \frac{\Delta Q}{\Delta t}$$

doi:10.1088/978-0-7503-6477-5ch4

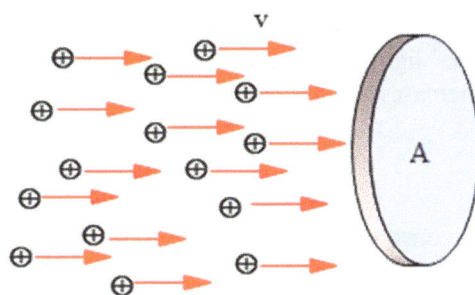

Total charge ΔQ passes through the area
A during a time Δt constituting an
average current $i_{av} = \Delta Q/\Delta t$

For a continuous charge distribution (even though the charge carriers such as electrons, protons are discrete), the instantaneous current can be given as

$$i = \frac{dQ}{dt} = \frac{dq}{dt}$$

So, electric current is defined as a time rate of flow of charge. In general, it is a scalar quantity.

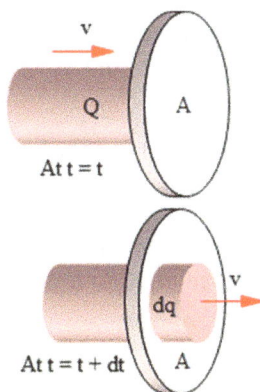

Current = rate of flow of charge = dq/dt

4.3 Types of current

Current in vacuum: If a charged particle moves in a vacuum (free space) it does not experience any resistance. If an electric field E acts on a charge $+q$, the speed of a charged particle at time t is:

$$v = \frac{qE}{m}t + v_0; \quad v_0 = \text{initial speed}$$

In the process of increasing the speed of the charge particle, if the speed v approaches the speed of light c, relativistic effect comes in to play. The mass of the charged particle increases with speed as

$$m = \frac{m_0}{\sqrt{1 - \frac{v^2}{c^2}}}$$

As the charge remains constant, the specific charge (q/m) decreases with increasing speed.

Current in fluid: Let us consider a fluid (gas) of positive charge particles. As each positive charged particle collides with the others in the fluid, it follows a zig-zag path and its net displacement is zero averaged over a long time in the absence of any electric field and other external forces. But, in the presence of an external electric field each positive charge carrier moves in the direction of the applied electric field E with a constant *drift speed* v_d which is given as:

$$\vec{v_d} = \mu \vec{E},$$

where μ = mobility of the charge particles. The *mobility* depends on the nature of the fluid and the charge carried.

Conduction current: The flow of charge, that of electrons in conductors under an electric field is called conduction current. It occurs in the same way as in liquids. We will discuss more about it in later sections.

Convection current: When a medium (solid, liquid or gas) containing charged particles scattered in it moves with a velocity v, the electric current is given as

$$i = \frac{dQ}{dt},$$

where dQ = amount of charge in the cloud that moves during time dt through the vertical plane.

Conventional current: The convection current metals is taken in the opposite direction of the drift flow of electrons, which is defined as the rate of flow of protons (even through they do not move) relative to the electrons.

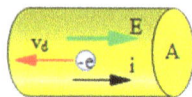

Conventional corrent is opposite to drift velocity of
the electrons or electron current in a metal

Polarisation current: If an electric field is applied, a charge separation occurs in a piece of dielectric. This is called polarization, as discussed in earlier chapters. During the process of polarization, the +ve charges move in the direction of the external or applied field and the −ve charges move in the opposite direction of the applied field. This type of relative motion of the charge carriers in a dielectric during polarization is known as polarization current. It is different from a conduction current because conduction does not take place in an ideal dielectric.

Displacement current: If there is no dielectric in a capacitor, polarization of vacuum (or air) gives rise to a current called displacement current. It was hypothesized by Maxwell while forming the equations of electromagnetic waves. In essence, it is a time-varying electric field. We will explain it in detail in the chapter of *electromagnetic waves*.

Direct current (DC): When the conduction electrons at a given point are driven in the same sense in a conductor by a constant electric field, it causes a direct current whose sense (direction) of flow does not change with time at a given position.

Alternating current (AC): A sinusoidal electric field (or applied voltage) of a generator pushes the conduction electrons of the external circuit back and forth resulting in a conduction current whose magnitude and direction change with time with a frequency equal to the frequency of rotor of the generator. This conduction current is known as alternating current. If the frequency of the AC supply is 50 Hz, the current reverses its direction 50 times in one second. The simple alternating current wave is sinusoidal.

Transient current: A current which exists for a brief time (in the order of milliseconds, say) is called a transient current. Polarization current is a transient current that decays exponentially. This occurs in charging and discharging R–C circuits, and building and decaying of curents in R–L circuits which will be discussed in later chapters.

4.4 Current density

When we say current, it does not make precise sense. We have to mention the area through which the current flows. Hence, neither the current nor the area of cross-section can specifically explain the flow of charge. If we divide the total current by the cross-sectional area, we get 'current per unit cross-section'. It is known as current density, a vector quantity in order to show the direction of current.

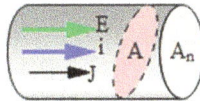

Current, current density under the
action of an electric field in a metal

The current density $\overrightarrow{J}$ is defined as the ratio of current i and the cross-sectional (effective) area A_n through which the charge passes or crosses perpendicularly.

$$J_{av} = \frac{i}{A_n}$$

At any given point, the current density can be given as

$$J = \frac{di}{dA_n}.$$

Current density is defined as the current per unit perpendicular area; it is a vector, whereas current is a scalar quantity. In a special case, filamentary current can be treated as a vector.

4.4.1 Convection current density

Let a cylindrical stream or cloud of charge having volume charge density ρ cross a vertical plane. If it moves through a distance dx during a time dt, the amount of charge dQ crossing the plane is

$$dQ = \rho dV = \rho A dx,$$

where A = effective area of the plane through which the cloud passes = area of cross-section of the cloud but need not be area of the given plane.

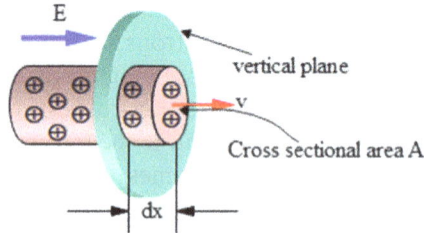

A cylindrical cloud of cross sectional area A and volume charge density ρ crosses a vertical plane with a velocity v = dx/dt

Then, the current density

$$J = \frac{dQ/dt}{A} = \frac{\rho A dx}{A dt}, \quad \text{where} \quad \frac{dx}{dt} = v$$

$$\Rightarrow \vec{J} = \rho \vec{v}$$

The current density $\vec{J}$ is a point function; it is parallel to the velocity $\vec{v}$ if ρ is +ve and anti-parallel to the velocity $\vec{v}$ if ρ is −ve.

4.4.2 Conduction current density

The above formula can also be used for conduction current which occurs due to the movement of the electrons in a conductor under an applied electric field. Just substitute the velocity of the charge cloud with a drift speed v_d of the electron cloud inside the conductor.

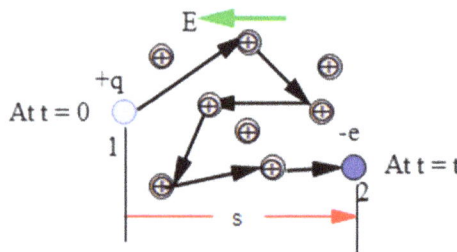

The charge q undergoes a displacement s
in a uniform elctric field E during a time t
constituting an average velocity called
drift velocity v$_d$=s/t

4-5

Then the density of conduction current is

$$\overrightarrow{J} = \rho \overrightarrow{v_d}, \text{ where } \rho = \text{density of negative charge}$$

By putting charge density $= \rho = $ (number of electrons per unit volume) (electronic charge) $= ne$, we have

$$\overrightarrow{J} = ne \overrightarrow{v_d}$$

Example 1 Find the net momentum of the conduction electrons in a straight conductor of length $l = 500$ m carrying a current $i = 60$ amp. Assume $n = 8.5 \times 10^{28}$.

Solution

The net momentum of the conduction electrons $= P = nmv_d Al$ where,

$$v_d = \frac{i}{neA} \qquad (4.1)$$

$$\Rightarrow P = \frac{mil}{e} = \frac{(9.1 \times 10^{-31})(60)(500)}{1.6 \times 10^{-19}} = 1.7 \times 10^{-7} \text{N s Ans.}$$

Example 2 How much time will be taken by a conduction electron to move a distance $l = 1$ km in a copper wire of cross-section $A = 1$ mm^2 if it carries a current $i = 5$ A? Assume $n = 8.5 \times 10^{28}$.

Solution

The time taken by a conduction electron to travel a distance l is

$$t = \frac{l}{v_d}, \text{ where } v_d = \frac{J}{ne}$$

$$\Rightarrow t = \frac{l}{J/ne}, \text{ where } J = \frac{i}{A}$$

$$\Rightarrow t = \frac{neAl}{i}$$

$$= \frac{8.5 \times 10^{28} \times 1.6 \times 10^{-19} \times 10^{-6} \times 10^3}{5} = 2.72 \times 10^6 \text{s} = 31.48 \text{ days}$$

N.B: It takes a long time to travel that distance! But this drift speed must not be misunderstood as the speed of transmission of the electromagnetic wave which can be equal to that of light (3×10^8 m s^{-1}). Ans.

Example 3 A thin beam of proton after accelerating through a potential difference of $V = 600$ kV, produces a current $i = 5 \times 10^{-3}$ A. Find the (a) linear charge density, and (b) electric field at a distance of $r = 1$ m from the beam.

Solution

(a) The electric current is given as

$$i = \frac{dq}{dt} = \frac{\lambda \, dx}{dt} = \lambda \, v$$

Or, $\lambda = \frac{i}{v} = \frac{5 \times 10^{-3}}{6 \times 10^{5}} = 0.833 \times 10^{-8} \, \text{C m}^{-1}$ Ans.

(b) The beam of protons behaves as an infinite and uniform line charge moving with a velocity equal to that of the protons. So, the electric field at a radial distance $r = 1m$ is

$$E = \frac{\lambda}{2\pi\varepsilon_0 r}$$

$$= 2 \times 9 \times 10^9 \times \tfrac{5}{6} \times 10^{-8}$$

$$= 150 \, \text{V m}^{-1} \, \text{Ans.}$$

4.5 Relation between current and current density

(i) For a flat surface, the total current passing through the surface is

$$i = JA_n = JA \cos\theta = \overrightarrow{J} \cdot \overrightarrow{A}$$

where θ = angle between $\overrightarrow{J}$ and $\overrightarrow{A}$.

(ii) For an elementary patch of area of dA, the current is $di = \overrightarrow{J} \cdot \overrightarrow{dA}$.

For a curved surface, integrating di over the total area, we have

$$i = \int \overrightarrow{J} \cdot \overrightarrow{dA}$$

(iii) For a closed-curved surface, integrating di over the total surface, we have,

$$i = \oint \overrightarrow{J} \cdot \overrightarrow{dA}.$$

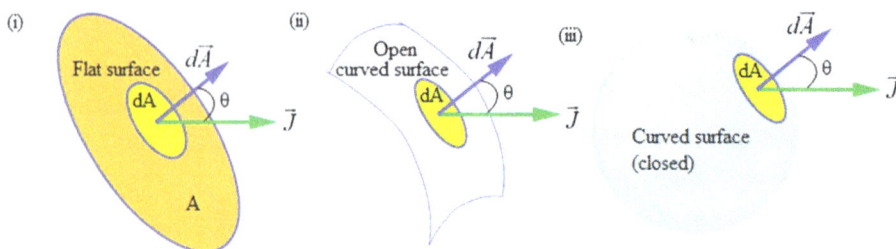

Take an elmentary patch of area dA on the given surface (flat, open-curved and closed-curved);then find the elementary or differential current di flowing through the elmentary patch by using the formula $di=Jds\cos\theta$, and finally,integrate it to find the total current i passing through the given surface.

4.5.1 Surface current density

Sometimes, charge flows on the surface of a conductor. This is called surface current. The density of surface current is given as

$$J_s = \frac{di}{dl} \text{ A m}^{-1}$$

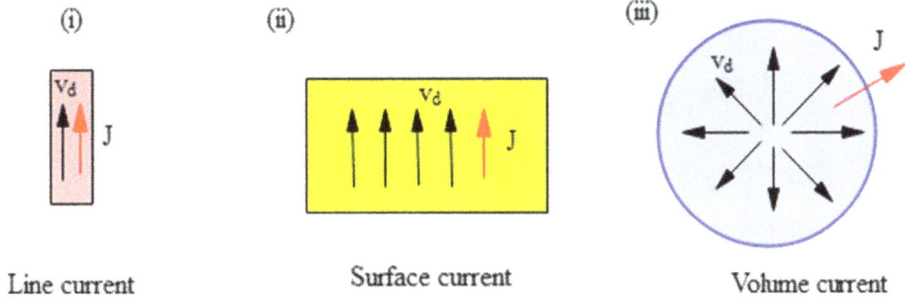

(i)   (ii)   (iii)

Line current   Surface current   Volume current

Example 4 Find the surface current density of a thin cylindrical shell having uniform surface charge density σ moving with velocity v.

Solution

Take an element of the cylindrical shell of length dl; let it contain a charge dQ. Then the surface current density is

$$J_s = \frac{i}{l} = \frac{\frac{dQ}{dt}}{2\pi R} = \frac{\left(\frac{2\pi R dl\sigma}{dt}\right)}{2\pi R} = \sigma v \text{ Ans.}$$

4.6 Equation of continuity

Let the charge dq pass through the cross-section A during time dt. Then, the current is

$$i = \frac{dq}{dt},$$

where $dq = n(Adl)e$; $n =$ density of conduction electrons.

Then, we have

$$i = \frac{nAedl}{dt} = nAev_d,$$

or

$$i = \frac{dq}{dt} = nev_d A$$

Example 5 Find the drift velocity of the electrons in a copper wire of cross-sectional area 1 mm^2 carrying a current of 1 A. Assume that each Cu atom gives one free electron; $\rho_{copper} = 9 \times 10^3$ km^{-3} and $Z = 63.5$ g.

Solution

The drift speed of an electron is

$$v_d = \frac{i}{neA} = \frac{1}{neA},$$

where

$$n = \frac{N}{Z}\rho = \frac{6.02 \times 10^{23}}{63.5} \times 9 \times 10^3 = 8.5 \times 10^{28} \text{ m}^{-3}.$$

Hence, the drift speed is

$$v_d = \frac{\frac{1}{8.5 \times 1.56}}{8.5 \times 10^{28} \times 1.6 \times 10^{-19} \times (10^{-3})^2}$$

$$= 10^{-5} \text{ m s}^{-1} \text{ Ans.}$$

N.B: *Electrons move with a snail's pace inside the conductor, whereas the electric power moves outside the conductor with great speeds between 0.6 and 0.99c, where c = speed of light.*

Example 6 Find the value of τ if the applied electric field of $E = 2$ V m^{-1} generates a drift speed $v_d = 0.8 \times 10^{-2}$ m s^{-1}.

Solution

As derived in section 4.8.5,

$$\tau = \frac{mv_d}{eE}$$

$$= \frac{9.1 \times 10^{-31}}{1.6 \times 10^{-19}} \times \frac{0.8 \times 10^{-2}}{2}$$

$$= \frac{7.28}{3.2} \times 10^{-31+17}$$

$$= 2.28 \times 10^{-14} \text{ s Ans.}$$

4.7 Theory of conduction

4.7.1 Drift speed

When an electric field is applied (maintained) inside the conductor, it drives the electrons against the electric field. Since the electrons collide with the ions in the metals, their motion will be zig-zag. However, each electron will gain a momentum (or velocity) in the opposite direction of the electric field. Consequently, each electron will undergo a displacement s along the conductor in certain time t. Then, average speed of the electron is called 'drift speed' v_d given as:

$$v_d = \frac{s}{t}$$

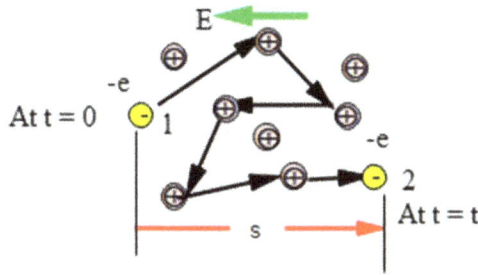

The electron (cloud) -e undergoes a
displacement s in a uniform elctric field E
during a time t constituting an average
velocity called drift velocity $v_d = s/t$

It is very small in metals in the order of centimetres per hour.

4.7.2 Relation between $\vec{J}$ and $\vec{v_d}$

$$\text{Since } J = \frac{i}{A_n} \text{ and } i = nev_d A_n$$

we have, $J = nev_d$

Vectorially we can write $\vec{J} = ne\,\vec{v_d}$

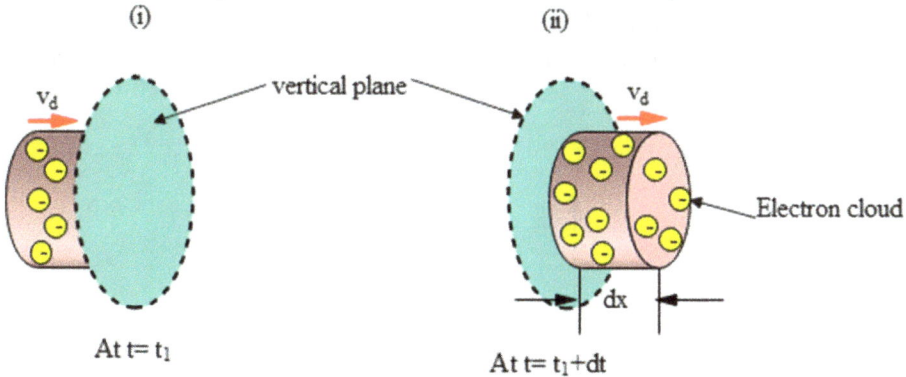

The free cloud of free electrons having electron density n
moves in a cylindrical conductor with a drift velocity $v_d = dx/dt$

4.7.3 Relation between $\vec{v_d}$ and $\vec{E}$

Let the electrons mean time between two collisions be τ. As each electron has
constant acceleration $\vec{a}$, the average increment of its velocity is:

$$\vec{v_d} = \vec{a}\,\tau, \quad \text{where } \vec{a} = -\frac{e\vec{E}}{m}$$

So, the drift velocity is given as $\vec{v_d} = -\frac{e\vec{E}}{m}\tau$.

4.8 Electrical resistance and Ohm's law

4.8.1 Definition

It is the property of any conductor by the virtue of which each electron requires an external electric field to move with a drift speed (corresponding to the applied electric field) against the electron cloud and metallic kernels (lattice atoms and ions). It means that some work must be done by an external agent in pushing an electron with a constant drift speed v_d along the wire (conductor). In other words, a potential is dropped along the conductor in the direction of the electric current.

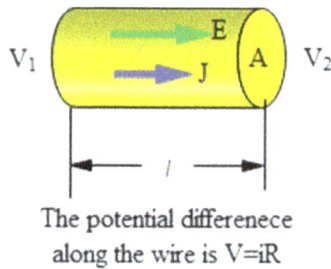

The potential differenece
along the wire is V=iR

4.8.2 Ohm's law (macroscopic form)

In 1827, German physicist Georg Ohm experimentally verified that current flowing through a conductor is directly proportional to the potential difference V between the terminals of the conductor, in a specified temperature range. It can be given as

$$i \propto V$$

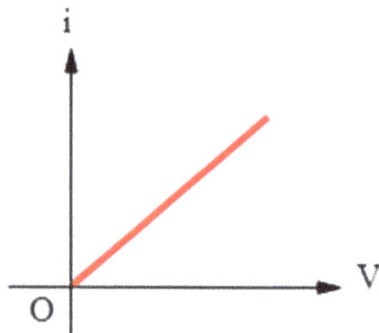

The slope of i-V graph is a
constant for Ohmic resitance

Ohm's law states that the electric current i in a resistor is directly proportional to the potential difference V across the resistor of resistance R; i = V/R; it is valid when appropriate conditions are met such as moderate temperature, electric and magnetic fields etc.

So, $\frac{V}{i}$ = constant, which is defined as the *resistance* of a conductor denoted by the letter R.

$$\text{or, } \frac{V}{i} = R$$

The above relation does not hold good at extreme conditions such as a very low and very high temperature, very high voltage or strong electric field and very high magnetic field etc.

Below certain critical (low) temperature, the conductor exhibits zero resistance (super conductivity) and quantum effect prevails. At high temperature, the material of the conductors may melt, inonize or go to plasma state; so, dynamics or principles of conduction change due to the physical and chemical changes. At high magnetic fields (Hall effect comes in to play due to strong Lorentz magnetic force); the current density J will no longer be linear to E because $F = e(E + v \times B)$. So, Ohm's law holds good up to a certain temperature for some conductors.

Example 7 Find the ratio of the resistance of conductors 1 and 2.

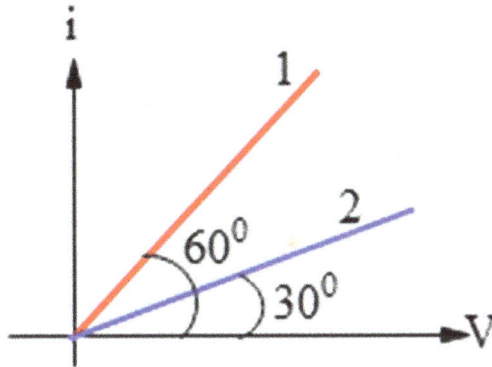

Solution

Using Ohm's law, $R = V/I$; so, resistor 1 is

$$R_1 = \left(\frac{V}{i}\right)_1 = \cot 60°$$

$$R_2 = \left(\frac{V}{i}\right)_2 = \cot 30°$$

Similarly, resistor 2 is

$$\text{So, } \frac{R_1}{R_2} = \frac{\cot 60°}{\cot 60°} = \frac{1/\sqrt{3}}{\sqrt{3}} = 1/3 \text{ Ans.}$$

4.8.3 Dependence of the resistance

Resistance of a conductor does not depend upon the current i flowing through it and the potential difference (p.d.) along the conductor. However, it depends upon

(a) Length as $R \propto l$ (b) area as $R \propto \frac{1}{A}$ (c) resistivity ρ: $R \propto \rho$

(i) **Dependence on length**

To push the electrons with a constant drift speed through double the distance (length) double work will be done. Then, V will be double for maintaining same current.

Hence $\frac{V}{i}(=R)$ will be double

$$\text{or, } R = \frac{V}{i} = \frac{qE}{i}l \text{ ; so } R \propto l \tag{4.2}$$

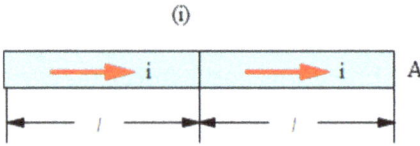

(i)

(ii)

For greater length, greater potential difference (= work per unit charge along the wire) for same current (drift speed)

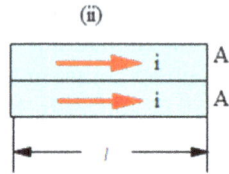

For greater area of cross section, greater current (= charge crosses per unit time) for same potential difference

(ii) **Dependence on area**

If the area of cross-section is doubled, double the amount of charge passes through the wire. So, the current is doubled for the same p.d. (or $\overrightarrow{E}$).

In the others, half of the previous p.d. (voltage) is required to set up the same current i. Then $R(=\frac{V}{i})$ will be halved.

$$\text{or, } R \propto \frac{1}{A} \tag{4.3}$$

Using equations (4.2) and (4.3), we have $R \propto \frac{l}{A}$

$$\text{So, } R = \rho \frac{l}{A},$$

where $\rho =$ resistivity of the conductor; R is in ohm (Ω).

$$\frac{1}{R} = G \text{ (Conductance−mho } \mho)$$

and

$$\frac{1}{\rho} = \sigma \text{ (Conductivity)}$$

Example 8 A wire is pulled through a die so that its length is doubled. What happens to its resistance?

Solution

The resistance of a conductor for an isotropic and uniform cross-sectional area is $R = \rho\frac{l}{A} = \rho\frac{l^2}{Al}$. Since, volume ($=Al$) remains constant $R \propto l^2$. As the length is doubled, resistance becomes four times the initial resistance.

If the length of the wire is doubled by joining an identical wire end to end (in series), the net resistance of the total wire will be doubled.

4.8.4 Point form of Ohm's law

The macroscopic or scalar form of Ohm's law is given as

$$\frac{V}{i} = R \tag{4.4}$$

Taking a conductor of length l and uniform cross-section A, we have

$$R = \rho\frac{l}{A} \tag{4.5}$$

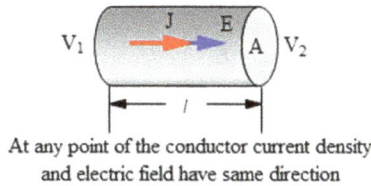

At any point of the conductor current density and electric field have same direction

The potential difference V along the conductor is

$$V = El \tag{4.6}$$

The current flowing through the conductor is

$$i = JA \tag{4.7}$$

Substituting R from equation (4.5), V from equation (4.6) and I from equation (4.7) in equation (4.4), we have

$$\frac{El}{JA} = \rho\frac{l}{A}$$

or,

$$\frac{E}{J} = \rho = \frac{1}{\sigma}; \quad \sigma = \text{conductivity and } \rho = \text{resistivity}$$

Vectorially, $\vec{J} = \sigma\vec{E}$

Since $\vec{J}$, σ and $\vec{E}$ are defined at a point, the above form is called point form/ differential form/microscopic form/vector form of Ohm's law.

Example 9 Find the resistivity of a metal carrying an electric field $E = 10 \ \text{V m}^{-1}$ causing a current density $J = 3 \times 10^5 \ \text{A m}^{-2}$.

Solution

The resistivity of a metal is given as

$$\rho = \frac{E}{J} = \frac{10}{3 \times 10^5} = 3.33 \times 10^{-5} \ \Omega - \text{m Ans.}$$

In a good electrical conductor, even a small electric field can produce a huge current because of large conductivity.

4.8.5 Microscopic interpretation of σ (or ρ)

Since $J = nev_d$ and $v_d = \frac{eE}{m}\tau$ as derived earlier, we can write

$$J = \frac{ne^2\tau}{m}E$$

Since the ratio of current density and electric field at any point of the conductor is

$$\frac{J}{E} = \sigma,$$

the conductivity of a conductor is given as $\sigma = \dfrac{ne^2\tau}{m}$.

It tells us that conductivity σ of a conductor depends on free electron density n and relaxation time or mean time τ between two successive collisions of an electron.

For the materials called insulators $\sigma < 10^{-5} \ (\Omega - \text{m})^{-1}$ m for metals (conductors) $\sigma > 10^{+5} \ (\Omega - \text{m})^{-1}$ and $10^{-5} < \sigma < 10^{+5}$ for the materials called semiconductors.

Example 10 An electric field $E = 5 \times 10^{-3} \ \text{V m}^{-1}$ sets a current $i = 1 \ \text{A}$ along a wire of radius $= 10^{-3}$ m. Find τ.

Solution

the electrical conductivity is

$$\sigma = \frac{J}{E}$$

$$= \frac{[1/\pi(10^{-3})^2]}{5 \times 10^{-3}} = 0.064 \times 10^9 \ (\Omega - \text{m})^{-1}$$

Then, the relaxation time is

$$\tau = \frac{m\sigma}{ne^2}$$

$$= \frac{(9.1 \times 10^{-31})(0.064 \times 10^9)}{(3 \times 10^{28})(1.6 \times 10^{-19})^2} \simeq 7.5 \times 10^{-14} \ \text{s Ans.}$$

4.8.6 Temperature dependence of resistance

If the temperature of a conductor increases, the atoms of the lattice vibrate with greater amplitudes and velocities. Furthermore, the conduction electrons move with greater speeds. Since $\sigma \propto \tau(=\frac{\lambda}{v})$ and $\bar{v}$ increases with temperature, we can say that σ decreases or ρ increases with temperature, hence the rate of collision of the conduction electrons with the lattice sites increases, meaning that the resistivity of the conductor increases.

It is experimentally verified that the resistivity of a conductor varies linearly with temperature up to a certain temperature. If ρ_0 = resistivity at 0 °C, the resistivity at θ °C is given as

$$\rho_\theta = \rho_0(1 + \alpha\theta),$$

where α = temperature coefficient of resistivity given as

$$\alpha = \frac{\rho_\theta - \rho_0}{\rho_0\theta} \text{ and its unit is } K^{-1} \text{ or } °C^{-1}.$$

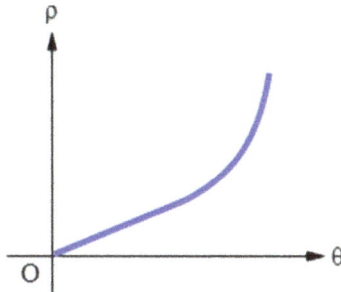

For copper, resistivity increases linearly with temperature up to certain temperature

In differential form,

$$\alpha = \frac{d\rho}{\rho d\theta}$$

Then, resistance R_θ at any temperature θ can be given as

$$R_\theta = R_0(1 + \alpha\theta),$$

where R_0 = resistance at 0 °C and α = average temperature coefficient of resistance (or resistivity).

As the alloys have very small value of α, their resistance does not change appreciably with increase (or decrease) in temperature. So, the alloys can be used for making resistances of constant value.

4.8.7 Superconductivity

At a very low temperature, the resistivity of a metal is considerably less than that at room temperature. Some metals lose their resistances completely at temperatures near 0 K (absolute zero). This property of a conductor is called superconductivity and the material is called a 'superconductor'. The temperature at which a material becomes a superconductor is called critical temperature T_C.

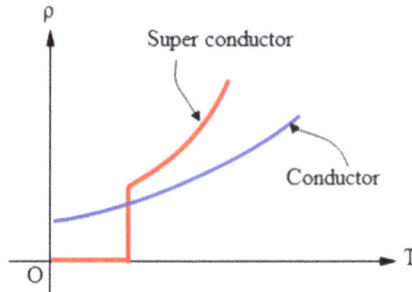

Resistivity of a normal conductor (copper, say) is not zero at zero Kelvin but the resistivity of a superconductor drastically falls to zero at certain critical temperature close to absolute zero.

A superconducting ring can retain electric currents of hundreds of amperes for a year without any external source. The nature of superconductors can be explained by quantum physics.

4.9 Calculation of resistance of arbitray shaped conductors

The expression $R = \rho \frac{l}{A}$ is valid for a conductor of uniform cross-section and conductivity. Let us find the resistance of a conductor of arbitrary shape and size between its terminals 1 and 2.

Let the field distribution inside the conductor be known. Then, the potential difference between points 1 and 2 can be given as

$$V = \int \overrightarrow{E} \cdot d\overrightarrow{l}$$

If the current density is not uniform, the total current passing through the conductor is:

$$i = \int \overrightarrow{J} \cdot d\overrightarrow{A}$$

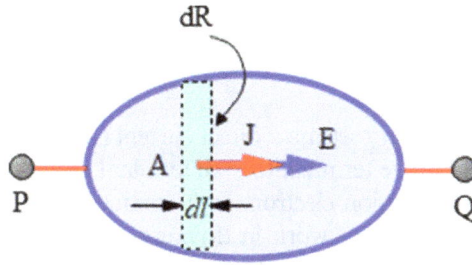

The resistance between P and Q is
given as $R = \int \dfrac{dl}{\sigma A}$

Then, the ratio of V and i, that is, $\dfrac{V}{i}$, gives the resistance R according to Ohm's law.

or, $R = \dfrac{V}{i}$

$= \dfrac{\int \vec{E} \cdot d\vec{l}}{\int \vec{J} \cdot d\vec{A}}$, where $J = \sigma E$

or, $R = \dfrac{\int \vec{E} \cdot d\vec{l}}{\int \sigma \vec{E} \cdot d\vec{A}}$

If areas of cross-section A and ρ remain uniform throughout the material, the above expression can be written as

$$R = \dfrac{\rho l}{A},$$

where A = area through which current flows.

Example 11 Find the resistance between the ends of a wire of cross-sectional area A, assuming that the conductivity of the wire increases linearly with distance (length) from σ_0 to $2\sigma_0$.

Solution

The resistance between P and Q is

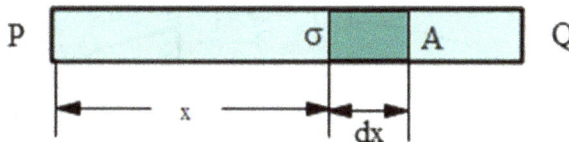

$$R = \dfrac{1}{A} \int_0^l \dfrac{dx}{\sigma}, \quad \text{where } \sigma = \sigma_0\left(1 + \dfrac{x}{l}\right)$$

or

$$R = \dfrac{1}{\sigma_0 A} \int_0^l \dfrac{dx}{1 + \frac{x}{l}} = \dfrac{l}{\sigma_0 A} \ln 2 \text{ Ans.}$$

4.10 Electromotive force

4.10.1 Definition

When we close the key, a battery sets up a direct current (DC) in the conductor. So, each charge carrier moves from +ve terminal to −ve terminal of the battery. In practice, the charge carriers are the conduction electrons in a conductor and ions in a battery. The electrostatic field that does postive work in driving a conventional +ve charge carrier outside the battery (inside the conductor), and it does equal negative work while the charge carrier moves back to the +ve pole (electrode) from −ve pole (electrode) of the battery against the electric field inside the battery. As a whole, the static electric field does zero work in a round trip; $\oint \vec{E} \cdot d\vec{l} = 0$.

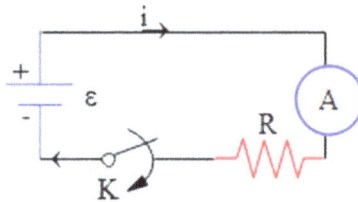

A seat of emf ε does a positive work dW by pushing an elementary charge dq; ε=dW/dq.

Hence, the electric field cannot take the credit of circulating a charge carrier along the closed loop (circuit). Then, it must be the battery that does a +ve work in pushing the charges from −ve to +ve terminal of the battery to set up a permanent potential difference across the terminals.

This is accomplished by different principles:
  (a) electrochemical action;
  (b) electromagnetic induction (electromechanical action);
  (c) thermionic emission;
  (d) photo-voltaic action etc.

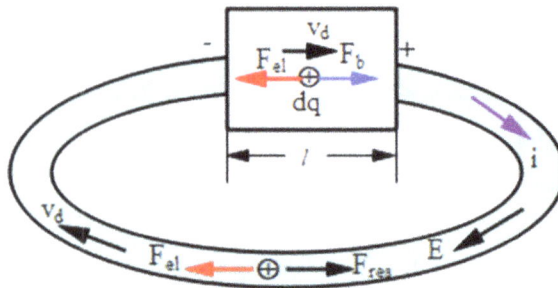

The elementary charge dq moves from negative to positive terminal of the battery against the electrostatic field inside the battery

The work done by the battery in pushing the positive charges from its negative terminal to positive terminal through a distance l with a force F_b is given as

$$W = F_b \cdot l = F_{el}l = qE_{el} \cdot l = qV_b$$

(F_b = force acting on the test charge $+q$ inside the battery due to the electrochemical action of the battery)

Then, the work done per unit charge is

$$\frac{W}{q} = V_b$$

which is called electromotive force (emf) of the battery denoted as ε.

$$\varepsilon = \frac{W_b}{q}$$

1. Emf is numerically equal to the work done by the battery in circulating a unit +ve charge.
2. A cell (seat of emf) generates a potential difference between its terminals across the circuit which is numerically equal to its emf when the circuit is open.

4.10.2 Internal resistance

Since the positive charge flows from −ve to +ve pole against the electrostatic field inside the battery, opposition of the electrolyte to the flow of charge causes an internal resistance of the battery denoted as r.

1. ε remains the same, whereas r increases with time to reduce the current.
2. If the current flows from +ve to −ve pole of the battery, (the battery) is said to be discharging and vice versa.

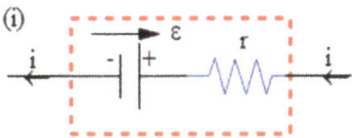

Battery is charging

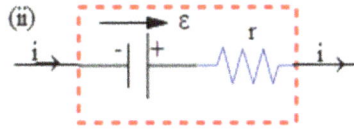

Battery is discharging

3. A practical cell of emf ε and internal resistance r is imagined as an ideal battery of emf ε and an external resistance r connected in series with the cell.

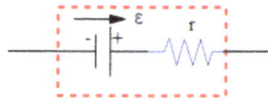

A practical (given) battery of emf ε is a series combination of an ideal battery and an internal resistance r of the given battery.

4.11 Electric field of a current-carrying conductor

Ohm's law tells us that $J = \sigma E$. Hence, an electric field is felt at every point of a current-carrying conductor. Then a question arises, what generates the electric field inside a conductor? You know that a charge is ultimately responsible for an electric field. Then, where do the charges stay in a conductor? Let us see.

4.11.1 Mechanism of DC

If you switch on a source (a DC generator or a battery) at $t = 0$ (say), due to the emf of the source it pulls the electrons from terminal 1 to the other terminal 2 causing a charge separation between the terminals.

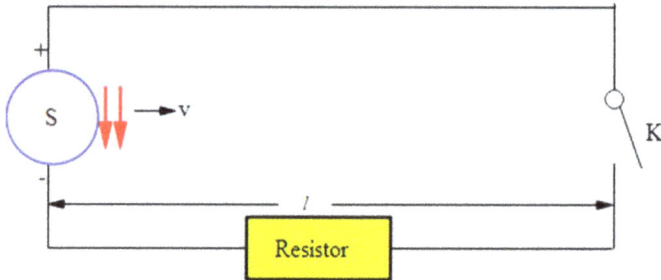

Just after the source S appeared at t=0, its electric field propagates to right as an electromagnetic wave being guided by the conductors, with a velocity v (= 0.6 to 0.9c); c = velocity of light in vacuum.

The charge separation generates an electric field which interacts with the electrons of the connecting wires (conductors) via an electromagnetic wave (force) which moves with a speed within $0.6c$ to $0.9c$, where $c =$ speed of light in a vacuum $= c (=3 \times 10^8$ m s^{-1}). If $v = 2c/3$, even a long conductor of length $l (=2$ km, say) will be charged completely after a very small time $t(=\frac{l}{v} = 0.1$ μs). When the circuit is open, the steady current in the circuit is zero. Hence the electric field inside the conductors is zero. It means, the potential difference V equal to potential difference across the source, that is, equal to emf ε of the source.

$$V = \varepsilon, \text{ open circuit}$$

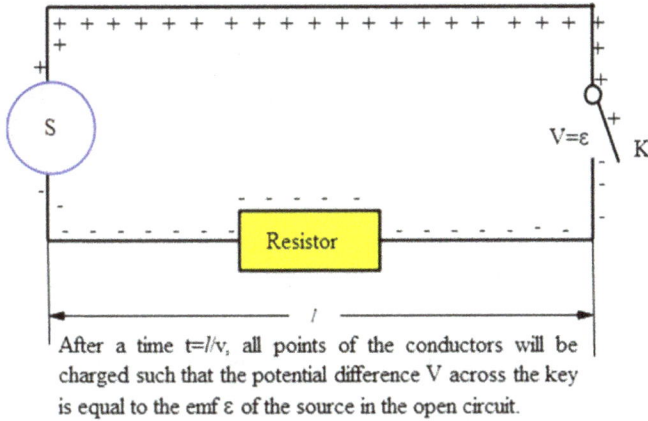

After a time t=l/v, all points of the conductors will be charged such that the potential difference V across the key is equal to the emf ε of the source in the open circuit.

After closing the key, a transient current flows, then a steady current i flows such that the potential difference V_R across the resistor is equal to the emf of the source. So, we can write

$$V_R = \varepsilon = iR$$

But, the potential difference across the key will be zero. It is because the connecting wires along the key K are assumed as zero resistors.

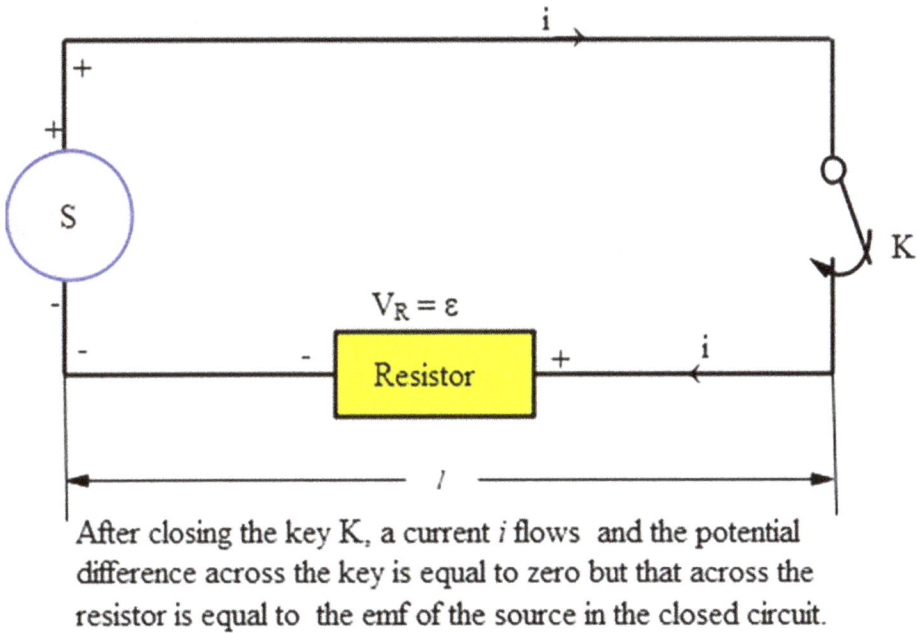

After closing the key K, a current i flows and the potential difference across the key is equal to zero but that across the resistor is equal to the emf of the source in the closed circuit.

4.11.2 Surface charges

As explained earlier, even through no current flows, a potential difference exists between the conductors. This is because of the equal and opposite charges induced at the surface of the conductor in the process of charging. But, in general, when $i \neq 0$, the surface charges can be redistributed so as to ensure the flow of electric current after closing the key.

4.11.3 External electric field of current-carrying conductor

Practically, the connecting copper or aluminium wires have negligible resistance $(0.1 \ \Omega \ \mathrm{km}^{-1})$. Hence, a very small electric field exists inside the conductor to maintain the current. This small electric field is always tangential to the surface in order to maintain the continuity of tangential electric field which can be given as

$$E_t = \frac{J}{\sigma}$$

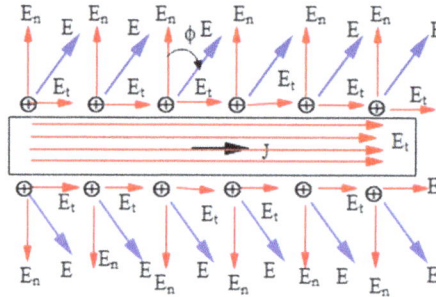

The net electric field E just outside the current carrying conductor is equal to the resultant or vector sum of tangential electric field E_t and normal electric field E_n.

Also, the surface charges generate a normal component of an electric field as discussed earlier, given as

$$E_n = \frac{\sigma_s}{\varepsilon_0}$$

So, the net electric field outside the conductor is

$$E = \sqrt{E_t^2 + E_n^2}$$

The angle made by $\overrightarrow{E}$ with the normal to the conductor surface is

$$\phi = \tan^{-1} \frac{E_t}{E_n}$$

Since, in general $E_t \ll E_n$, ϕ is very small. For an ideal conductor $\phi = 0$ (because $E_t = 0$ as $\sigma \to \infty$).

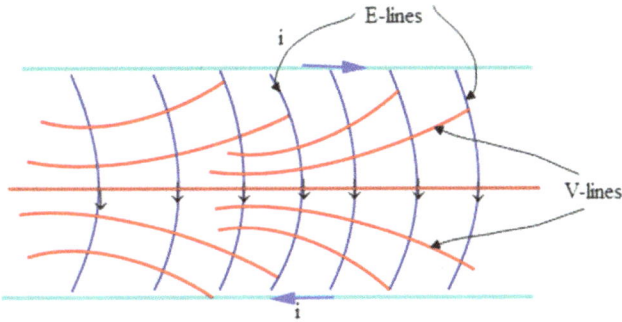

The electric field pattern (E-lines) and equipotential (V-lines) in between two parallel current carrying conductors;the E-lines and V-lines must intersect perpendicularly

The curved pattern of electric field between the parallel current-carrying conductors can be viewed by sprinkling the dielectric powder on a sheet of paper kept over the conductors. You can see that $\vec{E}$-field is not uniform and corresponding equipotential lines are drawn so that they are perpendicular to E-lines at each point.

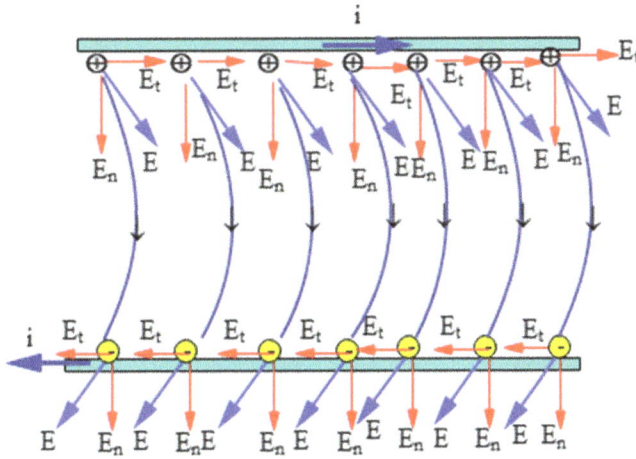

The field pattern outside (in between) two parallel current carrying conductors;the E-lines are curved due to the presence of both tangential and normal conponents; this curved pattern can be produced by the dielctric power sprinkled on a plane containing these current carrying conductors.

4.11.4 Electric field inside a current-carrying conductor

For uniform cross-section, homogeneous and isotropic conductors, if the current does not vary with time, the electric field inside the conductor is given as

$E = J/\sigma$ which remains constant inside the conductor.

It means that E-field is uniform inside the straight conductor. Hence, equipotential lines are uniform and parallel to the cross-section of the conductor.

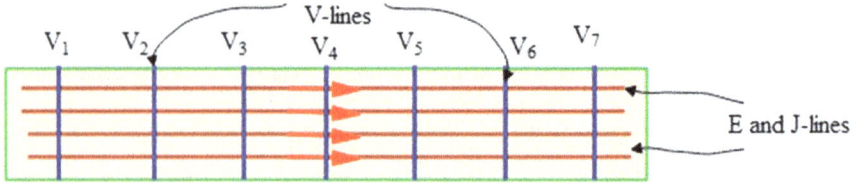

The red lines respresent electric field pattern (E-lines) and J-lines (current distribution); blue lines represent equipotential (V-lines) in a current carrying conductor; these lines are uniform for an isotropic, homogeneous conductor of uniform cross-section; the E-lines and V-lines intersect perpendicularly.

If the cross-sectional area varies in a truncketed conductor or bent conductor, the current density cannot be uniform. Since, $\vec{J}\,(=\frac{i}{An}) = \sigma\vec{E}$, the electric field $\vec{E}$ will be non-uniform inside the conductor.

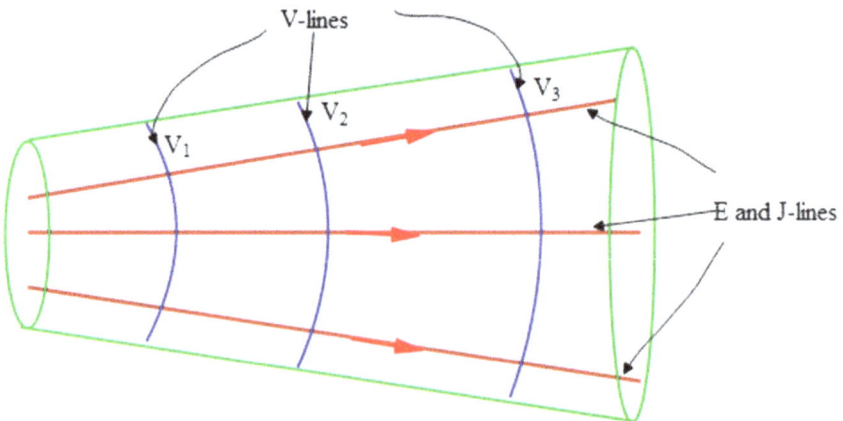

The electric field pattern (E-lines) and J-lines are shown as red lines and equipotential (V-lines) are shown as blue lines in a current carrying conductor; these lines are nonuniform for a conductor of nonuniform cross section;the E-lines and V-lines must intersect perpendicularly.

In the case of a bent wire, J-lines are curved; so, E-lines are curved because $\vec{J}$ in parallel to $\vec{E}$. The curved E-lines are produced by the deposited plus and minus surface charges at the bent. The surface charges are distributed so as to guide the current (at the bending) in the conductor.

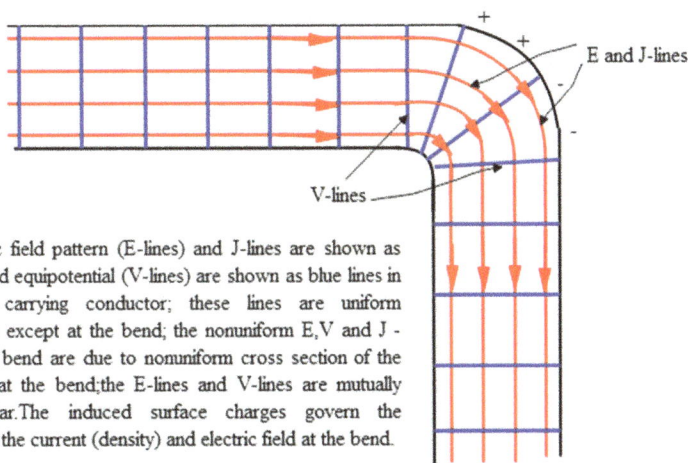

The electric field pattern (E-lines) and J-lines are shown as red lines and equipotential (V-lines) are shown as blue lines in a current carrying conductor; these lines are uniform everywhere except at the bend; the nonuniform E, V and J - lines at the bend are due to nonuniform cross section of the conductor at the bend; the E-lines and V-lines are mutually perpendicular. The induced surface charges govern the direction of the current (density) and electric field at the bend.

4.11.5 Volume charge

If the conductivity σ of the material changes, to maintain the same current density $J(=\sigma E)$, E must change. Hence, E will be more if σ is less (or ρ is more). Thus, the field distribution is denser in the region of greater resistivity. So, the excess charges will appear at the interface which accounts for the discontinuity of the field across it.

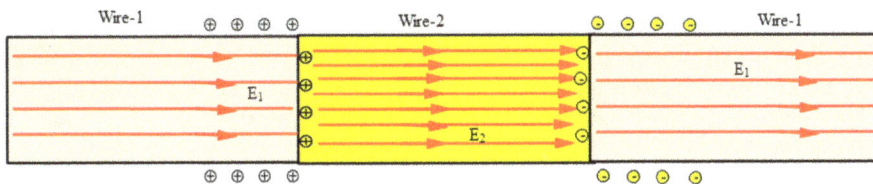

Electric field pattern is discontinuous at the interface of two different materials; the excess charge deposition at the junction will account for the discontinuty of the E-field; for uniform cross section, the current and current density remain equal along the composite wire

4.11.6 Apparent confusion between static, stationary, potential and moving field

For a constant current, even though the surface and volume charges move with the same drift speed, at any point their densities do not change with time. Hence, E-field patterns inside and outside the conductor remain unchanged. Even though the electrons move, we can call this field static, stationary or potential field in contrast to the moving $\vec{E}$ (electric) field of the electromagnetic wave which originates from a time-varying magnetic field $\vec{B}$.

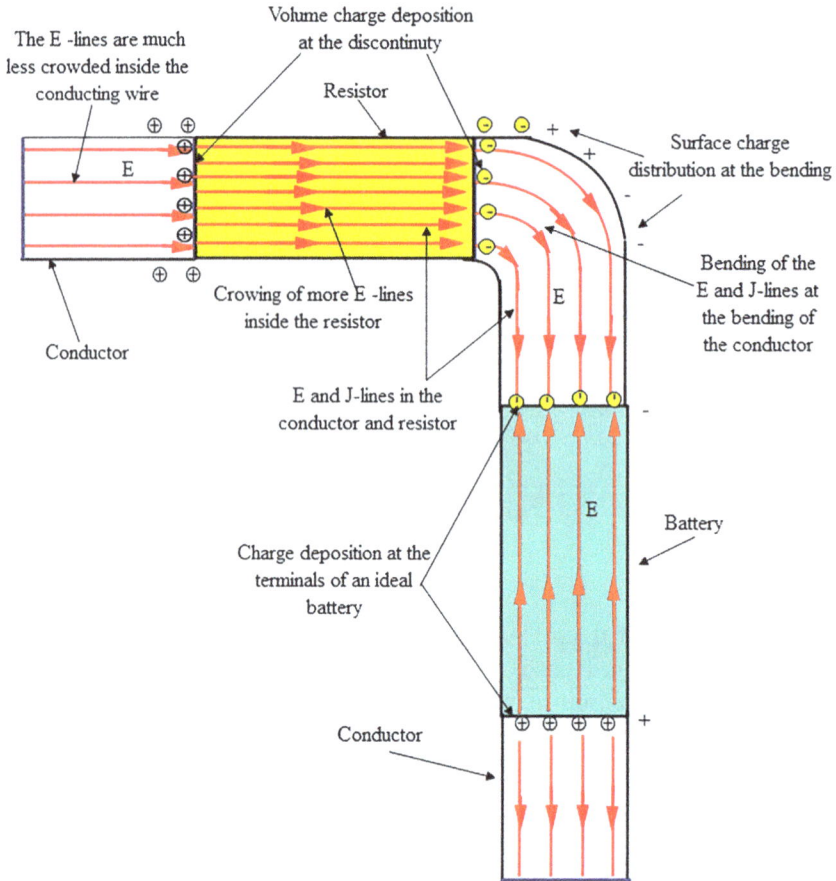

Electric field, current density and the (surface and volume) charge deposition;
Electric field inside the battery is in opposite sense to that inside the conductor, but
the sense of current flow is same in conductors, resistor and battery.

The charge distribution at the surface and bulk of the conductor are not directly but ultimately governed by the battery (or source of current). The charge distribution depends on shape, size, curvature, bending, change in area and conductivity of the conductor.

4.12 Energy conversion and electrical power

4.12.1 Input electrical energy

Let us consider a length l of the straight conductor of uniform cross-section A and conduction electron density n. Then, the total number of conduction electrons in the considered segment is $N = nAl$.

Since the uniform electric field E pushes each electron with a constant drift speed v_d against the resistance (offered by the fixed atoms in the lattice), the total work done by the field during a time dt in shifting the electrons by a distance ds is

$dW =$ (Work done on each electron) $\times$ (No. of electrons present in the segment)

$= (F_{el} \cdot ds)(N)$

$= (eEds)(nAl) \ (\because F_{el} = eE)$

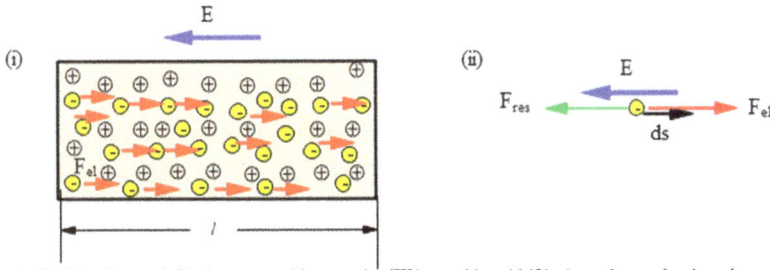

Electric field (acting to left) does a positive work dW in pushing (drifting) each conduction electron in its opposite direction (to right) through a very small (elementary)distance ds by a force $F_{el} = eE$.

or, $dW = (eEv_d dt)(nAl) \ (\because ds = v_d dt)$

$= (nev_d A)(El)dt$

By putting $nev_d A = i$ and $El = V_1 - V_2(=V)$, we have $dW = iVdt$.

Then, the total work done by the electric field on the assumed portion of the conductor during a time t is

$$W = \int_0^t iV \, dt$$

where $V =$ potential difference between terminals 1 and 2 of the given portion of the conductor.

4.12.2 Input electrical power

The electrical power of a voltage V while sending a current i can be given as rate of electrical work done.

$$\text{or } P_{el} = \frac{dW}{dt} = d(qV)/dt = Vdq/dt$$

Putting $dq/dt = i$, we have $P_{el} = i \, V$.

4.12.3 Heat dissipated

As the electrons travel from lower potential V_1 to higher potential V_2 they must lose their electrostatic potential energy or excess kinetic energy while accelerating in the applied electric field. This appears in the form of heat, light etc, due to the resistance offered by the conductor. Hence, the amount of heat liberated in the considered portion of the conductor is

$$Q = \int_0^t i \, V dt$$

$$= \int_0^t i^2 R dt \quad (\because V = iR)$$

$$= \int_0^t \frac{V^2}{R} dt \quad \left(\because i = \frac{V}{R} \right).$$

4.12.4 Thermal power

The rate of heat is liberated, that is, power loss in the resistor is called Ohmic heating, or Joule heating or copper-loss or thermal power or $i^2 R$ loss, which can be given as

$$\frac{dQ}{dt} (= P_R) = iV = i^2 R = \frac{V^2}{R}$$

We can use thermal energy in a room heater, toaster, electric iron etc, and in other electric circuits (power distribution and transmission) where power lost cannot be used.

4.12.5 Joule–Lenz law

The above expression is called the macroscopic form of Joule–Lenz law.

Substituting $i = JA$, $R = \rho \frac{l}{A}$ in the formula $\frac{dQ}{dt} = i^2 R$, we have $\frac{dQ}{dt} = (JA)^2 (\rho \frac{l}{A})$ $= \rho J^2 (Al)$, where $Al = V$ (volume of the segment)

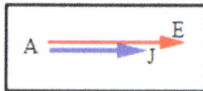

Electric field E and current density J are parallel (pointing to same direction) at any point inside the conductor.

Then, the power loss (rate of heat generated) per unit volume is

$$\frac{dQ}{dt} / V = Q_v = \rho J^2 = J. E = E^2 / \sigma \quad (\because J = \sigma E)$$

This expression is valid for any point of the conductor. Hence, we call it '*point (or differential) form*' of Joule–Lenz law.

4.12.6 Micro-interpretation of heat dissipation

The emf (battery) sets an electric field which pushes the electrons in the conductor. As a result, the electrons gain kinetic energy or lose electrostatic potential energy. The gained kinetic energy is lost in their repeated collisions with the site atoms of the lattice. The exchange in kinetic energy and momenta of the electrons causes the lattice atoms to vibrate with greater amplitudes. The vibrating metallic kernels of the lattice radiate electromagnetic energy in the form of heat, light etc, obeying the principle of electromagnetic radiation.

Thus, the excess kinetic energy of the electrons received from the electric field (ultimately from the battery) is spent in exciting the atoms of the lattice, which eventually will be lost as heat, light etc.

4.12.7 Power of an emf

A battery is ultimately responsible for setting an electric field inside and outside of the conducting wires. Hence, the battery does positive work in circulating the charges.

The rate of work done by a seat of emf (battery) to establish a current is defined as electrical power of a battery.

$$P_{el} = \frac{dW_b}{dt}$$

Electric field (acting to left) does a negative work while pushing each elementary charge dq (to left) while the ideal battery pushes the elementary charge(electron) by its electromotive action by applying a force $F_{battery} = F_{el} = eE$ to right. So, the battery does a positive work in shifting the charge from -ve terminal to +ve terminal of the battery against its own electrostatic field arising from the charge separation at the elecrodes by electrochemical action in case of a chemical cell and electromechanical action in the electric generators.

As discussed earlier, the work done by a battery to push the conventional +ve elementary charge dq from its −ve terminal to its +ve terminal against the emf can be given as

$$dW_b = \varepsilon \, dq$$

Then, the power delivered by the battery in setting a current i is

$$P_{el} = \frac{dW_b}{dt} = \varepsilon \frac{dq}{dt} = \varepsilon i$$

or, $P_{el} = \varepsilon i$

If current (or dq) flows in the direction of the emf, work done and power delivered by the battery is +ve and vice versa.

In practice, all batteries have small internal resistance due to the presence of electrolytes. Let us assume that the internal resistance of the battery be r.

(i) ε (ii) ε

If the current opposes the emf, work done by the battery is negative (or a positive work is done on the battery to charge it);if the current favors the emf, work done by the battery is positive as the battery is discharged in supplying energy to the external circuit(load)

When a current i flows through the battery, due to its internal resistance r, the heat is dissipated at a rate of

$$P_r = i^2 r$$

The surplus power, that is, $P_{el} - P_r$, of the battery is utilized in the external circuit (load). This is called output power of the battery or power input to the external load given as

$$P_{output}(=P_b) = P_{el} - P_r$$

$$\text{or,} \quad P_b = \varepsilon i - i^2 r = (\varepsilon - ir)i$$

Let the effective potential difference (or terminal voltage) across the battery be V_b. Since, it supplies a current i to the load,

$$P_{output} = V_b i$$

By using the last two equations, the terminal voltage of the battery is

$$V = V_b = \varepsilon - ir$$

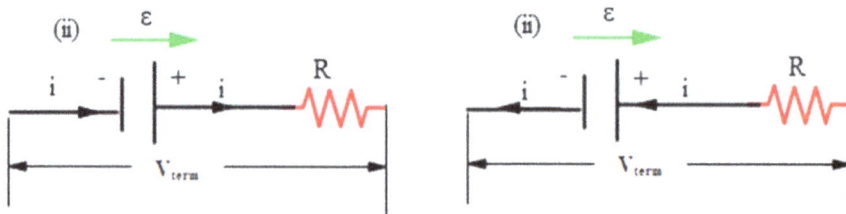

(ii) ε R   (ii) ε R

If the current opposes the emf, $V_{term} = \varepsilon + iR$; if the current favors the emf, $V_{term} = \varepsilon - iR$

Example 12 (Charging battery)
Describe the concept of charging of a battery and prove that the terminal voltage across a charging battery of emf ε, internal resistance r is

$$V_b = \varepsilon + ir, \quad \text{where } i = \text{charging current.}$$

Solution
While charging, the external agent (source battery) must do positive work εi while setting a charging current i through the charging battery. Furthermore, the source

battery will have to perform additional work in generating a power i^2r in the charging battery, where $r =$ internal resistance of the charging battery. Then, the total power supplied to the charging battery is

$$P = \varepsilon i + i^2 r$$

If the equivalent voltage (terminal voltage) V_b can impart the same current, we can write

$$P = Vi$$

Comparing the last two equations, we have

$$V = V_t = \varepsilon + ir$$

This tells us that the terminal voltage across a battery of emf ε and internal resistance r at the time of charging with a current i can be given as

$$V_t = \varepsilon + ir \quad \text{Proved}$$

N.B: During the charging, the accumulated (mixed) +ve and −ve ions can be displaced back to their respective electrodes clearing their paths. Thus, the internal resistance which increases in the course of discharging can be reduced to its original value.

Problem 1 A battery of emf 20 V internal resistance $r = 2\,\Omega$ is connected to a bulb by two wires of negligible resistance. If the current in the circuit is 0.2 A find the (i) rate at which the chemical energy of the battery changes (or power input of the battery), (ii) power loss in the battery, (iii) power output in the external circuit, (iv) power loss in the external circuit, (v) total power loss, and (vi) terminal voltage V of the battery.

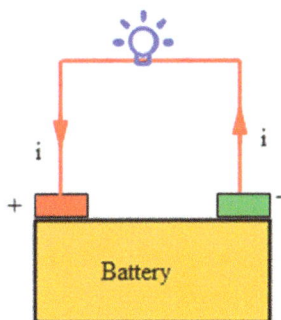

Solution

   (i) The rate of change in chemical energy of the cell

$$= (P_{el})_{\text{input}} = \varepsilon i = 20 \times 0.2 = 4 \text{ J Ans.}$$

   (ii) The power loss in the cell $= P_r = i^2 r = (0.2)^2(2) = 0.08$ J Ans.
   (iii) The power output in the external circuit $= P_{\text{output}} = 4 - 0.08 = 3.92$ J Ans.

(iv) The power loss in the external circuit $P_R = 3.92$ J Ans.

(v) The total power loss $= 3.92 + 0.08 = 4$ J Ans.

(vi) The terminal voltage $= V = \varepsilon - ir = 20 - (0.2)(2) = 19.6$ V Ans.

The electrical power input to any electrical equipment containing the circuit element (R, L and C) can be given as

$$P_{input} = iV,$$

where $i =$ current flowing through the equipment and $V =$ potential difference between the terminals of the equipment.

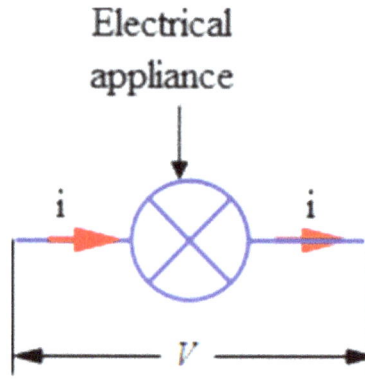

Problem 2 n_1 electron/s passes through a given cross-section to the right with velocity v_1 and n_2 proton/s passes through the same cross-section with velocity v_2 to the right. Find the current through a given cross-sectional area. Put $n_1 = 1.5 \times 10^{10}$ and $n_2 = 10^{10}$.

Solution

There are two currents in this problem; one is electron current moving to the right and the other is proton current moving to the left. Let us find one by one and then we will add them algebraically.

The direction of current is assumed as the opposite direction of motion of electrons; so, the current of the electron beam is given as

$$i_1 = \frac{\Delta q}{\Delta t} = \frac{\Delta N_1 q_1}{\Delta t} = \frac{dN_1}{dt} q_1 \text{ (points to the left)}$$

Similarly, the current of the proton beam is given as

$$i_2 = \frac{dN_2}{dt} q_2 \text{ (points to the left)}$$

So, the net current is

$$i = i_1 + i_2$$

$$= \left(\frac{dN_1}{dt}\right) - e + \left(\frac{dN_2}{dt}\right) e$$

Or, $i = (n_1 + n_2)e$
$$= (1 \times 10^{10} + 1.5 \times 10^{10})1.6 \times 10^{-19} = 4 \times 10^{-9} \text{ amp Ans.}$$

N.B: At first, fix the actual direction; then convert into the current direction and finally add them considering their directions.

Problem 3 Find the current associated with an electron revolving with a speed $v = 10^6$ m s^{-1} in an orbit of radius $R = 1$ Å.

Solution

The charge $\Delta q(=-e)$ flows (passes) through a fixed point during a time $\Delta t = T$.

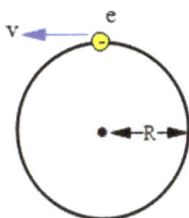

Then $i = \dfrac{\Delta q}{\Delta t} = \dfrac{e}{T}$, where $T = \dfrac{2\pi R}{v}$

or, $i = \dfrac{ev}{2\pi R} = \dfrac{(1.6 \times 10^{-19})(10^6)}{2 \times \frac{22}{7} \times (10^{-10})} \simeq 0.26 \times 10^{-3} A$ Ans.

Problem 4 (Convection current)

Find the current associated with a moving straight wire of linear charge density $\lambda = 2$ μC m^{-1} and of cross-section $A = 2$ mm^2, when the wire is pulled with a speed $v = 2$ m s^{-1}. (b) What is the current density over the given cross-sectional area?

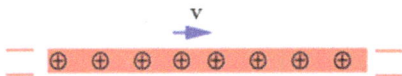

Solution

Let $dq(=\lambda \, dl)$ pass through a given vertical plane in time dt.

Then, the convection current is $i = \dfrac{dq}{dt} = \dfrac{\lambda dl}{dt} = \lambda v$ ($\because v = \dfrac{dl}{dt}$)
$$= 2 \times 10^{-3} \times 2 = 4 \text{ mA Ans.}$$

(b) The convection current density is $j = i/A = 0.0004/0.000004 = 1000$ Am^{-2}. Ans.

Problem 5 A charge cloud of length $l = 0.1$ m charge density $\rho = 3$ C m^{-3} crosses a plane perpendicularly with a velocity $v = 0.01$ m s^{-1}. Find the current density and draw the graph versus time.

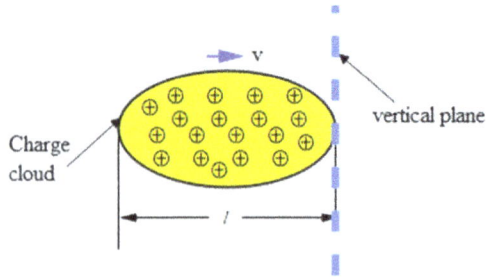

Charge cloud

v

vertical plane

l

A cloud of uniform charge volume density of positive charge crosses the vertical plane

Solution

Let the cloud touch the plane at $t = 0$ and leave the plane at $t(=\frac{l}{v})$.

$$\text{Then, } J = \begin{cases} 0\,; & t < 0 \\ \rho v\,; & 0 < t < \frac{l}{v} \\ 0\,; & t > \frac{l}{v} \end{cases}$$

The value of J during the time $t = \frac{l}{v} = \frac{0.1}{0.01} = 10$ s is given as

$$J = \rho v = (3)(0.01) = 0.03 \text{ A m}^{-2} \text{ Ans.}$$

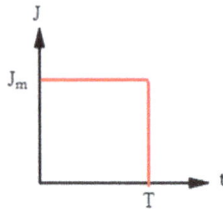

The current may vary because of non-uniform cross-section of the cloud but the current density remains constant. This is because the convection current density depends upon the volume charge density and the velocity of transportation of the charges.

Problem 6 A homogeneous beam of protons accelerated through a potential difference $V = 500$ kV has a circular cross-section of radius $R = 4$ mm. Assuming beam current $i = 32 \times 10^{-3}$ A. Find the:

(i) number of protons passing through a cross-section per second;
(ii) electric field at the surface of the beam;
(iii) potential difference between the surface and axis of the beam.

Solution

(i) The number of protons/second

$$\frac{i}{e} = \frac{32 \times 10^{-3}}{1.6 \times 10^{-19}} = 2 \times 10^{16} \text{ Ans.}$$

(ii) $E = \frac{\lambda}{2\pi\varepsilon_0 R}$, where $\lambda = \frac{i}{v}$

$$\text{or, } E = \frac{i}{2\pi\varepsilon_0 R v}$$

Since $\frac{1}{2}mv^2 = eV$, substituting $v = \sqrt{\frac{2eV}{m}}$ in equation (i),

$$E = \frac{i}{2\pi\varepsilon_0 R}\sqrt{\frac{m}{2eV}}$$

$$= \frac{2 \times 9 \times 10^9 \times 32 \times 10^{-3}}{4 \times 10^{-3}}\sqrt{\frac{1.6 \times 10^{-27}}{2 \times 1.6 \times 10^{-19} \times 500 \times 10^3}} \left(\because \frac{1}{4\pi\varepsilon_0} = 9 \times 10^9\right)$$

$$= 144 \times 10^9 \times 10^{-7} \text{ V m}^{-1}$$

$$= 14.4 \text{ kV m}^{-1} \text{ Ans.}$$

(iii) Applying Gauss's law, the flux of the electric field is

$$E \cdot 2\pi r l = \left(\frac{Q}{\pi R^2 l}\right)\left(\frac{\pi r^2 l}{\varepsilon_0}\right)$$

$$\text{or, } E = \frac{Qr}{2\pi\varepsilon_0 R^2 l} = \frac{\lambda r}{2\pi\varepsilon_0 R^2} \quad (Q/l = \text{linear charge density})$$

Then, $\Delta V = \int_0^R E \, dr = \frac{\lambda}{2\pi\varepsilon_0 R^2}\int_0^R r \, dr = \frac{\lambda}{4\pi\varepsilon_0}$, where $\lambda = i\sqrt{\frac{m}{2eV}}$

$$\text{or, } \Delta V = \frac{i}{4\pi\varepsilon_0}\sqrt{\frac{m}{2eV}} = \frac{ER}{2} = \frac{14.4 \times 10^3 \times 4 \times 10^{-3}}{2} = 28.8 \text{ V. Ans.}$$

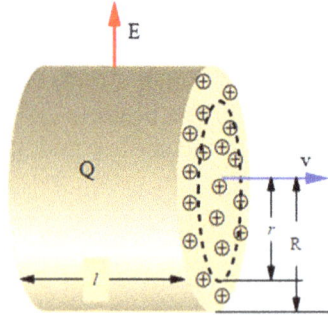

Problem 7 A current i flowing in a straight vertical wire touches the ground at point O. Then, the current flows radially away in the ground from O. Assume that the resistivity of the soil of the ground is $\rho = 110\ \Omega\ \mathrm{m}^{-2}$. If the current density of $\vec{J} = 2\hat{r}$ is measured at a radial distance $\mathrm{r} = 1$ m from O, find the (a) value of i, (b) the current density at a radial distance r from O, (c) the electric field at a radial distance r from O, (d) potential difference between any two points at a radial distance $a = 30$ cm and $b = 50$ cm from O.

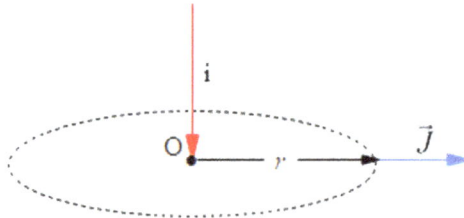

Solution

(a) The current will flow in the ground obeying spherical symmetry because we assume a uniform resistivity of the ground. Take an elementary area dA on a sphere of radius r which is vectorially given as $d\vec{A} = dA\hat{r}$. The total current flowing through the hemispherical surface of radius r drawn from O in the ground is

$$i = \oint \vec{J} \cdot d\vec{A}$$

Putting $\vec{J} = 2\hat{r}$ and $d\vec{A} = dA\hat{r}$ in the last expression, we have

$$
\begin{aligned}
i &= \oint \vec{J} \cdot d\vec{A} = \oint (2\hat{r}) \cdot (dA\hat{r}) \\
&= 2\oint dA = 2A_{\text{hemisphere}} = 2(2\pi R^2) \\
&= 4\pi(1)^2 = 12.57\ \text{A} \quad \text{Ans.}
\end{aligned}
$$

(b) The current density at a radial distance r is

$$\vec{J} = \frac{i}{2\pi r^2}\hat{r} = \frac{25.15}{2\pi r^2}\hat{r} \quad \text{Ans.}$$

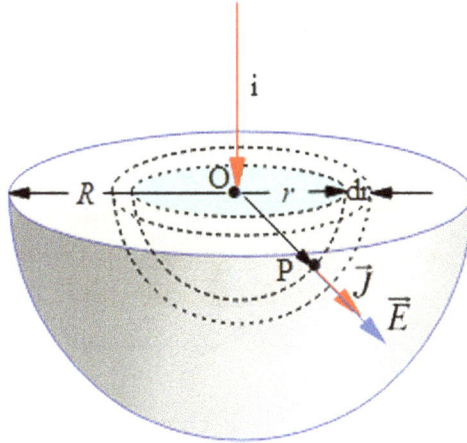

(c) The electric field at a radial distance is

$$\vec{E} = \rho\vec{J} = \frac{i\rho}{2\pi r^2}\hat{r} \quad \text{Ans.}$$

(d) The potential difference between the thin spherical shell of thickness dr at a radial distance is

$$dV = -Edr = -\frac{i\rho}{2\pi r^2}dr$$

The potential difference between these two points on the ground is

$$\Delta V = \int_0^V dV = V = -\int_a^b \frac{i\rho}{2\pi r^2}dr$$

$$= -\frac{i\rho}{2\pi}\int_a^b \frac{dr}{r^2} = -\frac{i\rho}{4\pi}\left(\frac{1}{a} - \frac{1}{b}\right)$$

$$\Rightarrow \Delta V = -\frac{(5)(110)}{2\pi}\left\{\frac{1}{(20/100)} - \frac{1}{50/100}\right\}$$

$$\Rightarrow \Delta V = -26.24 \text{ V} \quad \text{Ans.}$$

Problem 8 The current density $\vec{J} = 20000\hat{i}$ A m^{-2} is measured at a given area $A = 0.1$ cm^2 of the dotted section of the conductor of length $l = 200$ m. The normal to this area makes an angle $\theta = 60°$ with the conductor, as shown in the figure. The length of the conductor is $l = 200$ m. Find (a) the current passing through the area A, (b) the current flowing in the conductor, (c) the electric field in the conductor, (d) the potential difference between the ends of the conductor, (e) the power loss per unit volume in the conductor, (f) power loss in the conductor,(g) heat loss in the conductor per day.

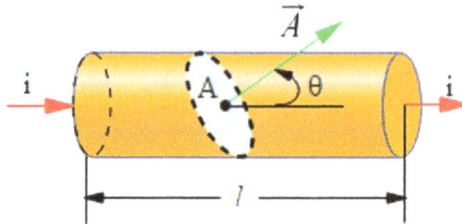

Solution

(a) The current flowing through the given area is

$$i' = \vec{J} \cdot \vec{A} = JA \cos 60^o = (2 \times 10^4)(10^{-5})\left(\frac{1}{2}\right) = 10 \text{ A} \quad \text{Ans.}$$

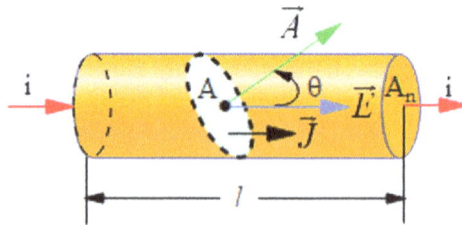

(b) The current flowing through the wire is

$$i = \vec{J} \cdot \vec{A}_n = JA_n \cos \theta° = JA_n,$$

where $A_n = A/\cos 60° = 2A$

$$\Rightarrow i = 2JA = 2(2 \times 10^4)(10^{-5}) = 40 \text{ A} \quad \text{Ans.}$$

(c) The electric field in the conductor is

$$\Rightarrow E = \rho J = (1.5 \times 10^{-6})(2 \times 10^4) = 0.03 \text{ V m}^{-2}$$

(d) The potential difference in the conductor is

$$\Delta V = El = \left(0.03 \text{ V m}^{-2}\right)(200 \text{ m}) = 3 \text{ V} \quad \text{Ans.}$$

(e) Power loss per unit volume in the conductor is

$$dP/dv = \rho J^2 = (1.5 \times 10^{-6})(2 \times 10^4)^2 = 600 \text{ W m}^{-3} \quad \text{Ans.}$$

The volume of the conductor is

$$v = A_n l = \left(2 \times 10^{-5}\right)(200) = 4 \times 10^{-3} \text{ m}^3$$

(f) So, the total power loss in the conductor is

$$P = \left(600 \text{ W m}^{-3}\right)\left(4 \times 10^{-3}\right) = 2.4 \text{ W} \quad \text{Ans.}$$

(g) Then, the total heat loss per hour is

$$Q = Pt = \left(2.4W\right)(3600s) = 8640 \text{ J} \quad \text{Ans.}$$

Problem 9 The charge q in a spherical region of radius R changes with time as $q = q_0 e^{-kt}$. Find the initial current density at the surface of the spherical region.
Solution
The total current flowing out is

$$i = \oint \overrightarrow{J} \cdot d\overrightarrow{A} = -\frac{dq}{dt}$$
$$= -\frac{d}{dt}(q_0 e^{-kt}) = +q_0 k e^{-kt}$$

The current = flux of the current density is

$$J. \, 4\pi R^2 = +q_0 k e^{-kt}$$

Then, the initial current density is

$$\overrightarrow{J}\big|_{t=0} = \frac{+q_0 k}{4\pi R^2}\hat{r} \quad \text{Ans.}$$

The direction of current is the outward direction of motion of the positive charge because q decreases with time inside the spherical region.

Problem 10 Find the distance covered by a free electron in copper during a displacement $l = 10$ mm along the wire if $J = 1$ A mm^{-2} acts in the conductor.

Solution

The distance covered by the electrons

$$D = \bar{v}\, t,$$

where $t = \dfrac{l}{v_d}$ and $\bar{v}=$ average velocity between two successive collisions.

$$\text{Then, } D = \bar{v}\frac{l}{v_d} \tag{4.8}$$

$$\text{Since, } v_d = \frac{J}{ne}, \text{ we have}$$

$$D = \bar{v}\frac{nel}{J},$$

where $\bar{v} = \sqrt{\dfrac{k}{m}T}$

$$\text{or, } D = \frac{nel}{J} \times \sqrt{\frac{kT}{m}} \text{ Ans.}$$

Problem 11 (a) Find the resistance of a conductor of length $l = \frac{1}{2}$ m area of cross-section $A = 10^{-6}$ m^2, having a current density $J = 10\,000$ A m^{-2} when an electric field of 0.01 V m^{-1} is applied inside the conductor. (b) In (a), find the heat dissipation per unit volume of the conductor during one minute.

Solution

According to Ohm's law,

$$R = \frac{V}{i} \text{ where } V = El \text{ and } i = JA$$

$$\text{Then, } R = \frac{El}{JA} = \frac{(0.01)(10)}{10000 \times 10^{-6}} = 10\,\Omega \text{ Ans.}$$

(b) Heat dissipated per unit volume $= jEt = (10000)(0.01)(60) = 6000$ Jm^{-3}.

Problem 12 Find the resistance of the conductor of conductivity σ between points 1 and 2.

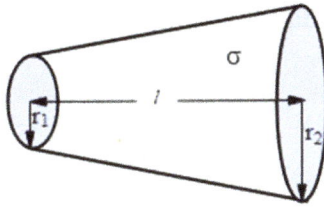

Solution

The potential difference across the thin strip is

$$dV = E \cdot dx$$

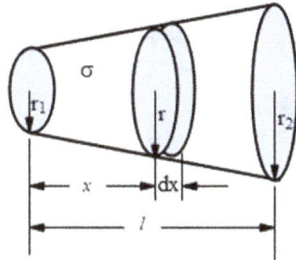

Integrating dV, the potential difference between terminals 1 and 2 is

$$V = \int_0^l E \cdot dx \tag{4.9}$$

The total current passing through the conductor is

$$i = \int J\pi r^2 = \int (\sigma E)\,\pi r^2$$

Putting $r = r_1 + \frac{r_2 - r_1}{l}$, we have

$$i = \pi\sigma \int_0^l E\left(r_1 + \frac{r_2 - r_1}{l}x\right)^2 \tag{4.10}$$

From equations (4.9) and (4.10) putting v and i in Ohm's law $R = \frac{v}{i}$, we have

$$R = \frac{\int_0^l E\,dx}{\int_0^l \pi\sigma E\left(r_1 + \frac{r_2 - r_1}{l}x\right)^2}$$

Since E is uniform in the conductor and σ is a constant pull, then out of the integral we obtain

$$R = \frac{1}{\pi\sigma} \int_0^l \frac{dx}{\left\{ r_1 + \frac{(r_2 - r_1)}{l}x \right\}^2}$$

or $R = \frac{l}{\pi\sigma r_1 r_2}$ Ans.

Problem 13 (Resistance of cylindrical conductor)

Find the resistance of the hollow cylindrical conductor between points 1 and 2. Assume $\sigma=$ conductivity of the conductor.

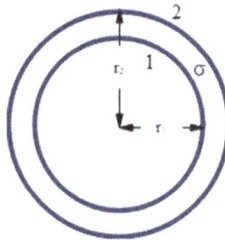

Solution

The potential difference across the thin cylindrical shell is

$$dV = E\ dr$$

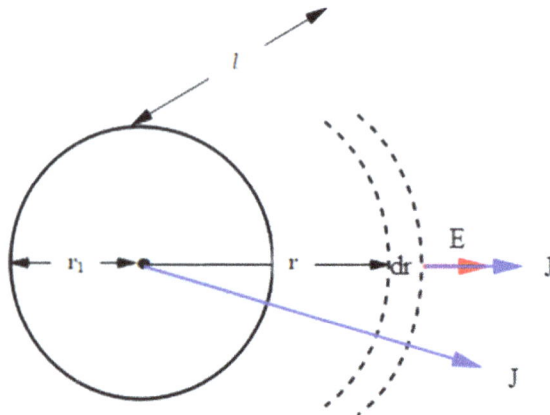

Integrating dV, the potential across the given cylinder is

$$V = \int_{r_1}^{r_2} E.\ dr \tag{4.11}$$

The current density at any point of the thin cylindrical shell is

$$J = \frac{i}{2\pi r l}$$

Then, the electric field in the thin cylindrical shell is

$$E = \frac{J}{\sigma} = \frac{i}{2\pi \sigma r l} \tag{4.12}$$

Substituting E from equation (4.12) in equation (4.11), we have

$$V = \frac{i}{2\pi \sigma l} \int_{r_1}^{r_2} \frac{dr}{r}$$

Then,

$$R = \frac{V}{i} = \frac{1}{2\pi \sigma l} \ln \frac{r_2}{r_1} \text{ Ans.}$$

Problem 14 An electric heater coil of 5 m long can radiate heat at 1000 K. If the resistivity of the coil at room temperature (27° C) is 1.1×10^{-6} Ω. m at zero degree Celsius and the diameter of the coil is 1 mm. Find (a) the rated power, (b) the rated current. Assume the coil is emissivity $=0.8$ and nearly 50% of the total heat is lost via radiation.

Solution

(a) The power radiated $=P = \varepsilon \sigma T^4 A$, where $\sigma=$ Stefan's constant and $A =$ surface area of the coil $A = \pi dl$. So, we have

$$P = \varepsilon \sigma \pi \mathrm{d} l T^4$$

Putting all values,

$$P = (0.8)(5.7 \times 10^{-8})\left(\frac{22}{7}\right)(10^{-3})(5)(2000)^4 = 712 \text{ W}$$

Since 50% of the heat generated is lost as radiation, the total power dissipation is

$$P' = 2 \times 712 \text{ W} = 1424 \text{ W Ans.}$$

(b) The resistivity of the coil at 1000 K is

$$\rho = \rho_0\{1 + \alpha(T - T_0)\} = 1.1 \times 10^{-6}\{1 + 0.0004(1000 - 300)\}$$
$$\approx 1.41 \times 10^{-6} \Omega. \text{ m}$$

Then, the resistance of the coil is given as

$$R = \rho \frac{l}{A_{cross-section}}$$

$$R = \rho \frac{l}{(\pi d^2/4)} = \left(1.41 \times 10^{-6}\right)\frac{(5)}{\pi(1 \times \frac{10^{-3})^2}{4}} \simeq 9.05 \, \Omega$$

As the efficiency of the heater is nearly 100%, almost total electrical energy is dissipated as heat. So, we can write

$$P' = i^2 R$$

Then, the rated current is

$$i = \sqrt{P'/R}$$

Using the above equations $i = \sqrt{\frac{1424}{9.05}} = 12.54$ A   Ans.

N.B: Here two areas are used; The first area (A) is surface area of the coil, the surface area used for calculating the thermal power radiation. The second area (A') is the cross-sectional area of the wire for finding its resistance.

Problem 15 What is the length of a heater wire (230 V, 1000 W) in a coil used for cooking? Assume that the diameter of the wire is $d = 1$ mm and the resistivity of nichrome is $\rho = 1.1 \times 10^{-6} \, \Omega$. m.
   Solution
   The rated power of the heater coil is

$$P = \frac{V^2}{R}$$

The resistance of the heater coil is

$$R = \frac{V^2}{P} = \frac{(230)^2}{1000} = 52.9 \, \Omega \tag{4.13}$$

We know that

$$R = \frac{\rho l}{A}, \text{ where } A = \frac{\pi d^2}{4}$$

$$\Rightarrow R = \frac{\rho l}{\left(\pi d^2/4\right)} = \frac{4\rho l}{\pi d^2} \tag{4.14}$$

Using last two equations,

$$\frac{4\rho l}{\pi d^2} = \frac{V^2}{P}$$

This gives

$$l = \frac{\pi d^2 V^2}{4\rho P}$$

Putting the given values, we have

$$l = \frac{(22/7)(10^{-3})^2(230)^2}{4(1.1 \times 10^{-3})(1000)}$$

$$\Rightarrow l \approx 37.8 \text{ m Ans.}$$

Problem 16 A current i varies with time in a coil of resistance R as shown in the graph. Find the (a) total charge flowing, (b) average current, and (c) heat dissipated in the resistor.

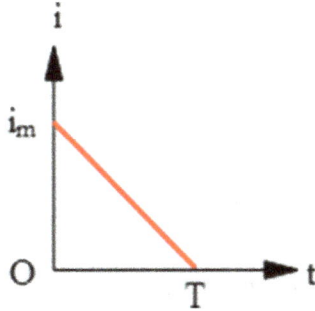

Solution

(a) The total charge flowing is

$$q = idt$$
$$= \text{Area under } i - t \text{ graph}$$
$$= \frac{1}{2}Ti_m = \frac{i_m T}{2} \text{ Ans.}$$

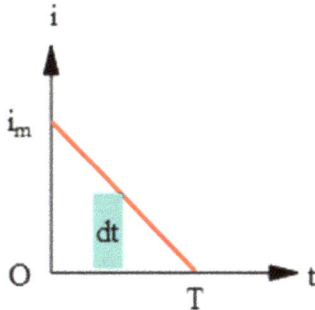

(b) The average current $= i_{av} = \frac{q}{T} = \frac{i_m T}{2}/T = \frac{i_m}{2}$ Ans.

(c) The heat dissipated $= \int P\,dt$

$$= \int i^2\,R\,dt, \text{ where } i = i_m(1 - \frac{t}{T})$$

$$= R \int_0^T \{i_m(1 - \frac{t}{T})\}^2\,dt$$

$$= \frac{1}{3}i_m^2\,RT \text{ Ans.}$$

Problem 17 Find the (a) drift speed of an electron in copper wire having a current density of 10 000 A m^{-2} and (b) electric field inside the wire.

Solution

The drift speed is

$$v_d = \frac{J}{ne} = \frac{10000}{8.49 \times 10^{28} \times 1.6 \times 10^{-19}} = 7.36 \times 10^{-7}\,\text{m s}^{-1}\,\text{Ans.}$$

The electric field is

$$E = \frac{J}{\sigma} = \frac{10\,000}{5.9 \times 10^7} = 1.695 \times 10^8\,\text{V m}^{-1}\,\text{Ans.}$$

N.B: Electrons drift slowly but the signal (electric field) propagates nearly at the speed of light.

Problem 18 A leaky parallel plate capacitor has two dielectric slabs of thickness d_1, d_2, permittivities ε_1, ε_2 and conductivities σ_1, σ_2, respectively. If the potential difference across the capacitor is V, find the (a) leakage current, (b) electric fields in the dielectrics, (c) total surface charge density,(d) free surface charge, deposited at the junction of the dielectrics.

Solution

(a) The potential difference across the capacitor is

$$V = V_1 + V_2 = E_1 d_1 + E_2 d_2 \tag{4.15}$$

Since the same current flows through the capacitors, the current density is equal in both dialectics. applying Ohm's law, we have

$$J = \sigma_1 E_1 = \sigma_2 E_2 \tag{4.16}$$

Solving the last two equations, the current density is

$$V = E_1 d_1 + E_2 d_2 = \frac{J}{\sigma_1}d_1 + \frac{J}{\sigma_2}d_2$$

$$\Rightarrow J = \frac{V}{\left(\frac{d_1}{\sigma_1} + \frac{d_2}{\sigma_2}\right)}$$

So the leakage current is

$$i = JA = \frac{\sigma_1\sigma_2 VA}{\left(d_1\sigma_2 + \sigma_1 d_2\right)} \quad \text{Ans.}$$

(b) Then the electric fields are

$$E_1 = \frac{V\sigma_2}{\sigma_1 d_2 + \sigma_2 d_1}, \quad E_2 = \frac{V\sigma_1}{\sigma_1 d_2 + \sigma_2 d_1} \quad \text{Ans.}$$

(c) Applying Gauss' law across the boundary, the total flux of electric field out is

$$\phi_E = \left(E_1 - E_2\right)(\delta A) = \frac{\sigma_s(\delta A)}{\varepsilon_0}$$

Then, the total (bound + free) surface charge density is

$$\sigma_s = \varepsilon_0\left(E_1 - E_2\right)$$

Putting the obtained values of electric fields, we have

$$\sigma_s = \frac{\varepsilon_0(\sigma_2 - \sigma_1) V}{\sigma_1 d_2 + \sigma_2 d_1} \quad \text{Ans.}$$

(d) Then, the free surface charge density is

$$\sigma_s = \left(D_2 - D_1\right) = \left(\varepsilon_2 E_2 - \varepsilon_1 E_1\right)$$

Putting the obtained values of electric fields,

$$\sigma_s = \left(\varepsilon_2 E_2 - \varepsilon_1 E_1\right)$$

$$= \varepsilon_2\left(\frac{V\sigma_1}{\sigma_1 d_2 + \sigma_2 d_1}\right) - \varepsilon_1\left(\frac{V\sigma_2}{\sigma_1 d_2 + \sigma_2 d_1}\right)$$

$$\sigma_{\text{free}} = \left(\frac{\sigma_1\varepsilon_2 - \sigma_2\varepsilon_1}{\sigma_1 d_2 + \sigma_2 d_1}\right) V$$

So, the free charge accumulated is

$$q_{\text{free}} = \left(\frac{\sigma_1\varepsilon_2 - \sigma_2\varepsilon_1}{\sigma_1 d_2 + \sigma_2 d_1}\right) VA \quad \text{Ans.}$$

N.B:
1. The conductivity of a practical dielectric is not zero, but very small. As the permittivities are small and finite, the free charge accumulation will be very small in the case of leakage current in the capacitors.
2. If the conductivity is very large for good conductors, the accumulation of free charges can be ignored.

Chapter 5

DC circuit and instrument

5.1 Introduction

In last chapter, we learnt how a battery works and delivers a current in a circuit. I hope that all of you have understood the mechanism of heat dissipation in a conductor in the last chapter. In this chapter we will calculate the current and voltage across a resistor and capacitors in a circuit. So, you have to apply Kirchhoff's circuital laws and Ohm's law. As we studied Ohm's law in the last chapter, we will discuss Kirchhoff's circuital laws (KCL) in this chapter. Using these two laws you will be able to find the current in any branch of an electric circuit. Then we will apply some basic techniques such as concept of symmetry, equipotential points etc, to find the equivalent resistance of complicated networks. Finally, we will explain the working principles of electrical instruments (galvanometer, ammeter, voltmeter, potentiometer etc) that measure the current, voltage and electromotive force (emf).

5.2 Kirchhoff's circuital law (KCL)

Kirchhoff's circuital Law has two parts, namely the first and second laws. The second law uses the term potential drop or potential difference (PD). We will see how to find PD across a battery, resistor and capacitor. As you know, the PD between two points 1 and 2 of a circuit of elements X (resistor, capacitor, battery etc), is always equal to the potential of the final (second) point minus the potential of the initial (first) point as we go along the wire. It can be given as

$$V_X = V_2 - V_1 = \Delta V. \tag{5.1}$$

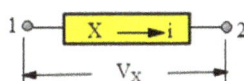

The potential difference across the circuit element X is
$V_X = V_2 - V_1$ when you are moving from terminal 1 to terminal 2.

5.2.1 PD across a battery (a group of cells)

For the sake of simplicity let us assume that the battery is ideal (having a zero internal resistance). If you go from the $-$ve to the $+$ve terminal inside the battery, you have to move against the electrostatic field of the battery. So, the potential rises; then you can write $V_b = +\varepsilon$. If you move from $+$ve to $-$ve terminal, you move in favour of the electrostatic field inside the battery. So, the potential drops and then you can write $V_b = -\varepsilon$. The measurement of V_b does not depend on the direction of current and resistance of the battery. However, a real battery is a series combination of an ideal battery and its internal resistance. So, the terminal PD while a current flows in a battery will no longer be equal to the emf of the battery. If we go from the negative to the positive terminal of a battery, the terminal potential difference is given as $V_b = +\varepsilon - ir$ (if current flows from $-$ve to $+$ve terminal inside the battery); and $V_b = \varepsilon + ir$ (if current flows from $+$ve to $-$ve terminal inside the battery). However, in an open circuit the battery does not carry a current; so, the potential drop will be equal to the emf of the battery.

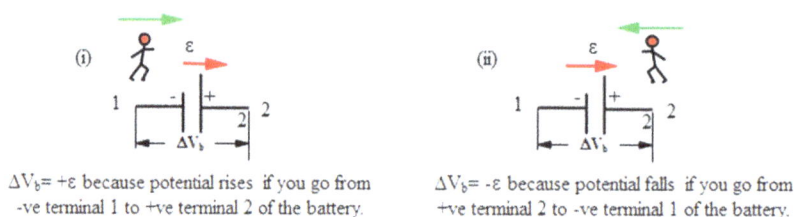

$\Delta V_b = +\varepsilon$ because potential rises if you go from $-$ve terminal 1 to $+$ve terminal 2 of the battery.	$\Delta V_b = -\varepsilon$ because potential falls if you go from $+$ve terminal 2 to $-$ve terminal 1 of the battery.

5.2.2 PD across a resistor

As we have learnt in the last chapter, the current flows (positive charge moves) in the direction of electric field inside the conductor. So, when you go in the direction of current, or electric field, the potential decreases and the PD along the direction of current is given as $V_R = -iR$, according to Ohm's law. Likewise, if you go against the direction of current, the PD is given as $V_R = +iR$. This means that the PD across a resistor depends upon both the magnitude and the direction of current in addition to the resistance.

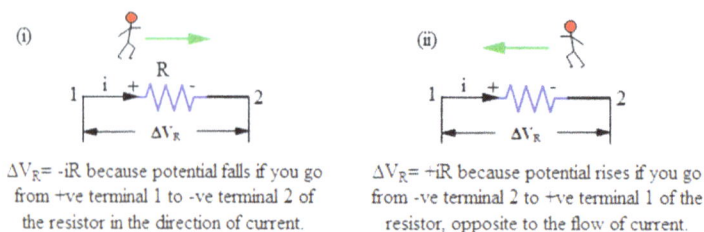

$\Delta V_R = -iR$ because potential falls if you go from $+$ve terminal 1 to $-$ve terminal 2 of the resistor in the direction of current.	$\Delta V_R = +iR$ because potential rises if you go from $-$ve terminal 2 to $+$ve terminal 1 of the resistor, opposite to the flow of current.

5.2.3 PD across a capacitor

If you move from the $-$ve plate to the $+$ve plate of a capacitor, you have to go in the opposite direction to the electric field produced by the capacitor. Hence, the potential increases as you go; then the difference is positive, which can be given as

$$V_c = \frac{q}{C}$$

If you move from the +ve plate to the −ve plate of a capacitor, you have to go in the same direction as the electric field produced by the capacitor. Hence, the potential decreases and the difference is negative, which can be given as

$$V_c = -\frac{q}{C}$$

Now, we can understand that the PD across the capacitor does not depend on the magnitude and direction of current. It depends upon the polarity of the capacitor along with the charge on each plate and the capacitance.

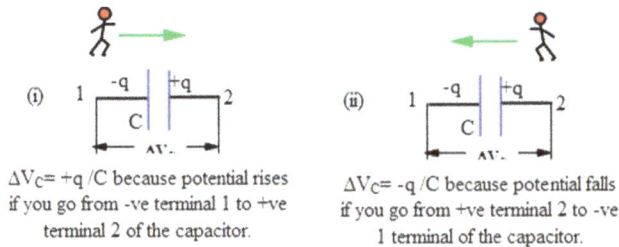

$\Delta V_C = +q /C$ because potential rises if you go from -ve terminal 1 to +ve terminal 2 of the capacitor.

$\Delta V_C = -q /C$ because potential falls if you go from +ve terminal 2 to -ve 1 terminal of the capacitor.

5.2.4 Kirchhoff's first law (Kirchoff's current law, KCL)

In a circuit we have so many branches and junctions (meeting point of the branches). In hydrodynamics, we know that the rate of flow or mass volume is called flux. In current electricity the flux is the electric current, that is, the flux of the J-field. We define an electric current as the rate of flow of electric charge. Because the charge is always conserved, we must conserve the current. Then, we can state the following.

The algebraic sum of current meeting at a junction is equal to zero.

$$\sum i = 0$$

Since current is an algebraic quantity, the net current = incoming current − outgoing current = 0; this means that the inward or incoming current is equal to the outward or outgoing current at a junction. The sign convention of the current in this case is arbitrary; if you choose incoming current as negative, the outward current will be positive or vice versa. This is Kirchhoff's first law or law of currents.

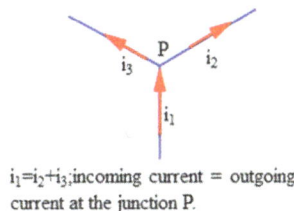

$i_1 = i_2 + i_3$; incoming current = outgoing current at the junction P.

5.2.5 Kirchhoff's second law $\int_a^b \vec{E}\cdot d\vec{l} = (V_b - V_a) = \Delta V_C$ (Kirchoff's voltage law, KVL)

This law talks about the conservation of energy. As we discussed in the last chapter, electrostatic field is present inside and outside the conductors, resistor, capacitor and batteries. We can call it electrostatic field which is a conservative field; the net work done by this field in a closed path is always zero, which is given as

$$\oint \vec{E}\cdot d\vec{l} = 0$$

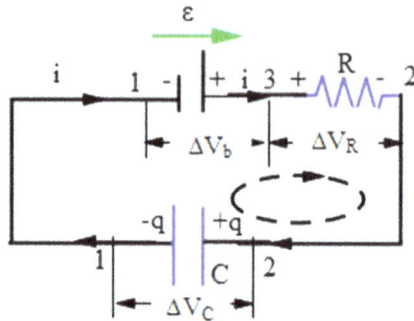

Sum of the voltage drop across the circuit
elements is zero taken in one sense (here
we have taken clockwise sense).

Splitting the closed integral from 1 to 2, from 2 to 3 and from 3 to 1 along the circuit, we have

$$-\int_1^2 \vec{E}\cdot d\vec{l} - \int_2^3 \vec{E}\cdot d\vec{l} - \int_3^1 \vec{E}\cdot d\vec{l} = 0$$

The PD along the capacitor is,

$$\int_1^2 \vec{E}\cdot d\vec{l} = \Delta V_R = V_R$$

The PD along the resistor is

$$\int_2^3 \vec{E}\cdot d\vec{l} = \Delta V_R = V_R$$

The PD along the battery is

$$\int_3^1 \vec{E}\cdot d\vec{l} = \Delta V_b = V_b$$

Using the last four equations, we have

$$\Delta V_C + \Delta V_R + \Delta V_b = 0$$

This tells us that the sum of PD across the circuit elements (R, C, source etc) in a closed circuit taken in one sense (either clockwise or anti-clockwise) is zero. So, we can write,

$$\sum \Delta V = \Delta V_R + \Delta V_b + \Delta V_c = 0$$

If you choose to write V instead of ΔV, the above expression can be given as

$$\sum V = V_R + V_b + V_c = 0.$$

This tells us that the algebraic sum of potential difference across all circuit elements in a closed loop taken in clockwise or anticlockwise sense will be equal to zero.

Example 1
   (a) Find the current in the single-loop circuit.
   (b) What is the terminal PD of the battery?

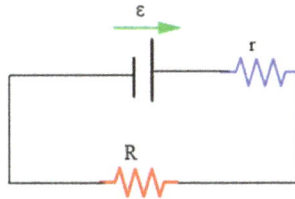

Solution
   (a) Applying KCL in the given circuit in the direction of current, we have

$$\Delta V_r + \Delta V_b + \Delta V_R = 0$$

$$\Rightarrow -ir + \varepsilon - iR = 0$$

$$\Rightarrow i = \frac{\varepsilon}{R + r} \text{ Ans.}$$

   (b) The terminal PD is

$$V_{\text{term}} = V_R = iR = \frac{\varepsilon R}{R + r} \text{ Ans.}$$

N.B: When $r \to 0$, $V_{\text{term}} = \varepsilon$; when $r \to \infty$, $V_{\text{term}} = \varepsilon$.

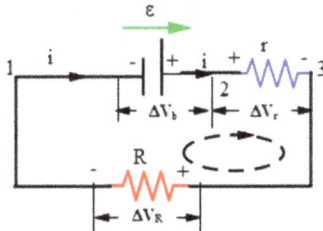

Example 2

(a) Discuss the effect of internal resistance on terminal voltage and emf of a battery.

(b) If we connect the terminals of a battery by a zero resistance, or short-circuit the battery, what will be the current and terminal voltage?

Solution

(a) Just after the switch is closed, the chemical reaction takes place in the battery producing a direct current. Due to the chemical action the electrolytes move. After a long time, due to the chemical degradation of electrolytes and electrodes and deposition of reaction by-products, the internal resistance of the battery tends to infinite (increases to a very large value). Eventually, the total resistance of the circuit becomes incredibly large and the current flowing in the circuit reduces to zero. You can say that the battery is dead (inactive). In this stage, the terminal PD will tend to zero, but the emf of the cell remains almost the same for a certain period. It is the internal resistance that increases to reduce the current to zero. Practically, emf of a battery also decreases over time due to the effect of temperature, polarization, electrode degradation etc. However, the emf of a battery decreases very slowly as compared with the change in internal resistance of the battery.

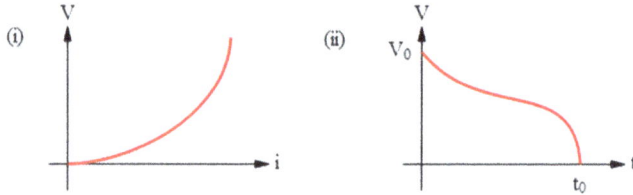

In this case, the internal resistance r increases, and Ohm's law

$$i = \frac{V}{R + r}$$

does not hold good (the system becomes non-ohmic). The V–i graph is given. After a time t_0, the terminal PD will be zero. If the battery is reversible, you can recharge it to restore its terminal voltage to its maximum value V_0. During the process of charging, the deposited electrolytes are again cleansed and the internal resistance decreases to its normal value. Ans.

(b) If we connect the terminals of a battery by a resistance R, the current in the circuit is

$$i = \frac{\varepsilon}{R + r} = \frac{\varepsilon}{r} \ (\because R = 0), \text{ where } R = 0$$

Then, the terminal voltage is

$$V_{\text{term}} = iR = \frac{\varepsilon R}{R + r}, \text{ where } R = 0$$

So, the terminal voltage is zero. In another way, the terminal voltage is given as

$$V_{term} = \varepsilon - ir$$

Putting $i = \frac{\varepsilon}{r}$, we can also find the same 'zero' terminal voltage. This means that due to a very small internal resistance of the cell or battery, a huge current flows such that the terminal voltage will be equal to zero. No potential will be dropped across the battery of zero internal reistance. Ans.

5.3 Grouping of resistors

In electrical circuits, we group the resistors to get a resistance suitable for our practical purpose. There are two basic ways we can group the resistors, namely series and parallel. The combination of series and parallel grouping is called mixed grouping.

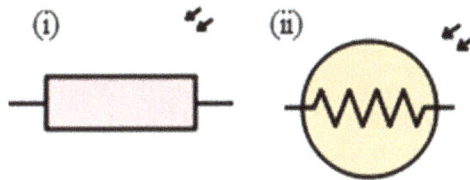

Symbols of a resistor

5.3.1 Series grouping

Let us connect the resistors end to end between terminals A and B. In this series connection the same current, i, say, flows through each resistor. Then, the PD between terminals A and B is

$$V_{AB} = V_1 + V_2 + \cdots + V_n$$

$$\Rightarrow V_{AB} = iR_1 + iR_2 + \cdots + iR_n$$

$$\Rightarrow V_{AB} = i(R_1 + R_2 + \cdots + R_n)$$

Let us substitute the combination by a single resistor R. If it carries the same current when the same PD V_{AB} is applied between A and B, we call it an *equivalent* resistance. So, we can write

$$V_{AB} = iR$$

Comparing the last two equations, we have

$$R_{AB} = V_{AB}/i = i(R_1 + R_2 + \cdots + R_n)$$

In a handy form, we can write

$$R = \sum R_i$$

This states that the equivalent resistance in a series grouping is equal to the sum of all resistances.

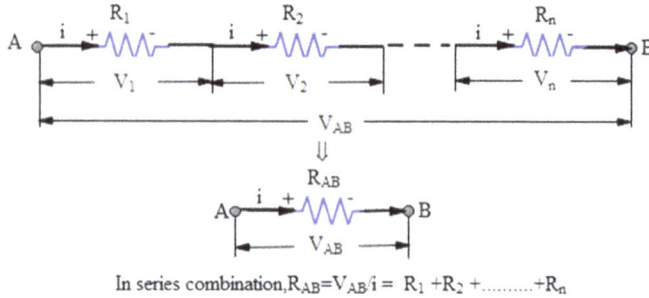

In series combination, $R_{AB} = V_{AB}/i = R_1 + R_2 + \ldots + R_n$

5.3.2 Parallel grouping

If you bring either end of each resistor to a single point having potential V_1 and other ends to a point B at potential V_2, the PD $V(=|V_2 - V_1|)$ is the same across all resistors. This is known as parallel combination.

The total current I drawn by the combination is

$$i = i_1 + i_2 + \cdots + i_n$$
$$= \frac{V}{R_1} + \frac{V}{R_2} + \cdots + \frac{V}{R_n}$$

$$\Rightarrow i = V\left(\sum \frac{1}{R_i}\right)$$

Let us substitute the combination by a single resistor R. If it carries the same current when the same PD V_{AB} is applied between A and B, we call it an *equivalent* resistance. So, we can write

$$i = V_{AB}/R = V/R$$

Comparing the last two equations, we have

$$\frac{1}{R} = \sum \frac{1}{R_i}$$

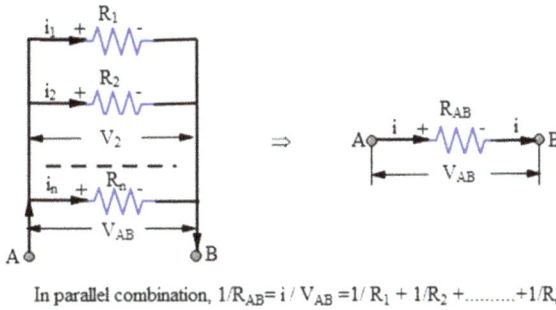

In parallel combination, $1/R_{AB} = i / V_{AB} = 1/R_1 + 1/R_2 + \ldots\ldots + 1/R_n$

Example 3 Four resistors each of resistance R and another resistor of resistance X are joined between A and B. If the equivalent resistance between A and B is equal to $2R$, find the value of X.

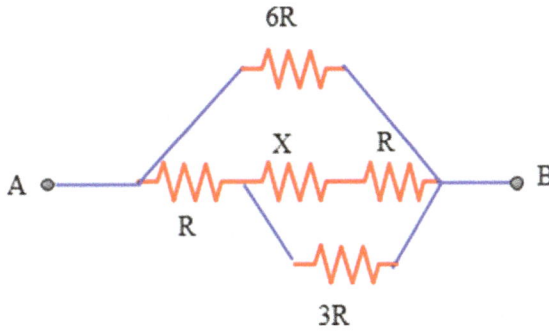

Solution

We can see that X and R are in series to give a resistance

$$r = X + R$$

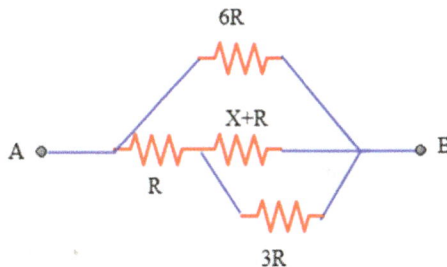

Now $3R$ and r are in parallel to give us an effective resistance

$$r' = \frac{3R(R + X)}{(R + X) + 3R} = \frac{3R(R + X)}{4R + X}$$

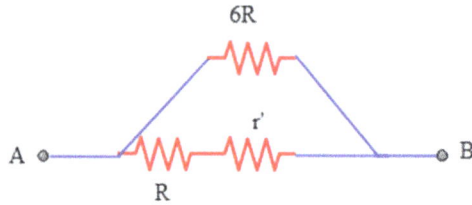

Now R and r' are in series to give us an effective resistance

$$r_1 = R + r' = R + \frac{3R(R + X)}{4R + X} = \frac{(7R + 4X)R}{4R + X} \tag{5.2}$$

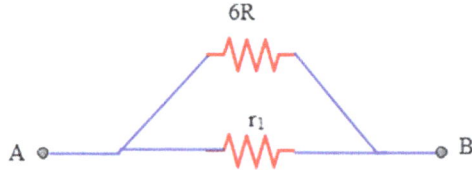

Since r_1 and $6R$ are in series, the effective resistance between A and B is

$$r_{AB} = \frac{r_1(6R)}{r_1 + 6R} = 2R \text{ (given)}$$

$$r_1 = 3R \tag{5.3}$$

Using the equations (5.2) and (5.3), we have

$$r_1 = \frac{(7R + 4X)R}{4R + X} = 3R$$

$$\Rightarrow X = 5R \text{ Ans.}$$

N.B: For any two resistors R_1 and R_2, the equivalent resistance in series combination is $R_{eq} = R_1 + R_2$ and parallel combination, $R_{eq} = \frac{R_1 R_2}{R_1 + R_2}$.

5.4 Finding equivalent resistance required for complex systems of resistors

In a complicated network there are too many resistors joined in an apparently complex manner. So, we cannot solve the problem easily by directly using the formulae of series and parallel combination of the resistors. In this case, we can simplify the circuit to a combination of series and parallel resistors using the concepts of equipotential points, electrical symmetry by the help of KCL and Ohm's law. Finally, the circuit will be reduced to a simpler form of series and parallel combinations so that it will be easier to find the equivalent resistance of the network between the given terminals.

5.4.1 Equipotential points

In a current-carrying electrical network, two points 1 and 2, say, are said to be equipotential if they are at the same potential ($V_1 = V_2$). So, the PD between points 1 and 2 is

$$\Delta V = iR = 0.$$

Now we have two cases, if $R = 0$, $\Delta V = 0$ ($i \neq 0$) and if $i = 0$ (R is finite) $\Delta V = 0$. The first case tells us that when we connect any two points by an ideal conductor (zero resistor), the PD between the points becomes zero. It is called *short circuiting*. The second case tells that, if we connect any two points by a non-zero resistor and find no current along the resistor, we can also term these points *equipotential*. After finding equipotential points, we join them to a single point to simplify the given circuit.

5.4.2 Electrical symmetry

If the branches AB and AC have the same resistance and the same current, we can conclude that the same potential $V = iR$ will be dropped along these branches. Then, the branches AB and AC are said to be electrically symmetrical. So, we can bring the equipotential points B and C to a single point to simplify the circuit diagram. Even though a resistor is connected between the equipotential points, no current will flow through the resistor. Then, we can remove this resistor from the circuit to simplify it.

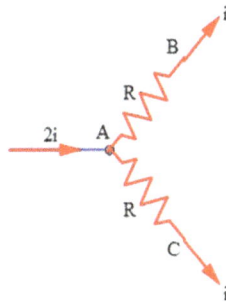

The branches AB and AC are electrically symmetrical as they have same resistance carrying equal current.

Example 4 Five resistors each of resistance R are connected between A and B. Find the equivalent resistance between A and B.

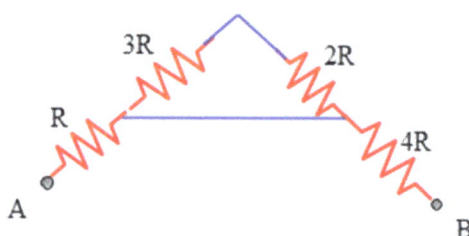

Solution

First of all, we name all the junctions as shown in figure (i). We can see that C and D are at the same potential, as these points are connected by zero resistance. Now, we can consider these two points C and D as one point. As a result, the resistors $3R$ and $2R$ will form a loop without having any emf. As this loop does not have any current due to the absence of a seat of emf, it is useless. So, you can remove these two resistors to get a most simplified circuit as shown in figure (iii). Now, we have two resistors R and $4R$ in series to form a net resistance of $5R$ between A and B.

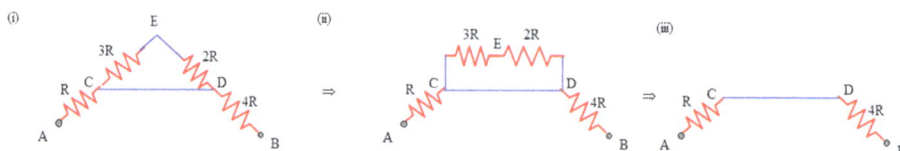

Alternative method: Alternately, consider the connecting wire CD as a zero resistor which is in parallel to a series combination of $3R$ and $4R$ between C and D. The presence of the zero resistor between C and D is shown in figure (ii). The effective resistance between these two points will be zero which is irrespective of any resistor between these points. Then redraw the circuit diagram as shown in figure (iii). Now we have a simple series circuit between A and B containing resistor R, $4R$ and zero resistance. So, we have

$$R_{AB} = R + 4R = 5R \text{ Ans.}$$

N.B:

1. If we get a closed loop of resistors without any battery, it carries no current. Then remove the total loop to get a simpler circuit or if the current in any branch is zero, remove it.

2. If a zero resistor is connected between any two points A and B, $R_{AB} = 0$. But, the total current passing through the zero resistor may not be zero.

3. $i_{CD} = \dfrac{\Delta V}{R} =$ finite because when $\Delta V \to 0$, the resistance will also tend to zero; $R \to 0$.

Example 5 Five resistors each of resistance R are connected between A and B. Find the equivalent resistance between A and B.

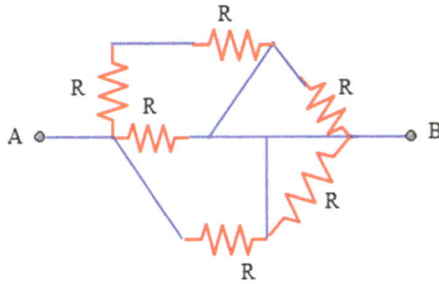

Solution

First of all, we name all the junctions as shown in figure (i). We can see that C and G are at the same potential, A, D and E are at the same potential as these points are connected by zero resistance (because the connecting copper wires have negligible resistance). So, consider these three points as one point, A, say. Similarly, points C, G, F, H and I are equipotential and consider these points as one point (C, say). Then redraw the circuit diagram as shown in figure (ii) below. Now we have a simple parallel circuit between A and C containing resistor $2R$, R and R. Similarly, there are three resistors of resistance $2R$, zero and R connected between B and C. Due to the presence of a zero resistor, $R_{BC}=0$. Now the circuit is simplified further as shown in figure (iii).

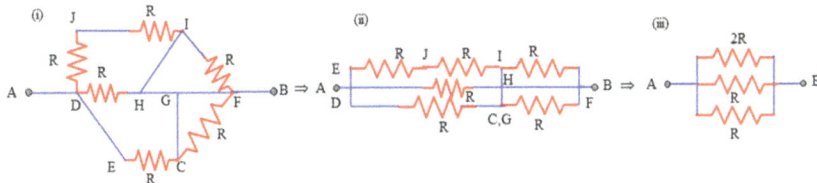

So, the final resistance between A and B is given by the formula

$$1/R_{AB} = 1/2R + 1/R + 1/R$$

Then, we have $R_{AB} = 2R/5$. Ans.

Example 6 Six resistors (shown as straight lines each of resistance R) form a pyramid. The base and the apex of the pyramid are, respectively, ABC and D. Find the effective resistance between A and B.

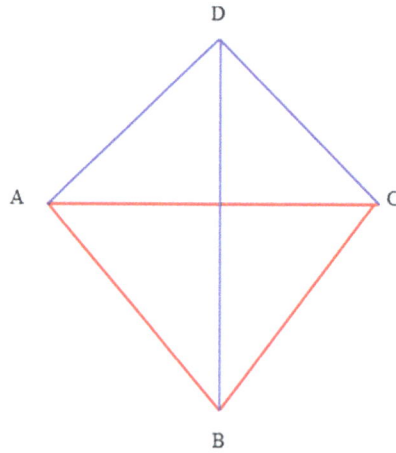

Solution

The branches ADB and ACB are symmetrical relative to terminals A and B. Hence, points D and C are equipotential. Since $R_{DC} \neq 0$, $i_{DC} = 0$. Then remove the branch DC and then the circuit is reduced to a simpler one, as shown in figure (iii).

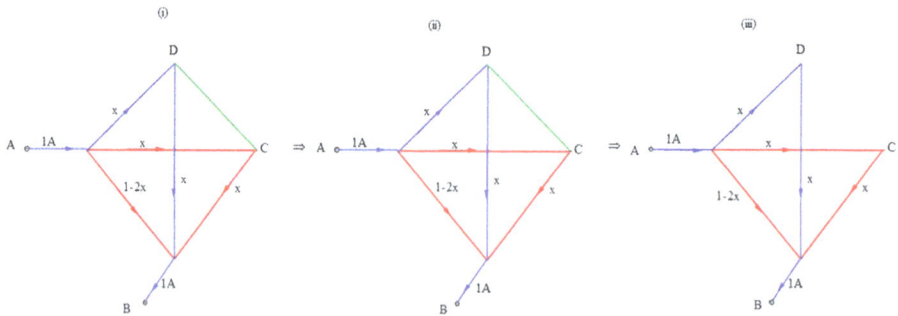

Alternative method: Press the apex D so that the pyramid becomes a *triangle* (ABC) plus a *star* (AD, CD and BD) system. After all resistors lie in the same plane it will be easier for us to work. You can see that equal current flows through branches ACB and ADB, hence C and D are equipotential. Please note that the circuit is electrically symmetrical about the dotted green line CD; the branch CD carries *no* current because C and D are equipotential. In this case it will satisfy the balanced *Wheatstone Bridge* condition given as $R_{BC}/R_{AC} = R_{BD}/R_{AD}$. We will talk about this in a later section.

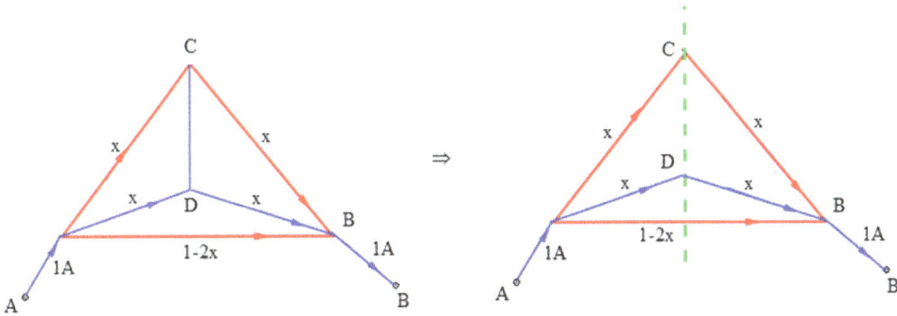

In both methods, you can prove that the branch CD carries zero current and therefore it is removed from the given network. Now we have three branches, namely ADB, ACB and AB connected in parallel between A and B. Both ADB and ACB have two resistors each of resistance R in series. So, we have the resistance $2R$, $2R$ and R in parallel. Then, the net resistance between A and B is given as

$$\frac{1}{R_{AB}} = \frac{1}{2R} + \frac{1}{2R} + \frac{1}{R} = \frac{2}{R}$$

$$\Rightarrow R_{AB} = \frac{R}{2} \text{ Ans.}$$

5.5 Grouping of cells

We can control the current as per our need by an appropriate mixed grouping of cells or batteries. Basically, we have two types of grouping such as (i) series and (ii) parallel. The combination of these two groupings is called mixed grouping.

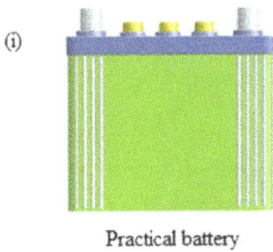

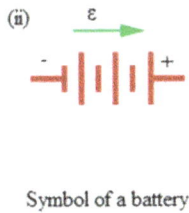

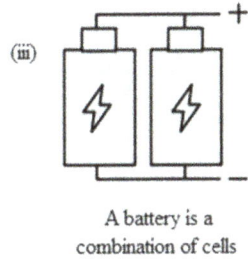

Practical battery

Symbol of a battery

A battery is a combination of cells

5.5.1 Series grouping

5.5.1.1 Correct connection

Let us connect two cells end to end so that the +ve terminal of one is connected with the −ve terminal of the other, we call it *correct* series connection. Let a current i flow through each cell. The emfs of the cells are ε_1 and ε_2, and internal resistances are r_1 and r_2, respectively. Let V_1 and V_2 be the potentials of points A and B, respectively. Then, applying the second KCL, we have

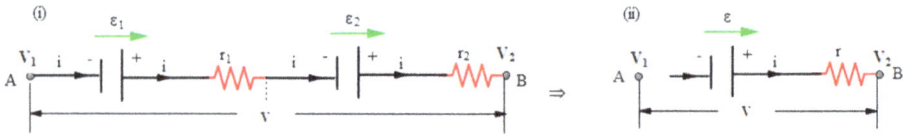

In the correct connection, the equivalent resistance is equal to the sum of all resistance $r = r_1 + r_2$ and the equivalent emf is equal to the sum of magnitude of all emf, $\varepsilon = \varepsilon_1 + \varepsilon_2$

$$V_1 + \varepsilon_1 - ir_1 + \varepsilon_2 - ir_2 = V_2$$

$$\Rightarrow V(=V_2 - V_1) = \varepsilon_1 + \varepsilon_2 - i(r_1 + r_2) \tag{5.4}$$

Let us substitute the given combination by a single cell of emf ε and internal resistance r. If it carries the same current i to have the same PD (terminal voltage), then you can call it an equivalent cell.

Applying KCL for the equivalent cell, we have

$$V = \varepsilon - ir \tag{5.5}$$

Comparing equations (5.4) and (5.5), we have the following results:

$$\varepsilon = \varepsilon_1 + \varepsilon_2$$

$$r = r_1 + r_2.$$

5.5.1.2 Wrong (incorrect) connection

If you connect the cell wrongly so that the +ve terminal of one is connected with the +ve terminal of the other, each cell tries to produce its own current in the direction of its emf. As the emf of the cells oppose each other in the wrong connection, the current will flow in the direction of the stronger cell having greater emf.

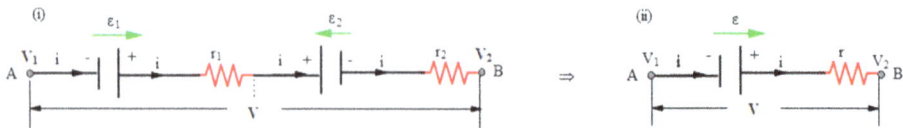

In the incorrect connection, the equivalent resistance is equal to the sum of all resistance $r = r_1 + r_2$ and the equivalent emf is equal to the subtraction of magnitude of two emf, $\varepsilon = \varepsilon_1 - \varepsilon_2$ assuming that 1st cell is stronger than the 2nd cell

Then, applying the second KCL, we have

$$V_1 + \varepsilon_1 - ir_1 - \varepsilon_2 - ir_2 = V_2$$

$$\Rightarrow V(=V_2 - V_1) = \varepsilon_1 - \varepsilon_2 - i(r_1 + r_2) \tag{5.6}$$

Let us substitute the given combination by a single cell of emf ε and internal resistance r. If it carries the same current i to have the same PD (terminal voltage), then you can call it an equivalent cell.

Applying KCL for the equivalent cell, we have

$$V = \varepsilon - ir \qquad (5.7)$$

Comparing equations (5.6) and (5.7), we have the following results:

$$\varepsilon = |\varepsilon_1 - \varepsilon_2|$$

$$r = r_1 + r_2$$

In general, for series combination of n cells, the equivalent cell can be given as

$$\varepsilon = \sum \varepsilon_i \quad \text{and} \quad r = \sum r_i$$

The net emf is equal to the 'algebraic' sum (but not always the sum of the magnitude) of emfs because emf is an algebraic quantity which can be +ve in one direction and −ve in the opposite direction. However, the net resistance remains the same as that in the case of correct connection which is equal to the scalar sum of all the resistances in series.

5.5.2 Parallel grouping

If right side terminals of all cells are connected to a point 1 and left side terminals are connected to another point 2, you may call it parallel combination, where the terminal PD across all cells will be equal (to V, say). For two cells of emf ε_1, ε_2 and internal resistances r_1 and r_2, applying KVL, we have

$$\varepsilon_1 - i_1 r_1 = V$$

and $\varepsilon_2 - i_2 r_2 = V$

Then, the total current

$$i = i_1 + i_2 = \frac{\varepsilon_1 - V}{r_1} + \frac{\varepsilon_2 - V}{r_2} \qquad (5.8)$$

If we substitute the total combination by a cell of emf ε and internal resistance r, if it carries the same current to have the same terminal voltage, we call it equivalent cell. Applying KVL, we have

$$V = \varepsilon - ir$$

or,

$$i = \frac{\varepsilon - V}{r} \qquad (5.9)$$

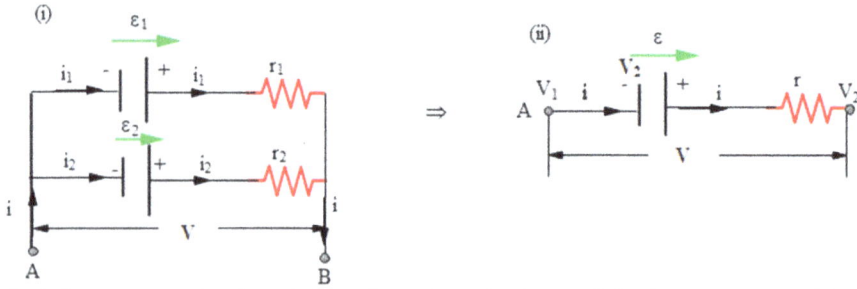

In the incorrect connection, the equivalent resistance is equal to the sum of all resistance $r = r_1 + r_2$ and the equivalent emf is equal to the subtraction of magnitude of two emf, $\varepsilon = r(\varepsilon_1/r_1 + \varepsilon_2/r_2)$, where $1/r = 1/r_1 + r_2$

Comparing equations (5.8) and (5.9), we have

$$\frac{\varepsilon_1 - V}{r_1} + \frac{\varepsilon_2 - V}{r_2} = \frac{\varepsilon - V}{r}$$

Then, the equivalent cell and its internal resistance can be given as follows:

$$\varepsilon = \frac{\frac{\varepsilon_1}{r_1} + \frac{\varepsilon_2}{r_2}}{\frac{1}{r_1} + \frac{1}{r_2}}$$

$$\frac{1}{r} = \frac{1}{r_1} + \frac{1}{r_2}$$

In general, for a parallel combination of n cells, the equivalent cell and its internal resistance can be given as follows:

$$\varepsilon = \frac{\sum \varepsilon_i / r_i}{\sum 1 / r_i}$$

$$\frac{1}{r} = \sum \frac{1}{r_i}$$

where ε_i and r_i are the emf and internal resistance of the ith cell, respectively. Please remember that emfs are algebraic scalars.

Example 7 Two cells of emf $\varepsilon_1 = 9$ V and $\varepsilon_1 = 4$ V are connected to an external load of resistance $R = 2$ Ω as shown in the figure below. The internal resistances of the cells are $r_1 = 1.5$ Ω and $r_2 = 1$ Ω. Find the current in the external resistor.

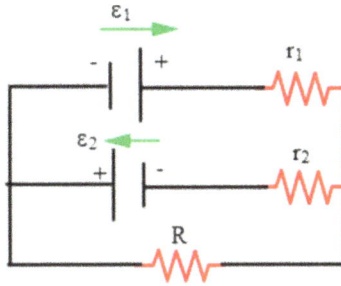

Solution

The effective emf of the parallel combination of the cells is

$$\varepsilon = \frac{\dfrac{\varepsilon_1}{r_1} + \dfrac{\varepsilon_2}{r_2}}{\dfrac{1}{r_1} + \dfrac{1}{r_2}}$$

$$= \frac{\varepsilon_1 r_2 + \varepsilon_2 r_1}{r_1 + r_2}$$

$$= \frac{9(1) + (-4)(1.5)}{1 + 1.5} = 2/3 \text{ V}$$

$$r = \frac{r_1 r_2}{r_1 + r_2} = \frac{(1.5)(1)}{1.5 + 1} = 0.6 \text{ Ohms.}$$

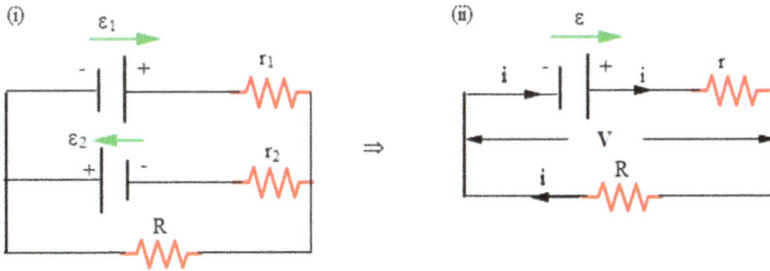

Then, the current flowing through the external resistor is

$$i = \frac{\varepsilon}{R + r} = \frac{2/3}{2 + 0.6} = 10/39 \text{ A Ans.}$$

5.6 Measuring instruments

5.6.1 Introduction

Electrical measuring instruments are all the devices used to measure the current, voltage, emf, resistance and power. We measure the unknown resistance by using Ohm's law as $R = \frac{V}{i}$. The voltage across any circuit element can be measured by a 'voltmeter' and current along any branch can be measured by an 'ammeter'. The emf of a cell can be measured by a 'potentiometer'. Generally, we measure the voltage and current in a circuit element X by connecting the voltmeter and ammeter in two possible ways, as described at the end of section 5.6.4. The basic electro-mechanical instrument that can measure current is called a 'galvanometer'.

5.6.2 Galvanometer

A galvanometer consists of a coil which is free to rotate in a permanent magnetic field. When a current passes through the coil, it experiences a torque by the magnetic field and deflects against a retarding torque of a torsional spring. The degree of deflection quantifies the magnitude of current. We will discuss its working principle in detail in the chapter 7 of this book. A galvanometer can be modified to measure any desired current. We can call it an ammeter. It can also be modified to be a voltmeter that can measure any desired voltage. These modifications can be done by connecting a suitable resistance with the galvanometer. The attached resistance is called 'shunt' denoted by S.

The maximum current that can flow through the galvanometer without damaging it is called maximum galvanometric current i_g (or full-scale deflection) and the corresponding voltage drop across the galvanometer is V_g. Then the ratio of the maximum voltage and current is called galvanometric resistance, denoted as

$$G = \frac{V_g}{i_g}$$

The deflection θ in a moving coil galvanometer is directly proportional to current i flowing through it. Since $i \propto \theta$, we can write

$$i = k\theta; \quad (k \text{ is called galvanometric constant}),$$

where $\frac{\theta}{i}(=\frac{1}{k})$ is called current resistivity, which is defined as current per unit deflection angle of the coil.

(i)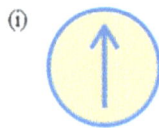

(ii)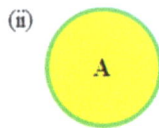

(iii)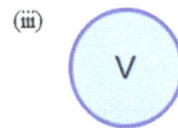

| Symbol of a galvanometer | Symbol of an ammeter | Symbol of a voltmeter |

5.6.3 Ammeter

This is a current measuring device. You take a galvanometer of full-scale deflection i_g. If you want to build an ammeter to measure a current, i, say, which is greater than i_g, you have to divert the surplus current $i - i_g$ through a low resistance (shunt) connected in parallel with the galvanometer so that the galvanometer remains protected. Since, G and S are in parallel, equal potential V_g is dropped across both; so,

$$V_g = i_g G = (i - i_g)S$$

(i)

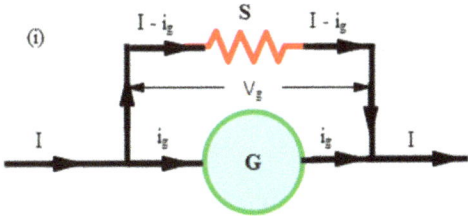

The maximum excess current I-i_g flows through the parallel shunt resistance S to form an ammeter

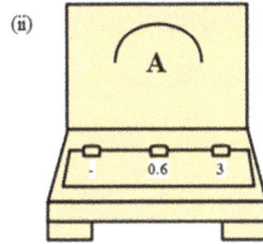

(ii)

Ammeter

Then, the required shunt resistance is given as

$$S = \frac{i_g G}{i - i_g}$$

5.6.4 Voltmeter

This is a device to measure voltage. You can convert a galvanometer to a voltmeter to measure a voltage V say, which is greater than the galvanometric voltage V_g. This is possible only when the excess voltage $V - V_g$ will be dropped across an appropriate shunt resistor S that is connected in series with the galvanometer such that maximum (rated) current i_g will flow through the galvanometer without damaging it.

As the galvanometer G and the shunt S are in series, the same maximum (rated) current i_g can flow. So, we can write

$$V = (G + S)i_g$$

(i)

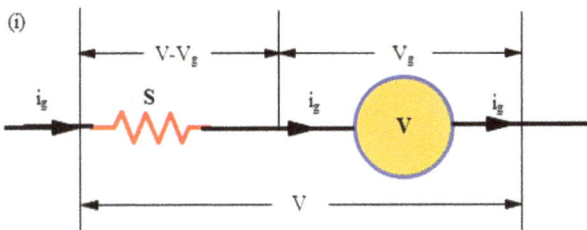

The maximum excess voltage V-V_g is dropped across the series shunt resistance S to form a voltmeter

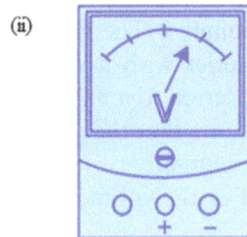

(ii)

Voltmeter

Then, the required shunt resistance is given as

$$S = \frac{V}{i_g} - G$$

N.B: We can connect the voltmeter and ammeter in two possible ways, with a very little difference, as shown in the following figures.

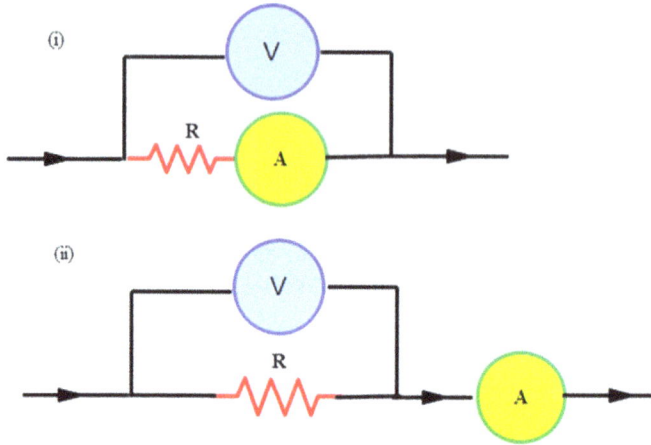

Two ways of connecting the ammeter and voltmeter in a circuit of resistance R

Example 8 What value of shunt resistance is required to measure a current of 10.006 A by using a galvanometer of rating 6 mA, 25 Ω?
Solution
The formula for the parallel shunt is

$$S = \frac{i_g G}{i - i_g}$$

$$= \frac{(0.006)(25)}{(10.006 - 0.006)} = 0.015 \ \Omega \ \text{Ans.}$$

Example 9 The galvanometer works as a 15 V full-scale voltmeter when connected in series with a resistance of 2950 Ω. Find the galvanometric resistance. If it is converted to a 5 A ammeter when connected in parallel with a resistance X, find X. Assume that the galvanometric current is 5 mA.
Solution
For the voltmeter, the galvanometric resistance is given as

$$G = \frac{V}{i_g} - S$$

$$= \frac{15}{5/1000} - 2950 \ \Omega = 50 \ \Omega$$

For the ammeter, the parallel shunt is given as

$$X = S = \frac{i_g\, G}{i - i_g}$$

$$= \frac{(0.005)(50)}{(5 - 0.005)} = 0.05\ \Omega \text{ (nearly) Ans.}$$

5.6.5 Potentiometer

This is a manually adjustable variable resistor with three terminals. It is used to control the flow of electricity in a circuit. It operates as a voltage divider and therefore it has many applications. It provides the desired PD between any two points of a circuit. It can also be used to measure the emf and internal resistance of a given cell by comparing the emfs of a standard cell. In general, a potentiometer can measure an unknown voltage by comparing it with a known voltage.

Symbol of potentiometer

This measures the voltage across the terminals of a source S of emf without drawing any current. Since no current passes through S at a null position of the slider at C which slides on the rheostat AB, the terminal voltage across the source S is equal to its emf, which can be given as

$$V_S(=\varepsilon) = i R_{AC}, \text{ where } i = \frac{V_{AB}}{R_{AB}}$$

Then, the emf of the unknown cell is

$$\varepsilon = V_{AB} \frac{R_{AC}}{R_{AB}}$$

As you know, the rheostat AB has uniform cross-sectional wire; so, resistance between any two points of the rheostat is directly proportional to the distance between the two points. Hence, the emf of the unknown cell is given as

$$\varepsilon = V_{AB} \frac{x}{l}, \text{ where } AC = x \text{ and } AB = l$$

Since, $i_s = 0$, internal resistance of the source (cell) does not play any role.

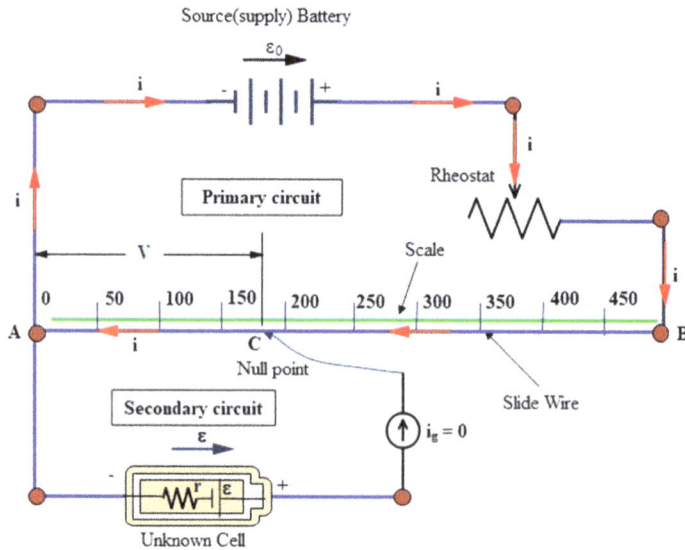

Source(supply) Battery

Primary circuit

Rheostat

Scale

Null point

Secondary circuit

$i_g = 0$

Slide Wire

Unknown Cell

At null position C of the contact the current flowing through the galvanometer is zero;so emf of the unknown cell is equal to the potential difference V between A and C

Example 10 A potentiometer wire of length 1 m has a resistance of 10 Ohm. It is connected in series with a resistance R and a cell of emf 5 V and negligible internal resistance. A source of emf of 10 mV is balanced against a length of 60 cm of the potentiometer wire. Find the value of R. Ans.

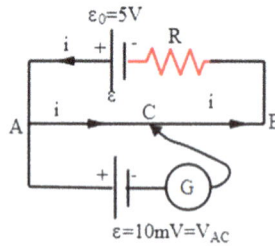

$\varepsilon_0 = 5V$

$\varepsilon = 10mV = V_{AC}$

Solution

Following the theory of a potentiometer, and the last figure, we have $AB = 100$ cm, $AC = 60$ cm

$$V_{AC} = iR_{AC}$$

$$= \left(\frac{\varepsilon_0}{R + R_{AB}}\right) R_{AC}$$

$\varepsilon_0 = 5$ V, $R_{AB} = 10$ Ω, $V_{AB} = 10 \times 10^{-3}$ V and $R_{AC} = \frac{AC}{AB} R_{AB} = \frac{60}{100} \times 10 = 6$ Ω,

we have $10 \times 10^{-3} = \left(\frac{5}{R + 10}\right) \times 6$ or, $R = 2990$ Ω Ans.

5.6.6 Wheatstone bridge

This is a network of four resistors R_1, R_2, R_3 and R_4, three of them are given and the fourth is unknown. When the network is connected to a battery between terminals A and B, if no current flows through the galvanometer, we call it a balanced Wheatstone bridge. When $i_g = 0$, we have

$$\frac{R_1}{R_2} = \frac{R_3}{R_4}$$

From the above expression, if any three resistances are given, the fourth can be calculated.

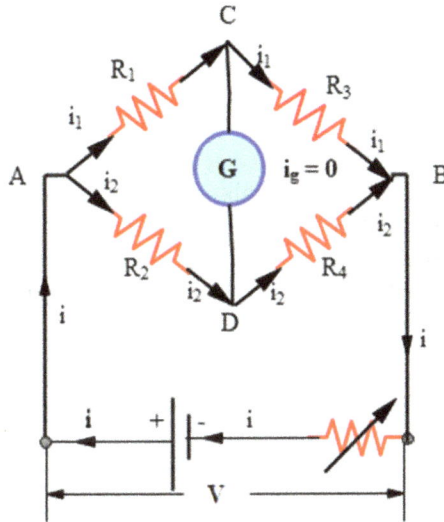

When no current flows through the galvanometer $i_g=0$, the wheatstone bridge is balanced; so, $R_1/R_2=R_3/R_4$

Proof:

If the current in the branch CD is zero, C and D are equipotential. Hence,

$$V_{AC} = V_{AD}$$

$$\Rightarrow i_1 R_1 = i_2 R_2$$

$$\Rightarrow \frac{i_1}{i_2} = \frac{R_2}{R_1} \tag{5.10}$$

Similarly, $V_{BC} = V_{BD}$

$$\Rightarrow i_1 R_3 = i_2 R_4$$

$$\Rightarrow \frac{i_1}{i_2} = \frac{R_4}{R_3} \tag{5.11}$$

From equations (5.10) and (5.11),

$$\frac{R_1}{R_2} = \frac{R_3}{R_4}$$

This is called a balanced Wheatstone bridge.

$$\text{If } \frac{R_1}{R_2} = \frac{R_3}{R_4} \text{ or } \frac{R_1}{R_3} = \frac{R_2}{R_4}, \quad V_{CD} = 0.$$

Since $R_{CD} = R_5 \neq 0$, $i_{CD} = \frac{V_{CD}}{R_{CD}} = 0$.

Using this condition we can find the possible equipotential points (C and D) in a given network.

5.6.7 Meter bridge

This is an instrument based on the balanced Wheatstone bridge to measure an unknown resistance, X, say, fitted between the tapping point B and C when a known resistance R is fitted between the tapping points A and B. The slider D is moved along the rheostat AC until i_g is zero at any point D, (say).

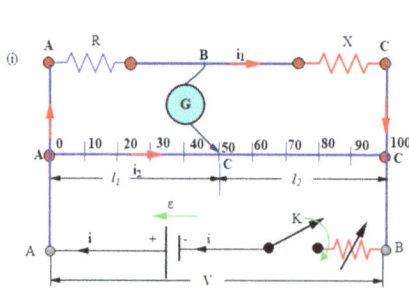

When the slider is at the null point C , no current flows through the galvanometer (i_g=0), then at the balanced wheatstone bridge condition, R/X = l_1/l_2

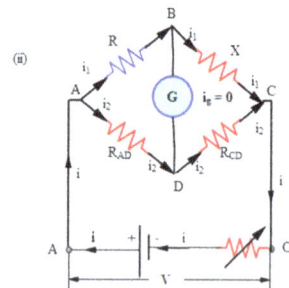

When i_g=0, the wheatstone bridge is balanced; so, R/X = l_1/l_2

Then, you can find that $\frac{R_{AD}}{R_{DC}} = \frac{l_1}{l_2}$, where l_1 and l_2 are the lengths of the resistance R_{AD} and R_{DC} measured by the scale. If $l_1 + l_2 = 1$ m, $l_1 = x$, we have $l_2 = (1 - x)$. After finding the value of $\frac{l_1}{l_2}$ equate it with the ratio R/X to obtain the unknown resistance

$$X = \frac{(l - x)}{x} R.$$

5.6.8 Post office box

The engineers working in the post office of the United Kingdom developed this technique to trace the faults in telecommunication cables. Therefore, it is known as post office box. This is a device to measure an unknown resistance by using the principle of the balanced Wheatstone bridge. It is a compact form of Wheatstone bridge having three arms AB, BC and AD with a range of resistors. The arms AB and BC have three resistances 10, 10^2 and 10^3 Ω, respectively. The arm AD has a wide range of resistance, ranging from 1 to 5000 Ω. The key K_2 is internally connected between A and C through a battery of emf ε. The unknown resistance X is connected between C and D which can be found by adjusting the circuit to a balanced Wheatstone bridge condition. Then, X can be given as

$$X = R\left(\frac{R_2}{R_1}\right)$$

where R_1/R_2 which can vary from 100:1 to 1:100R is the resistance adjusted (fitted) in the branch AD within the limit of $1 \rightarrow 5000$ Ω. After finding the unknown resistance X of the wire in the telecommunication cable we can find the length of the wire and hence the location of the fault (electrical breakdown) can be traced.

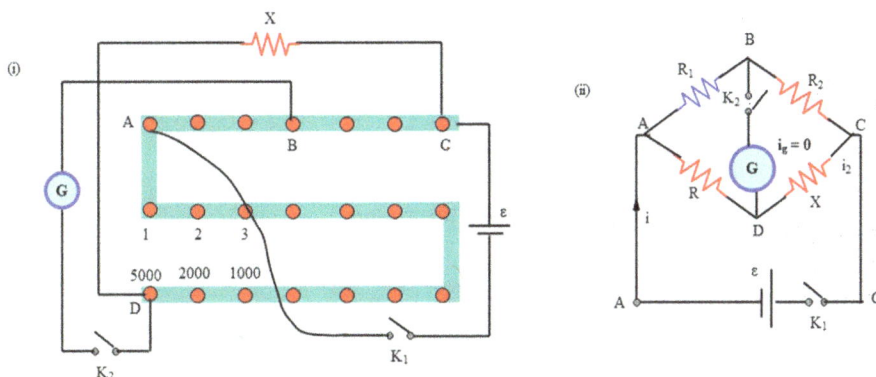

In figure (i),the post office box has three arms such as AB, BC and AD each containg some resistors. The galvanometer G is connected between B and D and the battery is connected between A and C.The unknown resistance X is fitted between C and D which can be calculated by the formula X=R(R₂/R₁) in the balanced Wheatstone Brige condition as shown in the figure(ii).

5.7 RC circuits

5.7.1 Charging of the capacitor

A circuit containing only resistors and capacitors is called an RC circuit. Let us take a single-loop RC circuit comprising a resistor R and a capacitor C connected in series with a battery of emf ε through a key K. If you close the key at $t = 0$, the battery will circulate a charge q, say, during a time t. If the capacitor is initially uncharged, its plates 1 and 2 receive charges $+q$ and $-q$, respectively, during the time t. So, the current that flows in the RC circuit at time t is given as

$$i = \frac{dq}{dt} \tag{5.12}$$

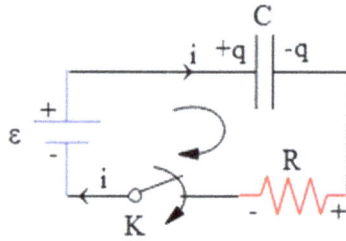

At time t the emf ε circulates a
charge q setting a current i=dq/dt

If you move in a clockwise sense, as discussed earlier, the PD across the capacitor, resistor and battery can be given as follows:

$$\Delta V_C = -\frac{q}{C} \tag{5.13}$$

$$\Delta V_R = -iR \tag{5.14}$$

$$\Delta V_b = +\varepsilon \tag{5.15}$$

Applying Kirchoff's voltage law, we have

$$\Delta V_C + \Delta V_R + \Delta V_b = 0 \tag{5.16}$$

Using the last four equations, we have

$$-\frac{q}{C} - iR + \varepsilon = 0 \tag{5.17}$$

Putting $i = \frac{dq}{dt}$ from equation (5.12) in equation (5.17), we have

$$-\frac{q}{C} - \frac{dq}{dt}R + \varepsilon = 0 \tag{5.18}$$

Then, the differential equation for the RC circuit is given as

$$R\frac{dq}{dt} + \frac{q}{C} - \varepsilon = 0 \tag{5.19}$$

and separating the variables, we have

$$\frac{dq}{C\varepsilon - q} = \frac{1}{RC}dt$$

5-28

Since a charge q flows during time t, integrating both sides,

$$\int_0^q \frac{dq}{C\varepsilon - q} = \frac{1}{RC} \int_0^t dt$$

$$\Rightarrow \ln(C\varepsilon - q)|_0^q = \frac{t}{RC}$$

$$\Rightarrow \ln\left(\frac{C\varepsilon - q}{C\varepsilon}\right) = -\frac{t}{RC}$$

$$\Rightarrow q = C\varepsilon\left(1 - e^{-\frac{t}{RC}}\right)$$

where RC is time constant of RC circuit denoted by τ(tau).

If $t \to \infty$, $q \to C\varepsilon$ $(=q_0)$

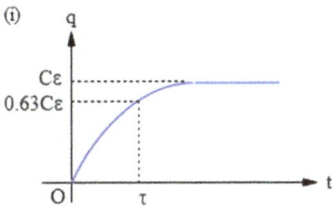

The charge of the capacitance builds up exponentially

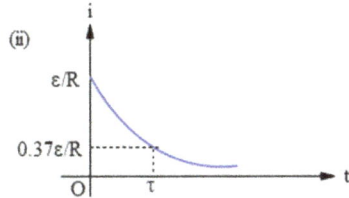

The current flowing through the capacitance decreases exponentially

This tells us that the charge on each plate builds up from zero to its maximum value exponentially (but not abruptly) after closing the key.

The current i at time t is given as

$$i = \frac{dq}{dt} = \frac{d}{dt} C\varepsilon(1 - e^{-t/RC})$$

$$\Rightarrow i = \frac{\varepsilon}{R} e^{-t/RC}$$

At $t = 0$, $i = \frac{\varepsilon}{R}(=i_0)$; at $t \to \infty$, $i \to 0$.

5.7.2 Time constant

The term 'RC' called time constant τ of an RC circuit given as.

$$\tau = RC$$

If $t = \tau$, $i = \frac{i_0}{e}$. Hence, the time constant of RC circuit is defined as the time after which the current drops by a factor e (to 36% of the maximum current). This means that the current in the circuit decays exponentially from its maximum value $i_0(=\frac{\varepsilon}{R})$ to zero after closing the key and the capacitor is fully charged to $V(=\varepsilon)$ after a long time (practically after a time period of five time constants). In the charging RC circuit, charge increases and current decreases exponentially.

5.7.3 Discharging of the capacitor

Let us open the key, remove the battery and then close the key. You can see that a current i flows in the circuit from +ve to −ve plate. This is because the capacitor acts as a battery temporarily due to its initial PD of q_0/C. Since the charge q of the capacitor plates decreases with time, the current in the circuit is given as $i = -\frac{dq}{dt}$. This current flows till the PD between the pates decreases to zero. In other words, the charge on each plate decreases to zero finally.

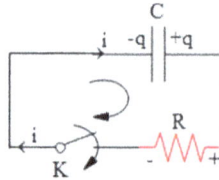

At time t the capacitor carries a charge q and a
current $i = dq/dt$ flows through the capacitor

Applying KCL in clockwise sense,

$$\Delta V_C + \Delta V_R = 0$$

$$\Rightarrow -\frac{q}{C} + (iR) = 0$$

$$\Rightarrow -\frac{q}{C} + \left(-\frac{dq}{dt}\right)R = 0 \left(\because i = -\frac{dq}{dt}\right)$$

$$\Rightarrow -\frac{dq}{dt} = \frac{q}{RC}$$

$$\Rightarrow \frac{dq}{q} = -\frac{1}{RC}dt$$

Since, the charge on the capacitor plates decreases from O to q during time t, integrating both sides,

$$\Rightarrow \int_0^q \frac{dq}{q} = -\frac{i}{RC}\int_0^t dt$$

$$\Rightarrow \ln q|_{C\varepsilon}^q = -\frac{t}{RC}$$

$$\Rightarrow \ln \frac{q}{C\varepsilon} = -\frac{t}{RC}$$

$$\Rightarrow q = C\varepsilon e^{-\frac{t}{RC}}$$

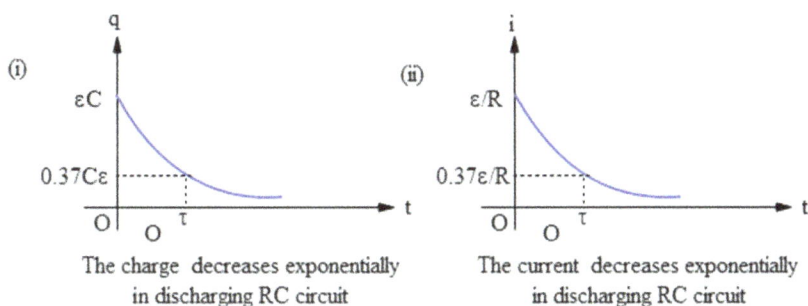

(i) The charge decreases exponentially in discharging RC circuit

(ii) The current decreases exponentially in discharging RC circuit

The current in the circuit is

$$\Rightarrow i = -\frac{dq}{dt} = -\frac{d}{dt}\left(C\varepsilon e^{-\frac{t}{RC}}\right)$$

$$\Rightarrow i = \frac{\varepsilon}{R}e^{-\frac{t}{RC}}$$

The above discussion tells us the following.

In a discharging RC circuit both charge and current decrease exponentially from their maximum values to zero. At the time $t = \tau(=RC)$, both charge and current decrease by 64%. After five to six time constants, the current will decay to zero nearly.

Example 11 If a single-loop RC circuit with an ideal battery of emf $\varepsilon = 100$ V and internal resistance $r = 2$ Ohm is connected to an external resistor $R = 98$ Ω, find the charge of a capacitor $C = 1$ μF after a time, $t = 2$ ms.

Solution

The external resistor R and the internal resistance r of the battery are in series. So, the total resistance of the circuit is $R_{eq} = R + r = 98 + 2 = 100$ Ω. Then, the charge of the capacitor at time t is

$$q = C\varepsilon(1 - e^{-\frac{t}{R_{eq}C}})$$

$$= (10^{-6})(100)\left(1 - e^{-\frac{2\times 10^{-3}}{(100)(10^{-6})}}\right)$$

$$= \left(1 - \frac{1}{e^2}\right) \times 10^{-4} = 86.46 \times 10^{-6} \text{ C Ans.}$$

Example 12 After how many constants does the charge of a discharging RC circuit decrease by 50%, 99%?

Solution

The charge of the discharging capacitor at time t is

$$q = C\varepsilon e^{-\frac{t}{RC}}$$

It is given that,

$$q = \frac{q_{max}}{2} = \frac{C\varepsilon}{2}$$

From the last two equations,

$$\frac{1}{2} = e^{-\frac{t}{RC}}$$

$$\Rightarrow -\ln 2 = -\frac{t}{RC} \ln e$$

$$\Rightarrow \frac{t}{RC} = \ln 2 \approx 0.693 \text{ Ans.}$$

Similarly, for 99% decrease in charge, $\frac{t}{RC} = \ln 100 = 4.6$ (nearly 5).   Ans.

5.8 Energy consideration and heat dissipated in *RC* circuits

5.8.1 Work done by the battery (W_b)

During the charging of the capacitor, the total charge sent by the battery is

$$q = \int_0^\infty i \, dt$$

The current during the charging of the capacitor decreases exponentially, given as

$$i = \frac{\varepsilon}{R} e^{-\frac{t}{RC}}$$

Then, the total charge flown through the battery is

$$q_{flowed} = C\varepsilon$$

You can directly get this value by putting $t \to \infty$ in the equation

$$q = C\varepsilon\left(1 - e^{-\frac{t}{RC}}\right)$$

Then, the total positive work done (or energy supplied) by the battery is

$$W_b = \varepsilon q = \varepsilon(C\varepsilon)$$

$$\Rightarrow W_b = C\varepsilon^2.$$

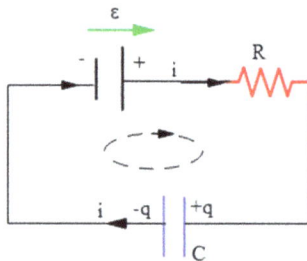

The total work done by the battery is W=qε,
where q = the total charge flown = Cε

5.8.2 Increase in potential energy (ΔU)

The charge of the capacitor increases from zero initially to a final (steady state) charge $q_0 = C\varepsilon$, the electrostatic potential energy stored in the capacitor increases from zero by

$$\Delta U = \frac{q_0^2}{2C} = \frac{C\varepsilon^2}{2}.$$

5.8.3 Total heat dissipated (Q)

As the current flows through the resistor, the total heat dissipated in the resistor is

$$Q = \int_0^\infty P\, dt$$
$$= \int_0^\infty i^2\, R dt$$
$$= \int_0^\infty \left(\frac{\varepsilon}{R}e^{-\frac{t}{RC}}\right) R dt$$

$$\Rightarrow Q = \frac{1}{2}C\varepsilon^2$$

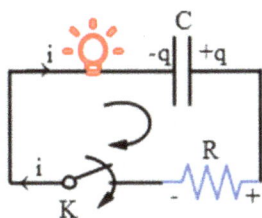

As the capacitor releases its stored electric energy exponentially, the brightness of the glowing bulb decreases gradually

This tells us that half of the total energy supplied by the battery is utilized in increasing the potential energy of the capacitor and the other half of the energy is dissipated in the form of heat, light, sound, etc. This is in accordance with the conservation of total energy (or the work–energy theorem) given as

$$W_b = \Delta U + Q$$

The above expression is useful for finding total heat dissipated in any complicated RC circuit.

In the aforementioned energy consideration in an RC circuit, we have ignored the internal resistance of the cell and that of the capacitor.

Example 13 In charging single-loop RC circuit, after how many time constants will the potential energy in the capacitor be 25% of its maximum value?

Solution

The potential energy of the capacitor at time t is

$$U = \frac{q^2}{2C}$$

The charge of the capacitor at any time t is

$$q = C\varepsilon\left(1 - e^{-\frac{t}{RC}}\right)$$

Using the last two equations, we have

$$U = \frac{C\varepsilon^2}{2}\left(1 - e^{-\frac{t}{RC}}\right)^2, \tag{5.20}$$

where $\frac{C\varepsilon^2}{2} = U_0$. Then, we have

$$U = U_0\left(1 - e^{-\frac{t}{RC}}\right)^2$$

The given condition is that

$$U = \frac{U_0}{4} \tag{5.21}$$

Using the last two equations

$$\left(1 - e^{-\frac{t}{RC}}\right)^2 = \frac{1}{4}$$

$$\Rightarrow 1 - e^{-\frac{t}{RC}} = \frac{1}{2}$$

$$\Rightarrow e^{-\frac{t}{RC}} = \frac{1}{2}$$

$$\Rightarrow -\frac{t}{RC}\ln e = -\ln 2$$

$$\Rightarrow \frac{t}{RC} = \ln 2 = 0.693 \text{ Ans.}$$

So, after 0.7 time constants (approximately), the energy stored in the capacitor decreases to 25% of its maximum value. Ans.

Example 14 A bulb is connected to a charging and a discharging capacitor separately. How does the brightness of the bulb change after closing the key?

Solution

We have seen that in both cases the current decreases exponentially to zero. So, the illumination I (or brightness) which is directly proportional to the square of the current can be given as

$$I = I_0 e^{-\frac{2t}{RC}}$$

This tells us that illumination of the bulb decreases exponentially as the current decreases exponentially. As the final current is zero, the bulb will be dimmer and after a long time (practically after few time constants of milli seconds) the bulb will stop glowing. In the charging RC circuit the capacitor offers an infinite resistance behaving as an open circuit after a long time (in steady state). But initially, in both cases, just after closing the key, the capacitor behaves as a zero resistor (short-circuit). This example tells us that in the discharging RC circuits the capacitors get discharged releasing their stored energy slowly, which will appear in the form of heat and light, etc. But in a charging RC circuit, a portion of the energy supplied by the battery (50% for the ideal battery) is dissipated in the form of radiation in the bulb. Ans.

Problem 1 Two cells of emf $\varepsilon_1 = 10$ V and $\varepsilon_2 = 6$ V, internal resistance $r_1 = r_2 = 1\ \Omega$, are connected with an external resistor $R = 8\ \Omega$ and a capacitor of capacitance $C = 2\ \mu F$ having an initial charge $q_0 = 1\ \mu C$, as shown in the figure below. Find (a) the initial current (b) initial terminal PD of the combination of two cells (c) total charge flowing in the circuit.

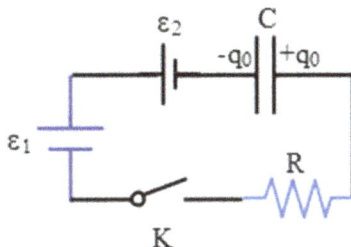

Solution

(a) Let $x =$ charge flown during a time t. So, the current in the circuit is given as $i = dx/dt$.

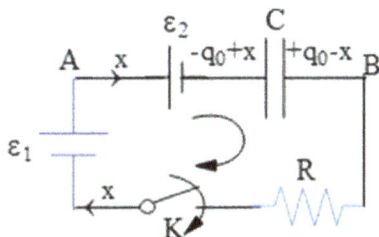

Applying Kirchoff's voltage law in the loop in a clockwise manner starting from A, we have

$$-\varepsilon_2 - ir_2 + (q_0 - x)/C + \varepsilon_1 - ir_1 - iR = 0$$

$$\Rightarrow \varepsilon_1 - \varepsilon_2 - i(r_1 + r_2 + R) - x/C + q_0/C = 0 \qquad (5.22)$$

Initially, putting the charge flowing $x = 0$ at $t = 0$ in equation (5.22), the initial current is given as

$$\varepsilon_1 - \varepsilon_2 - i_0(r_1 + R + r_2) + q_0/C = 0$$

$$\Rightarrow i_0(r_1 + R + r_2) = +(\varepsilon_1 - \varepsilon_2) + q_0/C$$

$$\Rightarrow i_0 = \{q_0/C + (\varepsilon_1 - \varepsilon_2)\}/(r_1 + R + r_2)$$

$$= \{1/2 + (10 - 6)\}/(1 + 8 + 1) = 0.45\,\text{A} \quad \text{Ans.}$$

(b) Using equation (5.22), the magnitude of initial terminal PD of the combination of the cells is given as

$$V_0 = |\varepsilon_1 - \varepsilon_2 - i_0(r_1 + r_2)| = q_0/C$$

$$-i_0 R = (10 - 6) - (0.45)(1 + 1) = -5\,\text{V} \quad \text{Ans.}$$

(c) Finally, putting the final current $i = 0$ in equation (5.22), the total charge flowing is given as

$$\varepsilon_1 - \varepsilon_2 - x/C + q_0/C = 0$$

$$\Rightarrow x = q_0 + (\varepsilon_1 - \varepsilon_2)C$$

$$= \{1 + (10 - 6)2\} = 9\,\mu\text{C} \quad \text{Ans.}$$

Problem 2 If two resistors R_1 and R_2 are connected in series and parallel, the effective resistances will be $R = 9\ \Omega$ and $R' = 2\ \Omega$, respectively. Find (a) R_1 and R_2, (b) ratio of power loss in the resistors if they are connected in (i) series (ii) parallel.
   Solution
   (a) For series combination of the resistors R_1 and R_2

$$R = R_1 + R_2 \tag{5.23}$$

For parallel combinations, of the resistors R_1 and R_2

$$R' = \frac{R_1 R_2}{R_1 + R_2} = 2 \tag{5.24}$$

Solving equations (5.23) and (5.24), we have

$$R_1 = 6\ \Omega \text{ and } R_2 = 3\ \Omega \text{ Ans.}$$

(i) In the series combination the current i is the same in each resistor. Using $P = i^2 R$, we have

$$P_1/P_2 = R_1/R_2 = 6/3 = 2 \text{ Ans.}$$

(ii) In the parallel combination the voltage V is the same in each resistor. Using $P = V^2/R$, we have

$$P_1/P_2 = R_2/R_1 = 3/6 = 1/2 \text{ Ans.}$$

N.B: If you take $R_1 = 3$ Ω and $R_2 = 6$ Ω, the answers will be 1/2 and 2, respectively.

Problem 3 Find (a) the equivalent resistance R_{AB} of the given network, (b) reading of the ammeter. Neglect the resistance of the ammeter.

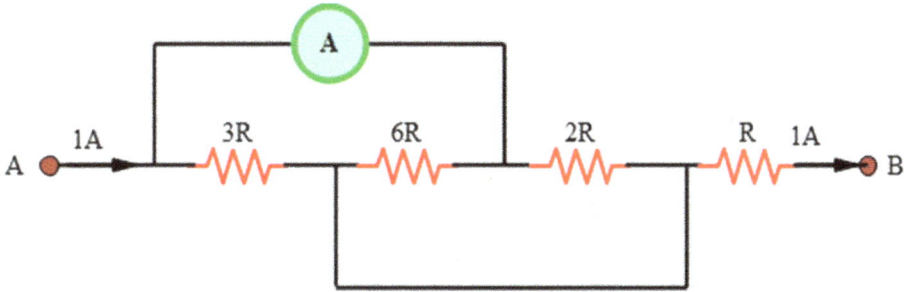

Solution
(a) **Method 1:**
Applying Kirchoff's voltage law for loop 1:

$$-3xR - 6yR = 0$$

$$\Rightarrow x + 2y = 0 \tag{5.25}$$

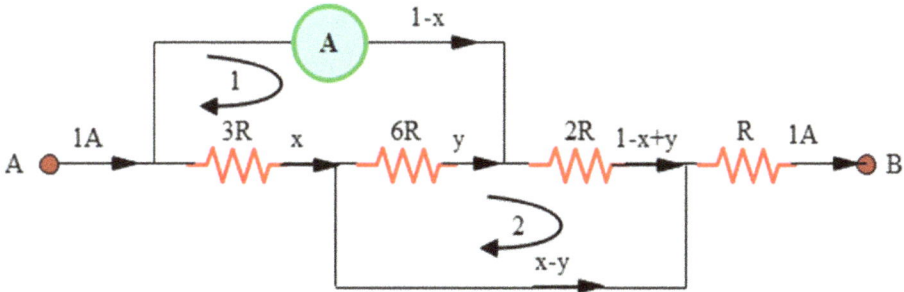

Applying Kirchoff's voltage law for loop 2:

$$-6yR - 2(1 - x + y)R = 0$$

$$\Rightarrow x - 4y = 1 \qquad (5.26)$$

Solving equations (5.25) and (5.26),

$$x = \frac{1}{3} \text{ and } y = -\frac{1}{6}A$$

Applying Kirchoff's voltage law from A to B, the PD between A and B is given as

$$V_A - 2(1 - x + y)R - R = V_B$$

$$V_A - V_B = V_{AB} = 2(1 - x + y)R + R = \{2(1 - x + y) + 1\}R$$

Putting $x = \frac{1}{3}$ and $y = -\frac{1}{6}A$, we have

$$V_{AB} = \{2(1 - x + y) + 1\}R = \left[2\left(1 - \frac{1}{3} - \frac{1}{6}\right) + 1\right]R = 2R$$

Then, the equivalent resistance

$$R = \frac{V_{AB}}{i} = \frac{2R}{1} = 2R \text{ Ans.}$$

(b) The reading of the ammeter $1 - x = 1 - \frac{1}{3}A = \frac{2}{3}A$ flows to the right. Ans.
Method 2 (Equipotential method):
Since A, C are connected with zero resistors (short-circuited), they are at the same potential; similarly, E and D are at the same potential. Bringing A and C to one point, and E and D to one point, we can redraw the circuit as below. We can see that three resistors $2R$, $3R$ and $6R$ are in parallel between A and D.

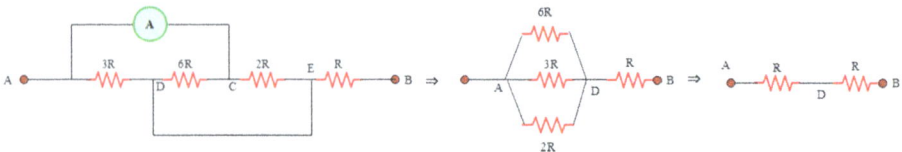

Then, $\frac{1}{R_{AD}} = \frac{1}{3R} + \frac{1}{6R} + \frac{1}{2R}$

$$\Rightarrow R_{AD} = R$$

Since R_{AD} is connected with the resistor R in series, the equivalent resistance between A and B is

$$R_{AB} = R_{AD} + R = R + R = 2R \text{ Ans.}$$

Even though A and C, E and D are connected with zero resistors, the currents in them are not zero. Hence, $i \neq 0$ even though $\Delta V = 0$.

The reading of the ammeter is given as

$$i' = 1 - x, \text{ where } x = \frac{1}{3}$$

Then, $i' = \frac{2}{3}A$ to right. Ans.

Problem 4 A Wheatstone bridge is connected with a battery of emf $\varepsilon = 6$ V and internal resistance $r = 1$ Ω. We assume the currents in different branches as shown in the figure below. By applying Kirchoff's voltage law for loops 1, 2 and 3, find (a) the current in the AD, mid branch CD and that following through the battery, (b) the effective resistance between A and C. Put $R = 1$ Ω. Put $G = 1$ Ω.

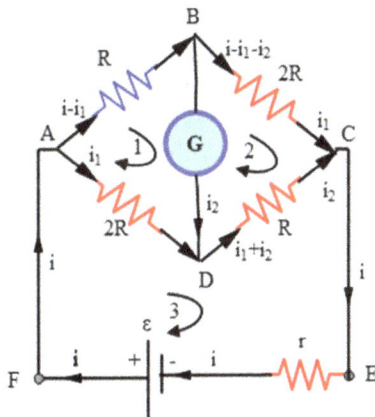

Solution
(a) Applying KCL for loop 1

$$-(i - i_1)(1) - i_2(1) + i_1(2) = 0$$

$$\Rightarrow 3i_1 - i_2 = i \tag{5.27}$$

Applying KCL for loop 2,

$$-2(i - i_1 - i_2) + 1(i_1 + i_2) + (i_2)(1) = 0$$

$$\Rightarrow 3i_1 + 4i_2 = 2i \tag{5.28}$$

Applying KCL for loop 3,

$$-2i_1 - (i_1 + i_2) + 6 - i = 0$$

$$\Rightarrow -3i_1 - i_2 + 6 = i \qquad (5.29)$$

Solving equations (5.27), (5.28) and (5.29), we have

$$i_1 = 1 \text{ A}, \quad i_2 = 0.5 \text{ A and } i = 2.5 \text{ A} \quad \text{Ans.}$$

(b) Applying KCL for loop from E to F through the battery,

$$V_{AC} = V_{EF} = 6 - i, \text{ where } i = 2.5 \text{ A}$$

$$\Rightarrow V_{AC} = V_{EF} = 3.5 \text{ V}$$

Since the network draws a current of 2.5 A, $V_{AC} = iR_{AC}$, where $i = 2.5$ A

$$\Rightarrow R_{eq} = R_{AC} = V_{AC}/i = 3.5/2.5 = 1.4 \ \Omega \quad \text{Ans.}$$

N.B: This is the general method of solving this network without any symmetry and equipotential points.

Problem 5 Twelve identical resistors each of resistance $R = 1 \ \Omega$ are connected to form a cube as shown in the figure below. Find (a) the equivalent resistance between the points A and B, (b) current drawn by the network if it is connected across a cell of emf of 6 V and internal resistance of $1/6 \ \Omega$.

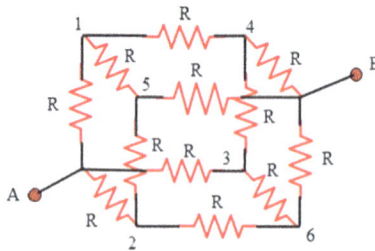

Solution

(a) **Method 1:**

The network is symmetrical about the body diagonal AB. Since equal currents flow in the branches between A and (1, 2 and 3), points 1, 2 and 3 are equipotential. Similarly, points 4, 5, 6 are equipotential. So, we can treat these three points 1, 2 and 3 as one point (C, say) and similarly, treat the three points 4, 5 and 6 as one point (D, say). Now you have three resistors between A and C, six resistors between C and D and three resistors between D and B.

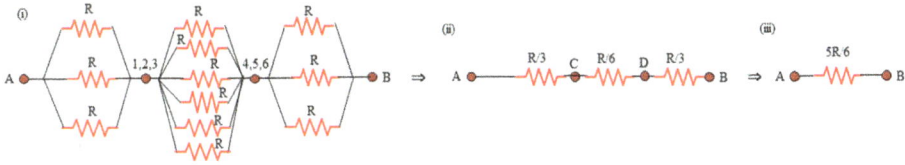

Then, the equivalent resistance between A and B is

$$R_{AB} = R_{AC} + R_{CD} + R_{DB}$$

$$= \frac{R}{3} + \frac{R}{6} + \frac{R}{3} = \frac{5R}{6} = \frac{5}{6}\,\Omega\ (\because R = 1\,\Omega)\ \text{Ans.}$$

Method 2:

(a) Following concept of symmetry, we have divided i-amp current in all branches as shown in the following figure.

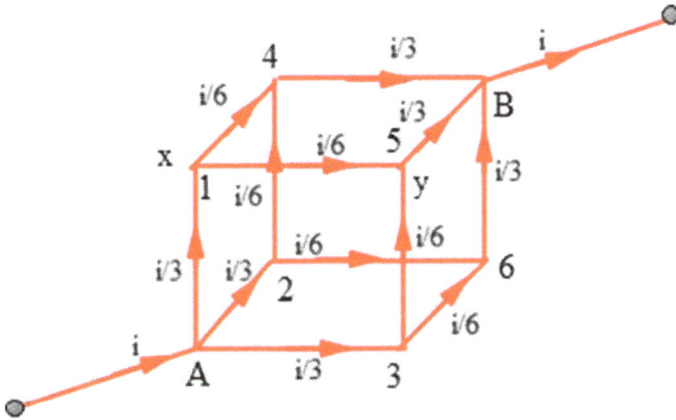

Then,

$$R_{AB} = V_{AB}\ (\because i_{AB} = 1),$$

where

$$V_{AB} = \left| V_{Ax} + V_{xy} + V_{yB} \right|$$

$$= \left(\frac{1}{3}R\right) + \left(\frac{1}{6}R\right) + \left(\frac{1}{3}R\right) = \frac{5}{6}R$$

$$\Rightarrow R_{AB} = \frac{5}{6}R\ \text{Ans.}$$

(b) Now the net resistance between A and B and the battery are connected in series to form a closed circuit. For the current i, applying Kirchoff's voltage law, in the loop taken in the clockwise sense, we have

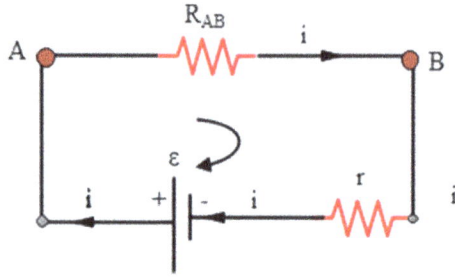

$$\varepsilon - ir - iR_{AB} = 0$$

So, the current drawn by the network from the cell is

$$i = \frac{\varepsilon}{R_{AB} + r} = \frac{6}{5/6 + 1/6} = 6 \text{ A Ans.}$$

Problem 6 In the last problem of cubic network, find the equivalent resistance between a two-face diagonal point of the cube.

Solution

Method 1:

We can solve this problem by equipotential method. For the sake of simplicity press the cube so that all vertices lie in the horizontal plane as shown in the following figure. We can see that the two-dimensional network is electrically symmetrical about the face diagonal AB as each side has equal resistance. So, we have two symmetrical triangular portions of the squares made by the diagonal AB as shown in the figure (i).

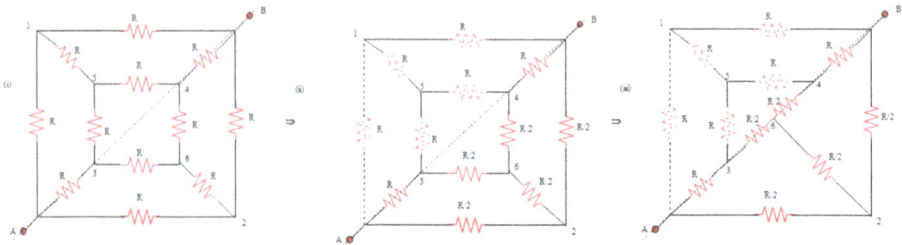

If you fold one triangle with the other, we can see that point 1 coincides with point 2. Similarly, points 5 and 6 coincide. While doing so, except the resistors between points 4 and B, and points A and 3, all other resistors overlap with their counterparts. As a result, the resistance of the respective branches will be each $R/2$ as shown in figure (ii).

Then it will be simplified to figure (v) through figures (iii) and (iv). Since this figure (v) is a balanced Wheatstone bridge, we can remove the middle branch between 2 and 6 because it does not carry any current. Then the circuit becomes simplified as the parallel combination of two series connections of resistance $3R/2$, $3R/2$ and $R/2$, $R/2$ as shown in figure (vi). Now we can solve the circuit to have a resistance $R_{AB} = 3R/4$ as these figures are step-wise drawn and self-explanatory.

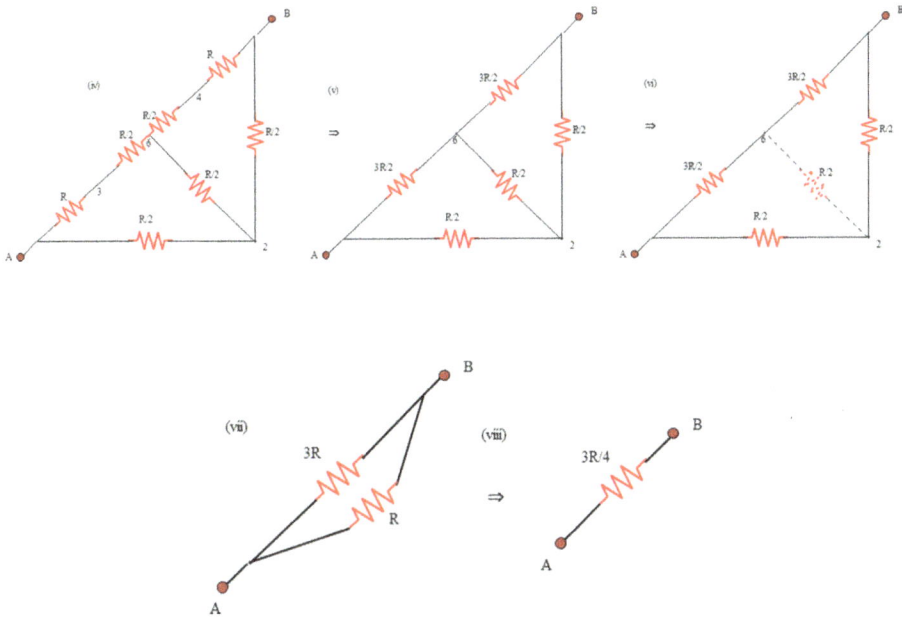

Method 2:

In this method, we can draw the current distribution as shown in figure (i) based upon the electrical symmetry. We can see that no current flows through the resistors between points 1,5 and 2,6 as shown in figure (i). So, we just remove these useless resistors in figure (ii) and now the circuit becomes simplified to the series and parallel combination, as shown in figure (iii). Then, by following the usual procedure, we can find the previous value of resistance between A and B. Ans.

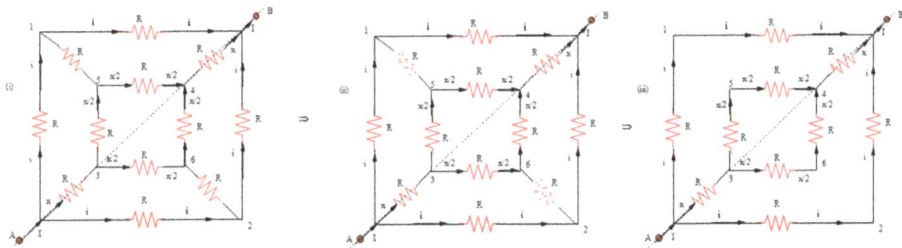

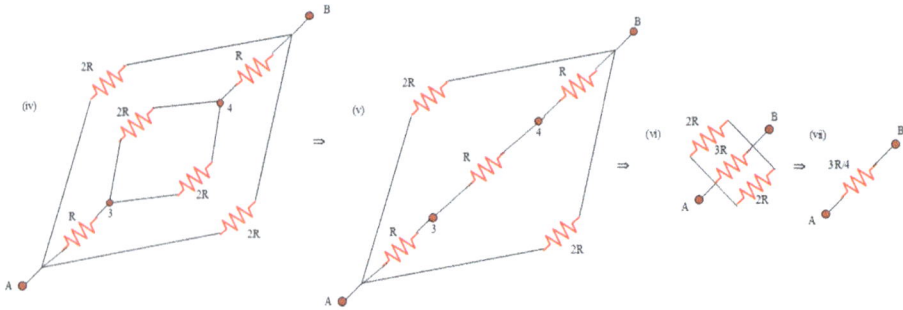

Problem 7 Six resistors are connected as shown in the figure below between A and B. Find the equivalent resistance R_{AB} in the network.

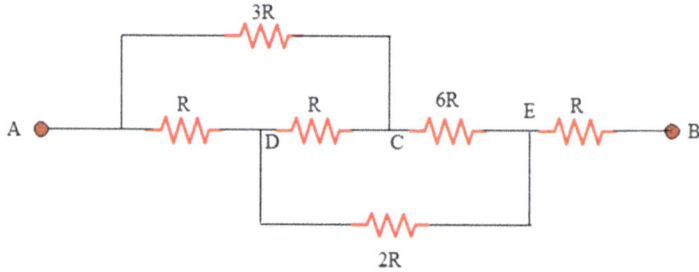

Solution

We can redraw the given diagram as shown in figure (i). It is now clear that the given network is a Wheatstone bridge. The ratio of the adjacent sides is equal, given as

$$\frac{R_{AC}}{R_{AD}} = \frac{R_{CE}}{R_{DE}} = 3$$

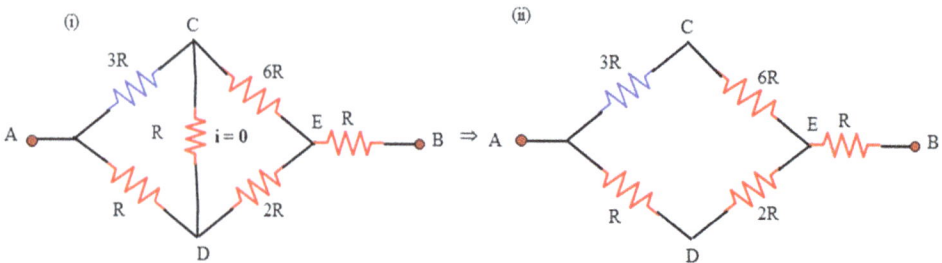

So, the given network is a balanced Wheatstone bridge $i_{CD} = 0$.

Then we remove the branch CD to obtain a simple circuit, as shown in figure (ii)

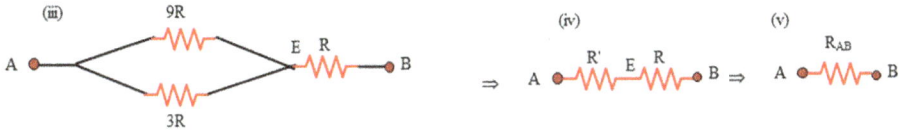

Hence, the equivalent resistance between A and E is

$$R_{AE} = R' = \frac{9R \times 3R}{9R + 3R} = 9R/4 \text{ Ans.}$$

Since R_{AE} and R are in series, the equivalent resistance between A and B is

$$R_{AB} = R + R' = R + 13R/4 = 13R/4.$$

Problem 8 Find the equivalent resistance R_{AB} between terminals A and B.

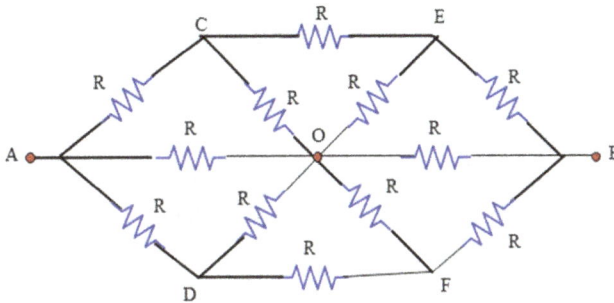

Solution

Method 1:

By inspection we can say that the lower half and the upper half of the given circuit is symmetrical about AB. Then, C and D are equipotential; E and F are equipotential. Superimposing D with C and F with E we have the following circuit. You can see that (AC and AD), (CE and DF), (EB and FB), (CO and DO) and (EO and FO) are superimposed.

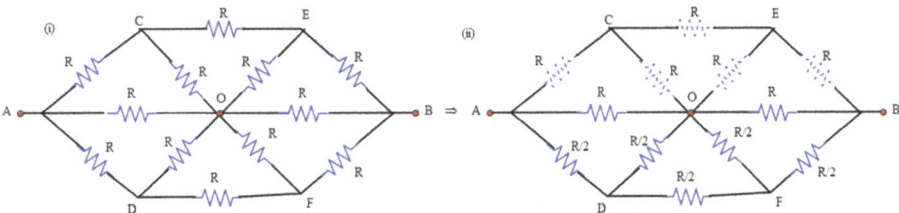

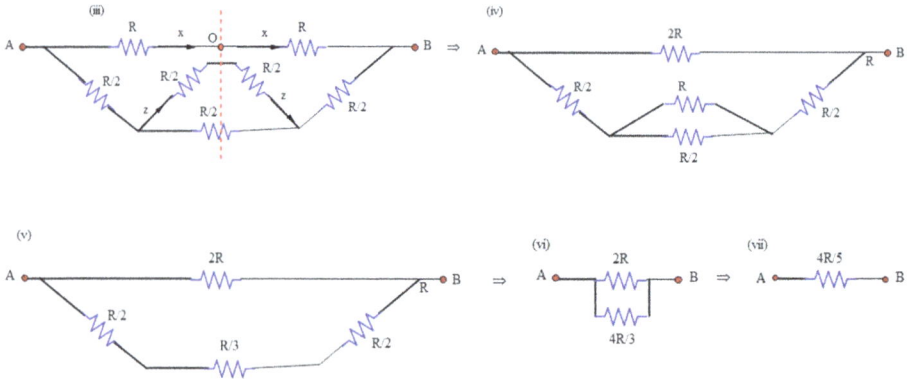

By current distribution following KCL, we understand that equal current passes through branches CO and OE. Then, you can separate branches COE and AOB as shown in the figure and solve it by the processes of series and parallel combination. Referring to figure (vi), the equivalent resistance between A and B is

$$R_{AB} = \frac{(2R)(4R/3)}{\frac{4}{3}R + 2R} = \frac{4}{5}R \text{ Ans.}$$

Method 2:
Following the symmetry of lower and upper halves, we can show that the same current flows in COE and DOF. Hence, you can separate these branches from AOB as shown.

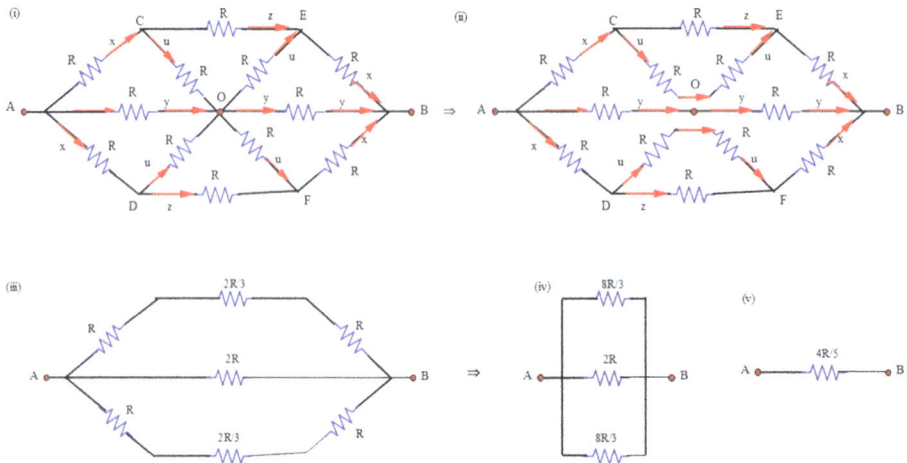

So, it is reduced to a simple network of three parallel resistors of resistances $8R/3, 2R$ and $8R/3$ to yield an equivalent resistance of $4R/5$, as shown in the above figures. Ans.

Method 3:

Based upon the circuit's electrical symmetry about XY and AB, applying Kirchoff's current law the current distribution is shown. Now applying Kirchoff's voltage law in loop 1, we have

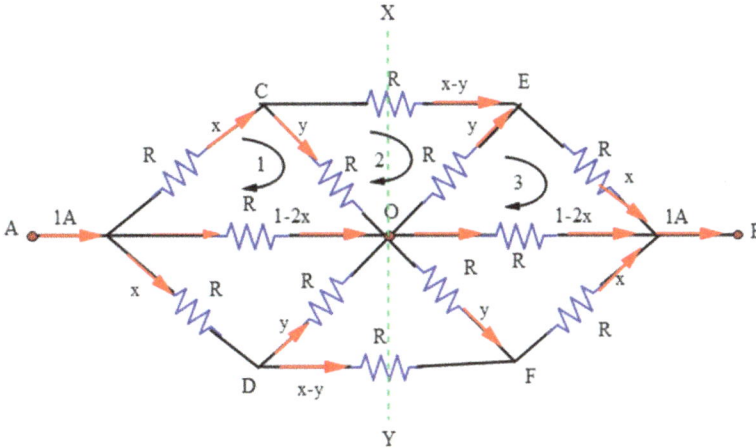

$$-xR - yR + (1 - 2x)R = 0$$

$$\Rightarrow 3x + y = 1 \tag{5.30}$$

Applying Kirchoff's voltage law in loop 2, we have

$$-yR - yR + (x - y)R = 0$$

$$\Rightarrow 3y = x \tag{5.31}$$

Solving equations (5.30) and (5.31), we have

$$x = 3/10 \text{ A} \quad \text{and} \quad y = 1/10 \text{ A}$$

Applying Kirchoff's voltage law in branch AOB, the voltage difference between A and B is given as

$$|V_{AB}| = (1 - 2x)R + (1 - 2x)R = 2(1 - 2x)R \tag{5.32}$$

If you replace the network by an effective resistance R_{eq}, it will drop the same voltage when carrying a current of 1 A. Then, you can write

$$|V_{AB}| = (1)R_{eq} = R_{eq} \tag{5.33}$$

Using equations (5.32) and (5.33), we have

$$R_{eq} = 2(1 - 2x) \tag{5.34}$$

Putting the obtained values of x in equation (5.34), we have

$$R_{eq} = R_{AB} = 2 - 4x = 2 - 4(3/10) = 4/5 \text{ A Ans.}$$

Problem 9 In the given infinite ladder network, find the (a) equivalent resistance between A and B (b) ratio of potential difference between A, B and C, D (c) current flowing through the branch CD.

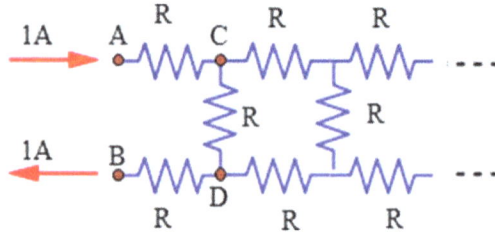

Solution

(a) Since infinite minus something is infinite, if you cut one cell from the given infinite network, the remaining network will still be an infinite network possessing same equivalent resistance, so we can write

$$R_{CD} = R_{AB} \tag{5.35}$$

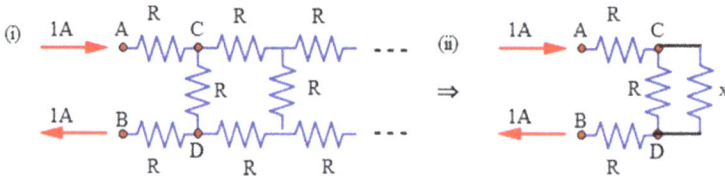

Substitute the infinite network from CD by an equivalent resistance $R_{CD} = x$, we obtain a simple network as shown in figure (ii). Let $R_{AB} = x$. Now, the simple series-parallel network in figure (ii) has the resistance

$$R_{AB} = \frac{(R)(x)}{x + R} + 2R$$

$$\Rightarrow x = \frac{(R)(x)}{x + R} + 2R$$

$$\Rightarrow x^2 + Rx = 3Rx + 2R^2$$

$$\Rightarrow x^2 - 2Rx - 2R^2 = 0$$

$$\Rightarrow x = (\sqrt{3} + 1)R \text{ Ans.}$$

(b) The PD between A and B is

$$V_{AB} = x(1)$$

The PD between C and D is

$$V_{CD} = (R_{CD})(1) = \frac{Rx}{(R + x)}$$

$$\Rightarrow V_{AB}/V_{CD} = \frac{R + x}{R} = \sqrt{3} + 2 \text{ Ans.}$$

(c) The current passing through the resistor R connected between C and D is given as

$$\Rightarrow i_{CD} = \frac{x}{R + x}$$

Putting the obtained value of x, we have

$$\Rightarrow i_{CD} = \frac{x}{R + (\sqrt{3} + 1)R} = \frac{(\sqrt{3} + 1)}{\sqrt{3} + 2} \text{ A Ans.}$$

Problem 10 Find R_{AB} in the infinite ladder network if the resistors are decreased by a factor 1/2 in each cell as shown.

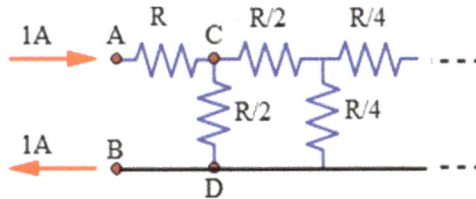

Solution

(a) In this infinite ladder network, we can see that the resistance of each resistor decreases by half in the next cell. Here, if we remove the first cell we will still get an infinite ladder network but it is a totally different one. We will get it if we connect two identical initial networks in parallel. So, the equivalent resistance between C and D will be equal to half of that between A and B. So, we can write

$$R_{CD} = R_{AB}/2 \tag{5.36}$$

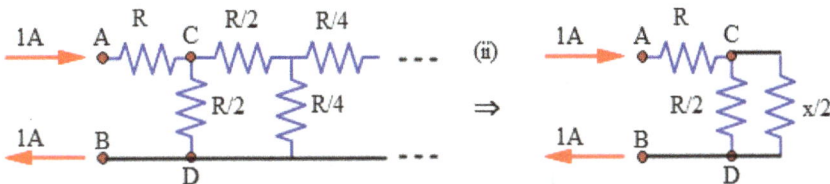

Substitute the infinite network from CD by an equivalent resistance $R_{CD} = x/2$, we obtain a simple network as shown in figure (ii). Let $R_{AB} = x$. Now, the simple series-parallel network in figure (ii) has the resistance

$$R_{AB} = \frac{(R/2)(x/2)}{x/2 + R/2} + R$$

$$\Rightarrow x = \frac{Rx}{2(x + R)} + R$$

$$\Rightarrow 2x^2 + 2Rx = 3Rx + 2R^2$$

$$\Rightarrow 2x^2 - Rx - 2R^2 = 0$$

$$\Rightarrow x = (\sqrt{17} + 1)R/4 \text{ Ans.}$$

(b) The potential difference between A and B is

$$V_{AB} = x(1) = x = (\sqrt{17} + 1)R/4$$

The potential difference between C and D is

$$V_{CD} = (R/2)(i_{CD}) = \frac{ix/2}{x/2 + R/2}(R/2)$$

$$= \frac{ixR}{2(x + R)} = \frac{RV_{AB}}{2(x + R)} \text{ (by assuming } x = 1)$$

$$\Rightarrow V_{AB}/V_{CD} = 2(R + x)/R$$

Putting the obtained value of x, we have

$$\Rightarrow V_{AB}/V_{CD} = 2\{R + (\sqrt{17} + 1)R/4\}/R = (\sqrt{17} + 5)/2 \text{ Ans.}$$

N.B: If each branch in the first cell (or unit) has the same resistance R, say, following the same method, you can show that the effective resistance is $1.4R$ approximately.

Problem 11 A resistance x is connected at the end of a ladder network such that the current drawn by the network between A and B does not depend on the number of square units of the network. (a) Find the value of x. (b) What is value of internal resistance r of the cell of emf ε so that the power loss in the external circuit will be maximum? (c) Find the maximum power loss in the external circuit.

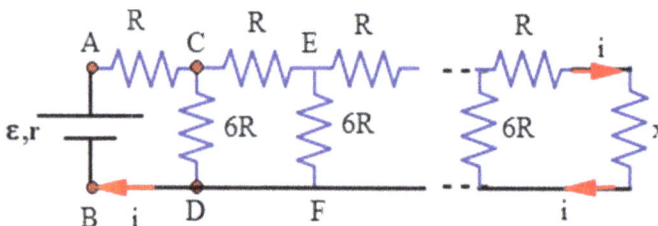

Solution

(a) As per the given condition, if the resistance x is joined at the beginning of the first cell, the effective resistance remains unchanged; so, if you place the resistance x in the first unit (between C and D), the effective resistance between A and B is

$$R_{AB} = R + x \tag{5.37}$$

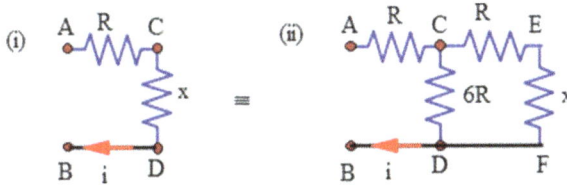

If the resistance x is connected at the beginning of the second unit (between E and F), the effective resistance is still equal to

$$R_{AB} = \frac{6R(R + x)}{6R + R + x} + R = \frac{6R(R + x)}{7R + x} + R \tag{5.38}$$

From equations (5.37) and (5.38),

$$R + x = \frac{6R(R + x)}{7R + x} + R$$

$$\Rightarrow x = \frac{6R(R + x)}{7R + x}$$

$$\Rightarrow 7Rx + x^2 = 6R^2 + 6Rx$$

$$\Rightarrow x^2 + Rx - 6R^2 = 0$$

$$\Rightarrow x = 2R \text{ Ans.}$$

(b) Applying Kirchoff's voltage law in the simplified circuit as shown in figure (ii), we have

$$-ir - i(R + x) + \varepsilon = 0$$

Then, the current in the circuit is

$$i = \frac{\varepsilon}{R + r + x} = \frac{\varepsilon}{R + r + 2R}(\because x = 2R)$$

$$\Rightarrow i = \frac{\varepsilon}{3R + r} \tag{5.39}$$

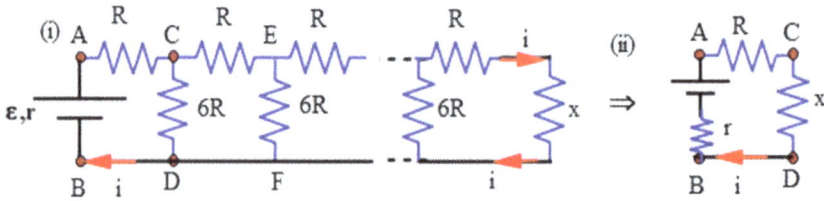

So, the power developed in the external circuit is

$$P = i^2 R_{AB} = \left(\frac{\varepsilon}{3R + r}\right)^2 (3R) (\because R_{AB} = 3R)$$

$$\Rightarrow P = \frac{3\varepsilon^2 R}{(3R + r)^2} \tag{5.40}$$

For the power to be maximum

$$\frac{dP}{dR} = \frac{d}{dR}\left\{\frac{2\varepsilon^2 R}{(3R + r)^2}\right\} = 0$$

$$\Rightarrow \frac{d}{dR}\left\{\frac{R}{(3R + r)^2}\right\} = 0$$

$$\Rightarrow \frac{(3R + r)^2 - 6R(3R + r)}{(3R + r)^4} = 0$$

$$\Rightarrow 3R + r = 6R$$

$$\Rightarrow r = 3R \text{ Ans.}$$

(c) Putting $r = 3R$ in the equation (5.40), the maximum power loss is

$$\Rightarrow P_{max} = \frac{3\varepsilon^2 R}{(3R + 3R)^2} = \frac{\varepsilon^2}{12R} \text{ Ans.}$$

Problem 12 N identical cells are connected in series with an external resistor carrying a current of i A. If n cells are connected wrongly in series with the same external resistor the current flowing through the cells will reduce by 25%. Find the value of n/N.

Solution

Assuming $r =$ internal resistance of each cell, for correct connection, the effective emf of the combination is

$$\varepsilon_{\text{eff}} = N\varepsilon$$

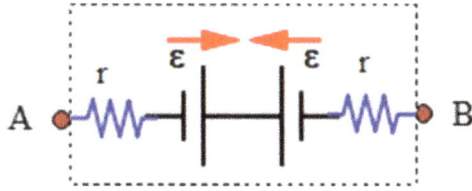

Then the current in the series circuit is

$$i = \frac{N\varepsilon}{r_{\text{eff}} + R} = \frac{N\varepsilon}{Nr + R} \quad (\because r_{\text{eff}} = Nr) \tag{5.41}$$

For wrong connection of one cell, the effective emf decreases by 2ε because emf of one cell cancels the emf of another cell. So, the effective emf of the combination is

$$\varepsilon'_{\text{eff}} = N\varepsilon - 2n\varepsilon = (N - 2n)\varepsilon$$

However, the effective internal resistance of the combination remains the same; $r'_{\text{eff}} = Nr$

Then the current in the series circuit is

$$i' = \frac{(N - 2n)\varepsilon}{Nr + R} \tag{5.42}$$

Using equations (5.41) and (5.42), the ratio of the currents is

$$\frac{i}{i'} = \frac{N}{N - 2n} \tag{5.43}$$

As the current drops by 25%, you can write

$$\frac{i}{i'} = \frac{i}{i - i/4} = \frac{4}{3} \tag{5.44}$$

Solving equations (5.43) and (5.44), we have

$$\frac{N}{N - 2n} = \frac{4}{3}$$

$$n/N = \frac{1}{8} \text{ Ans.}$$

N.B: Both N and n must be whole numbers. If $N = 8$, $n = 1$; if $N = 16$, $n = 2$ and so on.

Problem 13 n identical cells each of emf 6 V are connected in series with an external resistor of 5 Ω carrying a current of 10 A. If one cell is connected wrongly in series with the same external resistor the current flowing through the cells will be 8 A. Find the value of n and internal resistance of each cell.

Solution

In the correct series connection, the effective emf of the combination is

$$\varepsilon_{\text{eff}} = n\varepsilon = 6n$$

If r = internal resistance of each cell, the effective internal resistance of the combination is

$$r_{\text{eff}} = nr$$

Then, the current flowing in the circuit is

$$i = \frac{n\varepsilon}{nr + R} = \frac{6n}{nr + 5} = 10 \qquad (5.45)$$

For the wrong connection of a cell, the effective emf decreases by 2ε because emfs of two identical cells counteract with each other.

Then, the new effective emf of the combination is

$$\varepsilon'_{\text{eff}} = n\varepsilon - 2\varepsilon = (n - 2)\varepsilon = 6(n - 2)$$

The effective internal resistance of the combination remains the same, that is, $r'_{\text{eff}} = nr$

So, the new current flowing in the circuit is

$$i' = \frac{(n - 2)\varepsilon}{nr + R} = \frac{(n - 2)6}{nr + 5} = 8 \qquad (5.46)$$

By solving equations (5.45) and (5.46), we have

$$n = 10 \text{ and } r = 0.1 \ \Omega \text{ Ans.}$$

N.B: If we connect N cells wrongly, the emf of the equivalent cell is $\varepsilon_{\text{eff}} = (n - 2N)\varepsilon$ and internal resistance of equivalent cell is $r_{\text{eq}} = nr$, because wrong connection of a pair of identical cells reduces the net emf of 2ε as two emfs completely counteract each other to yield zero emf.

Problem 14 Three cells (1, 2 and 3) of emf 6, 8 and 12 V, respectively, are connected in parallel. The internal resistances are 1 Ω, 2 Ω and 3 Ω, respectively. (a) Find the current in each branch if the external load $R = 2$ Ω. (b) If the cell of emf 8 V in part (a) reverses its polarity, find the current through the external load.

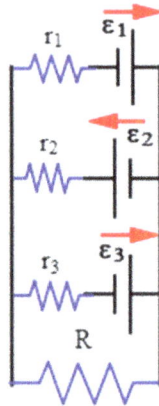

Solution

(a) Let us assume that the emf that is pointing to the right is taken as positive. So, the first and third emfs are positive and the second emf is negative. In the parallel combination of the cells, the equivalent cell emf is

$$\varepsilon_{\text{eff}} = \frac{\Sigma \varepsilon_i / r_i}{\Sigma 1 / r_i}$$

$$= \frac{\frac{6}{1} + (-\frac{8}{2}) + \frac{12}{3}}{\frac{1}{1} + \frac{1}{2} + \frac{1}{3}} = \frac{6 - 4 + 4}{\frac{1}{1} + \frac{1}{2} + \frac{1}{3}} = 36/11 \text{ V} \rightarrow$$

The internal resistance of the cell is

$$\frac{1}{r_{\text{eff}}} = \Sigma \frac{1}{r_i} = \frac{1}{1} + \frac{1}{2} + \frac{1}{3}$$

$$\Rightarrow r_{\text{eff}} = \frac{6}{11} \ \Omega$$

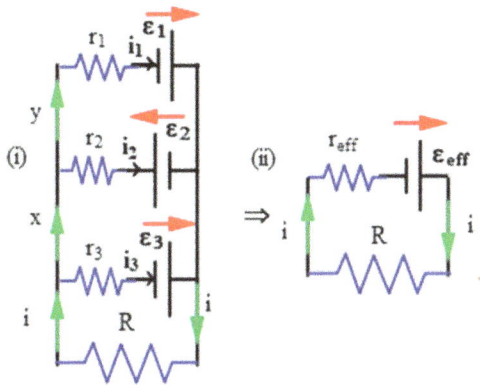

The current in the circuit is

$$i = \frac{\varepsilon}{R + r}$$

$$= \frac{36/11}{\frac{6}{11} + 2} = 9/7 \text{ A Ans.}$$

The terminal voltage across each cell is given as

$$V = \varepsilon_1 - i_1 r_1 = -\varepsilon_2 - i_2 r_2 = \varepsilon_3 - i_3 r_3 = iR = (9/7)(2) = 18/7 \text{ V}$$

Then, the current y flowing through the top cell is

$$i_1 = (\varepsilon_1 - V)/r_1 = (6 - 18/7)/1 = 24/7 \text{ A} \rightarrow \text{ Ans.}$$

The current flowing through the middle cell is

$$i_2 = (-\varepsilon_2 - V)/r_2 = (-8 - 18/7)/2 = -37/7 \text{ A} \leftarrow \text{ Ans.}$$

The negative sign signifies that the actual direction of the current is opposite to the assumed direction. So, the direction of i_2 is leftward. The current flowing through the bottom cell is

$$i_3 = (\varepsilon_3 - V)/r_3 = (12 - 18/7)/3 = 22/7 \text{ A} \rightarrow \text{ Ans.}$$

So, the value of x is

$$x = i - i_3 = 9/7 - 22/7 = -13/7 \text{ A} \downarrow \text{ Ans.}$$

The negative sign signifies that the actual direction of the current is opposite to the assumed direction. So, direction of x is downward.

(b) After reversing of the polarity of the cell, putting $\varepsilon = +8$ V, the net emf of the equivalent cell is

$$\varepsilon_{\text{eff}} = \frac{\Sigma \varepsilon_i / r_i}{\Sigma 1 / r_i}$$

$$= \frac{\frac{6}{1} + (+\frac{8}{2}) + \frac{12}{3}}{\frac{1}{1} + \frac{1}{2} + \frac{1}{3}} = \frac{6 + 4 + 4}{\frac{1}{1} + \frac{1}{2} + \frac{1}{3}} = 84/11 \text{ V}$$

Since the effective internal resistance remains constant, that is, $r_{\text{eff}} = 6/11$ Ω, The current in the circuit is

$$i = \frac{\varepsilon}{R + r}$$

$$= \frac{84/11}{\frac{6}{11} + 2} = 3 \text{ A Ans.}$$

Problem 15 The ends A, B and C of cells are maintained at potentials 20 V and -15 V. End B is earthed. One end of each cell is connected to a point D, find (a) the potential V of their function, (b) currents flowing in the branches.

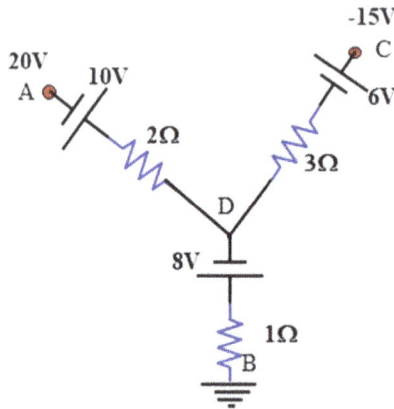

Solution

(a) Let i_1, i_2 and i_3 be the currents flowing in the branches AD, CD and BD, respectively. Assume that $V_D =$ potential at junction $D = V$. Let.

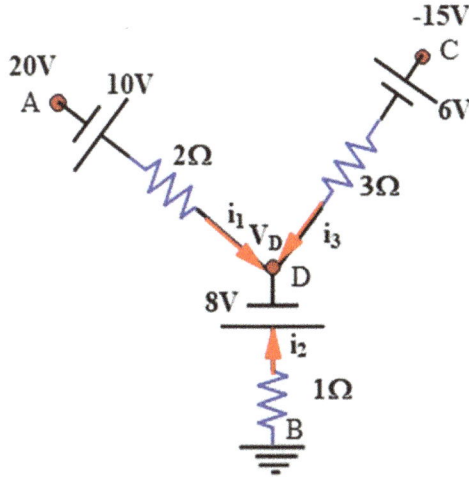

Applying Kirchoff's voltage law for the individual branch AD,

$$20 + 10 - i_1(2) = V$$

$$\Rightarrow i_1 = (30 - V)/2 \tag{5.47}$$

Applying Kirchoff's voltage law for the individual branch BD,

$$0 - i_2(1) - 8 = V$$

$$\Rightarrow i_2 = -(8 + V) \tag{5.48}$$

Applying Kirchoff's voltage law for the individual branch CD,

$$-15 - i_3(3) - 6 = V$$

$$i_3 = -(V + 21)/3 \tag{5.49}$$

Applying Kirchoff's current law at junction D,

$$i_1 + i_2 + i_3 = 0 \tag{5.50}$$

Putting i_1, i_2 and i_3 from equations (5.47), (5.48) and (5.49) in equation (5.50), we have

$$\frac{30 - V}{2} + \frac{V + 8}{-1} + \frac{V + 21}{-3} = 0$$

$$\Rightarrow V = 0 \text{ Ans.}$$

(b) Putting $V = 0$ in equation (5.47), we have

$$i_1 = \frac{30 - V}{2} = \frac{30 - 0}{2} = 15 \text{ A} \searrow$$

Putting $V = 0$ in equation (5.48), we have

$$i_2 = \frac{V + 8}{-1} = \frac{0 + 8}{-1} = -8 \text{ A} \uparrow$$

Putting $V = 0$ in equation (5.49), we have

$$i_3 = \frac{V + 21}{-3} = \frac{0 + 21}{-3} = -7 \text{ A} \nearrow \text{ Ans.}$$

N.B: We can see that the terminal potential difference between B and D is zero while a current of 8 A flows up through the cell.

Problem 16 (Mixed grouping of cells)

Among N identical cells each of emf ε, internal resistance r are connected in series to form a row and there are m numbers of such rows; so, this mixed grouping of cells drives a maximum current in an external load of resistance R. Find (a) the values of m and n, (b) maximum current flowing in the external resistor R. Put $N = 32$, $\varepsilon = 6$ V, $r = 1$ Ω and $R = 2$ Ω.

Solution

(a) Let n cells be connected in series and there are m series connections (rows) connected in parallel.

Then, the emf and internal resistance of the equivalent cell in a row are $\varepsilon_1 = n\varepsilon$ and $r_1 = nr$, respectively.

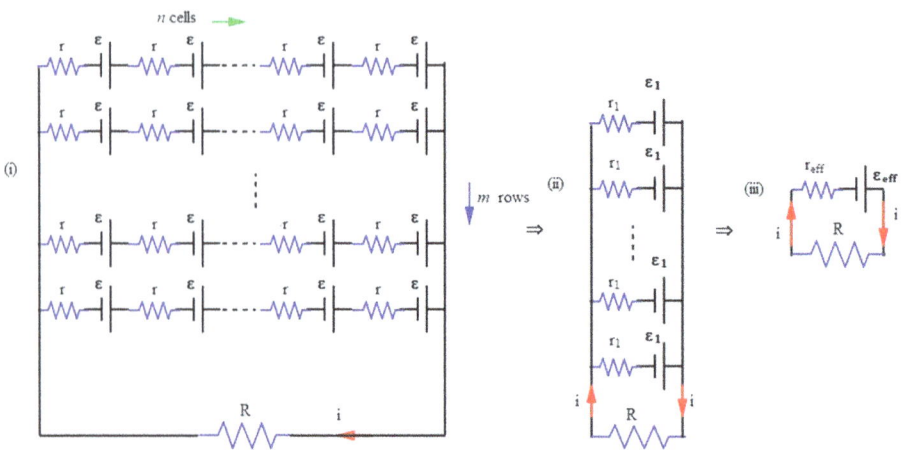

Since there are m identical rows each of equivalent emf ε_1 and internal resistance $r_1 = nr$, the effective internal resistance is

$$r_{\text{eff}} = \frac{nr}{m}$$

Hence, the current through the load is

$$i = \frac{\varepsilon_{\text{eff}}}{r_{\text{eff}} + R} = \frac{n\varepsilon}{\frac{nr}{m} + R} = \frac{mn\varepsilon}{nr + mR}$$

$$= \frac{nm\varepsilon}{\sqrt{(mR - nr)^2 + 4mnRr}} \tag{5.51}$$

$$= \frac{N\varepsilon}{\sqrt{(mR - nr)^2 + 4NRr}}$$

The current i is maximum,

$$mR - nr = 0 \text{ or } \frac{m}{n} = \frac{r}{R}$$

Since $mn = N$, we have

$$\frac{m}{(N/m)} = \frac{r}{R}$$

$$\Rightarrow m = \sqrt{N\frac{r}{R}} \tag{5.52}$$

Putting $N = 32$, $r = 1$ Ω and $R = 2$ Ω in equation (5.51), we have

$$\Rightarrow m = \sqrt{N\frac{r}{R}} = \sqrt{32\frac{1}{2}} = 4 \text{ Ans.}$$

Putting $m = N/n$ in equation (5.52), we have

$$n = N/\sqrt{N\frac{r}{R}} = \sqrt{N\frac{R}{r}} = \sqrt{32\frac{2}{1}} = 8 \text{ Ans.}$$

(b) For maximum current, putting $\frac{m}{n} = \frac{r}{R}$ and $m \times n = N$ in equation (5.51), we have

$$i_{\max} = \varepsilon\sqrt{\frac{N}{4Rr}} = 6\sqrt{\frac{32}{4(2)(1)}} = 12 \text{ A Ans.}$$

Problem 17 This RC circuit is comprised of a leaky capacitor of capacitance C and resistance R and a battery of emf ε and internal resistance r. If the switch S is closed at $t = 0$, find (a) the current i at (i) $t = 0$, (ii) $t = \infty$, and (b) the potential energy stored in the capacitor at $t \to \infty$.

Solution

A leaky capacitor is equivalent to an ideal capacitor of capacitance C connected with resistance R in parallel between terminals A and B of the capacitor, as shown in the figure below. Similarly, we know that a cell is equivalent to its emf connected with series with its internal resistance r in series between terminals D and E as shown in the figure.

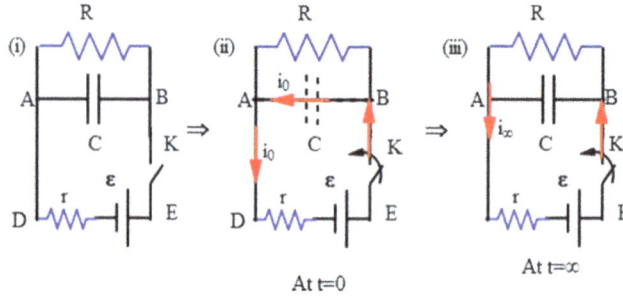

(a)

(i) At $t = 0$, capacitor acts as short-circuit (zero resistance). Then, the current does not flow in the resistor R. In other words, the voltage drop inside the capacitor is zero. Therefore, the initial current is

$$i_0 = \frac{\varepsilon}{r} \text{ Ans.}$$

(ii) At $t \to \infty$, the capacitor acts as an open circuit (infinite resistance). Then, the current does not flow in the resistor R. In other words, the voltage drop inside the capacitor is zero as no current flows through it. Therefore, the final or steady-state current is

$$i_\infty = \frac{\varepsilon}{R + r} \text{ Ans.}$$

(b) the potential energy stored in the capacitor at $t \to \infty$.

$$U = \frac{CV^2}{2}, \tag{5.53}$$

where $V =$ final potential dropped across the capacitor, given as

$$V = i_\infty R = \frac{\varepsilon R}{R + r} \tag{5.54}$$

Using equations (5.53) and (5.54), we have

$$U = \frac{C\varepsilon^2}{2}(\frac{R}{R + r})^2 \text{ Ans.}$$

Problem 18 In this RC circuit, find the (a) total charge flown, (b) energy stored in the capacitor, (c) total heat dissipated in the resistor R, (d) total heat dissipated in the circuit, and (e) total work done by the battery.

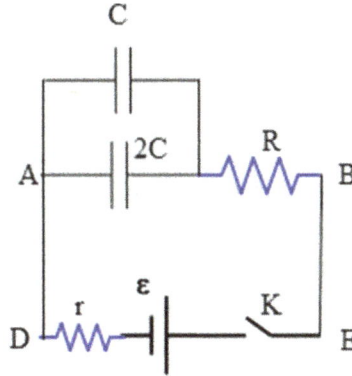

Solution

(a) First we find the equivalent capacitance and equivalent resistance between the terminals of the battery. As the capacitors of capacitance C and $2C$ are connected in parallel, the equivalent capacitance is $C_{eq} = 3$ C. Since the resistors R and r are connected in series between two terminals of the battery, the equivalent resistance is $R_{eq} = R + r$. Then the equivalent time constant is

$$\tau_{eq} = R_{eq} C_{eq}$$

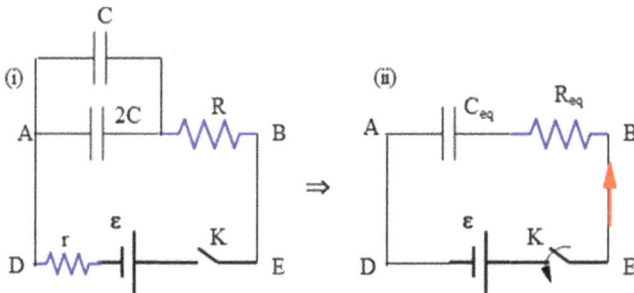

The growth of charge in the equivalent charging capacitor is

$$q = C_{eq}\varepsilon\left(1 - e^{-\frac{t}{\tau_{eq}}}\right) = C_{eq}\varepsilon\left(1 - e^{-\frac{t}{R_{eq}C_{eq}}}\right)$$

The total charge flowing is

$$q = C_{eq}\varepsilon = 3C\varepsilon \text{ Ans.} \tag{5.55}$$

(b) The energy stored in the capacitor is

$$U_C = \frac{q^2}{2C_{eq}}$$

$$= \frac{1}{2C}\left\{C_{eq}\varepsilon(1 - e^{-\frac{t}{R_{eq}C_{eq}}})\right\}^2 \quad (\because q = C_{eq}\varepsilon)$$

$$= \frac{C_{eq}\varepsilon^2}{2}(1 - e^{-\frac{t}{RC}})^2 \quad (\because R_{eq} = R \text{ and } C_{eq} = 3 \text{ C})$$

At $t \to \infty$, $U = \frac{C_{eq}\varepsilon^2}{2} = \frac{3C\varepsilon^2}{2}$ $(\because C_{eq} = 3 \text{ C})$ Ans.

(c) The total heat dissipated in the external circuit is

$$Q = \int_0^\infty P \, dt$$

$$= \int_0^\infty i^2 R \, dt$$

$$= R\int_0^\infty \left(\frac{\varepsilon}{R+r}e^{-\frac{t}{R_{eq}C_{eq}}}\right)^2 dt$$

$$= R\left(\frac{\varepsilon}{R+r}\right)^2 \int_0^\infty e^{-\frac{2t}{R_{eq}C_{eq}}} dt$$

$$= R\left(\frac{\varepsilon}{R+r}\right)^2 \frac{R_{eq}C_{eq}}{2}$$

$$= R\left(\frac{\varepsilon}{R+r}\right)^2 \frac{(R+r)(3C)}{2}$$

$$= \frac{3\varepsilon^2 RC}{2(R+r)} \text{ Ans.}$$

(d) The total heat dissipated in the entire circuit (including the internal resistance r of the battery itself) is

$$Q = \int_0^\infty P \, dt$$

$$= \int_0^\infty i^2 R_{eq} \, dt$$

$$= R_{eq}\int_0^\infty \left(\frac{\varepsilon}{R+r}e^{-\frac{t}{R_{eq}C_{eq}}}\right)^2 dt$$

$$= (R+r)\left(\frac{\varepsilon}{R+r}\right)^2 \int_0^\infty e^{-\frac{2t}{R_{eq}C_{eq}}} dt$$

$$= (R+r)\left(\frac{\varepsilon}{R+r}\right)^2 \frac{R_{eq}C_{eq}}{2}$$

$$= (R+r)\left(\frac{\varepsilon}{R+r}\right)^2 \frac{(R+r)(3C)}{2}$$

$$= \frac{3C\varepsilon^2}{2} \text{ Ans.}$$

(e) The total work done by the battery is

$$W_b = \int_0^{3C\varepsilon} \varepsilon \, dq$$

$$= 3C\varepsilon^2 \text{ Ans.}$$

Alternative method:
To find the total heat dissipated in the entire circuit but not in the external resistor R you can also use the formula

$$W_b = \Delta U + Q,$$

where $W_b = 3C\varepsilon^2$ and $\Delta U = U - 0 = \frac{3C\varepsilon^2}{2}$

$$Q = \frac{3C\varepsilon^2}{2}$$

This means that half of the *total* energy $3C\varepsilon^2$ supplied by the battery is stored in the form of electrostatic potential energy $C\varepsilon^2$ in the capacitors and other half of the energy $C\varepsilon^2$ is dissipated as heat in the circuit (*all* resistors). Ans.
N.B: Total heat dissipated in the external resistors $= QR/(R + r)$.

Problem 19 In the given circuit, find (a) the ammeter reading by neglecting the resistance of the ammeter (b) the current flowing in the battery (c) energy stored in the capacitor.

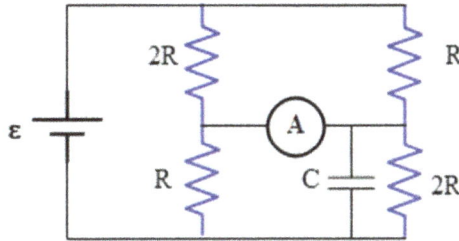

Solution
(a) Applying Kirchoff's voltage law KCL for loop 1, we have

$$-2Ry - R(y + z) + \varepsilon = 0$$

$$\Rightarrow 3y + z = \frac{\varepsilon}{R} \tag{5.56}$$

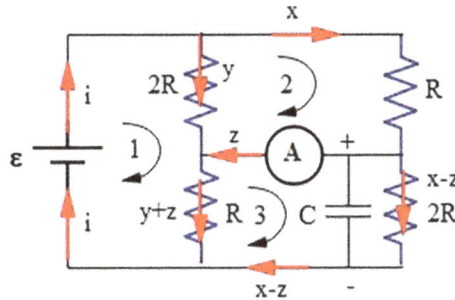

Applying Kirchoff's voltage law KCL for loop 2, we have
$$-xR + 2Ry = 0$$

$$\Rightarrow x = 2y \qquad (5.57)$$

Applying Kirchoff's voltage law KCL for loop 3, we have
$$-2R(x - z) + (y + z)R = 0$$

$$\Rightarrow 3z + y = 2x \qquad (5.58)$$

Solving these three equations, the ammeter reading is

$$y = z = \frac{\varepsilon}{4R} \left(\text{and } x = 2y = \frac{\varepsilon}{2R} \right) \text{Ans.}$$

(b) The current flowing through the battery is

$$i = x + y = \frac{\varepsilon}{2R} + \frac{\varepsilon}{4R} = \frac{3\varepsilon}{4R} \text{ Ans.}$$

(c) The energy stored in the capacitor is

$$U = \frac{CV^2}{2}, \qquad (5.59)$$

where $V = (x - z)(2R)$. Putting $x = \frac{\varepsilon}{2R}$ and $z = \frac{\varepsilon}{4R}$, we have

$$V = (x - z)2R = \left(\frac{\varepsilon}{2R} - \frac{\varepsilon}{4R} \right)(2R) = \frac{\varepsilon}{2} \qquad (5.60)$$

Using equations (5.59) and (5.60), we have
$$U = \frac{CV^2}{2} = \frac{C(\varepsilon/2)^2}{2} = \frac{C\varepsilon^2}{8} \text{ Ans.}$$

Problem 20 The branch AB, fitted with the resistor $R = 18 \ \Omega$ absorbs 60 W of power while carrying a current $i = 2$ A. If the battery has an internal resistance $r = 2 \ \Omega$, find the (a) values of x and y, (b) emf of the cell, (c) energy stored in the capacitor, (d) total power loss in the circuit. Put $R' = 4$ Ohms.

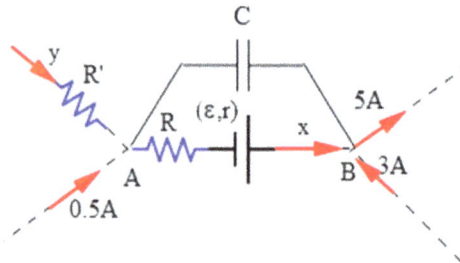

Solution

(a) Referring to the above figures and applying Kirchhoff's current law at junction B, we have

$$x + 3 = 5$$

$$\Rightarrow x = 2A$$

Applying Kirchhoff's current law at junction A,

$$0.5 + y = x = 2$$

$$\Rightarrow y = 1.5 \text{ Ans.}$$

(b) Applying Kirchhoff's voltage law for branch AB,

$$V_A - i(R + r) + \varepsilon = V_B$$

$$V_A - V_B = V = -\varepsilon + i(R + r)$$

$$\Rightarrow V = -\varepsilon + (2)(18 + 2) = -\varepsilon + 40 \tag{5.61}$$

The power absorbed by branch AB is

$$P = Vi$$

$$\Rightarrow 60 = (V)(2)(\because P = 60 \text{ W})$$

$$\Rightarrow V = 30 \tag{5.62}$$

Using equations (5.61) and (5.62) $\varepsilon = 10$ V Ans.

(c) Then, the energy stored in the capacitor is

$$U = \frac{CV^2}{2} = \frac{(20 \times 10^{-6})(30)^2}{2} = 9 \times 10^{-3} \text{ J Ans.}$$

(d) Total power loss is

$$P_{\text{loss}} = y^2 R' + P_{AB}$$

$$= (1.5)^2(4) + 60 = 69 \text{ W Ans.}$$

Problem 21 To measure the value of the resistance R, we have connected the voltmeter and ammeter as shown in the figure below. Can the ratio of voltmeter and ammeter reading $\frac{V}{i}$ give the correct value of R? Explain.

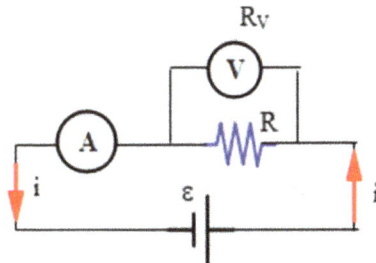

Solution

Let us assume that the ratio of voltmeter and ammeter reading is

$$\frac{\text{Voltmeter reading}}{\text{Ammeter reading}} = R_m,$$

where, R_m = meter reading of resistance

$$\text{or, } R_m = \frac{V}{i} \tag{5.63}$$

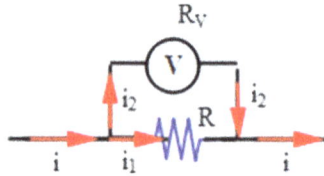

Since, the resistors R_V and R are connected in parallel,

$$i_2 R_V = i_1 R \tag{5.64}$$

According to Kirchoff's first law,

$$i = i_1 + i_2 \tag{5.65}$$

Using equations (5.63), (5.64) and (5.65), we have

$$\frac{1}{R} = \frac{1}{R_m} - \frac{1}{R_V}$$

If $R_V \to \infty$, $R \to R_m$. Hence, the ratio of voltmeter and ammeter reading cannot give the exact value of the resistance R. Ans.

Problem 22 The deflection of a moving coil galvanometer falls from 60 divisions to 10 divisions when a shunt of 8 Ω is connected. What is the resistance of the galvanometer?

Solution

In a moving coil galvanometer, the current i in the galvanometer is directly proportional to the angle of deflection ($i \propto \theta$)

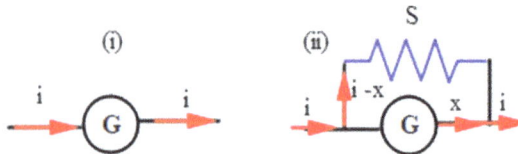

When the deflection of the galvanometer is 60 small divisions, let the current flowing through it be i. As the deflection decreases from 60 small divisions to 10 small divisions, the current in the galvanometer also decreases from i to x

proportionately. If we assume that the supply current is a constant quantity, the surplus supply current $i-x$ will be diverted and flow through the shunt resistance S. Then, the ratio of the initial and final (after introducing the shunt) currents in the galvanometer is given as

$$\frac{x}{i} = \frac{10}{60} = \frac{1}{6}$$

$$\Rightarrow x = \frac{i}{6} \tag{5.66}$$

For a shunted galvanometer,

$$(i - x)S = xG$$

or,

$$G = (i - x)\frac{S}{x} \tag{5.67}$$

Putting $x = \frac{i}{6}$ from equation (5.66) in equation (5.67) and $S = 8$ Ohm,

$$G = 40 \, \Omega \text{ Ans.}$$

Problem 23 The deflection of a moving coil galvanometer falls from $\beta = 50$ divisions to $\theta = 10$ divisions when a parallel shunt of $G = 5 \, \Omega$ is connected with the galvanometer. Assume that the battery has internal resistance $r = 5 \, \Omega$. What is the shunt resistance?

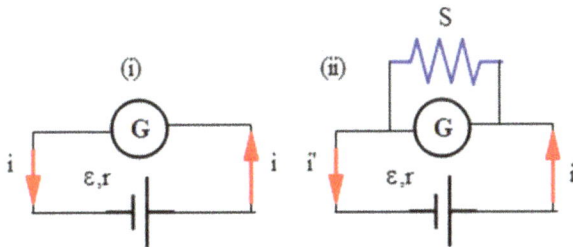

Solution
When the deflection of the galvanometer is $N = 50$ small divisions, let the current flowing through it be i. As the deflection decreases from $N = 50$ small divisions to $n = 10$ small divisions, the current in the galvanometer also decreases from i to i' proportionately. If we assume that the source battery is kept constant, a new current i' will be delivered by the source, the surplus supply current $i' - x$ will be diverted and flow through the shunt resistance S. Then, the ratio of the initial and final (after introducing the shunt) currents in the galvanometer is given as

$$\frac{x}{i} = \frac{\theta}{\beta} = \frac{10}{60} = \frac{1}{6} \tag{5.68}$$

For a shunted galvanometer,

$$(i' - x)S = xG$$

$$\Rightarrow x = \frac{i'S}{(G + S)} \tag{5.69}$$

The current drawn from the supply by the galvanometer without shunt as shown in figure (i) is

$$i = \frac{\varepsilon}{G + r} \tag{5.70}$$

The current drawn from the supply by the galvanometer without shunt as shown in figure (ii) is

$$i' = \frac{\varepsilon}{\{GS/(G + S)\} + r} \tag{5.71}$$

Using equations (5.68) and (5.71), we have

$$\Rightarrow x = \frac{S}{(G + S)} \frac{\varepsilon}{\{GS/(G + S)\} + r}$$

$$\Rightarrow x = \frac{\varepsilon S}{GS + (G + S)r} \tag{5.72}$$

Using equations (5.70) and (5.72), we have

$$\Rightarrow x/i = \frac{\frac{\varepsilon S}{GS + (G + S)r}}{\frac{\varepsilon}{G + r}}$$

$$\Rightarrow x/i = \frac{S(G + r)}{GS + (G + S)r} \tag{5.73}$$

Using equations (5.68) and (5.73), we have

$$\Rightarrow \frac{S(G + r)}{GS + (G + S)r} = \frac{\theta}{\beta}$$

$$\Rightarrow \frac{S(5 + 5)}{5S + (5 + S)5} = \frac{10}{50}$$

$$\Rightarrow \frac{S}{5 + 2S} = \frac{1}{10}$$

$$\Rightarrow S = 5/8 \; \Omega \; \text{Ans.}$$

Problem 24 (Multistage voltmeter)

The galvanometer G has internal resistance $G = 50$ Ω and full-scale deflection occurs at $i = 1$ mA. Find the series resistors R_1, R_2 and R_3 needed to use the arrangement as a voltmeter with different ranges as shown in the figure.

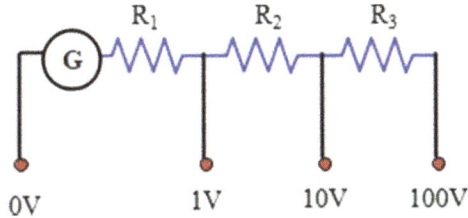

Solution

For the range of $V_1 = 1$ V,

$$i_g = \frac{V_1}{G + R_1}$$

$$\Rightarrow 10^{-3} = \frac{1}{50 + R_1}$$

$$\Rightarrow R_1 = 950 \ \Omega \text{ Ans.}$$

For the range of $V_2 = 10$ V

$$i_g = \frac{V_2}{G + R_1 + R_2}$$

$$\Rightarrow 10^{-3} = \frac{10}{50 + 950 + R_2}$$

$$\Rightarrow R_2 = 9 \times 10^3 \ \Omega \text{ Ans.}$$

For the range of $V_3 = 100$ V

$$i_g = \frac{V_3}{G + R_1 + R_2 + R_3}$$

$$\Rightarrow 10^{-3} = \frac{100}{50 + 950 + 9000 + R_3}$$

$$\Rightarrow R_3 = 90 \times 10^3 \ \Omega \text{ Ans.}$$

Problem 25 A potentiometer of length 1 m has a resistance of 25 Ω. It is then connected with a battery of emf 10 V and internal resistance of 2 Ω and resistor of

8 Ω in series with the wire. Calculate the emf of the primary cell when it gives a balance point at 75 cm.

Solution

Referring to the potentiometer diagram in section 5.6.5, the emf of the unknown cell is given as

$$\varepsilon = V_{AB}\frac{x}{l}$$

where the $l = 100$ cm $= 1$ m and $x = 0.75$ m;

$$\Rightarrow \varepsilon = V_{AB}\frac{0.75}{1} = 0.75V_{AB} \tag{5.74}$$

The terminal voltage V_{AB} of the supply battery is given as

$$V_{AB} = iR_{\text{Pot}} = \frac{\varepsilon_0 R_{\text{Pot}}}{R_{Rh} + R_{\text{Pot}} + r} = \frac{12 \times 25}{8 + 25 + 2} = 25/3 \text{ V} \tag{5.75}$$

Using equations (5.74) and (5.75)

$$\Rightarrow \varepsilon = 0.75V_{AB} = 0.75 \times 25/3 = 6.25 \text{ V Ans.}$$

Problem 26 Two balls of powers P_1 and P_2 are connected in series. If the supply voltage is equal to the rated voltage, find the power of the combination.

Solution

Let their resistances be R_1 and R_2, respectively.

For a rated voltage V, the power of the combination is

$$P = \frac{V^2}{R_1 + R_2} \quad (\because \text{ the resistors are connected in series})$$

Putting $R_1 = \frac{V^2}{P_1}$ and $R_2 = \frac{V^2}{R_2}$ we obtain

$$P = \frac{P_1 P_2}{P_1 + P_2} \text{ Ans.}$$

N.B: In a parallel combination, the combined power is equal to the sum of rated powers.

Problem 27 A 1000-watt heater coil can be cut into two equal parts and when each part is used in the rated supply voltage, it gives more power as $P \propto \frac{1}{R}$; but why is this not recommended?

Solution

Since, the power dissipated in the coil is $P = \frac{V^2}{R}$ and R decreases by two-fold if we cut it into two equal halves (say), the current flow will be doubled; so, the power dissipation will be doubled. As the heat liberation is doubled, the overheating damages the coil by reducing its life. Furthermore, the wires and connections may not be able to handle the higher current. The overheating may appear as an advantage for cooking, but it will burn out the shortened coil much quicker than the original coil. Overheating may melt insulation or cause sparks. It could also damage the power supply and start a fire. As a heater coil is designed for a specific power and heat dissipation, if we alter it, the balance between its safety and efficiency will be seriously hampered. Ans.

Problem 28 A fuse wire made of lead has radius of cross-section r. The short-circuit current of the wire is i. If density is d, melting point θ_m, specific heat C and resistivity ρ, assuming initial (room) temperature θ_0, find the time after which the fuse melts. Neglect the variation of resistance with the temperature, heat radiation and the non-uniform heating etc., assuming ideal case.

Solutions

Let t = time after which the fuse melts.

The heat developed in $Q = \frac{i^2 Rt}{J}$, where $R = \rho\frac{l}{A} = \rho\frac{l}{\pi r^2}$

$$\text{Then, we have } Q = \frac{i^2 \rho l t}{\pi r^2 J} \tag{5.76}$$

As the rise in temperature is $\Delta\theta = \theta_m - \theta_0$, the heat required to raise this temperature up to melting is

$$Q = mC(\theta_m - \theta_0) + mL = (\pi r^2 l d)[C(\theta_m - \theta_0) + mL] \tag{5.77}$$

where L = latent heat of vaporization of the fuse wire.
Using equations (5.76) and (5.77), we have

$$\frac{i^2 \rho l t}{\pi r^2 J} = \pi r^2 l d[C(\theta_m - \theta_0) + L]$$

$$t = \frac{J\pi^2 r^4 d[C(\theta_m - \theta_0) + L]}{\rho i^2} \text{ Ans.}$$

N.B: If we change the length of the fuse wire, the time after which it melts remains unchanged.

Problem 29 A capacitor C is connected to the cell of emf ε as shown in the figure below. After the capacitor is fully charged, let us disconnect the capacitor from the cell. Let us reverse the polarity of the connection by flipping the cell through $180°$ and then connect it to the capacitor. If the key is closed, find the
   (a) total charge flowing;
   (b) total heat dissipated.

Solution

(a) The quantity of initial charge on each plate of the capacitor is $q_0 = C\varepsilon$. When the capacitor is again connected after reversing the polarities of the connection, let $q = $ total charge flowing. Then the polarities of charges of plates 1 and 2 are $-q_0 + q$ and $+q_0 - q$, respectively. Since the final current $i = 0$ finally, applying KCL, we have

$$-\frac{q - q_0}{C} + \varepsilon = 0$$

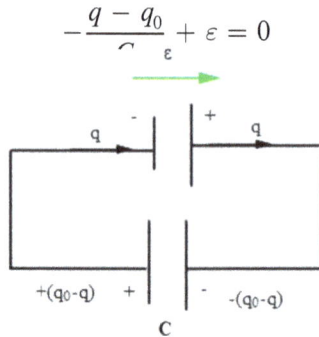

$$\Rightarrow q = q_0 + C\varepsilon = C\varepsilon + C\varepsilon = 2C\varepsilon \quad \text{Ans.}$$

(b) The heat dissipated is

$$Q = W_b - \Delta U$$

As the magnitude of initial and final charge on each plate does not change, which is equal to $C\varepsilon$, the change in electrostatic potential energy $=$

$$\Delta U = \frac{q_f^2 - q_i^2}{2C} = 0 \text{ and the work done by the battery (cell) is } W_b = \varepsilon q$$

$$= \varepsilon \cdot (2C\varepsilon) = 2C\varepsilon^2,$$

Then, we have

$$Q = 2C\varepsilon^2 \text{ Ans.}$$

Problem 30 Each link of the two-dimensional infinite grid of (a) squares as shown in figure (i), and (b) triangles, as shown in figure (ii), has resistance R. Using the principle of superposition and concept of symmetry, find the equivalent resistance of the grid between A and B.

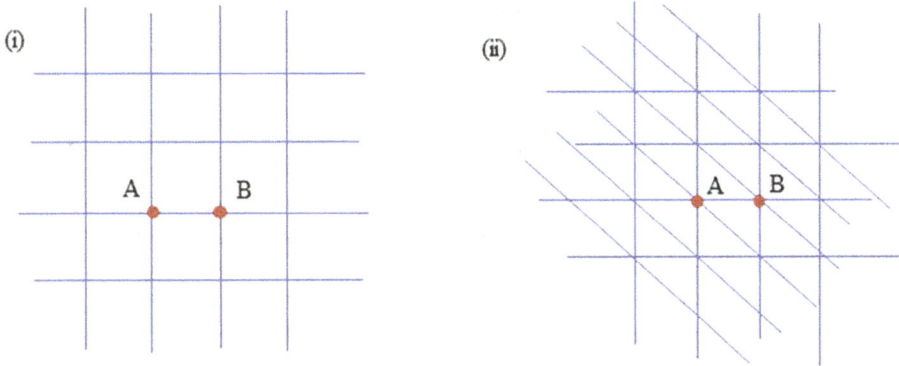

(i)   (ii)

Solution

(a) Let us try to solve these two problems by using the concept of electrical symmetry and superposition principle. Since each unit in both cases is repeating to infinity, the equivalent resistance between two nearest junctions remains the same, which is independent of the position of these two points. Due to infinite nature, the given network is symmetrical about any junction. This means, if you inject a current i, say, at the point or junction A, due to electrical symmetry because each link has the same resistance, the input current will be equally divided among all four branches connected to point A. So, a current of magnitude $x = i/4$ will flow in each branch as shown in figure (i). Similarly, if you extract a current of i out of point B, an equal amount of current $x = i/4$, will be collected through each of the four connecting branches at B, as shown in figure (ii).

As we know, it is the function of the battery to supply or inject a current i at A and simultaneously collect or extract the same current at B. So, this is a superposition of two effects such as injecting and withdrawing the current. While doing so, the net current in branch AB will be equal to $z = x + x = 2x$, and in the other branches the currents will be $y = (i-2x)/3$, where $x = i/4$. We can see that the symmetry of current distribution is disturbed after superposition which is quite natural. Then, the voltage drop between A and B is given as

$$V = (2x)R = 2xR = 2(i/4)R = iR/2 \qquad (5.78)$$

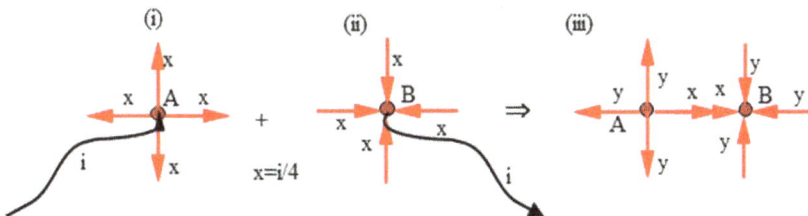

If we substitute the entire network (infinite grid) by a single resistor R_{eq}, the same current i will drop the same voltage, then we can write

$$V = iR_{eq} \tag{5.79}$$

Using these two equations, we have

$$R_{eq} = R/2 \text{ Ans.}$$

N.B: The exact answer will be $0.318R$, which can be found by following complex integration, which is beyond the scope of this book.

(b) Due to the infinite nature, the given network (two-dimensional grids of triangles) is symmetrical about any junction. This means, if you inject a current, i, say, at the point or junction A, due to electrical symmetry because each link has the same resistance, the input current will be equally divided among all six branches connected to point A. So, a current of magnitude $x = i/6$ will flow in each branch, as shown in figure (i). Similarly, if you extract a current of i from point B, the same amount of current $x = i/6$, will be collected through each of the six connecting branches at B. As explained earlier, the battery injects a current i at A and simultaneously collects or extracts the same current at B. So, this is a superposition of two effects such as injecting and withdrawing the current. While doing so, the net current in the branch AB will be equal to $z = x + x = 2x$, and in the other branches the currents will be $y = 5(i-2x)/6$, where $x = i/6$. We can see that the symmetry of current distribution is disturbed after superposition which is quite natural. Then, the voltage drop between A and B is given as

$$V = (2x)R = 2xR = 2(i/6)R = iR/3 \tag{5.80}$$

If we substitute the entire network (infinite grid) by a single resistor R_{eq}, the same current i will drop the same voltage, then we can write

$$V = iR_{eq} \tag{5.81}$$

Using these two equations, we have

$$R_{eq} = R/3 \text{ Ans.}$$

N.B: Resistance of infinite grid (2D and 3D) can most accurately be solved by complex integrations. This simplified method gives an approximated answer in part (a); but an exact answer in part (b), which may be a coincidence. Similarly, if we take an infinite honeycomb (uniform hexagonal) grid, this method gives a correct equivalent resistance ($=2R/3$). But in 3D, the answers will be approximated.

Problem 31 With reference to problem 17, the leaky capacitor of capacitance C and resistance R is connected to a battery of emf ε and internal resistance r. If the switch S is closed at $t = 0$, find (a) the charge flown through the capacitor, (b) current flowing through the capacitor (c) current flowing through the battery at time t.

Solution

As mentioned earlier, the leaky capacitor is equivalent to a pure capacitor and a resistor R (which is equal to the internal resistance of the leaky capacitor) in parallel. This combination is connected with the battery of emf ε and internal resistance r. Let the charge x flow through the capacitor during a time t after closing the key. Then the current through the capacitor is equal to dx/dt. If a current i is drawn from the battery, the current $I - dx/dt$ will flow through the resistor R.

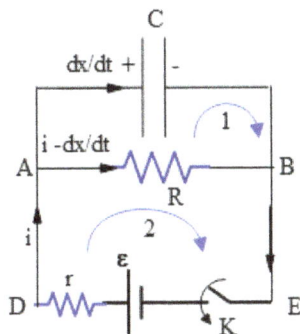

In this RC circuit, let us apply KVL in loop 1,

$$-\frac{x}{C} + \left(i - \frac{dx}{dt}\right)R = 0 \tag{5.82}$$

Then, applying KVL in loop 2,

$$\varepsilon - ir - \frac{x}{C} = 0 \tag{5.83}$$

Solving the last two equations,

$$-\frac{x}{C} + \left\{\left(\varepsilon - \frac{x}{C}\right)\frac{1}{r} - \frac{dx}{dt}\right\}R = 0$$

$$\Rightarrow -\frac{x}{C}\left(1 + \frac{R}{r}\right) + \frac{R\varepsilon}{r} - R\frac{dx}{dt} = 0$$

$$\Rightarrow -\frac{(R + r)x}{RrC} + \frac{\varepsilon}{r} = \frac{dx}{dt}$$

Separating the variables,

$$dt = \frac{dx}{-\dfrac{(R + r)x}{RrC} + \dfrac{\varepsilon}{r}}$$

During time t the charge flown through the capacitor is equal to x.

Integrating both sides,

$$\Rightarrow \int_0^t dt = \int_0^x \frac{dx}{-\dfrac{(R+r)x}{RrC} + \dfrac{\varepsilon}{r}}$$

$$\Rightarrow -\frac{RrC}{R+r}\left\{\ln\left(-\frac{(R+r)x}{RrC} + \frac{\varepsilon}{r}\right) - \ln\left(\frac{\varepsilon}{r}\right)\right\} = t$$

$$\Rightarrow \left\{\ln\left(-\frac{(R+r)x}{RrC} + \frac{\varepsilon}{r}\right) - \ln\left(\frac{\varepsilon}{r}\right)\right\} = -\frac{R+r}{RrC}t$$

$$\Rightarrow \left(-\frac{(R+r)x}{RrC} + \frac{\varepsilon}{r}\right)\Big/\left(\frac{\varepsilon}{r}\right) = e^{-\frac{R+r}{RrC}t}$$

$$\Rightarrow x = \left(\frac{\varepsilon CR}{R+r}\right)\left(1 - e^{-\frac{R+r}{RrC}t}\right) \text{ Ans.}$$

The current flowing through the capacitor is

$$\Rightarrow i_C = \frac{dx}{dt} = \frac{\varepsilon}{r}e^{-\frac{R+r}{RrC}t} \text{ Ans.}$$

Then the total current is

$$i = \frac{1}{r}\left(\varepsilon - \frac{x}{C}\right)$$

Putting the obtained value of x, we have

$$i = \frac{1}{r}\left\{\varepsilon - \frac{\varepsilon R}{R+r}\left(1 - e^{-\frac{R+r}{RrC}t}\right)\right\}$$

$$\Rightarrow i = \frac{\varepsilon}{R+r} + \frac{\varepsilon R}{(R+r)r}e^{-\frac{R+r}{RrC}t} \text{ Ans.}$$

IOP Publishing

Problems and Solutions in Electricity and Magnetism

Pradeep Kumar Sharma

Chapter 6

Magnetic field and its calculation

6.1 Introduction

A piece of magnetic ore first found in magnesia had the property of attracting iron and some materials. This ore was named magnesite (Fe_3O_4) and more popularly called a natural 'magnet'. The behaviour (properties) of a magnet is called magnetism. Magnetism has diverse applications in electrical engineering and electronics. In electrical engineering we have three types of electrical machines, namely motor, generator and transformer; each is based upon electromagnetic action. The use of magnets in medical electronics and compasses in civil engineering is crucial. In this chapter we will discuss the types of magnets and the ways of calculating the magnetic field by using Biot–Savart law and Ampère's circuital law.

Until 1820, electricity and magnetism had been treated as separate branches. But, after Oersted's discovery in 1820, magnetism and electricity were proved to be interrelated and merged as a single branch called electromagnetism. The notable scientists such as Ampère, Biot, Savart, Gauss, Weber and Faraday etc, advanced the knowledge of electromagnetism. In 1832, Faraday introduced the concept of magnetic lines of force to understand and interpret the electromagnetic phenomena more efficiently. In this chapter, we will explain the magnetic field by using Faraday's concept of lines of force.

6.2 Magnets and some factors (characteristics)

6.2.1 Temporary and permanent magnets

Permanent magnets do not require any external source to produce a magnetic field. They are natural and also artificial. The ferromagnetic materials like iron, cobalt and nickel can be made permanent magnets artificially. When a ferromagnetic metal is melted in a constant high magnetic field and slowly cooled down, it acquires a permanent magnetism. In this process, the atoms align and the electrons spin in the same direction. This is known as exchange of coupling of the electron spin, which is a quantum effect. The retaining of the ordering of the magnetic domains 'spin dipole

doi:10.1088/978-0-7503-6477-5ch6

moment of the electrons' is the root cause of ferromagnetism. If it exists permanently, we call it a permanent magnet. However, when a ferromagnetic material is rubbed in the same direction with a magnet or kept in a magnetic field for some time, it will get magnetized for a period of time. This is called temporary magnetism. There are several types of permanent magnets such as bar magnet, horseshoe magnet, needle magnet, cylindrical magnet, ring magnet, spherical and oval-shaped magnet etc.

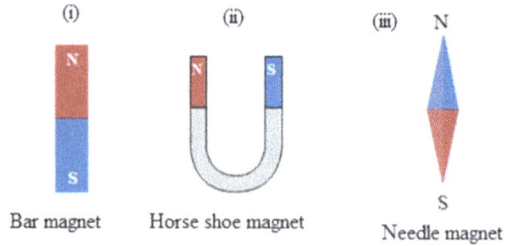

Bar magnet   Horse shoe magnet   Needle magnet

6.2.2 Electromagnets

Some electromagnets consist of a soft iron core having high permeability and low retentivity wrapped by a wire. When a current flows in a coil, it behaves as a magnet by producing a magnetic field which magnetizes the soft iron that is capable of confining and magnifying the magnetic fields produced by the coil to a thousand times due to its high permeability. This is called an electromagnet. The use of electro-magnets is all encompassing from **MRI** machines to a memory storage device, space craft propulsion system, amplifiers etc. In an electrical system the electromagnets are required for generators and motors and in many devices where we need a time-varying magnetic flux such as induction heating, on and off switching devices etc.

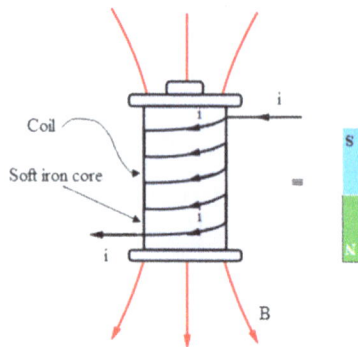

An electromagnet can behave as a bar magnet

6.2.3 Poles

If we move any magnet near to the powder of iron filings, most of it will stick to the ends of the magnet, called 'poles'. This means, a magnet has two poles, where its attraction is maximum and at its mid-point the attraction is minimum, called the 'neutral region'. A magnet when suspended by a silk thread always orients in a north–south (N–S) direction. The poles in N and S directions are called N-pole and S-pole, respectively. This means a magnet has two poles, i.e., N and S poles.

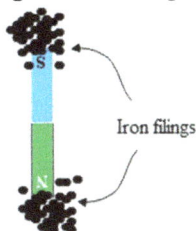

At the magnetic poles (N and S), the accumulation of iron filings is maximum.

6.2.4 Magnetic axis and magnetic meridian

The axis passing through the poles is called the magnetic axis, and the vertical plane passing through the magnetic axis is called the magnetic meridian.

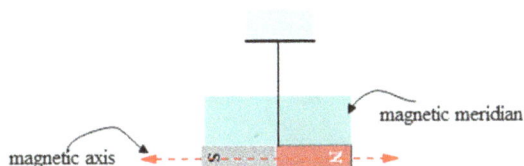

The vertical plane that passes through the magnetic axis is called magnetic meridian.

6.2.5 Magnetic length

The shortest distance between the poles is called magnetic length. This may be different from the actual length of a magnet.

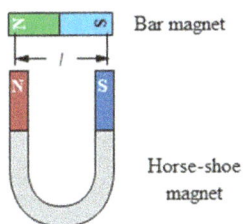

The shortest distance between the magnetic poles (N and S) is the length of a magnet.

6.2.6 No monopole, no-pole and consequent poles

If we cut a magnet, we cannot separate (isolate) the poles. Hence, monopoles do not exist. A magnetized ring (toroid carrying a current), called a ring magnet does not have any pole. It is not possible to tell which is north pole and which is south pole.

The separation of the magnetic poles
(N and S) of a magnet is impossible

Some faulty magnets may have two identical poles (consequent poles). This may be a temporary effect due to faulty magnetization of the material.

6.2.7 Attraction and repulsion of poles

Practically, we can see that unlike poles of magnets attract each other and like poles repel. So, a repulsion confirms the polarity of a magnet.

The like poles (N-N and S-S) repel
and unlike poles (N-S) attract

6.2.8 Pole strength

The strength of the poles is given as 'm', which is directly proportional to the force of attraction between the poles and any magnetic material like a piece of iron (or any other pole of a magnet). If we cut the magnet so as to decrease the area, the pole strength decreases proportionally. However, m does not depend effectively upon the length of the magnet. Following the conventions of opposite electrical charges q and $-q$, we can represent the pole strength of the magnet as $+m$ and $-m$.

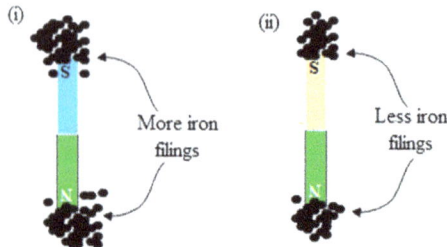

The accumulation of iron filings is more for
a magnet of stronger pole strength.

6.2.9 Magnetic dipole moment

We defined the electric dipole moment as a vector which is equal to the product of charge and the distance of separation between two poles (equal and opposite charges). It is given as

$$\vec{P_{el}} = q\,\vec{l}$$

The electric dipole moment points from $-q$ to $+q$.

Following the above logic, we can also define the magnetic dipole moment or magnetic moment as a vector quantity, given as

$$\vec{P_m} = ml\,\hat{n},$$

where $\hat{n}$ = unit vector of the area of cross-section of the N-pole. $\vec{P_m}$ points from S-pole to N-pole. The electric and magnetic moment can also be denoted as $\vec{M}$ or $\vec{\mu}$.

The magnetic dipole moment points from S- to N-pole;it is analogous to electric dipole moment that points from -ve to +ve charge

Example 1 If we bend a bar magnet of dipole moment M into a semi-circular arc, find the dipole moment of the new magnet.

Solution

For the bar magnet, the given dipole moment is

$$M = ml \tag{6.1}$$

For a semi-circular magnet the dipole moment is

$$M' = ml' \tag{6.2}$$

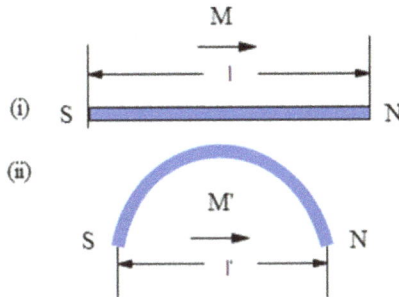

Using equations (6.1) and (6.2), $\frac{M'}{M} = \frac{l'}{l}$ where $\frac{\pi l'}{2} = l$

$$\Rightarrow \frac{M'}{M} = \frac{2}{\pi}$$

$$\Rightarrow M' = \frac{2M}{\pi} \text{ Ans.}$$

N.B: The dipole moment decreases; $M' < M$.

Example 2 As the magnetic moment point from S-pole to N-pole, their directions are shown in the following figures. Find the magnetic dipole moment of the combination of the identical bar magnets each of dipole moment p_0 as shown in figures (i) and (ii).

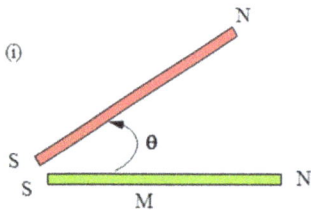

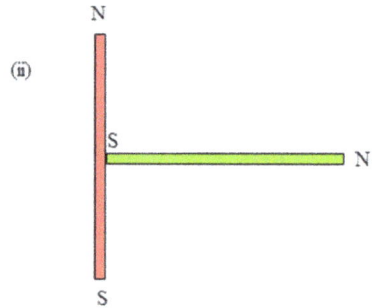

Solution
Referring to figure (i), applying the parallelogram law of vectors, the net magnetic moment is given as

$$p = \sqrt{p_0^2 + p_0^2 + 2p_0 p_0 \cos \theta}$$

$$= 2p_0 \cos \frac{\theta}{2} \text{ Ans.}$$

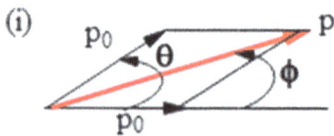

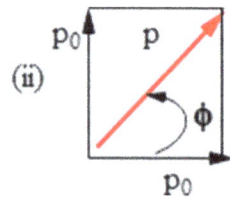

Referring to figure (ii), applying the parallelogram law of vector addition, the net magnetic moment is

$$p = \sqrt{p_0^2 + p_0^2} = \sqrt{2} p_0 \text{ Ans.}$$

6.3 Faraday's Concept of Field impressed Maxwell and Einstein

Faraday modified the Newtonian concept of aether to explain the action-at-a-distance—one body acting instantly through vacuum or aether with others. Between 1831 and 1837, while experimenting on electromagnetic induction, Michael Faraday introduced the idea of lines of magnetic and electric force to visualize how fields act in space. He developed the concept of field of a force and applied this idea to describe all forces such as electric, magnetic and gravitational forces produced by electric charges, magnets and ordinary neutral matter, respectively. Between 1961 and 1865, James Clerk Maxwell developed the theory of electromagnetic wave by using Faraday's concept of electric and magnetic field lines [1].

Faraday replaced Newtonian subtle granular aether by an electromagnetic field. As an experimentalist, he was interested in the physical conception of forces by his innovative field ideas. He proposed that the invisible electric and magnetic lines of forces are the physical entities that can create a strain or tension in physical vacuum or space (aether) surrounding the current-carrying conductors, magnets and charged objects etc.

Where Newton left off, Faraday undertook Newton's work to unify all forces. Newton explained the interactions by using his idea of action-at-a-distance and all-pervading granular aether. But, Faraday proposed the ideas of lines of force that showed the physical presence of the fields. Faraday's sincere attempt to visualize the interactions in a mechanical way of introducing the concept of field, is one of the revolutionary ideas in the history of physics. As a gifted experimentalist, he established this hypothesis from the experiments.

For instance, iron filings sprinkled on a piece of paper or cardboard placed above a magnet, align or arrange themselves in a fixed pattern around a magnet; when the magnet is removed, the pattern will disappear. What does this pattern represent? This inspired Faraday to imagine the presence of a kind of strain or tension generated in the space surrounding a magnet due to its presence. The presence of this strain will be independent of the presence of the iron filings.

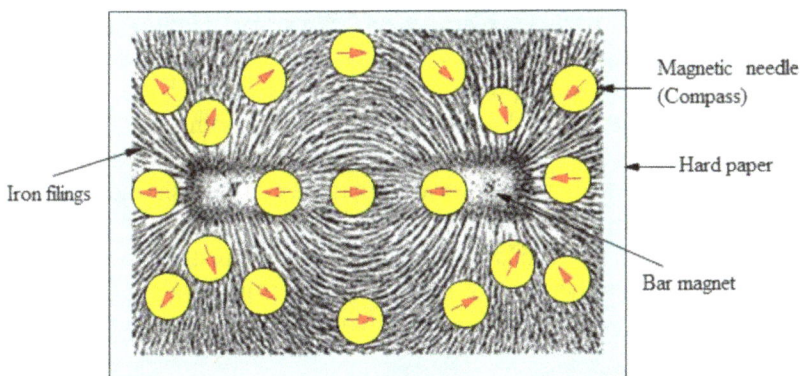

Alignment of iron filings on a thin hard paper placed over a bar-magnet tells us about the physical presence of magnetic the field of the magnet and alignment of magnetic needles proves the directional nature of the magnetic field.Magnetic lines or B-lines point from N-pole to S-pole outside the magnet and from S-pole to N-pole inside the magnet.So,B-lines are closed and three-dimensional.

So, Faraday had to hypothesize that a magnet can generate a permanent strain around it. He called this strain a *field* of the magnet or magnetic field. Similarly, in 1837, he proposed the existence of fields around charged bodies, called electric field, and later, he proposed the field surrounding a neutral object such as Earth, called gravitational field. So, the concept of field arises from the physical presence of matter and its effect on the others in the form of force etc. Thus, Faraday visualized the invisible field as a bunch of lines drawn along the line of alignment of the probes (iron filings, magnetic needle etc).

The implication of the idea of lines of force is all-encompassing without which we cannot even interpret any electric or magnetic effect. We can imagine how difficult it would have been for Oersted, Ampère and Arago to understand and analyse the electric and magnetic interactions without the field-lines which was developed by Faraday eleven years later. Professor Magnusson narrates the significance of Faraday's line of force as follows:

> We are so accustomed to use the lines of force when thinking of electromagnetic phenomena that without this concept we are entirely at sea. Conceive, if you can, some rational explanation of the strange action of the magnetic needle in the presence of a conductor carrying an electric current without using Faraday's concept of magnetic lines of force [2].

Interconversion of forces and field as a mechanical stress: This *field* idea impressed Maxwell who published his monumental work with a series of papers titled 'On Faraday's Lines of Force'. Recapitulating, Einstein followed Maxwell's concept of *constancy of light speed*. Maxwell's achievement is rooted in Faraday's experimental findings and his field idea of electromagnetism. So, Einstein glorifies Faraday in the following words:

> Faraday must have grasped with unerring instinct the artificial nature of all attempts to refer electromagnetic phenomena to actions-at-a-distance between electric particles reacting on each other. How was each single iron filing among a lot scattered on a piece of paper to know of the single electric particles running round in a nearby conductor? All these electric particles together seemed to create in the surrounding space a condition which in turn produced a certain order in the filings. These spatial states, today called fields, would, he was convinced, furnish the clue to the mysterious electromagnetic interactions.
>
> He conceived these fields as states of mechanical stress in an elastically distended body (ether/space). For at that time this was the only way one could conceive of states that were apparently continuously distributed in

space. The peculiar type of mechanical interpretation of these fields remained in the background - a sort of placation of the scientific conscience in view of the mechanical (Newtonian) tradition of Faraday's time. (Albert Einstein, 1940) [3].

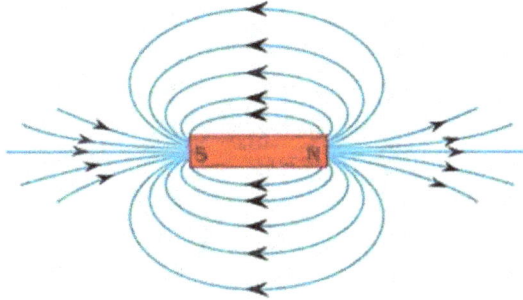

In 1845, Faraday discovered magneto-optical effect where the plane of polarization of a beam of linearly polarized light rotated when travelling along the direction of an external magnetic field. So, he understood the fact that the lines of force are real physical entities and dynamic in nature as these lines curve and turn just like a rubber band or plastic plate under a mechanical tension.

6.4 Magnetic field and lines of force

6.4.1 Magnetic field

The space surrounding a magnet where any other magnet or magnetic material experiences a magnetic force and torque is called magnetic field. We cannot see the magnetic, electric and the gravitational fields. However, we can see the effect of these fields on material particles very easily.

6.4.2 Lines of force

As mentioned earlier, Michael Faraday introduced the concept of lines of force to visualize the electric and magnetic field graphically. When we keep a magnet below a thin hard horizontal paper and sprinkle iron filings on it, we can see a spectacular pattern (array or arrangement) of the iron particles to form innumerable curves in between the poles. It seems as if something flows from north pole to south pole. If we bring a magnetic needle from N-pole to S-pole and draw its path, we can also get similar curves made by the iron filings. These lines are called 'lines of force'. Since the magnetic needle points from N-pole to S-pole outside the magnet, we can say that the magnetic lines of force come from N-pole to S-pole outside the magnet and from S-pole to N-pole inside the magnet.

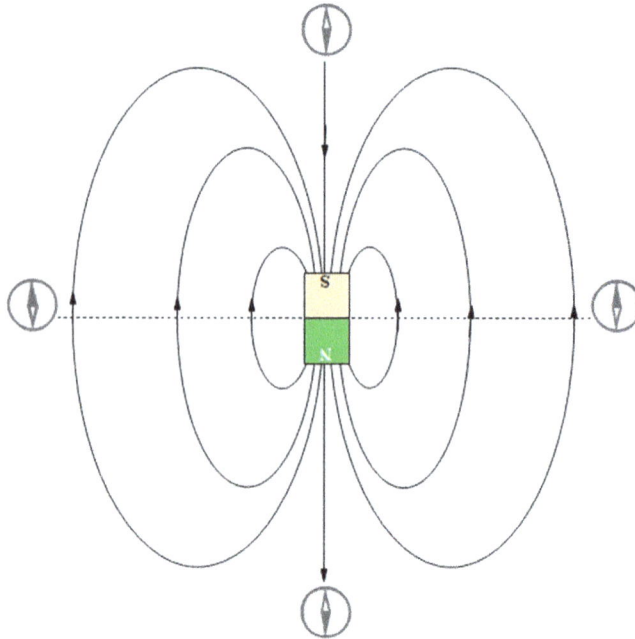

The orientation of iron filings creates a 2-D pattern of magnetic field of a bar magnet.Practically, magnetic field is three-dimensional.The lines of forces are more crowded near the poles of the magnet and least crowded along the perpendicular bisector of the magnet

So, magnetic lines of force are closed unlike electrostatic lines of force. The magnetic lines do not intersect like electric field lines because we cannot have two directions of each line of force of the same magnet at the point of their intersection.

6.4.3 Flux and flux density

The magnetic flux is proportional to the number of lines of force given as ϕ. Just like a in a tube of fluid, we can just imagine a line by observing the path followed by a dust or ink particle but we cannot count these lines. Similarly, we cannot count the magnetic field lines. Magnetic flux can be made numerically equal to m.

In hydrodynamics, according to the equation of continuity the density of v-lines of force is given by

$$v = \frac{\phi_v}{A_n}$$

Following this logic, in electrostatics, the electric density vector is defined s

$$D = \frac{q(=\phi_{el})}{A_n}$$

6-10

In current electricity the density of J-vector is defined as

$$J = \frac{\phi_J(=dq/dt)}{A_n}$$

Similarly, in magnetostatics, we define the magnetic field (flux) density B as numbers of lines of force passing through a unit cross-sectional area normally. It is given as

$$B = \frac{m(=\phi_m)}{A_n}$$

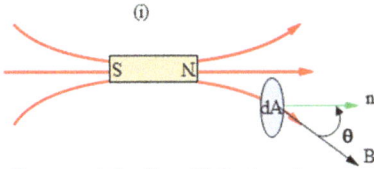

The magnetic flux (flux of B-lines) passing through the elementary area dA is $\phi=BdA\cos\theta$

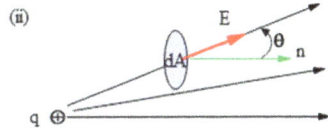

The flux of E-lines passing through the elementary area dA is $\phi=EdA\cos\theta$

The magnetic field induction, magnetic flux density and magnetic field density are the same thing. It is denoted by the letter B whose unit is Tesla ($=$ weber m^{-2}). So, the magnetic flux is given as

$$\phi_m = BA_n$$
$$= BA \cos \theta$$
$$= \vec{B} \cdot \vec{A}$$

$\phi_B = BA$

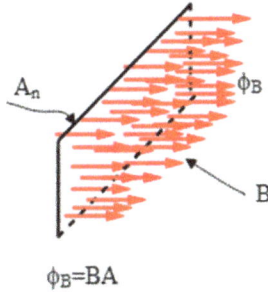

The magnetic flux (numbers of lines of force) passing through any cross-sectional area A is given as $\phi = \vec{B} \cdot \vec{A}$, where $\vec{B} =$ average flux density.

At any point, the flux density can be given as

$$B = \frac{d\phi}{dA_n}$$

Then, the flux passing through the curved area (surface) is

$$\phi = \int_S \vec{B} \cdot d\vec{A},$$

$\vec{B}$ is also called magnetic induction or magnetic field density. The magnetic lines of force are called B-lines. The tangent drawn at any point P of the lines of force from N to S-pole gives the direction of $\vec{B}$. If the density of B-lines is equal, the field is said to be uniform. If the density of B-lines changes from point to point, the field is non-uniform. Flux is positive if it is outward at N-pole and negative for inward flux at S-pole, outside the magnet.

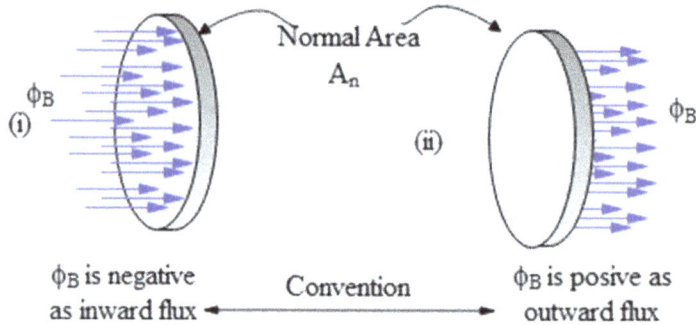

ϕ_B is negative as inward flux   Convention ϕ_B is posive as outward flux

6.5 Superposition of $\vec{B}$

If many magnets are shown near a piece of iron placed at P, each magnet tries to pull the piece of iron by its own force. The, the resultant force acting on the iron-piece is equal to the vector sum of all forces applied by all the magnets.

$$\vec{F} = \Sigma \vec{F_i}$$

Since the force of a magnet is directly proportional to the magnetic field density of its poles, the resultant field induction $\vec{B}$ is equal to the vector sum of the individual field inductions of the magnets at the point under consideration

$$\vec{B} = \Sigma \vec{B_i}$$

This is known as the superposition of $\vec{B}$-field like (magnetic field). We have applied the superposition theory for electric field in the first chapter (electric field and potential).

6.6 Gauss's law of magnetism

As discussed earlier, the magnetic field lines are closed. This means that the number of lines of force going out of the north pole towards the south pole outside the magnet is equal to the same number of lines of forces coming from S-pole towards N-pole inside the magnet.

Then, the net magnetic flux (which is proportional to the number of field lines) is

$$\phi_{net} = \phi_{out} + \phi_{in} = +\phi + (-\phi) = 0$$

Since $\phi_{net} = \int \vec{B} \cdot d\vec{A}$, for a closed surface enclosing the N-pole or any region space, we can write that the net flux passing through the closed surface is

$$\oint \vec{B} \cdot d\vec{A} = 0,$$

where $\vec{B}$ = net magnetic field at the point under consideration. This is called Gauss's law of magnetism. This states that the closed surface integral of magnetic induction is zero. This physically signifies that magnetic lines are closed, unlike electric fields. As we cannot separate them and isolate N-pole from S-pole, magnetic monopoles are impossible.

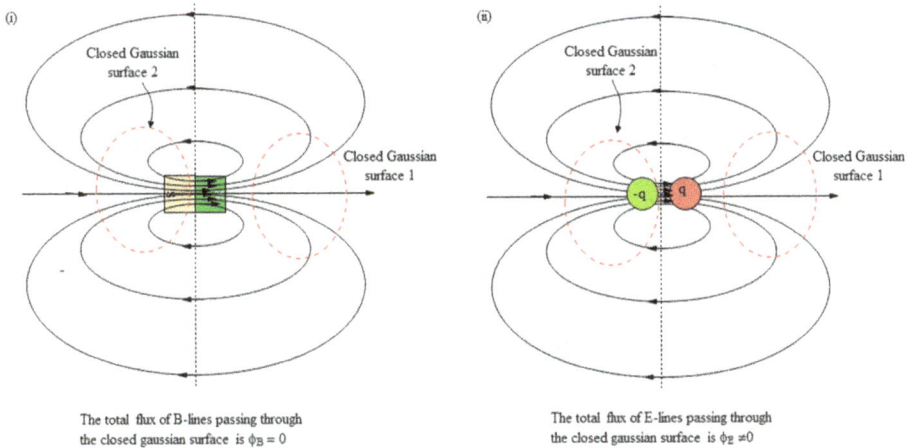

(i)

Closed Gaussian surface 2

Closed Gaussian surface 1

The total flux of B-lines passing through the closed gaussian surface is $\phi_B = 0$

(ii)

Closed Gaussian surface 2

Closed Gaussian surface 1

The total flux of E-lines passing through the closed gaussian surface is $\phi_E \neq 0$

Closed line integral or circulation of magnetic field induction B is zero, whereas the circulation of electric field E need not be zero. So, unlike isolated electric monopoles (charges), magnetic monopoles (single isolated magnetic poles) do not exist.

6.7 Modern view of magnetism

In 1820, Hans Christian Oersted discovered the magnetic effect of an electric current. This means that when a current flows in a wire, it can affect (rotate) a magnetic needle. A few days after, French physicist André-Marie Ampère argued that, since a magnetic needle is a magnet, only a magnetic field can influence it. So, a current-carrying conductor can establish a magnetic field in its surrounding space. In other words, a magnetic field is caused by a moving charge; it forms the basis of Ampére's laws. Ampère hypothesized molecular currents as the cause of magnetism in a permanent magnet. However, the modern theory of magnetism states that the magnetism in matter mostly arises from the motion of electrons. The spin and orbital motion of the electrons create magnetic dipoles (moments). When the atomic dipoles are randomly oriented, the net magnetic effect is zero. But the alignment of these dipoles in certain materials like iron causes ferromagnetism, creating a permanent magnetic field; if it is the soft iron core of an electromagnet, we get a magnetic field so long as current flows in the coils of the electromagnet. In addition to the motion of the electrons, the spin and orbital motion of protons and neutrons cause nuclear magnetism; but its effect is much weaker than the magnetism caused by the motion of electrons.

Example 3 A magnetic needle is placed by the side of a straight current-carrying conductor. In figure (i), current is up and figure (ii), the current is outward. Which is the correct direction of the magnetic needle?

Solution

By using the right-hand thumb rule (RHTR), we can show that the magnetic lines of forces in the surrounding space of a long straight current-carrying wire are circular surrounding the wire. So, option (i) is wrong and option (ii) is correct. The north pole of the magnetic needle aligns along the perimeter of the circle in anticlockwise direction. If we reverse the direction of current, the pointing of the needle will also get reversed. This establishes the fact that a magnetic field has a direction or the magnetic field induction is a vector.

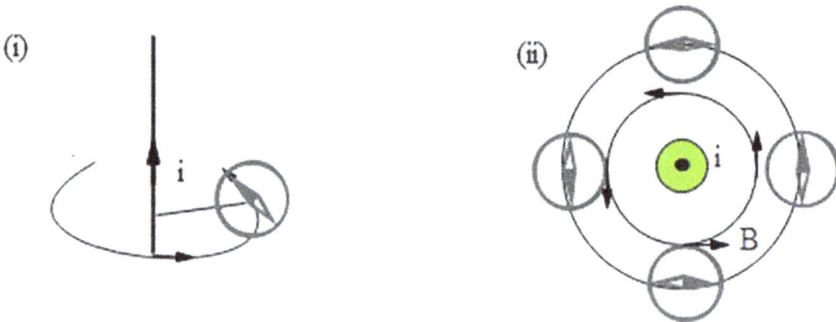

6.8 Right-hand thumb rule

This rule was introduced by James Clerk Maxwell. It helps us to find the direction of magnetic field due to a straight current-carrying conductor(infinitely long, finite and infinitesimal or elementary). If the extended thumb represents the direction of an upward current, the direction of magnetic field induction will be in the direction in which our fingers curl (in an anticlockwise sense when viewed from above) as shown in figure (i).

For current loops, if the sense or direction of curling of the fingers (anticlockwise when viewed from above) as shown in figure (ii), represents the direction of current in the loop (anticlockwise as viewed from above), the extended thumb pointing up will represent the direction of magnetic field along its axis or magnetic dipole moment of the current loop.

Following the RHTR, when we view from above, the clockwise current gives downward (or –ve) $\vec{B}$ and $\vec{\mu}$, whereas anticlockwise current gives upward (or +ve) $\vec{B}$ and $\vec{\mu}$. *If the thumb points towards you, this direction is considered outward from the writing board (black or white or smart board or computer screen etc); then, the sense of wrapping or curling of the fingers will be anticlockwise. So, an outward current has an anticlockwise magnetic field, and inward current produces a clockwise magnetic field; for the loops, an anticlockwise current has an outward magnetic field and outward magnetic moment, and a clockwise current will produce an inward magnetic field and inward magnetic moment.*

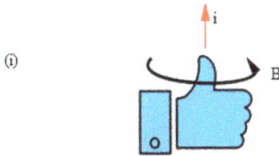

If the extended thumb of the right hand represents the current direction in a straight conductor/current element, the wrapping or curling of the fingers will represent the direction of magnetic field (B-lines)

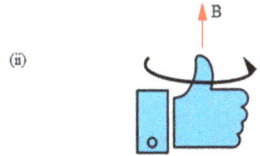

If the wrapping or curling of the fingers of the right hand represents the direction of current in a loop, the extended thumb represents the direction of the magnetic field (B-lines)

If we place a magnetic needle, it points perpendicular to the wire in the horizontal plane. If we reverse the direction of current, the pointer of the magnetic needle orients in the opposite direction. So, a magnetic field is directional, which can be characterized by the magnetic induction vector $\vec{B}$.

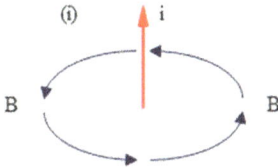

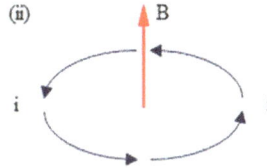

The symbolic representation of the direction of the magnetic field (B-lines) due to (i) a current element (ii) a current loop according to right hand thumb rule

Magnetic field induction has a magnitude, and it has a direction at any given point of space.

6.9 Biot–Savart law

In 1820, just after Oersted's discovery of magnetic effect of a current, the French physicists Jean Baptiste Biot and Felix Savart derived a relation to express the magnetic field of a current element. Also, French polymath Laplace's contribution to the Biot–Savart law is also considered. So it is known as the Biot–Savart–Laplace law. It suggests that each element of a current-carrying conductor, that is, current element $i\,\vec{dl}$, produces its magnetic field $\vec{dB}$ at any point P in space. The net magnetic field at point P can be given as the summation (integral) of the elementary magnetic field $\vec{dB}$.

$$\vec{B} = \int \vec{dB}$$

Experimentally, it is verified that $dB \propto \frac{1}{r^2}$, $dB \propto i\,dl$ and $dB \propto \sin\theta$. So, we can write

$$dB = \frac{\mu_0 i \; dl \; \sin \theta}{4\pi r^2},$$

where $\frac{\mu_0}{4\pi}$ is a constant of proportionality. Vectorially, it can be given as

$$d \overrightarrow{B} = \mu_0 \frac{i \; \overrightarrow{dl} \times \overrightarrow{r}}{4\pi r^3}$$

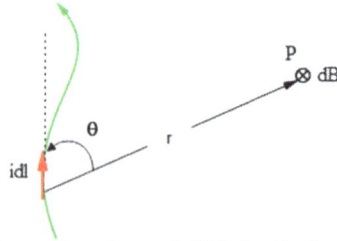

The current element idl produces an inward magnetic field dB at the given position (point P) whose direction can be given by right hand thumb rule or by the vector product in Biot-Savart law given as $d\overrightarrow{B} = \frac{\mu_0}{4\pi r^3} id\overrightarrow{l} \times \overrightarrow{r}$

Example 4 Derive an expression for magnetic induction due to a point charge q moving with a velocity $\overrightarrow{v}$ at a position $\overrightarrow{r}$ from the charge.

Solution

Biot–Savart law for a current element is

$$d\overrightarrow{B} = \frac{\mu_0 i \; \overrightarrow{dl} \times \overrightarrow{r}}{4\pi r^3} \qquad (6.3)$$

If a charged particle q undergoes a displacement $\overrightarrow{\delta l}$ during a time δt, putting $i = \frac{q}{\delta t}$, we have

$$i \; \overrightarrow{dl} \times \overrightarrow{r} = \frac{q \; \overrightarrow{dl}}{\delta t} \times \overrightarrow{r} \qquad (6.4)$$

Putting $\frac{\delta l}{\delta t} = v$ in equation (6.4),

$$i \; \overrightarrow{dl} \times \overrightarrow{r} = q \; \overrightarrow{v} \times \overrightarrow{r} \qquad (6.5)$$

Using equations (6.3) and (6.5)

$$\delta \overrightarrow{B} = \frac{\mu_0 q (\overrightarrow{v} \times \overrightarrow{r})}{4\pi r^3} \; \text{Ans.}$$

N.B: This equation was first derived by English physicist Oliver Heaviside in 1888.

6.10 Application of Biot–Savart law

Using Biot–Savart's law we can find the magnetic field of straight and circular current-carrying wires. First of all, we find the magnetic induction $d\overrightarrow{B}$ due to elementary length dl of the conductor. If $d\overrightarrow{B}$ due to the elements of the given current-carrying conductor varies, then we have to split it into respective components. Finally, integrating the components of $\overrightarrow{dB}$ we will find B_x, B_y etc; finally, vectorially adding these three components, we can find the resultant magnetic field.

Example 5 (Finite/infinite straight current)

Find the magnetic induction at a point P, located at a perpendicular distance R from a straight wire carrying a current i if the ends of the wire make internal angles θ_1 and θ_2 at P.

Solution

Consider a current element $i\,dy$ at a distance x from the origin O. The magnitude of $\overrightarrow{dB}$ due to the current element $i\,dy$ is at the given point P

$$dB = \frac{\mu_0 i\,dy\,\sin\theta}{4\pi r^2}$$

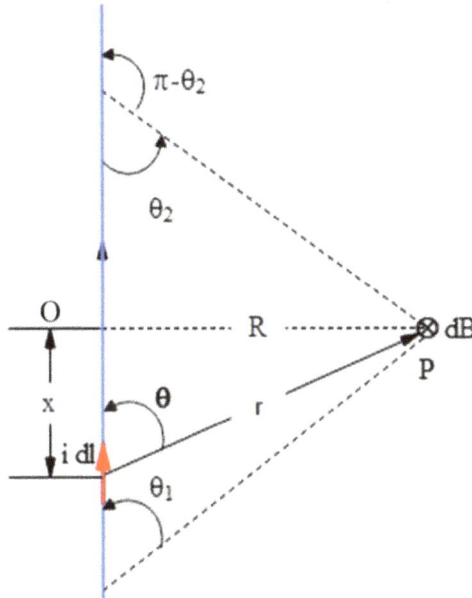

By applying the RHTR, the direction of $\overrightarrow{dB}$ due to each current element of the current-carrying wire/conductor P is inward. So, we need not split it into components. We can find the net field just by directly integrating dB, which is given as

$$B = \int dB$$

$$= \int \frac{\mu_0 i\,dy\,\sin\theta}{4\pi r^2}$$

$$= \frac{\mu_0 i}{4\pi} \int \frac{dy\,\sin\theta}{r^2},$$

where $y = R\cot\theta$, $dy = -R\,\mathrm{cosec}^2\theta\,d\theta$ and $r = R\,\mathrm{cosec}\,\theta$

$$\Rightarrow B = -\frac{\mu_0 i}{4\pi} \int_{\theta_1}^{\theta'_2} \sin\theta \, d\theta, \text{ where } \theta'_2 = \pi - \theta_2$$

$$\Rightarrow B = \frac{\mu_0 i}{4\pi R}(\cos\theta_1 + \cos\theta_2) \text{ Ans.}$$

N.B: For an infinite straight conductor, $\theta_1 = \theta_2 = 0$; so, we have

$$B = \frac{\mu_0 i}{2\pi R}$$

Since $B \propto \frac{1}{R}$, the pattern fades away hyperbolically. The lines of force are concentric circles forming coaxial cylindrical surfaces. The distance between two consecutive lines goes on increasing with increasing radial distance R or r.

Example 6 (Ring)

Derive an expression for $\overrightarrow{B}$ due to a ring of radius R carrying a current i, at an axial distance x from its centre.

Solution

Taking a current element $i \, dl$, its magnetic field at P is

$$dB = \frac{\mu_0 i \, dl \, \sin\theta}{4\pi r^2},$$

where $\theta = $ angle between $\overrightarrow{r}$ and $\overrightarrow{dl} = 90°$

$$\Rightarrow dB = \frac{\mu_0 i \, dl}{4\pi r^2}.$$

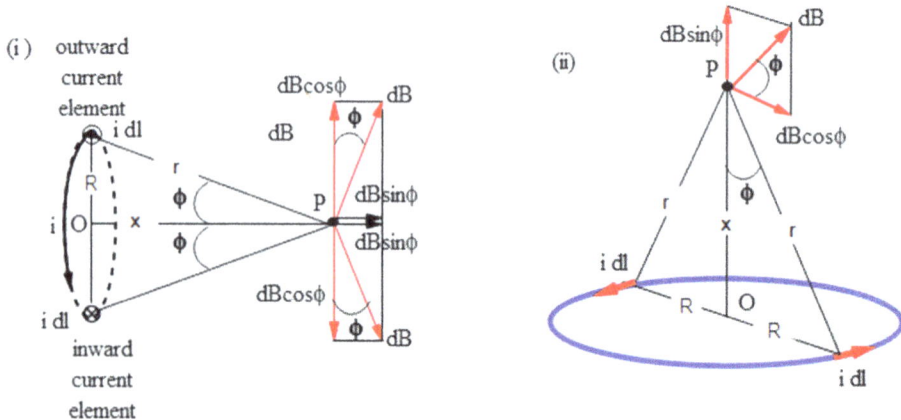

You can see that $d\overrightarrow{B}$ due to each element has the same magnitude but different directions at P. Since $d\overrightarrow{B} \perp \overrightarrow{r}$, it forms a cone; so, the net radial magnetic field will be zero; this is because the radial components of magnetic field dB due to all diametrically opposite pairs of current elements $i\,dl$ will cancel out. Then, the net magnetic field will be axial, which is given as

$$B = \int dB_x = \int dB \sin\phi$$

$$= \int \frac{\mu_0 i\,dl\,\sin\phi}{4\pi r^2}$$

$$= \frac{\mu_0 i \sin\phi}{4\pi r^2} \oint dl$$

$$= \frac{\mu_0 i \sin\phi}{4\pi r^2}(2\pi r)$$

$$= \frac{\mu_0 i R}{2r^2} \sin\phi,$$

where $\sin\phi = \frac{R}{r}$ and $r = \sqrt{R^2 + x^2}$

$$\Rightarrow B = \frac{\mu_0 i R^2}{2(R^2 + x^2)^{3/2}}$$

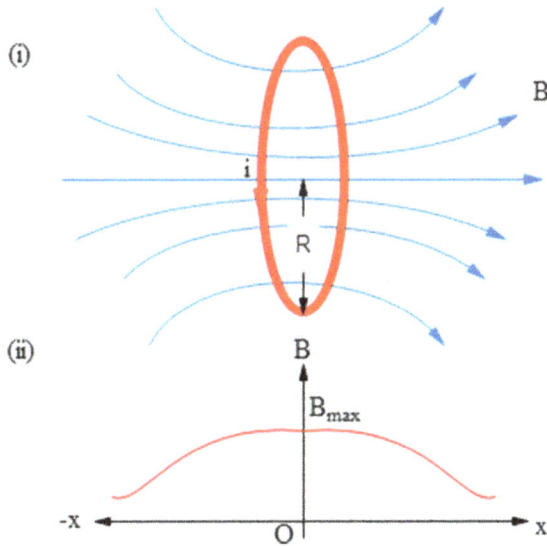

N.B: Since $B \propto \frac{1}{(R^2+x^2)^{3/2}}$ due to a ring, B is maximum at $x = 0$ (centre C) and minimum (zero) when $x \to \infty$. If $x = 0$, the magnetic field at the centre of the circular loop is given as

$$B_C = \frac{\mu_0 i}{2R} = B_{\text{max}}.$$

N.B: If we consider a circular arc that subtends an angle θ at the centre of curvature, its magnetic field at the centre C is given as

$$B = \frac{\mu_0 i \theta}{4\pi},$$

where θ is in radian.

Example 7 (Current-carrying solenoid)

Find $\overrightarrow{B}$ due to a finite solenoid of number of turns per unit length n and current i. The edges of the ends of the solenoid make internal angles θ_1 and θ_2 at the origin O lying on the axis of the solenoid.

Solution

A solenoid is comprised of many closely spaced rings. Each ring produces magnetic field in the same direction at any axial point O, say. Hence, the net field B of the solenoid at O is equal to the scalar sum of the fields due to all loops. Taking a coil of thickness dx at a distance x from O, its field is

$$B_{coil} = \frac{\mu_0 N_{coil} i R^2}{2(R^2 + x^2)^{3/2}} \tag{6.6}$$

Integrating the field of the thin coils, the net field at O due to the solenoid is

$$B_{solenoid} = \int B_{coil} \tag{6.7}$$

Using the last two equations,

$$B_{sol} = \frac{\mu_0 i R^2}{2} \int \frac{N_{coil}}{(R^2 + x^2)^{3/2}}, \tag{6.8}$$

where $N_{coil} = \frac{dN}{dx} \cdot dx = n\, dx$

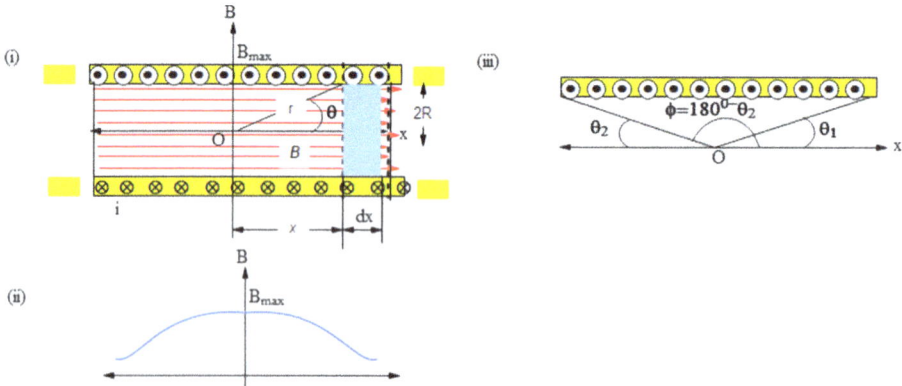

$$\Rightarrow B_{sol}(=B) = \frac{\mu_0 n i R^2}{2} \int \frac{dx}{(R^2 + x^2)^{3/2}} \tag{6.9}$$

Putting $x = R \cot \theta$, $dx = -R \operatorname{cosec}^2 \theta\, d\theta$ in equation (6.9),

$$B = \frac{\mu_0 n i R^2}{2} \int_{\theta_1}^{\theta_2' = 180 - \theta_2} \frac{-R \operatorname{cosec}^2 \theta\, d\theta}{R^3 \operatorname{cosec}^3 \theta}$$

$$\Rightarrow B = \frac{\mu_0 ni}{2}(\cos \theta_1 + \cos \theta_2) \text{ Ans.}$$

N.B: The magnetic field B at the end of a semi-infinite solenoid can be given as

$$B = \frac{\mu_0 ni}{2}$$

For an infinite solenoid, put $\theta_1 = \theta_2 = 0$ to obtain $B = \mu_0 ni$. This means that a uniform magnetic field exists in a long solenoid.

6.11 Magnetic dipole moment and its calculation

We have seen that a current loop produces a pattern of magnetic field similar to that of a bar magnet. So, a current loop behaves as a bar magnet. As you know, a bar magnet possesses a magnetic moment which points from south pole to north pole of the magnet. Likewise, a current-carrying loop possesses a magnetic moment that points from its south pole to north pole. Now it is essential to know the polarity of a current-carrying loop.

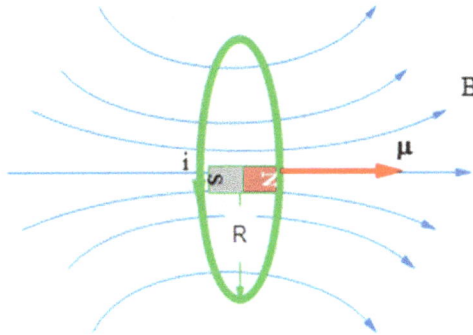

A current loop produces a magnetic field
pattern similar to that of a bar magnet

By using the right-hand thumb rule, we can see that the left side of the loop or coil is the north pole of the equivalent bar magnet as the lines of force emerge from this side. Likewise, you can imagine the right side of the loop as the south pole because the lines of force terminate at the right side of the loop. The polarities of the loop for the given direction of current are shown in the above figure.

6.11.1 Magnetic moment $\overrightarrow{\mu}$ plane loop

As mentioned earlier, the magnetic moment points in the direction of $\overrightarrow{B}$ (from S-pole to N-pole) along the axis of a magnet (coil). Its magnitude is given as

$$\mu = niA,$$

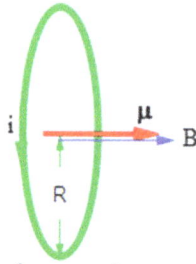

A current loop behaves as bar magnet possessing a magnetic dipole momenent $\mu=NiA$ whose direction is given by right hand thumb rule

where
n = Number of turns of the coil
i = Current flowing in the coil
A = Area bounded by the coil (area of the coil).

Example 8 Find the ratio of magnetic moment of a circular current loop and the magnetic induction at its centre.
Solution
The magnetic induction at the centre of the loop is

$$B = \frac{\mu_0 i}{2R}$$

The magnetic dipole moment of the loop is

$$\mu = niA = (1)(i)\pi R^2$$

Using the last two equations,

$$\frac{\mu}{B} = \frac{i\pi R^2}{\frac{\mu_0 i}{2R}} = \frac{2\pi R^3}{\mu_0} \quad \text{Ans.}$$

Example 9 What is the magnetic moment of current loop of radius R caused by an electron orbiting with an angular speed ω.
Solution
For a current loop the magnetic moment is
$$\mu = niA \tag{6.10}$$
As the electron passes through a given point at a regular time interval of T, the electric current produced by the orbiting electron is
$$i = e/T \tag{6.11}$$

The orbiting electron behaves as a circular current loop of area

$$A = \pi R^2 \qquad (6.12)$$

Using the last three equations, we have

$$\mu = (1)\left(\frac{e}{T}\right)\pi R^2 = e\frac{\omega}{2\pi} \cdot \pi R^2$$

$$\Rightarrow \mu = \frac{e\omega R^2}{2} \text{ Ans.}$$

6.11.2 $\overrightarrow{\mu}$ for a non-coplanar loop

For a non-coplanar current loop, we can divide the loop into several coplanar loops. Then find the dipole moment of each loop and add them vectorially to find the net moment

$$\overrightarrow{\mu} = i\Sigma \overrightarrow{A_i},$$

where $\overrightarrow{A_i}$ = vector of the ith loop.

If the loops are elementary, we can integrate (sum up) the elementary moments $d\overrightarrow{\mu}$ to obtain the net dipole moment. Then, we can write

$$\overrightarrow{\mu} = \int d\overrightarrow{\mu}, \text{ where } d\overrightarrow{\mu} = i\,d\overrightarrow{A};$$

$$\Rightarrow \overrightarrow{\mu} = i\int d\overrightarrow{A},$$

where $d\overrightarrow{A}$ = area vector of the elementary current loop taken in the direction of $d\overrightarrow{B}$ at its axis.

Example 10 Find the magnetic dipole moment of the current loop.

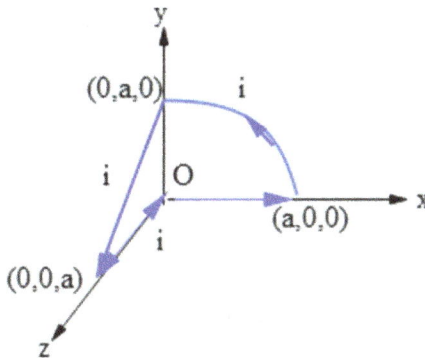

Solution

We split the areas of the given non-coplanar loop into two coplanar loops of area $\overrightarrow{A_1}$ and $\overrightarrow{A_2}$.

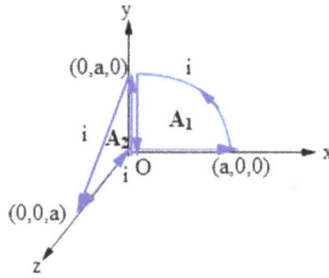

The net moment of the given loop is

$$\vec{\mu} = i\Sigma \vec{A_i},$$

where $\vec{A_1} = \frac{\pi a^2}{4}\hat{k}$ and $\vec{A_2} = \frac{a^2}{2}\hat{i}$

$$\Rightarrow \vec{\mu} = \frac{a^2}{4}i(2\hat{i} + \pi\hat{k}) \text{ Ans.}$$

6.12 Ampère's circuital law

In 1820, just after Oersted's discovery of magnetic effect of electric current, Ampère deduced a law to calculate the magnetic field of an electric current. This is known as Ampère's circuital law. This law relates magnetic fields to the electric currents that produce them. This states that the closed line integral (circulation) of magnetic induction in a closed loop is equal to μ_0 times the current enclosed by the loop. It is given as

$$\oint \vec{B} \cdot d\vec{l} = \mu_0 i$$

This law helps us to find the magnetic induction for some symmetrical current distribution.

Proof

Taking an elementary segment $d\vec{l}$ in an arbitrary loop (contour) enclosing the long straight current conductor, the right side of the expression is

$$\oint \vec{B} \cdot d\vec{l} = \oint B \, dl \cos \phi$$

$$= \oint \left(\frac{\mu_0 i}{2\pi R}\right)(dl \cos \phi)$$

$$= \frac{\mu_0 i}{2\pi} \oint \frac{dl \cos \phi}{R}$$

$$= \frac{\mu_0 i}{2\pi} \oint d\theta$$

$$= \frac{\mu_0 i}{2\pi} \cdot 2\pi$$

$$= \mu_0 i$$

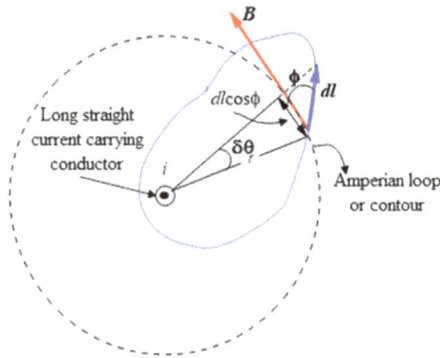

Along a closed Amperian loop the line intergral
of the magnetic field induction is equal to $\mu_0 i$.

If no (zero) current is enclosed by the contour,

$$\oint \vec{B} \cdot d\vec{l} = 0$$

This tells us that the magnetic field is conservative under this condition.

6.13 Application of Ampère's circuital law

Example 11 (Straight long current-carrying conductor)
Find the magnetic field distribution due to a long straight current-carrying conductor of circular cross-section. Assume R = radius of cross- section of the wire and i_0 = current flowing through the wire.
   Solution
   Taking a circular contour centered to a wire of radius $r < R$, to find B inside the conductor using Ampère's circuital law, we have

$$\oint \vec{B} \cdot d\vec{l} = \mu_0 i',$$

where $i' = J \cdot \pi r^2$.

$$\Rightarrow \oint B \, dl \cos 0 = \mu_0 (J \pi r^2)$$

$$\Rightarrow B \oint dl = \mu_0 (J \pi r^2)$$

$$\Rightarrow B (2\pi r) = \mu_0 (J \pi r^2)$$

$$\Rightarrow B = \frac{\mu_0 J}{2}; \ r \leqslant R, \ \text{where } J = \frac{i}{\pi R^2}$$

For outside the wire, $(r > R)$

$$\oint B \cdot dl = \mu_0 i',$$

where $i' = i_0$ because the contour encloses the total current

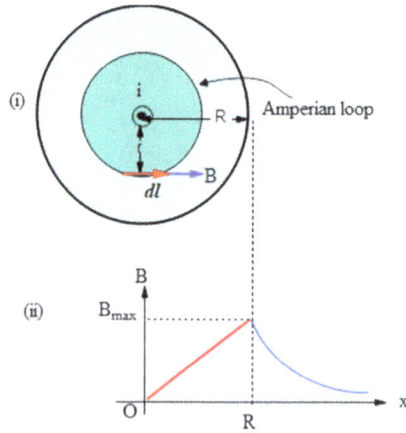

$$\Rightarrow B \oint dl = \mu_0 i_0$$

$$\Rightarrow B \, 2\pi r = \mu_0 i_0$$

$$\Rightarrow B = \frac{\mu_0 i_0}{2\pi r}; \ r \geqslant R$$

$$\Rightarrow B = \frac{\mu_0 i_0 r}{2R^2}; \ r \leqslant R$$

$$= \frac{\mu_0 i_0}{2\pi r}; \ r \geqslant R \text{ Ans.}$$

This means, B varies linearly inside the conductor and hyperbolically outside the conductor. Hence, the density of lines of force is maximum at the periphery of the wire, zero at the center and infinity.

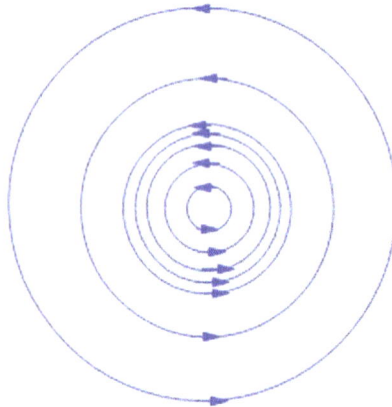

Example 12 (Infinite solenoid)

Find B due to a current-carrying infinite (long) solenoid. Assume $n =$ number of turns per unit length and $i =$ current per turn.

Solution

Take a rectangular contour *abcda* and apply Ampère's circuital law to obtain

$$\oint \vec{B} \cdot d\vec{l} = \mu_0 i$$

$$\Rightarrow \int_a^b \vec{B} \cdot d\vec{l} + \int_b^c \vec{B} \, d\vec{l} + \int_c^d \vec{B} \cdot d\vec{l} + \int_d^a \vec{B} \cdot d\vec{l} = \mu_0 Ni$$

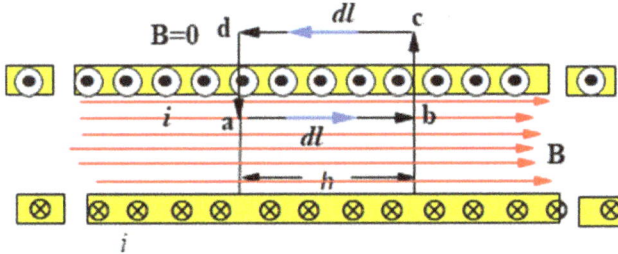

Since $\vec{B} \cdot d\vec{l} = 0$ from b to c and d to a because $\vec{B} \perp d\vec{l}$, and $\vec{B} \cdot d\vec{l} = 0$ from c to d because $B = 0$ outside the solenoid, we have

$$\int_a^b \vec{B} \cdot d\vec{l} = \mu_0 Ni$$

$$\Rightarrow \int_0^h B \, dl = \mu_0(nh)$$

$$\Rightarrow Bh = \mu_0 \, nhi$$

$$\Rightarrow B = \mu_0 ni$$

Magnetic field B is uniform inside a long (infinite) solenoid and does not depend on the radius of cross-section of the solenoid. Ans.

Example 13 (Infinite current sheet)

Find the field distribution due to an infinite (large) current sheet.

Solution

Applying Ampère's circuital law by taking the rectangular contour, we have

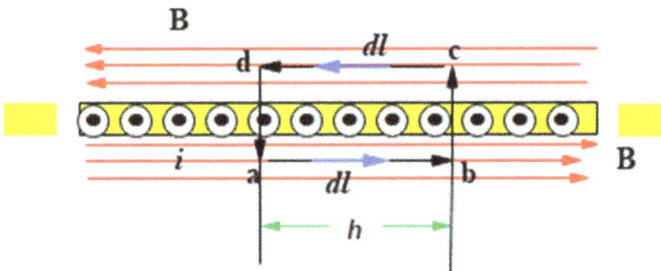

$$\oint \vec{B} \cdot \vec{dl} = \int_a^b \vec{B} \cdot \vec{dl} + \int_b^c \vec{B} \cdot \vec{dl} + \int_c^d \vec{B} \cdot \vec{dl} + \int_d^a \vec{B} \cdot \vec{dl} = \mu_0 i$$

Since $\vec{B} \| \vec{dl}$ along ab and cd, $\vec{B} \cdot \vec{dl} = B \, dl$. Since $\vec{B} \perp \vec{dl}$ for the path bc and da, $\vec{B} \cdot \vec{dl} = 0$.

Then, we have

$$\int_a^b B \, dl + \int_c^d B \, dl = \mu_0 i$$

$$\Rightarrow B \int_0^h dl + B \int_0^h dl = \mu_0 i$$

$$\Rightarrow 2Bh = \mu_0 i,$$

where $i = J \cdot h$

$$\Rightarrow B = \frac{\mu_0 J}{2},$$

where J = linear current density $= \frac{di}{dh}$. So, the B-lines are uniformly distributed on both sides of the plate (sheet). However, the directions of $\vec{B}$ are opposite in both sides of the sheet. Ans.

Problem 1 Find the magnetic induction at a point P on the perpendicular bisector of the current-carrying conductor.

Solution

Here, $\cos \theta_1 = \cos \theta_2 = \frac{l}{\sqrt{2}l} = \frac{1}{\sqrt{2}}$

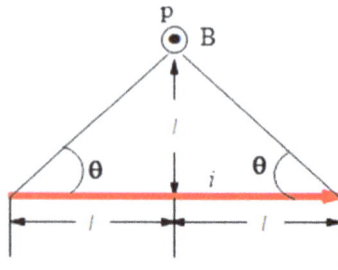

Then, the magnetic field induction at P is

$$B = \frac{\mu_0 i}{4\pi l}(\cos \theta_1 + \cos \theta_2)$$

$$= \frac{\mu_0 i}{4\pi l}\left(\frac{1}{\sqrt{2}} + \frac{1}{\sqrt{2}}\right)$$

$$= \frac{\mu_0 i}{2\sqrt{2}\pi l} \text{ (outward) Ans.}$$

Problem 2 A point positive charge $q = 22 \times 10^{-6}$ C at point $P\,(-2, 3, -4)$ m moves with a velocity $\vec{v} = \hat{i} - 2\hat{j} + \hat{k}$ m s^{-1}. Find the field induction $\vec{B}$ at a point $Q\,(-1, 2, -1)$ m due to the moving charge at the given position.

Solution

With reference to example 4, the magnetic field due to a moving charge is

$$\delta\vec{B} = \frac{\mu_0 q(\vec{v} \times \vec{r})}{4\pi r^3}$$

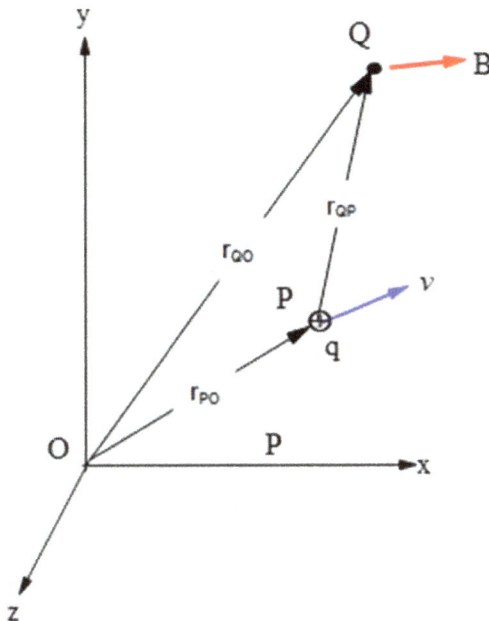

The position of the point Q relative to the current element at P is

$$\vec{r} = \vec{r}_{QP} = \vec{r}_{QO} - \vec{r}_{PO} = (-\hat{i} + 2\hat{j} - \hat{k}) - (-2\hat{i} + 3\hat{j} - 4\hat{k})$$
$$= -\hat{i} + 2\hat{j} - \hat{k} + 2\hat{i} - 3\hat{j} + 4\hat{k} = \hat{i} - \hat{j} + 3\hat{k}; \text{ so, } r = \sqrt{1 + 1 + 9} = \sqrt{11} \text{ m}$$

Putting the values of $\vec{r}$, $\vec{v}$ and r, in the above expression, we have

$$\delta\vec{B} = \frac{\mu_0 q(\vec{v} \times \vec{r})}{4\pi r^3} = \frac{4\pi \times 22 \times 10^{-13}}{4\pi(\sqrt{11})^3}(\hat{i} - 2\hat{j} + \hat{k}) \times (\hat{i} - \hat{j} + 3\hat{k})$$

$$= \frac{22 \times 10^{-13}}{11\sqrt{11}}(\hat{i} - 2\hat{j} + \hat{k}) \times (\hat{i} - \hat{j} + 3\hat{k})$$

$$= -0.603 \times 10^{-13}(5\hat{i} + 2\hat{j} - \hat{k}) \text{ T Ans.}$$

Problem 3 A large straight current-carrying conductor is bent in the form of an L shape. Find magnetic field induction $\vec{B}$ at point P referring to figure (i).

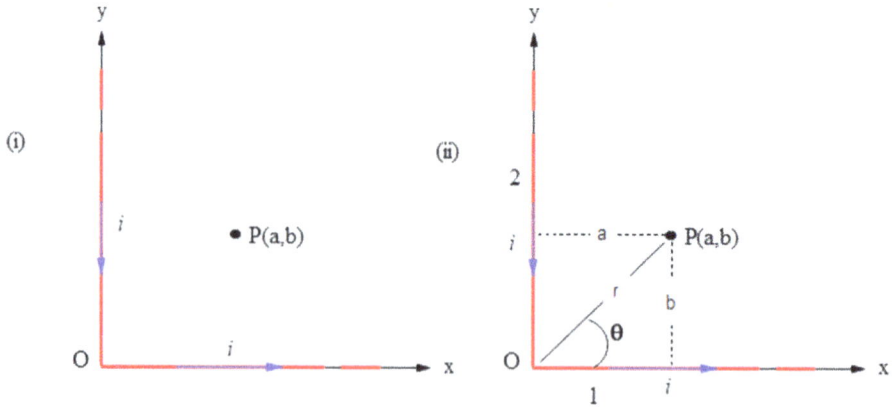

Solution

Referring to the above figure (ii), divide the conductor into two semi-infinite segments 1 and 2. Then, the magnetic induction at point P is

$$\vec{B} = \vec{B_1} + \vec{B_2} \tag{6.13}$$

The magnetic induction due to segment 1 is

$$\vec{B_1} = \frac{\mu_0 i}{4\pi b}(\cos\theta + \cos 0)\hat{k}$$

$$\vec{B_1} = \frac{\mu_0 i}{4\pi b}(\cos\theta + 1)\hat{k} \tag{6.14}$$

The magnetic induction due to segment 2 is

$$\vec{B_2} = \frac{\mu_0 i}{4\pi a}(\cos\theta'_1 + \cos 0)\hat{k}$$

$$= \frac{\mu_0 i}{4\pi a}\{\cos(90 - \theta) + \cos 0\}\hat{k}$$

$$\vec{B_2} = \frac{\mu_0 i}{4\pi a}(\sin\theta + 1)\hat{k} \tag{6.15}$$

Using the last three equations,

$$\vec{B} = \frac{\mu_0 i}{4\pi}\left(\frac{1}{b}\cos\theta + \frac{1}{a}\sin\theta + \frac{1}{a} + \frac{1}{b}\right)\hat{k},$$

where

$$\cos\theta = \frac{a}{\sqrt{a^2 + b^2}} \text{ and } \sin\theta = \frac{b}{\sqrt{a^2 + b^2}}$$

$$\Rightarrow \vec{B} = \frac{\mu_0 i}{4\pi}\left(\frac{1}{b}\frac{a}{\sqrt{a^2 + b^2}} + \frac{1}{a}\frac{b}{\sqrt{a^2 + b^2}} + \frac{1}{a} + \frac{1}{b}\right)\hat{k}$$

$$\Rightarrow \vec{B} = \frac{\mu_0 i}{4\pi}\left(\frac{a + b + \sqrt{a^2 + b^2}}{ab}\right)\hat{k} \text{ Ans.}$$

Problem 4 Find $\vec{B}$ at the origin due to the long wire carrying a current i.

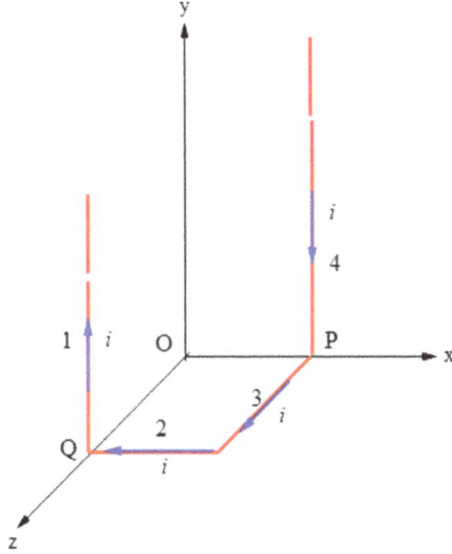

Solution

We divide the wire into four parts 1, 2, 3 and 4. The net magnetic induction at O is

$$\vec{B} = \vec{B_1} + \vec{B_2} + \vec{B_3} + \vec{B_4} \tag{6.16}$$

The magnetic induction due to 1 at O is

$$\vec{B_1} = -\frac{\mu_0 i}{4\pi a}(\cos 0 + \cos 90°)\hat{i} = -\frac{\mu_0 i}{4\pi a}\hat{i} \tag{6.17}$$

The magnetic induction due to 2 and 3 at O can be given as

$$\vec{B_2} = \vec{B_3} = -\frac{\mu_0 i}{4\pi a}(\cos 90° + \cos 45°)\hat{j} = -\frac{\mu_0 i}{4\sqrt{2}\,\pi a}\hat{j} \tag{6.18}$$

The magnetic induction due to 1 at O is

$$\vec{B_4} = -\frac{\mu_0 i}{4\pi a}(\cos 90° + \cos 0°)\hat{k} = -\frac{\mu_0 i}{4\pi a}\hat{k} \tag{6.19}$$

Using the above four equations,

$$\vec{B} = -\frac{\mu_0 i}{4\pi a}(\hat{i} + \sqrt{2}\hat{j} + \hat{k}) \text{ Ans.}$$

Problem 5

(a) Find the magnetic induction $\vec{B}$ at P due to two parallel long straight conductors each carrying an outward current i.

(b) Draw the graph of $\vec{B}$ versus y.

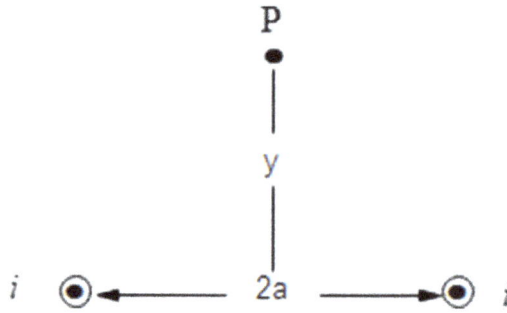

Solution

(a) The magnetic fields due to conductors 1 and 2 are $\vec{B_1}$ and $\vec{B_2}$, respectively, are perpendicular to their respective position vectors $\vec{r_1}$ and $\vec{r_2}$. In the given case, $r_1 = r_2 = r$ (say); then, we have

$$B_1 = B_2 = \frac{\mu_0 i}{2\pi r} \tag{6.20}$$

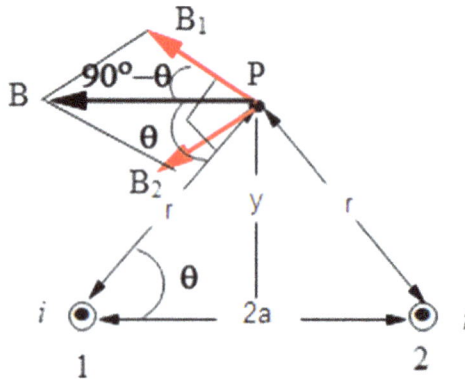

The net induction at P is

$$\vec{B} = \vec{B_1} + \vec{B_2} \tag{6.21}$$

Since, the vertical (y) component of $\vec{B_1}$ and $\vec{B_2}$ cancel each other, using the last two equations, the net field is given as

$$\vec{B} = 2B_1 \sin \theta (-\hat{i})$$
$$= 2\left(\frac{\mu_0 i}{2\pi r}\right)\left(\frac{y}{r}\right)(-\hat{i})$$

$$\Rightarrow \vec{B} = -\frac{\mu_0 i y}{\pi(a^2 + y^2)}\hat{i} \text{ Ans.}$$

(b) If we take the derivative of B with respect to y and set it to zero, we have

$$\Rightarrow \frac{dB}{dy} = \frac{d}{dy}\left\{\frac{\mu_0 i y}{\pi(a^2 + y^2)}\right\} = \frac{\mu_0 i}{\pi}\frac{d}{dy}\left\{\frac{y}{(a^2 + y^2)}\right\} = 0$$

$$\Rightarrow \frac{d}{dy}\left\{\frac{y}{(a^2 + y^2)}\right\} = 0$$

$$\Rightarrow \frac{(a^2 + y^2) - 2y^2}{(a^2 + y^2)^2} = 0$$

$$\Rightarrow y^2 - a^2 = 0 \Rightarrow y = \pm a$$

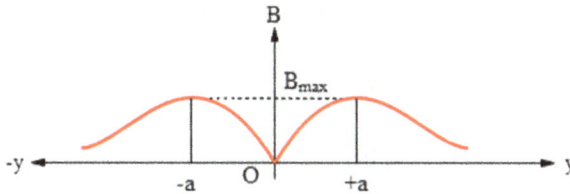

This means that the magnitude of $B(y)$ will be maximum at $y = +a$ and $y = -a$. At $y = 0$ and y tends to infinity, the magnetic induction will be zero. So, the curve is traced as in the figure below. Ans.

Problem 6 A charged particle of mass m and charge q revolves in a circular path. Find the ratio of its magnetic dipole moment and angular momentum about the centre of the circular path.

Solution

Let the angular frequency and radius of circular path be ω and R, respectively. The magnetic dipole moment

$$\mu = niA$$
$$= (1)\left(\frac{q}{T}\right)(\pi R^2) = \left(\frac{q\omega}{2\pi}\right)\pi R^2$$

$$\Rightarrow \mu = \frac{q\omega R^2}{2} \tag{6.22}$$

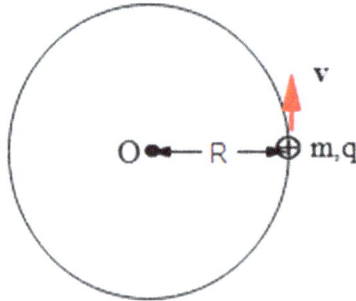

The angular momentum about C is

$$L = (mR^2)\omega \tag{6.23}$$

Using the last two equations, we have

$$\frac{\mu}{L} = \frac{q}{2m} \text{ Ans.}$$

N.B: The ratio of magnetic moment and angular momentum is called gyro-magnetic ratio, which is a constant for uniform charge distribution.

Problem 7 Find the magnetic dipole moment of a spinning electron.
   Solution
   Assuming the electron as a sphere of uniform volume charge density, for the spinning electron we can use

$$\frac{\mu}{L} = \frac{e}{2m} \tag{6.24}$$

The angular momentum of the spinning sphere is

$$L = I\omega = \left(\frac{2}{5}mR^2\right)\omega \tag{6.25}$$

Using the last two equations, we have

$$\mu = \frac{eR^2\omega}{5} \text{ Ans.}$$

N.B: You have to follow the basic procedure to calculate the magnetic moment if the charge distribution of a sphere is not uniform.

Problem 8 In the figure below, (i) is comprised of a circle and two current-carrying conductors, one semi-infinite and one finite. The current i flowing down in branch 1 is divided into two parts in branches 2 and 3 and again gets united in branch 4, as shown in figure (ii). Find the magnetic induction at O.

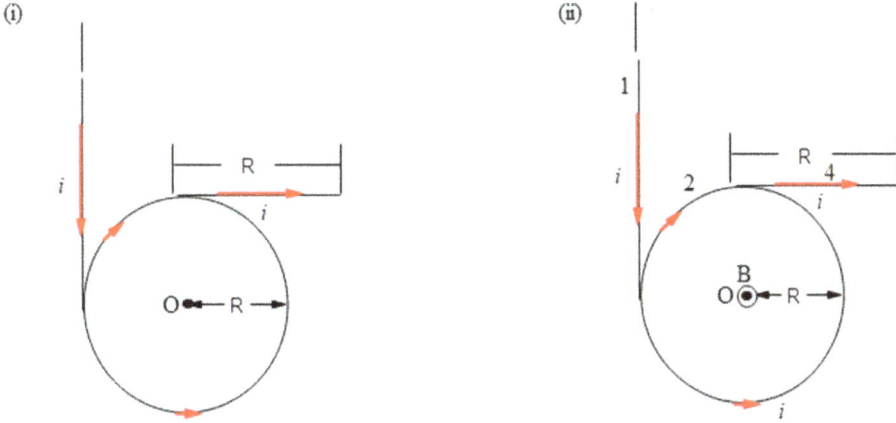

Solution

The magnetic induction at O is

$$\vec{B} = \vec{B_1} + \vec{B_2} + \vec{B_3} + \vec{B_4} \tag{6.26}$$

The magnetic induction due to conductor 1 at O is

$$\vec{B_1} = \frac{\mu_0 i}{4\pi R}\hat{k} \tag{6.27}$$

$$\vec{B_2} = \frac{\mu_0 i_1 \theta_1}{4\pi R}(-\hat{k}) \tag{6.28}$$

$$\vec{B_3} = \frac{\mu_0 i_2 \theta_2}{4\pi R}\hat{k} \tag{6.29}$$

The magnetic induction due to conductor 4 at O is

$$\vec{B_4} = -\frac{\mu_0 i(\cos 90° + \cos 45°)}{4\pi R}\hat{k} = -\frac{\mu_0 i}{4\sqrt{2}\,\pi R}\hat{k} \tag{6.30}$$

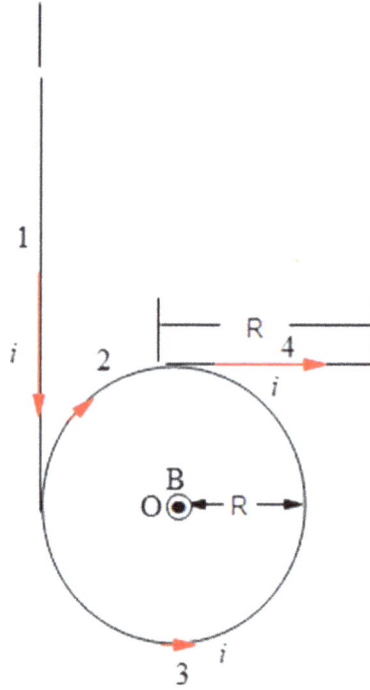

Using the above five equations, we get

$$\vec{B} = \frac{\mu_0 \hat{k}}{4\pi R}\left(i + i_2\theta_2 - i_1\theta_1 - \frac{i}{\sqrt{2}}\right)\hat{k} \tag{6.31}$$

Since branches 2 and 3 are in parallel,

$$i_1 R_1 = i_2 R_2$$

Since, $R_1 = \rho\frac{l_1}{A} = \frac{\rho}{A}(R\theta_1)$ and $R_2 = \frac{\rho}{A}(R\theta_2)$,

$$i_1\theta_1 = i_2\theta_2 \tag{6.32}$$

Using equations (6.31) and (6.32), we get

$$\vec{B} = \frac{\mu_0 i}{4\pi R}\left(1 - \frac{1}{\sqrt{2}}\right)\hat{k} \text{ Ans.}$$

Problem 9 Find the (a) dipole moment of the spiral of inner and outer radii a and b, respectively, and (b) magnetic induction at O. Assume $N =$ total number of turns, $i =$ current flowing in the spiral.

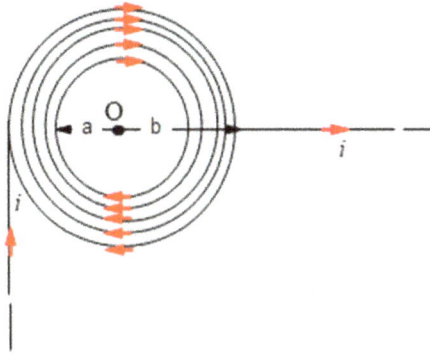

Solution

(a) Let us take a thin coil of thickness dr in which the number of turns of the coil is

$$dN = \frac{N}{b-a} \cdot dr$$

The dipole moment of the thin coil is

$$d\mu = (dN)(i)(A)$$
$$= \left(\frac{N\,dr}{b-a}\right)(i)(\pi r^2)$$

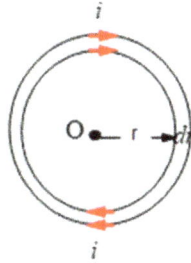

Then, the magnetic moment of the spiral is

$$\mu = \int d\mu$$
$$= \frac{\pi N i}{b-a} \int_a^b r^2 dr$$
$$= \frac{\pi i N}{3}(a^2 + ab + b^2) \text{ Ans.}$$

(b) Let us take a thin coil of thickness dr. Then the number of turns of the coil is

$$dN = \frac{N}{b - a} \cdot dr$$

The magnetic field at the centre of the coil is

$$dB = \frac{\mu_0(dN)(i)}{2r}$$

Using the last two expressions, we have

$$dB = \frac{\mu_0\left(\frac{N}{b-a}\right) \cdot dr(i)}{2r} = \frac{\mu_0 i}{2}\left(\frac{N}{b - a}\right)\frac{dr}{r}$$

Then, the net magnetic field due to the spiral at the centre of the coil is

$$B = \int_0^B dB = \frac{\mu_0 i}{2}\left(\frac{N}{b - a}\right)\int_a^b \frac{dr}{r}$$

$$\Rightarrow B = \frac{\mu_0 Ni}{2(b - a)} \ln \frac{b}{a} \text{ Ans.}$$

Problem 10 Two large parallel current sheets having linear current densities J (outward) and $-2\,J$ (inward) are arranged as shown in the figure below. (a) Find the fields in regions a, b and c. (b) Draw the field pattern.

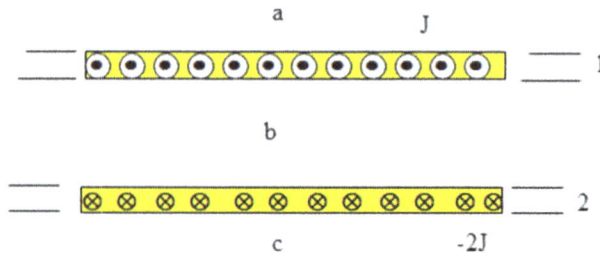

Solution

(a) Let the net field due to sheets 1 and 2 be $\vec{B_1}$ and $\vec{B_2}$, respectively. The field in region a is

$$\vec{B_a} = \vec{B_1} + \vec{B_2} = -B_1\hat{i} + B_2\hat{i} = -\frac{\mu_0 J}{2}\hat{i} + \frac{\mu_0(2J)}{2}\hat{i} = \frac{\mu_0 J}{2}\hat{i} \text{ Ans.}$$

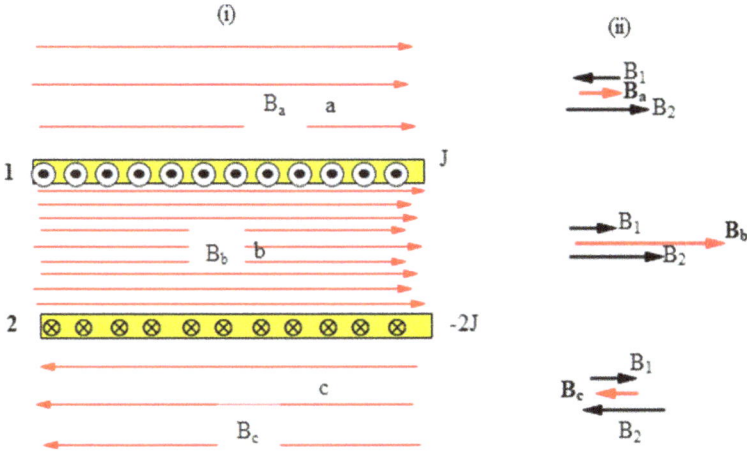

The field in region b is

$$\vec{B_b} = \vec{B_1} + \vec{B_2} = \frac{\mu_0 J}{2}\hat{i} + \frac{\mu_0(2J)}{2}\hat{i} = \frac{3\mu_0 J}{2}\hat{i} \ \text{Ans.}$$

The field in region c is

$$\vec{B_a} = \vec{B_1} + \vec{B_2}$$
$$= \frac{\mu_0 J}{2}\hat{i} - \frac{\mu_0(2J)}{2}\hat{i} = -\frac{\mu_0 J}{2}\hat{i} \ \text{Ans.}$$

(b) Please note that B_1 and B_2 oppose in regions a and c and favour each other in region b. We can see that the magnetic induction in the region a and c is one third of that in region b. So the number of B-lines in region b must be three times that in regions 2 and 3. If you chose nine B-lines in region b, the number of B-lines will be three in regions 1 and 2. Furthermore, the direction of B-lines in regions a, b and c will be right, right and left, respectively. Ans.

Problem 11 Two circular current-carrying loops, as shown in the figure below, procure a net zero magnetic field at the centre. Find the net magnetic field at the centre of the loops if the smaller loop is rotated about the vertical axis that passes through the center, by 90°. The currents and radii are $i_1 = 10$ A, $i_2 = 20$ A, $R_1 = 2$ m and $R_2 = 1$ m.

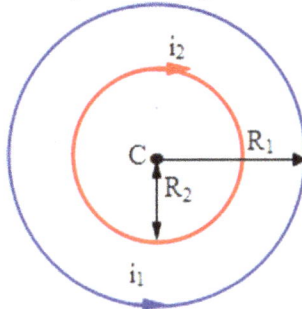

Solution

The net field at C is

$$\overrightarrow{B} = \overrightarrow{B_1} + \overrightarrow{B_2} \tag{6.33}$$

The magnetic fields of the loops are

$$\overrightarrow{B_1} = \frac{\mu_0 i_1}{2R_1}\hat{k} \tag{6.34}$$

$$\overrightarrow{B_2} = -\frac{\mu_0 i_2}{2R_2}\hat{k} \tag{6.35}$$

Then, the net field is

$$\overrightarrow{B} = \frac{\mu_0}{2}\left(\frac{i_1}{R_1} - \frac{i_2}{R_2}\right)\hat{k} = 0$$

$$\Rightarrow i_2 = \frac{i_1 R_2}{R_1} = \frac{(10)(1)}{2} = 5 \text{ A}$$

If the smaller loop is rotated by an angle $90°$, the magnetic fields will be perpendicular to each other at the centre. Then, the net field at C is

$$\overrightarrow{B} = \overrightarrow{B_1} + \overrightarrow{B_2} \tag{6.36}$$

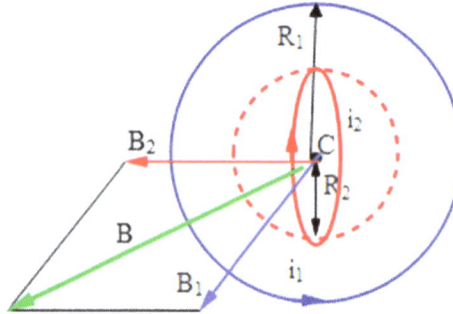

The magnetic fields of the loops are

$$\overrightarrow{B_1} = \frac{\mu_0 i_1}{2R_1}\hat{k} \tag{6.37}$$

$$\overrightarrow{B_2} = \pm\frac{\mu_0 i_2}{2R_2}\hat{i} \tag{6.38}$$

6-40

Then, the magnetic field is

$$\vec{B} = \frac{\mu_0}{2}\left(\frac{i_1}{R_1}\hat{k} \pm \frac{i_2}{R_2}\hat{i}\right)$$

$$= \frac{4\pi \times 10^{-7}}{2}\left(\frac{5}{1}\hat{i} + \frac{10}{2}\hat{k}\right)$$

$$= \pi \times 10^{-6}(\hat{i} + \hat{k})\text{ T Ans.}$$

Problem 12 A very long current-carrying wire passes along the x-and z-axes after twisting to form a circular loop while it passes through the origin O. Find $\vec{B}$ at C. Assume that the circular loop lies in the yz-plane.

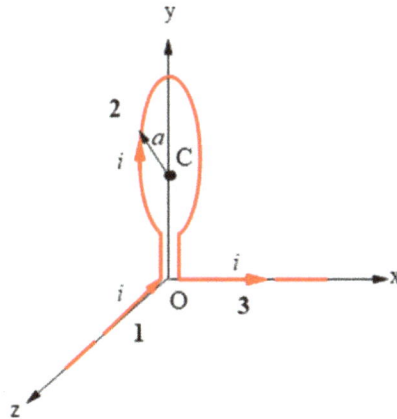

Solution

Let us divide the given conductor into three parts, namely 1, 2 and 3. The net magnetic field at C is the sum of two semi-infinite conductors 1 and 3, and the circular loop 2. So, we can write

$$\vec{B_C} = \vec{B_1} + \vec{B_2} + \vec{B_3} \tag{6.39}$$

The field due to conductor 1 at C is

$$B_1 = \frac{\mu_0 i}{4\pi a}\hat{i} \tag{6.40}$$

The field due to conductor 2 at C is

$$B_2 = -\frac{\mu_0 i}{2a}\hat{i} \tag{6.41}$$

The field due to conductor 3 at C is

$$\overrightarrow{B_3} = \frac{\mu_0 i}{4\pi a}\hat{k} \qquad (6.42)$$

Using the above equations,

$$B_C = \frac{\mu_0 i}{4\pi a}\{-(2\pi - 1)\hat{i} + \hat{k}\} \text{ Ans.}$$

Problem 13 (Hollow cylinder)

Find the variation of $\overrightarrow{B}$ as the function of radial distance r due to a hollow cylinder of inner radius a and outer radius b carrying a current i_0. Also draw the B–r graph.

Solution

Taking a circular Amperian loop of radius r ($>a$) and applying Ampère's circuital law,

$$\oint \overrightarrow{B} \cdot \overrightarrow{dl} = \mu_0 i$$

$$\Rightarrow B \int 2\pi r = \mu_0 i,$$

where

$$i = \frac{i_0}{\pi(b^2 - a^2)} \cdot \pi(r^2 - a^2)$$

$$= \frac{i_0(r^2 - a^2)}{b^2 - a^2} \text{ (due to a uniform distribution of current)}$$

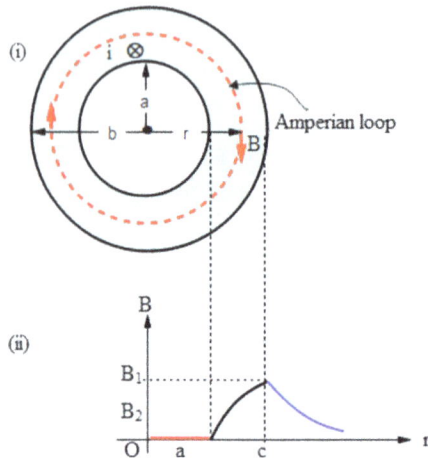

Then, the magnetic field in the tube is given as follows:

$$B = \frac{\mu_0 i_0 (r^2 - a^2)}{2\pi(b^2 - a^2)r} \text{ for } a \leqslant r \leqslant b$$

$B = 0$ for $r \leqslant a$ (as the current enclosed by the Amperian loop $i = 0$)

$$B = \frac{\mu_0 i_0}{2\pi r} \text{ (because } i = i_0) \text{ Ans.}$$

Problem 14 (Coaxial cable)

A long coaxial cable has a core of radius a and sheath of inner radius b and outer radius c. Find the variation of $\overrightarrow{B}$ as the function of radial distance r due to a coaxial cable while it carries a current i_0. Plot the B–r graph.

Solution

Taking a circular Amperian loop (contour) of radius r ($>a$) and applying Ampère's circuital law,

$$\int \overrightarrow{B} \cdot d\overrightarrow{l} = \mu_0 i$$

$$\Rightarrow B \int 2\pi r = \mu_0 i,$$

where

$$i = i_0 - \frac{i_0}{\pi(c^2 - b^2)} \cdot \pi(r^2 - b^2) = \frac{i_0(c^2 - r^2)}{(c^2 - b^2)}$$

$$= \frac{i_0(c^2 - r^2)}{c^2 - b^2}$$

So, we have

$$B = \frac{\mu_0 i(c^2 - r^2)}{2\pi r(c^2 - b^2)} \text{ for } b \leqslant r \leqslant c.$$

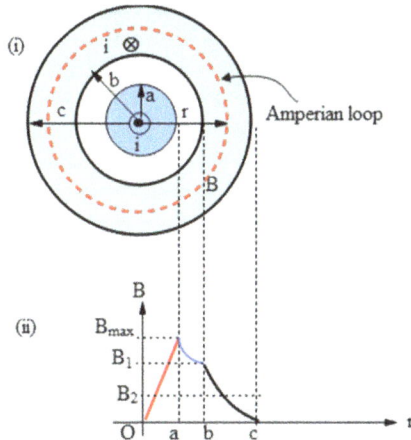

As derived earlier, the magnetic field inside the core is

$$B = \frac{\mu_0 i}{2\pi a^2} r \text{ for } r \leqslant a,$$

In between the core and sheath, the magnetic field inside the core is

$$B = \frac{\mu_0 i}{2\pi r} \text{ for } a \leqslant r \leqslant b,$$

Since the current enclosed by the Amperian loop outside the coaxial cable is zero, we can write

$$B = 0 \text{ for } r \geqslant c.$$

The variation of magnetic field with radial distances is shown in figure (ii). Ans.

References

[1] Thompson S P 1898 *Michael Faraday: his life and work* (Cassell)
[2] Magnusson C E 1920 The centennial of the discoveries of Oersted, Arago, and Ampère *J. Am. Inst. Electr. Eng.* **39** 1031–3
[3] Lambert K 2011 The uses of analogy: James Clerk Maxwell's 'On Faraday's lines of force' and early Victorian analogical argument *Br. J. Hist. Sci.* **44** 61–88

IOP Publishing

Problems and Solutions in Electricity and Magnetism

Pradeep Kumar Sharma

Chapter 7

Magnetic forces, torques and energy

7.1 Introduction

In 1820, Danish physicist Hans Christian Oersted discovered that an electric current produced deflection in a magnetic needle. This discovery established a link between electricity and magnetism. French mathematician-physicist André-Marie Ampère was independently trying to establish a connection between electricity and magnetism just a week after hearing of Ørsted's discovery. Ampère conducted rigorous experiments with the current carrying conductors, and founded a new force law that governed the interaction of electric currents. This is known as Ampère's force law which will be discussed in in this chapter. Later Ampère deduced another law to find the magnetic field due a current carrying conductor which is called Ampère's circuital law. We have explained it in the previous chapter. Thus, within a couple of months after Oesterd's discovery, Ampère presented his laws on 18th September 1820, he presented his findings to the French academy of science. Ampère argued that since a magnetic needle is a permanent magnet which can only be deflected by another magnetic field, a current-carrying conductor must produce a magnetic field to influence the magnetic needle. So, a current-carrying conductor should behave as a magnet. If this is true, two current-carrying elements must behave as two magnets. Hence, they should either attract or repel each other. This was the thought experiment conducted by Ampère. Then he conducted a series of experiments to establish these laws. It is interesting to learn that the discovery of Oersted and Ampère, in 1821 inspired British scientist Michael Faraday to invent the homopolar motor. We will discuss it in the next chapter.

7.2 Ampère's force

As a current element $i\,\overrightarrow{dl}$ behaves as a magnet, it must experience a force $d\,\overrightarrow{F}$ called 'Ampère's force' when placed in a magnetic field $\overrightarrow{B}$ which can be given as

$$d\,\overrightarrow{F} = i\,\overrightarrow{dl} \times \overrightarrow{B}$$

doi:10.1088/978-0-7503-6477-5ch7

This law is experimentally verified, which tells us that the magnetic force acting on the current element is directly proportional to the current element idl, magnetic field induction B and the sin θ, where θ is the angle between $\vec{B}$ and the current element $i\,\vec{dl}$. Hence, Ampère's force is perpendicular to the plane containing current element and magnetic field vector, as shown in the figure below.

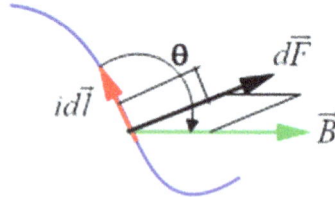

The Ampere's force on a current element is given as $d\vec{F} = i\vec{dl} \times \vec{B}$

Example 1 Find the force of interaction between two parallel currents.
Solution

The Ampère's force acting on the current element $i_2\,\vec{dl_2}$ due to the magnetic field B_1 of the current i_1 is

$$d\,\vec{F_2} = i_2\,\vec{dl_2} \times \vec{B_1} = i_2\,dl_2\hat{j} \times \frac{\mu_0 i_1}{2\pi r}(-\hat{k})$$

$$= -\frac{\mu_0 i_1 i_2\,dl_2}{2\pi r}\hat{i}$$

$$\Rightarrow \frac{\vec{dF_2}}{dl_2} = -\frac{\mu_0 i_1 i_2}{2\pi r}\hat{i}$$

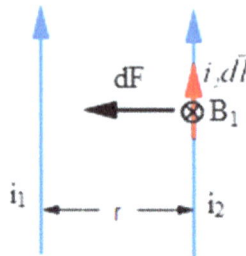

In the above discussion we can see that parallel currents attract each other, and reversing the direction of either current, we can show that anti-parallel currents repel. Ans.

7.3 Force acting on any arbitary current-carrying conductor in a uniform magnetic field

The Ampère's force on the current element $i\,\overrightarrow{dl}$ of the finite current-carrying conductor is

$$\overrightarrow{dF} = i\,\overrightarrow{dl} \times \overrightarrow{B}$$

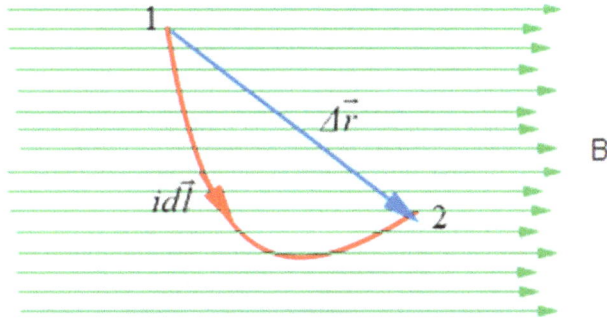

In a uniform magnetic field the Ampere's force is
independent on the shape of the conductor.

The direction of $\overrightarrow{dF}$ may not be same on each element in an arbitrary shaped current-carrying conductor. However, the net force acting on the conductor can be given as,

$$\overrightarrow{F} = \int \overrightarrow{dF} = \int i\,\overrightarrow{dl} \times \overrightarrow{B}$$

$$= -i\left(\int_{r_1}^{r_2} d\,\overrightarrow{l}\right) \times \overrightarrow{B} = i(\overrightarrow{r_2} - \overrightarrow{r_1}) \times \overrightarrow{B}$$

$$= i\,\overrightarrow{r_{21}} \times \overrightarrow{B}$$

This tells us that the net Ampère's force on the conductor does not depend on its total length and its shape in a uniform magnetic field. It depends upon the relative position vector of the end points of the conductor.

7.4 Force acting on a current loop in a magnetic field

The Ampère's force acting on the current element is

$$\overrightarrow{dF} = i\,\overrightarrow{dl} \times \overrightarrow{B}$$

Then, the total (net) force acting on the current loop is

$$\overrightarrow{F} = \int \overrightarrow{dF}$$

$$\Rightarrow \overrightarrow{F} = i \oint \overrightarrow{dl} \times \overrightarrow{B}.$$

7.4.1 Case I: $\vec{B}$ is uniform

If $\vec{B}$ is uniform, pull out of the integral to obtain

$$\vec{F} = i\left(\oint d\vec{l}\right) \times \vec{B} = 0 \left(\because \oint d\vec{l} = 0 \text{ for a closed loop}\right)$$

So, the Ampère's force acting on a closed loop placed in a uniform magnetic field is always zero. It does not depend upon the shape and size of the loop.

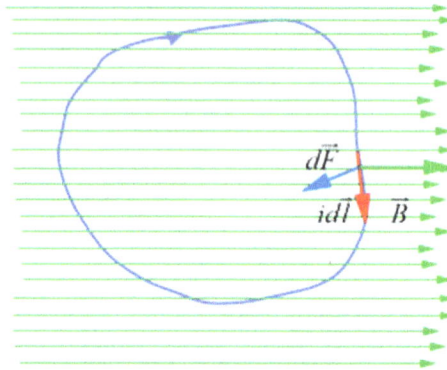

In a uniform magnetic field the Ampere's
force acting on a lcurrent oop is zero

Example 3 An arbitrary shaped wire in figure (i) and a semi-circular wire in figure (ii) carrying a current i are placed in a uniform magnetic field $\vec{B}$. Find the Ampère's force acting the wire. Put $PQ = 2R$ in figure (i).

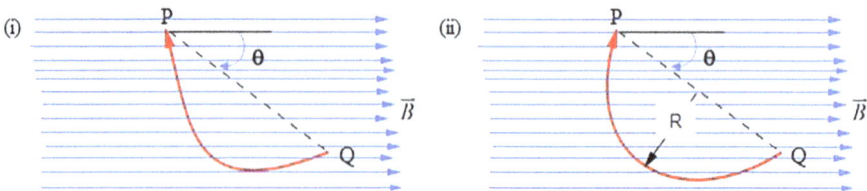

Solution

Mentally make it a closed loop by mentally joining the ends P and Q. Let the forces acting on the curved and straight portion of the loop be $\vec{F_1}$ and $\vec{F_2}$, respectively. Since the net Ampère's force acting on the loop is $\vec{F}_{\text{net}} = \vec{F_1} + \vec{F_2} = 0$, we have $\vec{F_1} = -\vec{F_2}$, where $\vec{F_2} = i(2R)B(-\hat{k}) = -2iRB\hat{k}$ by Ampère's law.

Then, we have

$$\vec{F_1} = 2iRB\,\hat{k} \text{ Ans.}$$

N.B: Both cases have the same Ampère's force because it does not depend on the shape of the current-carrying wire in a uniform magnetic field.

7.4.2 Case II: $\bar{B}$ is non-uniform

An element $\vec{dl}$ of the loop experiences an Ampère's force

$$d\vec{F} = i\,\vec{dl} \times \vec{B}$$

The net Ampère's force acting on the loop can be given as

$$\vec{F} = \int d\vec{F} = i \oint \vec{dl} \times \vec{B}$$

For a non-uniform magnetic field, $\vec{B}$ is not the same for all points of the conductor; so, we cannot pull $\vec{B}$ out of the integral. When a current loop is kept in a non-uniform magnetic field, a net non-zero magnetic force can act on the loop.

7.4.3 Case III

If the loop is very small and the magnetic field is not-uniform, by using the expression

$$\vec{F} = i \oint d\,\vec{l} \times \vec{B}$$

followed by a rigorous calculation, we have

$$\vec{F} = \mu \frac{\partial \vec{B}}{\partial n},$$

where $\frac{\partial \vec{B}}{\partial n}$ is the directional derivative of $\vec{B}$ in the direction of axis (positive normal) of the small current loop (dipole) and $\vec{\mu} =$ magnetic moment of the dipole.

In a Cartesian co-ordinate system,

$$\vec{F} = \left(\frac{\partial \vec{B}}{\partial x} \cdot \vec{\mu} \right) \hat{i} + \left(\frac{\partial \vec{B}}{\partial y} \cdot \vec{\mu} \right) \hat{j} + \left(\frac{\partial \vec{B}}{\partial z} \cdot \vec{\mu} \right) \hat{k}$$

You can prove the force equation by using the expression

$$\vec{F} = -\frac{\partial U}{\partial x}\hat{i} - \frac{\partial U}{\partial y}\hat{j} - \frac{\partial U}{\partial z}\hat{k},$$

where

$$U = -\vec{\mu}.\vec{B} - (= -\mu_x B_x - \mu_y B_y - \mu_z B_z)$$

which will be derived in a later section.

7.4.4 Fleming's left-hand rule

This is used to find the direction of Ampère's force. To understand it, extend the thumb, forefinger and middle finger of your left hand so that they remain perpendicular to each other. If the forefinger and middle finger represent the direction of the magnetic field and current, respectively, the extended thumb will represent the direction of Ampère's force.

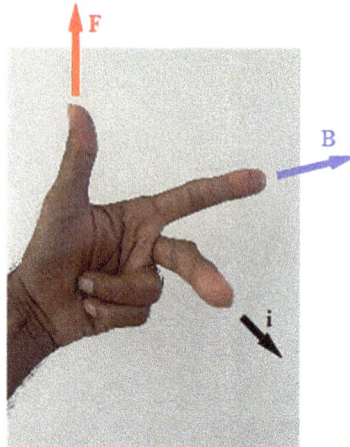

Fleming's Left Hand Rule (FLHR): Thumb represnts
Force (F),fore finger shows magnetic induction (B)
and midddle finger represents current (i).

Example 4 A square loop carrying a current i is placed in a magnetic field of a long straight conductor carrying a current i_0 as shown in the figure (i) below. Find the force acting on the wire loop. Put $x = 1 = r$.

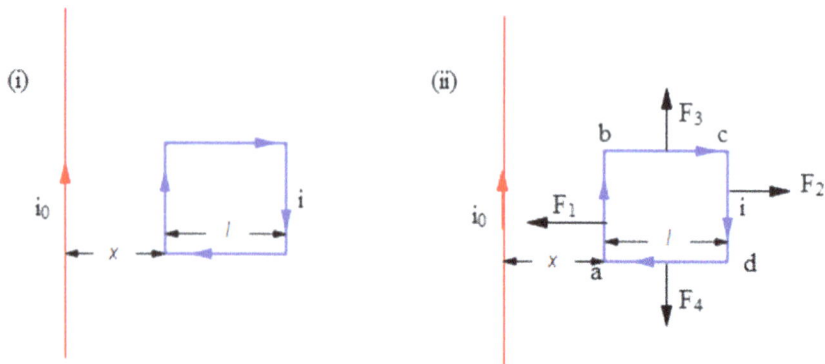

Solution
Referring to the above figure (ii), the force acting on the loop can be given as

$$\vec{F} = i \oint d\,\vec{l} \times \vec{B}$$

$$= i\left[\int_a^b d\,\vec{l} \times \vec{B} + \int_b^c d\,\vec{l} \times \vec{B} + \int_c^d d\,\vec{l} \times \vec{B} + \int_d^a d\,\vec{l} \times \vec{B}\right]$$

The directions of forces $\vec{F_1}$, $\vec{F_2}$, $\vec{F_3}$ and $\vec{F_4}$ are given by using Ampère's force law. It is also easy to find the direction of Ampère's force by Fleming's left-hand rule (FLHR). We can see that due to the concept of symmetry, the force acting on each element of bc is counterbalanced by the force acting on the opposite element of ad. Then, $\vec{F_2} + \vec{F_4} = 0$. Now we have

$$\vec{F} = \vec{F_1} + \vec{F_2}$$
$$= i\left[\int_a^b d\,\vec{l} \times \vec{B} + \int_c^d d\,\vec{l} \times \vec{B}\right]$$
$$= i\left[\int_0^l (dl\,\hat{j}) \times \left(-\frac{\mu_0 i_0 \hat{k}}{2\pi x}\right) + \int_0^l (-dl\,\hat{j}) \times \left(-\frac{\mu_0 i_0 \hat{k}}{2\pi(l+x)}\right)\right]$$
$$= -\frac{\mu_0 i i_0 l}{2\pi x(l+x)}\hat{i}.$$

Now you can put $x = l = r$. Ans.

Alternative method:

The net force acting on the frame is

$$\vec{F} = (-F_1\hat{i} + F_2\hat{i})$$
$$= -(F_1 - F_2)\hat{i}$$
$$= (ilB_1 - ilB_2)\,\hat{i} \quad (\because F_1 = ilB_1 \text{ and } \because F_2 = ilB_2)$$
$$= -il(B_1 - B_2)\hat{i}$$
$$= -il\left\{\frac{\mu_0 i}{2\pi r} - \frac{\mu_0 i}{2\pi(2r)}\right\}\hat{i}$$
$$= -\frac{\mu_0 i i_0 l}{4\pi r}\hat{i} \text{ Ans.}$$

Example 5 Prove qualitatively that a small loop will experience a force when placed coaxially with a magnet.

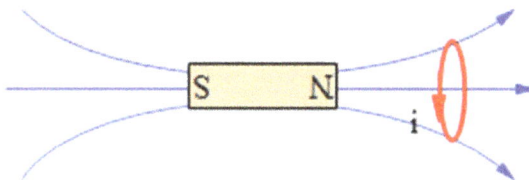

Solution

Let the magnetic field be B at any point P of the current loop. Since $\vec{B}$ forms a cone with the loop it changes its direction from point to point on the perimeter of the loop. Then, you cannot treat $\vec{B}$ as uniform. Now resolve $\vec{B}$ into two components one along the x-axis, that is, B_x, and the other is along the radial direction, that is, B_r. Due to the axial component of B, that is, B_x, the net force acting on the loop is zero because each element of the loop will experience outward radial force dF_r of equal magnitude. On the other hand, due to the radial component B_r, the loop experiences an axial force dF_x at each point of the loop. Hence, the loop will experience an axial force in the negative x-direction in this case.

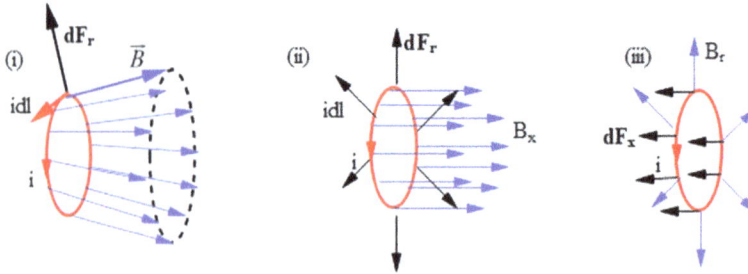

In figure (i), we can see that the Ampère's force acting at each element of the ring or loop is equally inclined; so, these elementary forces form a cone. Their radial components will cancel out. So, the net force acting on the loop points to the left towards the stronger magnetic field. Ans.

7.5 Magnetic torque

In the last section we learnt that, in general, a magnetic force acts on each element of a current-carrying loop. Each force $d\vec{F}$ will produce a torque $d\vec{\tau}$ about a fixed point O given as

$$d\vec{\tau} = \vec{r} \times d\vec{F},$$

where

$$d\vec{F} = i(d\vec{l} \times \vec{B})$$

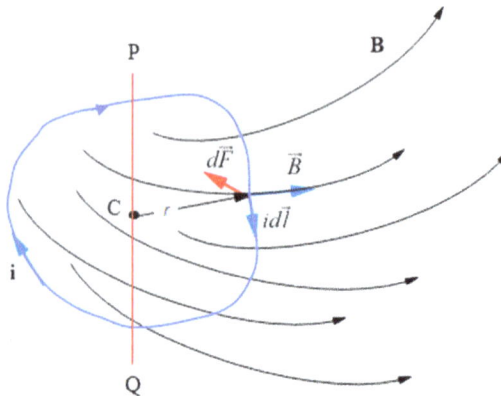

The current loop experiences a torque in a non-uniform magnetic field

Then, the total torque experienced by the current loop about O can be given as

$$\vec{\tau} = \int d\vec{\tau} = \vec{r} \times (i\,d\vec{l} \times \vec{B})$$

$$\Rightarrow \vec{\tau} = i \int \vec{r} \times (d\,\vec{l} \times \vec{B})$$

Example 6 Derive an expression for the torque experienced by a coplanar current loop placed in a uniform magnetic field.

Solution

First of all, resolve the applied or external field into two components: parallel to the plane of the loop, i.e. $\vec{B}_{\parallel}$; and perpendicular to the plane of the loop, i.e. $\vec{B}_{\perp}$. Due to $B_{\perp}$ ($\otimes$ or inward $\vec{B}$) each element of the current loop experiences a force perpendicular to $d\,\vec{l}$ in the plane of the loop. In consequence, the loop gets stretched into a circular shape or tends to be squeezed if it is flexible. However, the net force acting on the loop is zero due to $B_{\perp}$. Now we consider the effect of $\vec{B}_{\parallel}$ by taking two elements of lengths dl_1 and dl_2. We can see that the forces acting on dl_1 and dl_2 are

$$d\,\vec{F_1} = +\,i\,dl_1\,B_{\parallel}\sin\theta_1\hat{k}, \text{ and } d\,\vec{F_2} = -i\,dl_2\,B_{\parallel}\sin\theta_2\hat{k} \text{ respectively.}$$

Putting, $dl_1 \sin\theta_1 = dl_2 \sin\theta_2 = dy$, we have

$$d\,\vec{F_1} = +iB_{\parallel}dy\hat{k} = -d\,\vec{F_2} = -iB_{\parallel}dy\hat{k}$$

Referring to rotational dynamics, the net torque due to two equal and opposite forces ($d\,\vec{F_1}$ and $d\,\vec{F_2}$) does not depend upon the position of the reference axis. So, we can choose any axis parallel to the y-axis. The torque about that axis due to the above two equal and opposite forces is

$$d\,\vec{\tau} = +\{x_1(dF_1) + x_2 \times (dF_2)\}\hat{j} = \{x_1(-iB_{\parallel}dy) + x_2(iB_{\parallel}\,dy)\}\hat{j} = (x_1 + x_2)(iB_{\parallel}dy\,)\hat{j}$$

$$\Rightarrow d\,\vec{\tau} = iB_{\parallel}xdy\,\hat{j}$$

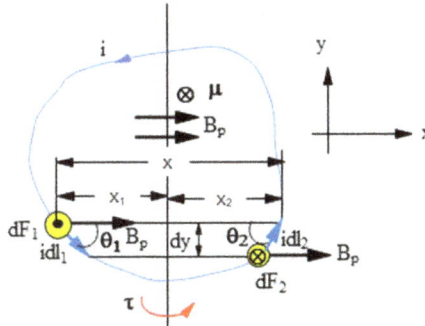

The torque experienced by the current loop in a uniform magnetic field does not depend on the reference axis which is given as $\vec{\tau} = \mu \times \vec{B}$

The net torque experienced by the loop is

Then, the net torque acting on the loop is

$$\vec{\tau} = \int d\,\vec{\tau} = \int iB_{\parallel}\,x\,dy\,\hat{j}$$

$$= i\,B_{\parallel}\hat{j}\int x\,dy = i\,B_{\parallel}\hat{j}\,A\;(\because \int x\,dy = \text{area } A \text{ bounded by the loop})$$

$$\Rightarrow \vec{\tau} = \mu B_\parallel \hat{j} \ (\because iA = \text{magnetic moment} = \mu)$$

Vectorially, the torque is given as

$$\vec{\tau} = \vec{\mu} \times \vec{B_\parallel}$$

Since, $\vec{\mu} \times \vec{B_\perp} = 0 \ (\because \vec{\mu} \| \vec{B})$, we can write

$$\vec{\tau} = \vec{\mu} \times (\vec{B_\parallel} + \vec{B_\perp})$$

Putting $\vec{B_\parallel} + \vec{B_\perp} = \vec{B}$, we have

$$\vec{\tau} = \vec{\mu} \times \vec{B} \ \text{Ans.}$$

N.B: In rotational mechanics we know that the net torque due to two equal and opposite forces does not depend on the reference frame. This expression holds good for non-coplanar loops of any shape and size placed in a uniform magnetic field. This expression is also valid for a small current loop in a non-uniform magnetic field.

Example 7 Find the torque experienced by the rectangular loop of length l and breadth b which carries a current i and is subjected to a horizontal magnetic field $\vec{B}$.

Solution

Assume that the loop is free to rotate about a horizontal axis perpendicular to the magnetic field B. The forces acting on the sides AD and BC are parallel to the axis OP; so, they will not be able to produce torques about the given axis OP. The forces acting on AB and CD are equal and opposite. Hence, they can produce a couple which does not depend on the reference axis. Therefore, you can take the torque due to the forces $-F$ and F acting on the sides AB and CD about the given axis OP to find the total torque or couple.

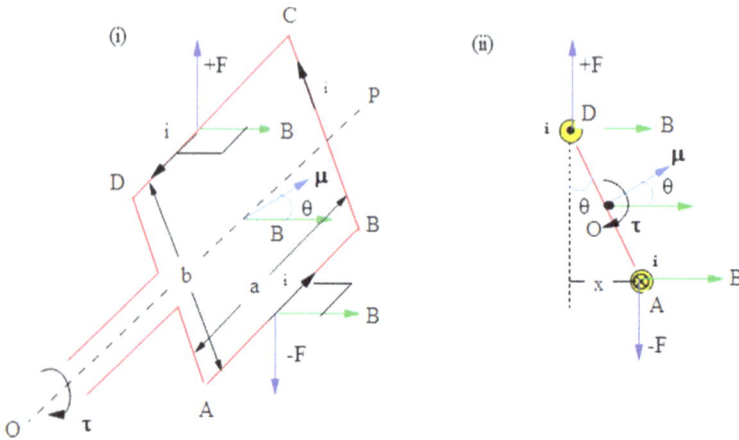

The net torque experienced by the loop is

$$\tau = F\frac{b}{2} \sin \theta \hat{k} + F\frac{b}{2} \sin \theta \hat{k} = Fb \sin \theta \hat{k}$$

$$= (ilB)b \sin \theta \hat{k} = i(lb)B \sin \theta \hat{k}$$

$$= iAB \sin \theta \hat{k}$$

For a coil of N loops, the torque acting on the coil is

$$\tau = NiAB \sin \theta$$

$$= \mu B \sin \theta \; (\because \mu = NiA)$$

Vectorially, we can write $\overrightarrow{\tau} = \overrightarrow{\mu} \times \overrightarrow{B}$. Ans.

7.6 Work done in displacing a current loop in a magnetic field

Ampère's force law tells us that whenever a straight current-carrying conductor of length l is kept perpendicular to a magnetic field B, a mechanical force acts on it, which can be given as

$$F_m = ilB$$

This mechanical force F_m does work dW if the conductor undergoes a displacement dx, which is given as

$$dW = F_m dx$$

$$= ilB \; dx = iBl \; dx$$

$$\Rightarrow dW = iB \; dA \; (\because \; dA = ldx),$$

where dA = area swept by the conductor during the displacement dx (or time dt).

During a time t, the work done by Ampère's force (the mechanical force), i.e. mechanical work done W_{mech} is given as $W_{\text{mech}} = \int iBdA$, where $BdA = d\phi$= change in flux or flux cut or swept by the conductor during time dt.

$$\text{or,} \quad W_{\text{mech}} = \int i \; d\phi$$

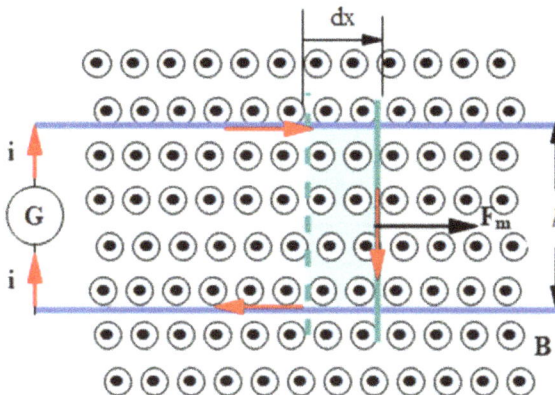

Mechanical work is done in displacing the current carrying conductor = (current) (change in flux enclosed by the loop)

When the current i is maintained constant (by a constant current generator) we can pull i out of the integral to obtain.

$$W_{\text{mech}} = i \int_{\theta_1}^{\theta_2} d\phi$$

$$\Rightarrow W_{\text{mech}} = i(\phi_2 - \phi_1) = i\phi,$$

where ϕ = flux cut (swept) by the conductor or change in flux passing through the loop formed by the conductor.

If $\phi_2 > \phi_1$, W_m is positive; if $\phi_2 < \phi_1$, W_m negative; if $\phi_1 = \phi_2$, $W_m = 0$.

Example 8 A constant current generator sends a current i through the smooth conducting rod of mass m and length l which is free to slide on the horizontal conducting rail. For an upward magnetic field $\vec{B}$, find the speed of the rod as the function of time.

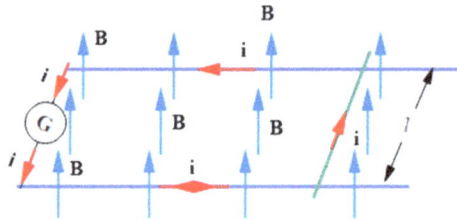

Solution
The Ampère's force acting on the rod is

$$F = ilB$$

This accelerates the rod with

$$a = \frac{F}{m} = \frac{ilB}{m}$$

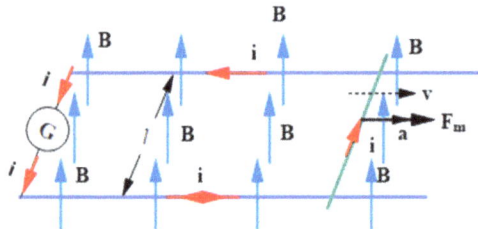

Then, velocity of the rod at any time t is

$$v = u + at$$
$$= 0 + \frac{ilB}{m}t \ (\because u = 0)$$
$$\Rightarrow v = \frac{ilB}{m}t$$

The speed of the rod varies linearly with time. However, it does not happen in practice due to the effect of induction, slight frictional loss and copper loss etc, which will be discussed in the next chapter 'Electromagnetic induction'. Ans.

7.7 Mechanical (potential) energy possessed by a current loop (or a tiny magnet)

Referring to the last section, the mechanical work done when a current loop carrying a constant current i, is moved in a magnetic field, is given as

$$W_{\text{mech}} = i(\phi_2 - \phi_1)$$

This means that a current loop possesses a mechanical energy in a magnetic field because of its ability of doing mechanical work. For a constant current i, the magnetic fluxes linked with the loop such as ϕ_1 and ϕ_2 depend solely on the positions of the conductor or loop. So, we can call this work or mechanical energy 'potential energy U of the loop' treating the magnetic field as conservative which satisfies the condition, given as

$$\oint \vec{B} \cdot d\vec{l} = 0$$

when the contour does not thread any current (or additional source of magnetic field). Under the above restricted condition, we can write the work–potential energy theorem as

$$W_{\text{mech.}} = -\Delta U$$
$$\Rightarrow -(U_2 - U_1) = i\phi_2 - i\phi_1$$

Comparing both sides, we have

$$U_2 = -i\phi_2 \text{ and } U_1 = -i\phi_1$$

This means that at any arbitrary position of current loop (or a tiny magnet), the potential energy possessed by it is given as

$$U = -i\phi,$$

where $\phi = \vec{B} \cdot \vec{A}$; A = area of the loop

$$\Rightarrow U = -i\vec{A} \cdot \vec{B}$$

Substituting $i\vec{A} = \mu$ (magnetic dipole moment of the loop), we have

$$U_{\text{mech}} = -\vec{\mu} \cdot \vec{B}.$$

7.8 Lorentz force

The Lorentz force equation appeared in Maxwell's 1861 and 1864 papers in Treatise of Electricity and Magnetism. But at that time it was not evident how this equation was related to the force on a moving charge by an electromagnetic field. For the first time, in 1895, J J Thomson derived the magnetic force on the cathode rays (a stream of fast-moving electrons, discovered by J J Thomson in 1897), but got a wrong multiplier of 1/2 which was corrected by Heaviside in 1885 and 1889 and gave its present vector form. In 1895, by using the Heaviside's version of Maxwell's equations and Lagrangian mechanics, Hendrik Lorentz derived a formula of the net force on a point charge in the presence of both electric and magnetic fields.

7.8.1 Magnetic force

Since a current element is a moving point charge, Lorentz argues that Ampère's force acting on a current element is a force that acts on a charged particle while it moves in a magnetic field, which is known as magnetic force F_m that not only depends upon the velocity of the charged particle but also on the magnetic field. Experimentally, Lorentz observed that

  (i) $\vec{F_m}$ acts on a charge perpendicular to both $\vec{v}$ and $\vec{B}$.
 (ii) $|\vec{F_m}|$ is directly proportional to the speed of the charged particle relative to the observer.
(iii) $|\vec{F_m}|$ is directly proportional to the applied magnetic field $\vec{B}$.
 (iv) $|\vec{F_m}|$ is directly proportional to sin θ, where θ = angle between $\vec{v}$ and $\vec{B}$.
  (v) $|\vec{F_m}|$ is directly proportional to the quantity of charge.
 (vi) Reversal of $\vec{v}$ or $\vec{B}$ reverses the direction of force $\vec{F_m}$.
(vii) Change in polarity (or nature) of the charge reverses the direction of force.

From the above observations Lorentz concludes that whenever a charge q moves with a velocity $\vec{v}$ in a magnetic field of induction $\vec{B}$ at any point, at that point the charged particle experiences a magnetic force.

$$\vec{F_m} = q\left(\vec{v} \times \vec{B}\right)$$

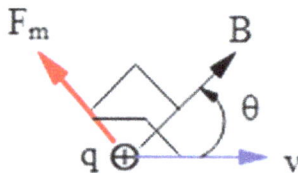

Magnetic force

N.B: If $\vec{v} \perp \vec{B}$, F_m is maximum; that is, qvB.
If $\vec{v} \parallel \vec{B}$, $\vec{F_m}$ is zero.
If $v = 0$, $\vec{F_m}$ is zero.

The magnitude of F_m can be given as

$$F_m = qv\,B \sin \theta,$$

where θ = angle between $\vec{v}$ and $\vec{B}$, q is algebraic (+ve and −ve).
Lorentz force is the sum of electric and magnetic forces; it tells us how charges respond to electric and magnetic fields; in the force formula, v = velocity of the charge relative to the observer, not necessarily relative to ground and the electric field E is the sum of induced and static electric fields (please refer to section 7.9).

7.8.2 Electric force

In addition to the magnetic field, if an electric field $\vec{E}$ is present, an additional force, that is, electric force $\vec{F_{el}}$ acts on the charge, which is given as

$$\vec{F_{el}} = q\vec{E}$$

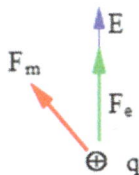

Electric force

7.8.3 Lorentz force

Then, the total force due to the electric and magnetic fields acting on the point charge is called Lorentz's force which is given as

$$\vec{F} = \vec{F_{el}} + \vec{F_m}$$
$$= q\vec{E} + q\vec{v} \times \vec{B}$$

$$\vec{F} = q\left(\vec{E} + \vec{v} \times \vec{B}\right),$$

where $\vec{E} = \vec{E}_{static} + \vec{E}_{ind}$ (sum of induced electric field and static electric field)

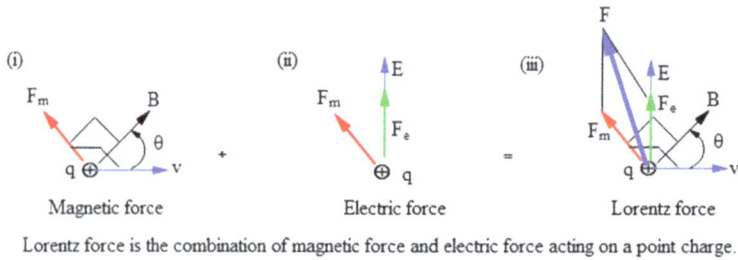

(i) Magnetic force   (ii) Electric force   (iii) Lorentz force

Lorentz force is the combination of magnetic force and electric force acting on a point charge.

We will talk about $\vec{E}_{\text{ind}}$ in the next chapter. Please remember that $\vec{v}$ is velocity of charge relative to the observer (may not be relative to ground). The above expression is called Lorentz force equation.

Example 10 An electron moving with $\vec{v} = 10^6$ m s^{-1} $\hat{k}$ is subjected to an electric field $\vec{E} = 10^6\hat{i}$ v m^{-1} and magnetic field $\vec{B} = 1\hat{j}$ Tesla. (a) Find the Lorentz force acting on the electron. (b) If we reverse the direction of $\vec{E}$, find the Lorentz force acting on the electron.

Solution

(a) The Lorentz force is

$$\vec{F} = q(\vec{v} \times \vec{B} + \vec{E})$$
$$= -(1.6 \times 10^{-19})\{(10^6\,\hat{k}) \times (1\hat{j}) + 10^6\,\hat{i}\}$$
$$= -(1.6 \times 10^{-19})\{-10^6\hat{i} + 10^6\,\hat{i}\} = 0\ \text{Ans.}$$

Hence, no net force acts on the electron.

(b) If we reverse the direction of $\vec{E}$, the net force acting on the electron is

$$\vec{F} = q(\vec{v} \times \vec{B} + \vec{E})$$
$$= -(1.6 \times 10^{-19})\{(10^6\,\hat{k}) \times (1\hat{j}) + (-10^6\,\hat{i})\}$$
$$= -(1.6 \times 10^{-19})\{-10^6\hat{i} - 10^6\,\hat{i}\} = 3.2 \times 10^{-13}\hat{i}\ N\ \text{Ans.}$$

7.9 Induced electric field

Referring to section 7.8, if a charge particle $+q$ moves with a velocity $\vec{v}$ relative to the observer O in a crossed electric field $\vec{E}$ and magnetic field $\vec{B}$, for an observer at rest (relative to ground or magnetic field), Lorentz force acting on the charge (as viewed by the observer) is

$$\vec{F_1} = q\left(\vec{E} + \vec{v} \times \vec{B}\right) \tag{7.1}$$

If the observer moves with a velocity v' relative to ground, the point charge appears to move to the observer O' with a velocity $\overrightarrow{v_r} = (\overrightarrow{v} - \overrightarrow{v'})$ relative to the observer, he detects the magnetic force

$$F_m = q\,\overrightarrow{v_r} \times \overrightarrow{B} = q\left(\overrightarrow{v} - \overrightarrow{v'}\right) \times \overrightarrow{B}$$

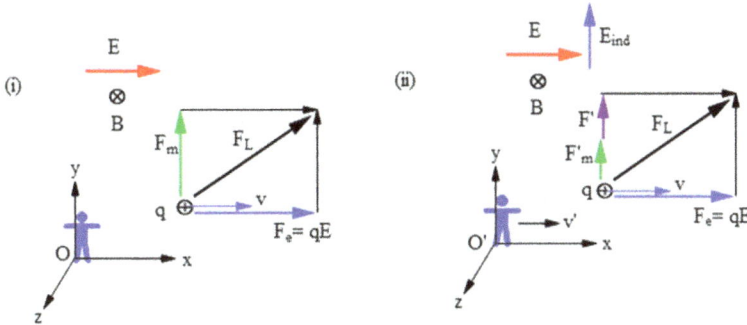

The observer moving relative to the magnetic field detects an additional electric force F' in addition to the electrostatic and magnetic forces. This additional (induced) electric force is associated with an induced electric field E_{ind}; $\vec{F'} = q\vec{E}_{ind} = qv'^r \times \vec{B}$

Let the observer detect (observe) an electric force $F_{el} = qE'$, where $E' =$ electric field observed by the moving observer so that the net force, that is, Lorentz force acting on the charge is

$$\overrightarrow{F_2} = q\left\{ E' + q(\overrightarrow{v} - \overrightarrow{v'}) \times \overrightarrow{B} \right\} \tag{7.2}$$

We know that Lorentz force remains nearly invariant in non-relativistic field transformations (when the velocity is much less that the speed of light). So, by equating $\overrightarrow{F_1}$ with $\overrightarrow{F_2}$ in equations (7.1) and (7.2), we have

$$E' = E + \overrightarrow{v'} \times \overrightarrow{B},$$

where $\overrightarrow{v'} \times \overrightarrow{B}$ is called induced electric field $\overrightarrow{E}_{ind}$ which arises from the motion of the observer relative to the magnet (or magnetic field). Then, we can write

$$\overrightarrow{E}_{ind} = \overrightarrow{v'} \times \overrightarrow{B}.$$

In general, the induced electric field is given as

$$\overrightarrow{E}_{ind} = \overrightarrow{v}_{om} \times \overrightarrow{B},$$

where $\overrightarrow{v}_{om} =$ velocity of the observer relative to the magnetic field but need not be relative to the ground.

From the foregoing discussions we can conclude that, a moving observer may observe an extra electric field called induced electric field $\overrightarrow{E}_{ind}$ due to their motion relative to the magnetic field. In consequence, there is an extra electric force $q\vec{E}_{ind}$. Then, the total electric field $\vec{E}'$ is a combination of static electric field $\vec{E}_{st}$. and induced electric field $\vec{E}_{ind}$.

$$\vec{E}' = \vec{E}_{st} + \vec{E}_{ind}$$

$$\Rightarrow \vec{E}' = \vec{E} + \vec{v}_{om} \times \vec{B}$$

N.B: Please note that the static electric field $\vec{E}_{st}$ is conservative, whereas the induced electric field $\vec{E}_{ind} = (\vec{v}_{om} \times \vec{B})$ is non-conservative because it drops to zero just outside the magnetic field. We will discuss more about the induced electric field in the next chapter.

1. $\vec{F}_{ind} = q\vec{E}_{ind} = q(\vec{v}_{om} \times \vec{B})$ *should not be misinterpreted as the Lorentz magnetic force* $\vec{F}_m = q(\vec{v}_{po} \times \vec{B})$, *where* $v_{po}(=v) =$ *velocity of the charged particle relative to the observer, whereas* $\vec{v}_{om} =$ *velocity of the observer relative to the magnetic field.*

2. *In the Lorentz force equation,* $\vec{F} = q(\vec{E} + \vec{v} \times \vec{B})$, *where* $\vec{E} = \vec{E}_{st} + \vec{E}_{ind}$ *in general. Hence, we call* $\vec{E}$ *electric (not electrostatic) field and* $v = v_{po}$ *(but not* v_{om}*).*

Example 11 A person is moving in a magnetic field $\vec{B} = -2\hat{k}$ tesla with a velocity $\vec{v} = 5\hat{i}$ m s^{-1}. Find the (i) magnetic force, (ii) induced electric force, and (iii) Lorentz force acting on a charge $q = 2$ μC stationary relative to ground.

Solution

(i) The magnetic force is

$$\vec{F}_m = q\left(\vec{v}_{PO} \times \vec{B}\right) = (2 \times 10^{-6})\{(-5\hat{i}) \times (-2\hat{k})\} = -2 \times 10^{-5}\hat{j} \text{ N Ans.}$$

(ii) The induced electric force is

$$\vec{F}_{ind} = q\vec{E}_{ind} = q\left(\vec{v}_{om} \times \vec{B}\right)$$

$$= (2 \times 10^{-6})\{(5\hat{i}) \times (-2\hat{k})\} = 2 \times 10^{-5}\hat{j} \text{ N Ans.}$$

(iii) Then the total force is

$$\vec{F} = \vec{F}_m + \vec{F}_{el} = (-10^5\hat{j}) + (+10^{-5}\hat{j}) = 0 \text{ Ans.}$$

In one reference frame, the Lorentz force F is purely magnetic, in another reference frame it may be purely electric, in general, in some reference frames the force is partly electric and partly magnetic. Hence, we would prefer to call it the electromagnetic field, which is ultimately caused by the relative motion between the charge, magnetic field and observer. These field transformations can be explained by relativity.

Problem 1 A current-carrying conductor of mass m, length l carrying a current i hangs from the ceiling by two identical springs each of stiffness k. If an inward magnetic field B is given, the spring will be found undeformed. (a) Find the current in the conductor. (b) What will the deformation of the springs be if the direction of magnetic field is reversed (made outward)? Put $m = 50$ g, $g = 10$ m s^{-2}, $l = \frac{1}{2}$ m and $B = 1$ T and $k = 50$ N m^{-1}.

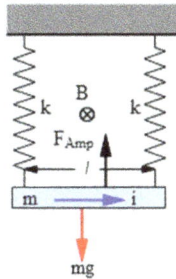

Solution

(a) The forces acting on the rod are $mg\downarrow$, $F_{amp}(=ilB\uparrow)$ and $F_{sp} = 2kx = 0$ (because the spring is undeformed).

Since the rod is in equilibrium, $F_{net} = 0$

$$\Rightarrow -mg + ilB = 0$$

$$\Rightarrow ilB = mg$$

$$\Rightarrow i = mg/lB = (1/20)(10)/(0.5)(1) = 1 \text{ A Ans.}$$

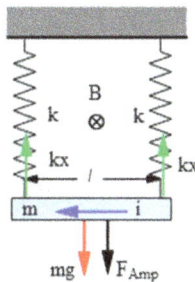

(b) When the magnetic field is reversed, the Ampère's force will be vertically down in addition to the weight of the rod. So, the spring will be stretched to develop an upward force to balance the resultant of weight and Ampère's force. Now the forces acting on the rod are $mg\downarrow$, $F_{amp}(=ilB\downarrow)$ and $F_{sp} = 2kx\uparrow$ (because the spring is elongated). Since the rod is in equilibrium,

$$F_{\text{net}} = 0$$

$\Rightarrow mg + ilB = 2kx$ (net spring force $= 2kx$ because two springs are in parallel)

$$\Rightarrow x = \frac{mg + ilB}{2k}$$

Putting $ilB = mg$, we have

$$\Rightarrow x = \frac{mg + mg}{2k} = \frac{mg}{k} = \frac{(1/20)(10)}{50} = 0.01 \ m \ \text{Ans.}$$

Problem 2 A conductor (rod) of mass m, length l carrying a current i is subjected to a magnetic field of induction B. If the coefficients of static friction between the conducting rod and rail is μ_s (a) find the value of i if the rod starts sliding. (b) Find the minimum possible value of i.

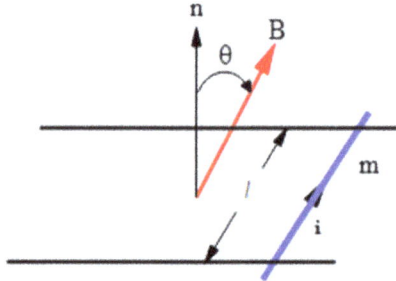

Solution

(a) The Ampère's force acting on the conductor is

$$F_A = ilB \sin \phi,$$

where $\phi=$ angle between current and magnetic field $= 90°$.

$$\Rightarrow F_A = ilB,$$

which acts on the rod at an angle θ with the horizontal, as shown in the figure below according to the FLHR.

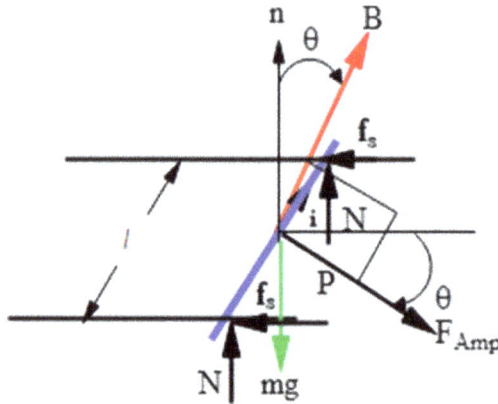

Referring to the free-body diagram, the net horizontal force acting on the rod is

$$F_x = 2f_s - F_A \cos \theta = 0$$

$$\Rightarrow f_s = \frac{1}{2} F_A \cos \theta \tag{7.3}$$

The net vertical force acting on the rod is

$$F_y = 2N - (mg + F_A \sin \theta) = 0$$

$$\Rightarrow N = \frac{1}{2}(mg + F_A \sin \theta) \tag{7.4}$$

Law of friction,

$$f_s \leqslant \mu_s N \tag{7.5}$$

By using the last four equations, we have

$$F_A \cos \theta \leqslant \mu_s(mg + F_A \sin \theta)$$

$$\Rightarrow F_A \leqslant \frac{\mu_s mg}{\cos \theta - \mu_s \sin \theta}$$

$$\Rightarrow ilB \leqslant \frac{\mu_s mg}{\cos \theta - \mu_s \sin \theta}$$

$$\Rightarrow i \leqslant \frac{\mu_s mg}{(\cos \theta - \mu_s \sin \theta)lB} \quad \text{Ans.}$$

(b) For the minimum possible value of i, the factor '$(\cos \theta - \mu \sin \theta)$' must be maximum. Its maximum value is equal to $\sqrt{1 + \mu_s^2}$. So, the minimum possible value of the current to slide the rod is

$$i_{min} = \frac{\mu mg}{\left(\sqrt{1 + \mu_s^2}\right)lB} \quad \text{Ans.}$$

Problem 3 A circular flexible current loop of radius R carrying a current i is placed in an outward magnetic field B. (a) Find the tension in the loop. (b) If we spin the loop with an angular speed ω, find the tension in the string. Assume the mass of the loop as m.

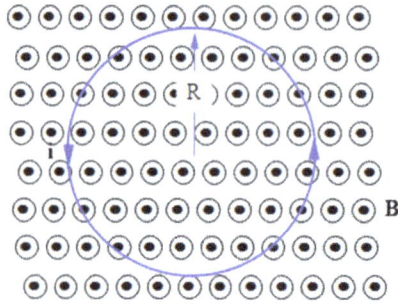

Solution

(a) The elementary segment of length dl of the loop is in equilibrium under the forces $T\swarrow$, $T\searrow$ and Ampère's force $dF\uparrow$.

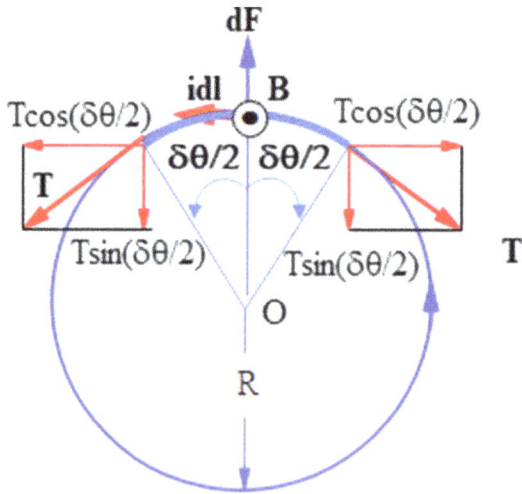

Resolving the forces radially,

$$\sum F_r = ma_r$$

$$\Rightarrow dF - 2T \sin \frac{d\theta}{2} = 0$$

$$\Rightarrow idl\, B - T\, d\theta = 0$$

$$\Rightarrow T = iB\frac{dl}{d\theta} = iB\frac{Rd\theta}{d\theta} = iBR \text{ Ans.}$$

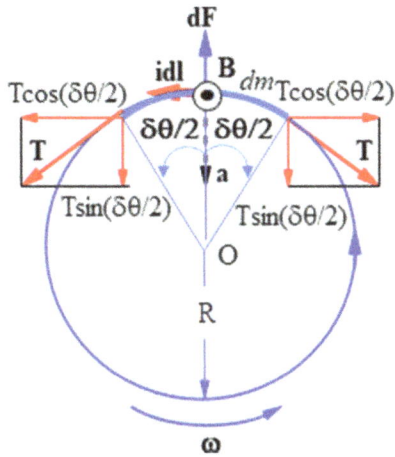

N.B: Tension is not zero but the net Ampère's force is zero.

(b) If we spin the loop with angular speed ω, the elementary segment of length dl and mass dm of the loop is accelerating towards the centre of the loop under the forces $T \nearrow$, $T \searrow$ and Ampère's force $dF \uparrow$.

Resolving the forces radially,

$$\sum F_r = ma_r$$

$$\Rightarrow dF - 2T \sin \frac{d\theta}{2} = dm(-a_r) = -dm R \omega^2$$

$$\Rightarrow idl\ B - T\ d\theta = -dm R \omega^2$$

$$\Rightarrow T\ d\theta = idl\ B + dm R \omega^2$$

$$\Rightarrow T = iB \frac{dl}{d\theta} + \frac{dm R \omega^2}{d\theta}$$

$$= iB \frac{R d\theta}{d\theta} + \frac{dm R \omega^2}{d\theta} = iBR + \frac{m R \omega^2}{2\pi} \text{ Ans.}$$

Problem 4 Two identical circular loops each of radius r carrying currents i_1 and i_2 are placed coaxially at a small distance of separation d. Find the force of interaction between them.

Solution

Since the loops are very close, we use the expression

$$\frac{dF_2}{dl_2} = i_2 B_1, \text{ where } B_1 = \frac{\mu_0 i_1}{2\pi d}$$

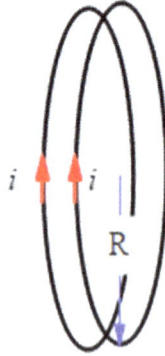

$$F_2 = \int i_2 B_1 dl_2$$
$$= \int i_2 \frac{\mu_0 i_1}{2\pi d} dl_2$$
$$= \frac{\mu_0 i_1 i_2}{2\pi d} \int_0^{2\pi R} dl_2$$
$$= \frac{\mu_0 i_1 i_2}{2\pi d} 2\pi R$$
$$= \frac{\mu_0 i_1 i_2 R}{2d} \text{ Ans.}$$

Problem 5 A long thin plate that is bent in the form of circular arc carries an outward current i which is distributed uniformly over its cross-section. A long straight rigid wire carrying a current i is kept along the axis of the thin plate. (a) What is the magnetic field of the current-carrying thin plate at its axis? (b) Find the Ampère's force acting on the wire. (c) Find the mass of the long thin wire if it floats under the action of the Ampère's force of the thin plate and its weight.

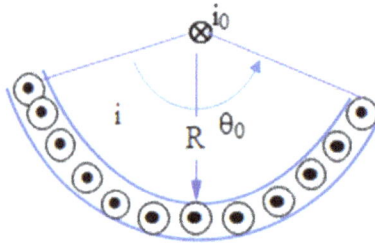

Solution

(a) Take the two symmetrical elementary segments of the cross-section of the tube which behave as two long straight current-carrying conductors 1 and 2. The vertical components of dB on the long current-carrying wire due to segments 1 and 2 of the thin plate (circular arc) will cancel each other. Then, their net field is

$$2dB_x = 2dB \sin(90 - \theta) = 2dB \cos \theta,$$

where

$$dB = \frac{\mu_0 di}{2\pi R}$$

$$\Rightarrow 2dB_x = \frac{2\mu_0}{2\pi R} di \cos \theta,$$

where

$$di = \frac{i}{R\theta_0} \cdot R d\theta = \frac{i d\theta}{\theta_0}$$

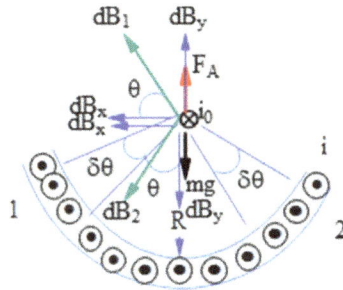

$$\Rightarrow 2dB_x = \frac{\mu_0 i}{\pi \theta_0} \cdot \cos \theta \, d\theta$$

The net magnetic field is

$$B = \frac{\mu_0 i}{\pi \theta_0} \int_0^{\theta_0.2} \cos \theta \, d\theta$$

$$\Rightarrow B = \frac{\mu_0 i \sin \frac{\theta_0}{2}}{\pi \theta_0} \text{ Ans.}$$

(b) Then, the upward magnetic (Ampère's) force is

$$F = i_0 l B = \frac{i_0 \left(\mu_0 i \sin \frac{\theta_0}{2} \right)}{\pi \theta_0}$$

$$\Rightarrow F = \frac{\mu_0 i i_0}{\pi \theta_0} \sin \frac{\theta_0}{2} \text{ Ans.}$$

(c) If the thin conductor remains at rest, the vertical Ampère's force balances the weight of the wire. So, we can write

$$F = \frac{\mu_0 i i_0}{\pi \theta_0} \sin \frac{\theta_0}{2} = mg$$

$$\Rightarrow m = \frac{\mu_0 i i_0 \sin \frac{\theta_0}{2}}{\pi \theta_0 g} \text{ Ans.}$$

Problem 6 A straight conductor of length r_0 carrying a current i is placed perpendicular to a long straight conductor which carries a current i_0, as shown in the figure below. (a) Find the force of interaction between them. (b) If the finite conductor is pivoted smoothly at the end P very close to the long conductor, find its angular acceleration. Put $r_0 = l$ for (a) and r_0 is very small for (b).

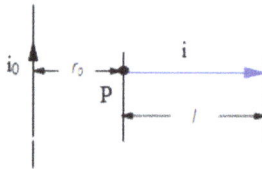

Solution

(a) The force on the current element $i dr$ of the conductor is

$$dF = i dr \, B \sin 90°$$
$$= i dr \, \frac{\mu_0 i_0}{2\pi r} \qquad (7.6)$$
$$= \frac{\mu_0 i i_0}{2\pi} \frac{dr}{r}$$

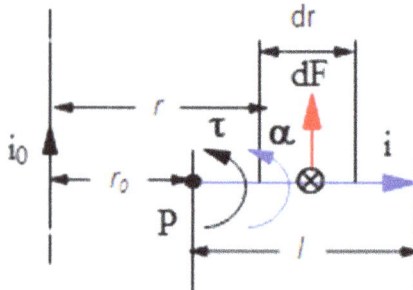

The net force is,

$$F = \int dF$$

$$= \int_{r_0}^{2r_0} \frac{\mu_0 i i_0}{2\pi r} dr = \frac{\mu_0 i i_0}{2\pi} \int_{r_0}^{2r_0} \frac{dr}{r}$$

$$= \frac{\mu_0 i i_0}{2\pi} \ln 2 \text{ Ans.}$$

(b) The torque of the force dF about the pivot P is given as

$$d\tau = r \, dF$$

Then, the net torque about P is

$$\tau = \int d\tau = \int_{r_0}^{2r_0} r \, dF \tag{7.7}$$

$$\tau = \int_0^l r \, dF = \int_0^l r \frac{\mu_0 i i_0}{2\pi r} dr$$

$$= \frac{\mu_0 i i_0}{2\pi} \int_0^l dr = \frac{\mu_0 i i_0 l}{2\pi}$$

The conductor will rotate about the pivot P by the action of the torque of the Ampère's force. So, the angular acceleration of the finite conductor is

$$\alpha = \tau/I = \frac{\mu_0 i i_0 l}{2\pi} / (ml^2/3) = \frac{3\mu_0 i i_0}{2\pi ml} \text{ Ans.}$$

Problem 7 A closed loop carrying a current i is placed so that its plane is perpendicular to the long straight conductor which carries a current i_0, as shown in the figure. (a) Find the torque acting on the current loop. (b) Find the net force acting on (i) each straight conductor and (ii) the current loop.

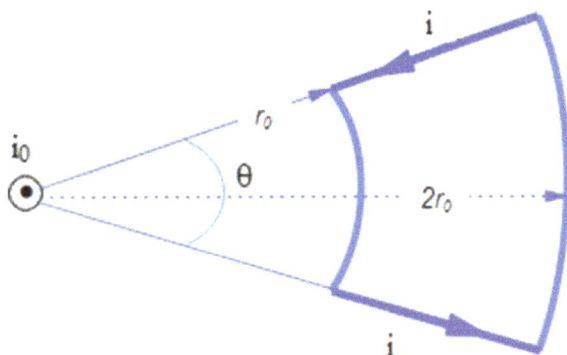

Solution

(a) The force acting on the curved segments is zero because $\overrightarrow{B}$ (B_1 and B_2) is parallel to the current i flowing in the curved portion (arcs) of the loop. The force acting on the current element idr is

$$d\overrightarrow{F} = iBdr \sin 90° \hat{k}$$
$$= i\left(\frac{\mu_0 i_0}{2\pi r}\right)dr\ \hat{k}$$
$$= \frac{\mu_0 i i_0\ dr}{2\pi r}\hat{k}$$

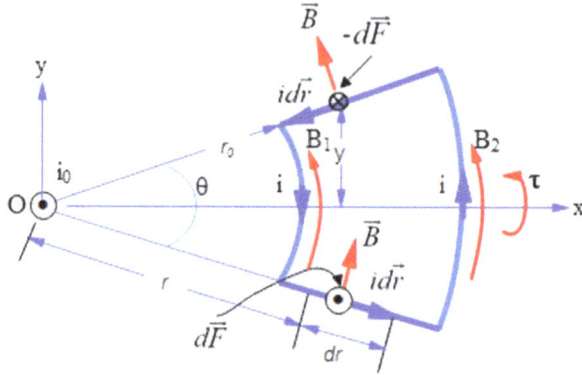

Take two elements idr on straight (top and bottom) segments of the loop. They experience an inward and upward force, respectively, of the same magnitude dF. These two equal and opposite forces will produce a couple and rotate the loop about the x-axis. The net torque due to $d\overrightarrow{F}$ and $-d\overrightarrow{F}$ about x-axis is

$$d\overrightarrow{\tau} = (ydF + ydF)\hat{i}$$
$$= 2ydF\hat{i}$$
$$= \{2(r \sin \tfrac{\phi}{2})\tfrac{\mu_0 i i_0}{2\pi r}dr\}\hat{i}$$
$$= \mu_0 i i_0 \sin \tfrac{\phi}{2}\int_{r_0}^{2r_0} dr$$
$$= \mu_0 i i_0 r_0 \sin \tfrac{\phi}{2}\ \text{Ans.}$$

(b)

(i) The net force acting on each straight conductor of the loop is

$$F = \int dF = \int \frac{\mu_0 i i_0\ dr}{2\pi r} = \frac{\mu_0 i i_0}{2\pi}\int_{r_0}^{2r_0} \frac{dr}{r} = \frac{\mu_0 i i_0}{2\pi} \ln 2\ \text{Ans.}$$

(ii) The Ampère's forces acting on the top and bottom straight conductor of the loop are inward and outward, respectively. The forces acting on the curved portions of the loop is zero as mentioned earlier. So, the net force acting on the current loop is zero. Ans.

Problem 8 A current i flows in a free rectangular rigid loop of mass m, width b and length a. The loop is placed in a horizontal magnetic field of induction B. At the given position, find the (a) torque acting on it and (b) angular acceleration of the loop at the given instant.

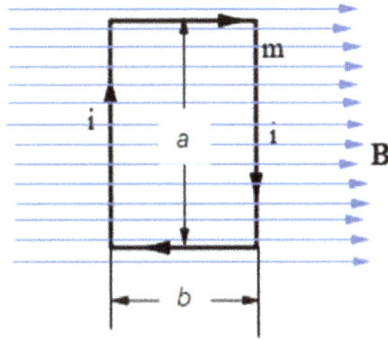

Solution

(a) The top and bottom branches do not experience any force or torques because the magnetic induction $\vec{B}$ is parallel with these branches. The other two vertical conductors experience equal and opposite (inward and outward) forces. The net torque acting due to two equal and opposite forces $\vec{dF_1}$ and $\vec{dF_2}$ does not depend on the reference point (or axes). However, for the sake of simplicity, choosing the y-axis and taking the torque $d\tau$ about the y-axis that passes through the center of mass of the loop, we have

$$d\,\vec{\tau} = \left(\tfrac{b}{2}d\,F + \tfrac{b}{2}d\,F\right)(-\hat{j})$$

$$= -\,bdF\hat{j}, \text{ where } d\,F = d\,F = idl\,B$$

$$\Rightarrow d\,\vec{\tau} = -ibB\,dl\,\hat{j}$$

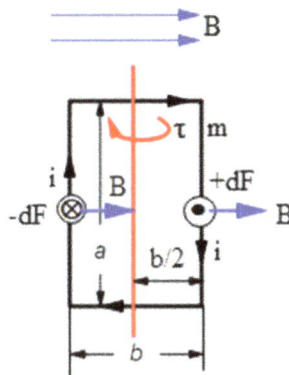

Then, the total torque experienced by the loop is,

$$\vec{\tau} = \int d\vec{\tau}$$

$$= -ibB\,\hat{j}\int_0^a dl$$

$$\vec{\tau} = -ibaB\,\hat{j} \quad \text{Ans.}$$

(b) Since, the loop is free, it will rotate about its centroidal y-axis with

$$\alpha = \frac{\tau}{I_c},$$

where I_c = the moment of inertia of the loop is given as

$$I_c = 2 \times \frac{1}{12}(\lambda b)b^2 + 2 \times (\lambda a)\left(\frac{b}{2}\right)^2$$

$$= \frac{2\lambda b^2}{4}\left(\frac{b}{3} + a\right),$$

Putting $\lambda = \frac{m}{2(b+a)}$ in the last expression, we have

$$I_c = \frac{mb^2(b + 3a)}{12(b + a)}$$

$$\Rightarrow \vec{\alpha} = -\frac{12(b + l)iaB}{mb(b + 3a)}\,\hat{j} \quad \text{Ans.}$$

Problem 9 A straight rod carrying current i having length l is made to rotate with a constant angular velocity ω about the y-axis in a radial magnetic field $\vec{B} = B_0\hat{r}$. Find the external power required for rotating the rod. Put $i = 2$ A, $l = 1/2$ m, $R = 1/2$ m, $B_0 = 1$ T and $\omega = 20$ rad s^{-1}.

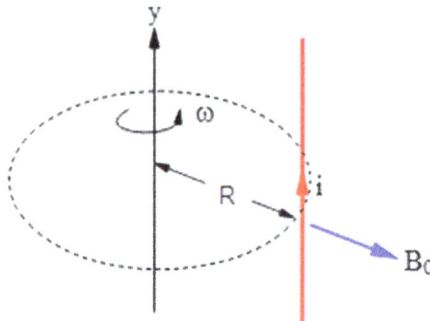

Solution

The Ampère's force acting on the rod is

$$F_a = ilB_0$$

To rotate the rod with constant angular velocity ω, we need to apply equal and opposite force $F_{ext}(=F_m)$ to generate a zero net torque about the y-axis. Then, the power delivered by the external agent is

$$P_{ext} = \vec{\tau}_{ext} \cdot \vec{\omega}$$
$$= \tau\omega \cos\pi = -(F_{ext} \cdot R)\omega = -F_a R\omega = -(ilB_0)R\omega \text{ Ans.}$$

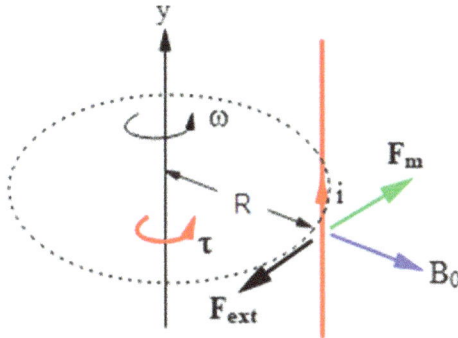

Putting $i = 2$ A, $l = 1/2$ m, $R = 1/2$ m, $B_0 = 1$ T and $\omega = 20$ rad s^{-1}, in the last expression, the magnitude of power delivered by the external agent is

$$P_{ext} = ilB_0 R\omega = (2)(1/2)(1)(1/2)(20) = 10 \text{ W Ans.}$$

N.B: The power delivered by the external agent is negative as its torque is opposite to the torque delivered by Ampère's force.

Problem 10 Two long straight parallel current-carrying conductors each of length l and current i are placed at a distance r_0. Find the total work done by the Ampère's force when an external agent is slowly reducing the distance of separation between the conductors to $\frac{r_0}{2}$. Assume that current flows in the same direction in both conductors.

Solution

As the current flows in the same direction, the conductors attract each other. The force acting on conductor 2 by conductor 1 is

$$F = ilB$$
$$= il\frac{\mu_0 i}{2\pi r}$$
$$= \frac{\mu_0 i^2 l}{2\pi r}$$

This force does work dW in displacing conductor 2 by a distance ds is

$$dW = F\ ds$$

$$= \frac{\mu_0 i^2 l}{2\pi r}(-dr)\ (\because ds = -dr)$$

Then, the total work done by Ampère's force is

$$W = \int dW$$

$$= -\frac{\mu_0 i^2 l}{2\pi} \int_{r_0}^{r_0/2} \frac{dr}{r}$$

$$= \frac{\mu_0 i^2 l}{2\pi} \ln 2\ \text{Ans.}$$

Problem 11 A square loop of mass m and side a carries a current i and is placed on a rough table in a magnetic field $\vec{B}$, as shown in the figure below. If $m = 50$ g, $a = 0.5$ m, $B = 2$ T, $\theta = 30°$ and $g = 10$ m s^{-2}, find the minimum magnitude of current so that the loop will be toppled.

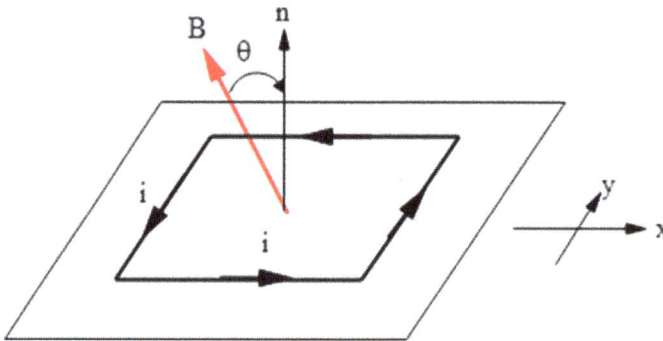

Solution

The magnetic dipole moment of the loop is

$$\vec{\mu} = iA\hat{k} = ia^2\hat{k}$$

Then, the torque experienced by the loop is

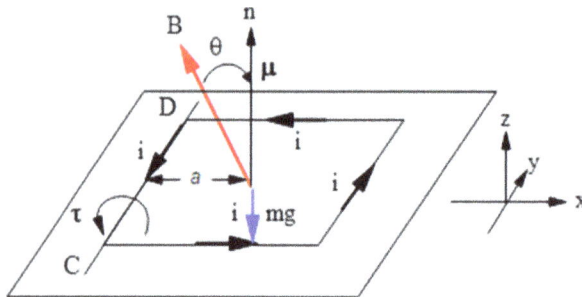

$$\vec{\tau}_m = \vec{\mu} \times \vec{B}$$
$$= (ia^2\hat{k}) \times (B\cos\theta\hat{k} - B\sin\theta\hat{i})$$
$$= -ia^2B\sin\theta\,(\hat{k} \times \hat{i})$$

$$\Rightarrow \vec{\tau}_m = -ia^2B\sin\theta\,\hat{j} \tag{7.8}$$

This torque must overcome the gravitational torque about the branch CD which is given as

$$\vec{\tau}_{gr} = mg\frac{a}{2}\hat{j} \tag{7.9}$$

Equating $|\vec{\tau}_m|$ and $|\vec{\tau}_{gr}|$ from equations (7.8) and (7.9), we have

$$i = \frac{mg}{2aB\sin\theta} = \frac{\left(\frac{1}{20}\right)(10)}{2\left(\frac{1}{2}\right)(2)\left(\frac{1}{2}\right)} = 0.5 \text{ A Ans.}$$

Problem 12 A wire carrying a current i is wrapped N times over a uniform solid of mass m and radius R which is placed on a (i) very rough, (ii) smooth horizontal surface. A horizontal magnetic field of induction $\vec{B}$ is present, as shown in the figure below. Find the (a) magnetic torque, (b) angular acceleration experienced by the sphere. Assume that the mass of the wire is negligible compared to the mass of the sphere. Neglect the mass of the wire when compared to the mass of the sphere.

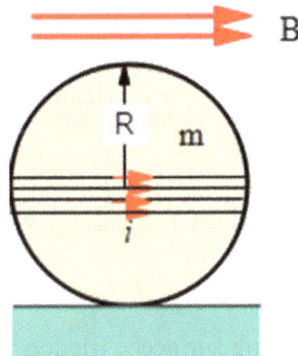

Solution

    (i) For very rough contacting surfaces:

        (a) The current loop experiences a magnetic torque which will tend to rotate the sphere in a clockwise direction. As a result, the lowest point P tends to slide to the left and the static friction acts to the right. As the static friction, normal reaction and weight of the sphere pass through P, the net torque due to these forces will be zero about P. Then, the net torque acting on the sphere is equal to the magnetic torque which does not depend upon the reference axis.

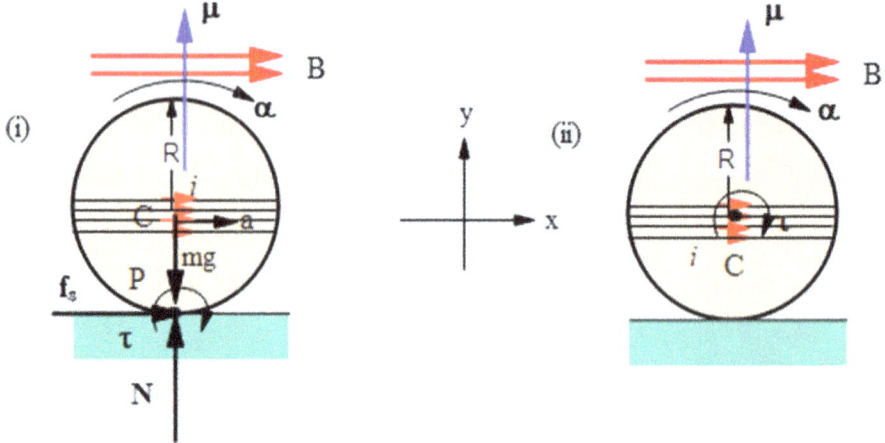

$$\vec{\tau} = \vec{\mu} \times \vec{B}$$
$$= (NiA\,\hat{j}) \times (B\,\hat{i})$$
$$= -NiA\,\hat{k}, \text{ where } A = \pi R^2$$

$$\Rightarrow \vec{\tau} = -N\pi R^2 iB\,\hat{k} \text{ Ans.}$$

(b) As the surface is very rough, the sphere will roll without sliding. So, the instantaneous axis of rotation of the sphere passes through its lowest point P. Then, the angular acceleration of the sphere about P is

$$\vec{\alpha} = \frac{\vec{\tau}}{I_P}$$
$$= \frac{-N\pi R^2 iB}{\frac{7}{5}mR^2}\hat{k} \left(\because I_P = \frac{7}{5}mR^2\right)$$
$$= -\frac{5N\pi iB}{7m}\hat{k} \text{ Ans.}$$

(ii) For the smooth contacting surface/s, friction is absent.
   (a) The magnetic torque toque does not depend upon the reference point/axis which was obtained earlier as

$$\Rightarrow \vec{\tau} = -N\pi R^2 iB\,\hat{k},$$

The other forces like weight and normal reaction cannot produce any torque about the centre of mass as they pass through it. So, the net the torque about the centre of mass is equal to the magnetic torque as it is frame independent. Ans.

(b) Then, the sphere will free to rotate about the centroidal axis parallel to the z-axis, with an angular acceleration which can be given as

$$\vec{\alpha} = \frac{\vec{\tau}}{I_c}$$

$$= \frac{-N\pi R^2 iB}{\frac{2}{5}mR^2}\hat{k} \left(\because I_c = \frac{2}{5}mR^2\right)$$

$$= -\frac{5N\pi iB}{2m}\hat{k}. \text{ Ans.}$$

Problem 13 A tiny magnet having a dipole moment μ' is placed at a distance $x(\gg R)$ on the axis of the current loop of dipole moment μ. Find the force acting on the tiny magnet.

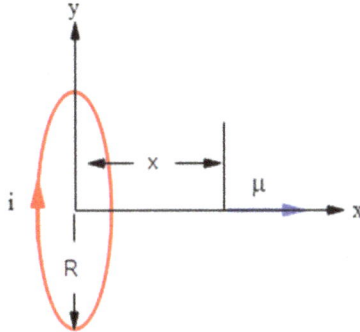

Solution
The magnetic field due to a ring at an axial distance x is

$$B = \frac{\mu_0 i R^2}{2(R^2 + x^2)^{3/2}}$$

$$= \frac{\mu_0 i \pi R^2}{2\pi x^3} \ (\because x \gg R)$$

$$= \frac{\mu_0 \mu}{2\pi x^3} \ (\because i\pi R^2 = \mu)$$

Then, the force acting on the tiny magnet is

$$F_x = \mu'\frac{\partial B}{\partial x} = \mu'\frac{\partial}{\partial x}\frac{\mu_0\mu}{2\pi x^3} = -\frac{3\mu\mu'}{2\pi x^4} \text{ Ans.}$$

Problem 14 Two particles of charge q_1 and q_2 are moving with velocities $\vec{v_1}$ and $\vec{v_2}$, respectively. Derive an expression for the (a) magnetic and (b) electrostatic force acting on the charge q_2 which is at a position $\vec{r}$ relative to the charged particle q_1.

Do not consider the relativistic effect as the velocities are much less than the speed of light.

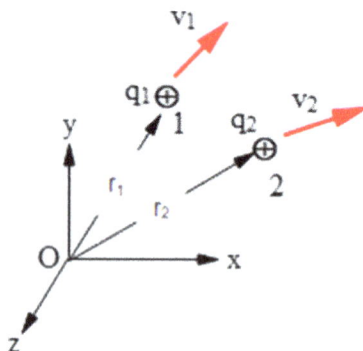

Solution

(a) Let $\vec{r}$ = position vector of particle 2 relative to particle 1 which is given as

$$\vec{r} = \vec{r_2} - \vec{r_1} \qquad (7.10)$$

As derived earlier, the magnetic field due to q_1 at q_2 is

$$\vec{B_1} = \frac{\mu_0 q_1}{4\pi r^3}(\vec{v_1} \times \vec{r}) \qquad (7.11)$$

Then, the magnetic force experienced by q_2 due to the magnetic field of q_1 is

$$\vec{F_{2_m}} = q_2(\vec{v_2} \times \vec{B_1})$$

$$= q_2\left\{\vec{v_2} \times \frac{\mu_0 q_1}{4\pi r^3}(\vec{v_1} \times \vec{r})\right\}$$

$$\Rightarrow \vec{F_{2_m}} = \frac{\mu_0 q_1 q_2}{4\pi r^3}[\vec{v_2} \times (\vec{v_1} \times \vec{r})] \qquad (7.12)$$

Using equations (7.10) and (7.11), we have

$$\vec{F_{2m}} = \frac{\mu_0 q_1 q_2}{4\pi |\vec{r_2} - \vec{r_1}|^3}[\vec{v_2} \times \{\vec{v_1} \times (\vec{r_2} - \vec{r_1})\}] \text{ Ans.}$$

(b) The electrostatic force acting on particle 2 due to the electric field of particle 1 is

$$\vec{F_{2el}} = q_1\vec{E_1}, \qquad (7.13)$$

where the electric field of particle 1 at particle 2 is

$$\vec{E_1} = \frac{q_1 \vec{r}}{4\pi\varepsilon_0 r^3} \qquad (7.14)$$

Using equations (7.13) and (7.14), we have

$$\vec{F}_{2el} = \frac{q_1 q_2}{4\pi\varepsilon_0 r^3}\vec{r} \qquad (7.15)$$

Using equations (7.10) and (7.15), we have

$$\vec{F}_{2el} = \frac{q_1 q_2 (\vec{r_2} - \vec{r_1})}{4\pi\varepsilon_0 \,|\vec{r_2} - \vec{r_1}|^3} \quad \text{Ans.}$$

Problem 15 Find the electric and magnetic forces force of interaction between two charges q_1 and q_2 moving with velocities $\vec{v_1}$ and $\vec{v_2}$ as shown in (a) figure (i) and (b) figure (ii). Discuss the validity of Newton's third law.

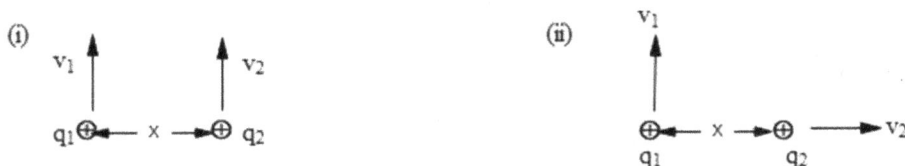

Solution

(a) In figure (i), as derived in the last problem, the electrostatic force acting on particle 2 due to the electric field of particle 1 is

$$\vec{F}_{2el} = \frac{q_1 q_2}{4\pi\varepsilon_0 r^3}\vec{r} = \frac{q_1 q_2}{4\pi\varepsilon_0 x^2}\hat{i}$$

Referring to the last problem, the magnetic force acting on particle 2 due to the magnetic field of particle 1 is

$$\vec{F}_{2m} = \frac{\mu_0 q_1 q_2}{4\pi r^3}[\vec{v_2} \times (\vec{v_1} \times \vec{r})]$$

$$\Rightarrow \vec{F}_{2m} = \frac{\mu_0 q_1 q_2}{4\pi r^3}[v_2\hat{j} \times (v_1\hat{j} \times x\hat{i})] = -\frac{\mu_0 q_1 q_2 v_1 v_2}{4\pi x^2}\hat{i}$$

Similarly, the force acting on particle 1 can be given as the combination of the following forces:

$$\vec{F}_{1el} = -\frac{q_1 q_2}{4\pi\varepsilon_0 x^2}\hat{i}$$

$$\vec{F}_{1m} = \frac{\mu_0 q_1 q_2 v_1 v_2}{4\pi x^2}\hat{i}$$

We can see that

$$\vec{F}_{1el} = -\vec{F}_{2el} \text{ and } \vec{F}_{1m} = -\vec{F}_{2m} \text{ Ans.}$$

(b) In figure (i), as derived in the last problem, the electrostatic force acting on particle 2 due to the electric field of particle 1 is

$$\vec{F}_{2el} = \frac{q_1 q_2}{4\pi\varepsilon_0 r^3}\vec{r} = \frac{q_1 q_2}{4\pi\varepsilon_0 x^2}\hat{i}$$

Referring to the last problem, the magnetic force acting on particle 2 due to the magnetic field of particle 1 is

$$\vec{F}_{2m} = \frac{\mu_0 q_1 q_2}{4\pi r^3}[\vec{v_2} \times (\vec{v_1} \times \vec{r})]$$

$$\Rightarrow \vec{F}_{2m} = \frac{\mu_0 q_1 q_2}{4\pi r^3}[v_2 \hat{i} \times (v_1 \hat{j} \times x\hat{i})] = \frac{\mu_0 q_1 q_2 v_1 v_2}{4\pi x^2}\hat{j}$$

Similarly, we can show that the electric and magnetic forces acting on particle 1 can be given as follows:

$$\vec{F}_{1el} = -\frac{q_1 q_2}{4\pi\varepsilon_0 x^2}\hat{i}$$

$$\vec{F}_{1m} = 0\hat{i}$$

In this case $\vec{F}_{1el} = -\vec{F}_{2el}$ but $\vec{F}_{1m} \neq -\vec{F}_{2m}$. It violates the Newton's third law! Try to find the answer for this anomaly. Ans.

Problem 16

(a) Show that magnetic force cannot change the speed of a charged particle.
(b) If a proton is moving in a magnetic field with a velocity $\vec{v} = (-\hat{i} + 2\hat{j} + 3\hat{k})10^6$ m s^{-1} at a point with an acceleration $(2\hat{i} - \hat{j} + c\hat{k})$, find the value of c.

Solution

(a) The magnetic force $\vec{F}_m$ is given as

$$\vec{F}_m = q(\vec{v} \times \vec{B})$$

It tells us that $\vec{F}_m$ is always perpendicular to $\vec{v}$. Hence, power delivered by a magnetic field is

$$P_m = \vec{F}_m \cdot \vec{v} = 0$$

As you know, power is the rate of change in kinetic energy given as

$$P = \frac{dK}{dt}$$

So, we have $\frac{dK}{dt} = 0$. This means that kinetic energy of a charged particle remains constant. In other words, speed of a charged particle remains constant in a magnetic field. Ans.

(b) Since, $\overrightarrow{F} \cdot \overrightarrow{v} = 0$,

$$\overrightarrow{a} \cdot \overrightarrow{v} = 0$$

$$\Rightarrow (2\hat{i} - \hat{j} + c\hat{k}). (-\hat{i} + 2\hat{j} + 3\hat{k}) = 0$$

$$\Rightarrow -2 - 2 + 3c = 0$$

$$\Rightarrow c = 4/3 \text{ Ans.}$$

Problem 17 An electric field of intensity E and a magnetic field of induction B are present, as shown in the figure. A charged particle of mass m is projected with a speed v_0 from the origin. Find its speed as the function of y.

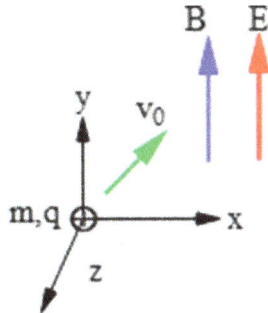

Solution
The magnetic field cannot change the speed of the charged particle because work done by a magnetic force is zero as discussed earlier; but, the speed of the charged particle can be changed by the electric field $\overrightarrow{E}$. The electrical force acting on q is

$$\overrightarrow{F_{el}} = qE\,\hat{j} \tag{7.16}$$

The work done by the electrical force is

$$W = \int dW = \int \overrightarrow{F_{el}} \cdot d\overrightarrow{y} = (qE \cdot \hat{j}) \cdot (dy\,\hat{j}) = qEy$$

This is equal to change in kinetic energy

$$\Delta K = \frac{1}{2} m \, (v^2 - v_0^2) \tag{7.17}$$

Using equations (7.16) and (7.17), we have

$$\frac{1}{2} m \, (v^2 - v_0^2) = qEy$$

$$\Rightarrow v = \sqrt{v_0^2 + \frac{2qEy}{m}} \text{ Ans.}$$

N.B: According to the work–energy theorem the sum of work done by electric and magnetic force is equal to the change in kinetic energy of the particle. As the magnetic force does not perform any work, the electric force is responsible for increasing the kinetic energy of the charged particle.

Problem 18 A negative charge of mass m and charge $-q$ is projected with a velocity v in an inward magnetic field $\overrightarrow{B}$. Find the (a) radius of the circular path, (b) period of revolution of the charge, and (c) kinetic energy of the particle in terms of radius of the circular path.

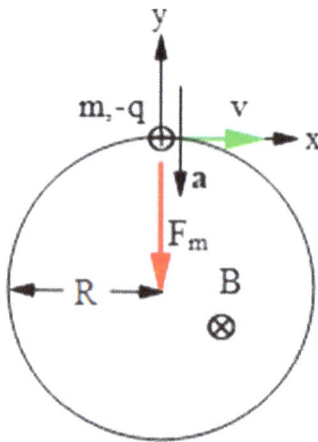

Solution
   (a) The force acting on q is

$$\overrightarrow{F_m} = (-e)v\hat{i} \times (-B\hat{k})$$
$$= - \, ev \, B \, \hat{j}$$

This force pulls the particle towards the centre of the circular path with an acceleration

$$a = \frac{v^2}{R}.$$

$$\Rightarrow \frac{F_m}{m} = \frac{v^2}{R}$$

$$\Rightarrow \frac{evB}{m} = \frac{v^2}{R}$$

$$\Rightarrow R = \frac{mv}{eB} \text{ Ans.}$$

(b) period of revolution of the charge is

$$T = \frac{2\pi R}{v}$$

$$= \frac{2\pi R}{\frac{eBR}{m}} = \frac{2\pi m}{eB} \text{ Ans.}$$

(c) The kinetic energy of the particle is given as

$$KE = \frac{mv^2}{2},$$

where the velocity of the particle in terms of radius of curvature is

$$v = \frac{eBR}{m}$$

Then we have,

$$KE = \frac{m\left(\frac{eBR}{m}\right)^2}{2} = \frac{e^2 B^2 R^2}{2m} \text{ Ans.}$$

N.B: Time period (or frequency) of revolution does not depend on the speed of the charge particle. The expression of KE is valid for the non-relativistic case where the mass of the particle remains constant.

Problem 19 An α-particle and a proton are projected with equal (i) KE, (ii) momenta, and (iii) speed. Find the ratio of (a) the radii of curvature and (b) the period of revolution.

Solution

(a) The radius of curvature is given as

$$R = \frac{mv}{qB} = \frac{\sqrt{2mK}}{qB}$$

(i) For a constant KE, the ratio of the radius of curvature is

$$\frac{R_\alpha}{R_p} = \sqrt{\frac{2m_\alpha K_\alpha}{q_\alpha B}} \bigg/ \sqrt{\frac{2m_p K_p}{q_p B}}$$

$$= \sqrt{\frac{m_\alpha}{m_p} \cdot \frac{q_p}{q_\alpha}} = \sqrt{4 \cdot \frac{1}{2}} = \sqrt{2} : 1 \text{ Ans.}$$

(ii) For a constant momentum, the ratio of the radius of curvature is

$$\frac{R_\alpha}{R_p} = \frac{m_\alpha v_\alpha / q_\alpha B}{m_p v_p / q_p B} = \frac{q_p}{q_\alpha} = 1 : 2 \; (\because m_\alpha v_\alpha = m_p v_p) \text{ Ans.}$$

(iii) For a constant speed, the ratio of the radius of curvature is

$$\frac{R_\alpha}{R_p} = \left(\frac{m_\alpha}{m_p}\right)\left(\frac{v_\alpha}{v_p}\right)\left(\frac{q_p}{q_\alpha}\right) = \left(\frac{4}{1}\right)\left(\frac{1}{1}\right)\left(\frac{1}{2}\right) = 2 : 1 \text{ Ans.}$$

(b) The period of revolution of the charged particle is given as

$$T = \frac{2\pi m}{eB}$$

It depends upon the specific charge q/m, irrespective of speed, momentum and KE. So, for (i), (ii) and (iii), the ratio of the radius of curvature is

$$\frac{T_\alpha}{T_p} = \frac{m_\alpha}{q_\alpha} \bigg/ \frac{m_p}{q_p} = \frac{m_\alpha q_p}{m_p q_\alpha} = \frac{m_\alpha}{m_p} \cdot \frac{q_p}{q_\alpha} = 4 \cdot \frac{1}{2} = 2 \text{ Ans.}$$

Problem 20

(a) A positive charge $+q$ is projected with a velocity $\vec{v}$ at an angle θ with vertical in the region of vertical magnetic field of induction $\vec{B}$. Describe the nature of the path followed by the charge ignoring the gravity of earth.

(b) If an electric field $\vec{E}$ is applied parallel to $\vec{B}$, find the y-displacement of the particle as the function of time.

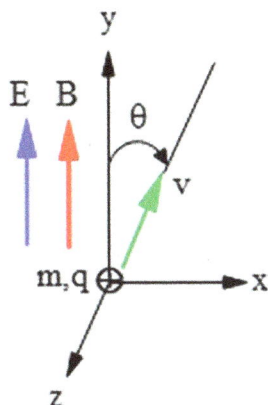

Solution

(a) Let us split the velocity $\vec{v}$ into two components $\vec{v_{\parallel}}$ and $\vec{v_{\perp}}$, where $\vec{v_{\parallel}}$ and $\vec{v_{\perp}}$ are the velocities parallel and perpendicular to $\vec{B}$, respectively. Due to $v_{\perp}(=v\sin\theta)$, the charged particle moves in a circle with a time period T and radius given as

$$T = \frac{2\pi m}{qB} \text{ and } R = \frac{qBR}{mv_{\perp}} \text{ as discussed in the previous example.}$$

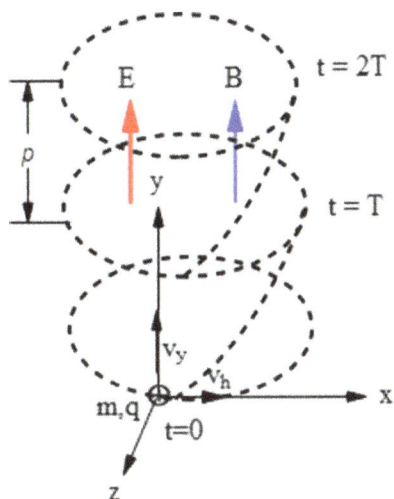

Due to $v_{\perp}(=v\cos\theta)$, the particle will drift along the direction of $\vec{B}$ while revolving with a constant time period and radius about the y-axis. As a result, the path of the particle is a helix with a pitch.

$$p = v_\parallel T$$

$$= (v \cos \theta)\left(\frac{2\pi m}{qB}\right)$$

$$= \frac{2\pi m v \cos \theta}{qB} \text{ Ans.}$$

(b) Due to the presence of vertical electric field, the particle keeps on accelerating vertically up. So, the pitch will go on increasing. However, the radius and time period remain the same. The path will be a helix with increasing pitch. The y-displacement can be given as

$$y = v\, t \cos \theta + \frac{1}{2}\left(\frac{qE}{m}\right)t^2 \text{ Ans.}$$

Problem 21 A current of two amperes flows from end 1 to end 2 of the wire. If the coordinates of ends 1 and 2 are $(1, 2, -3)m$ and $(-2\ -5,1)m$, respectively, and the wire is placed in a uniform magnetic field $\overrightarrow{B} = (\hat{i} + \hat{j} + \hat{k})$ T, find the Ampère's force experienced by the wire.

Solution

The force acting on the wire is

$$\overrightarrow{F} = i\ \overrightarrow{r_{21}} \times \overrightarrow{B} = -(\overrightarrow{r_2} - \overrightarrow{r_1}) \times \overrightarrow{B}$$

$$= 2[\{(-2) - (1)\}\hat{i} + \{(-5) - (2)\}\hat{j} + \{(1) - (-3)\}\hat{k} \times (\hat{i} + \hat{j} + \hat{k})]$$

$$= 2(-3\hat{i} - 7\hat{j} + 4\hat{k}) \times (\hat{i} + \hat{j} + \hat{k})$$

$$\overrightarrow{F} = 2 \begin{vmatrix} \hat{i} & \hat{j} & \hat{k} \\ -3 & -7 & 4 \\ 1 & 1 & 1 \end{vmatrix}$$

$$= 2[\hat{i}(-7 - 4) - \hat{j}(-3 - 4) + \hat{k}(-3 + 7)]$$

$$= 2(-11\hat{i} + 7\hat{j} + 4\hat{k}) \text{ N Ans.}$$

IOP Publishing

Problems and Solutions in Electricity and Magnetism

Pradeep Kumar Sharma

Chapter 8

Electromagnetic induction

8.1 Introduction

8.1.1 Faraday's electric motor—electromagnetic rotations

In the fall of 1820, just after Oersted's discovery, Faraday assisted Humphrey Davy and Willium Wollaston in a series of electromagnetic experiments for obtaining a continuous magnetic rotation but did not succeed. On 3rd September 1821, Faraday experimented with a vertical wire and suspended magnetic needle to obtain a continuous magnetic rotation of the wire around the magnetic pole. He suspended a conducting wire hinged smoothly from a fixed point so that it touched a conducting liquid (mercury). He put a permanent vertical magnet at the center of the container whose one face remained just above the liquid. When a current passed through the wire, it spun around the vertical magnet. The electric current converted to magnetism and then mechanical energy. On the next day, he conducted the same experiments by using mercury as the contacting liquid. His original electric motor was lost and he designed another model of his electric motor in 1822, which is shown in the following figure.

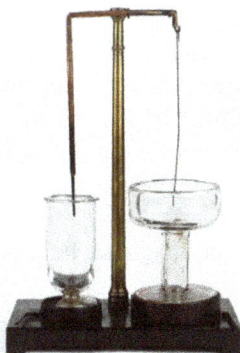

Faraday's electric homopolar motor (1821).

After one year he understood that the infinite series of concentric circular magnetic field around the wire was responsible for pushing the pole of the magnet around the wire. This is called a homopolar motor because the magnetic poles do not change, unlike AC motors and brushed (commutator) DC motors. In the homopolar motor, the wire rotates in one sense (direction) due to a constant torque produced by the Ampère force. Eventually, the current-carrying wire experiences a continuous rotation. The day of the discovery of the motor, Faraday noted:

> Arranged a magnet needle in a glass tube with mercury about it and by a cork, water, etc. supported a connecting wire so that the upper end should go into the silver cup and its mercury and the lower move in the channel of mercury round the pole of the needle. The battery arranged with the wire as before. In this way got the revolution of the wire round the pole of the magnet [1].

The continuous rotation of the wire around the pole of the magnet without any human intervention was a matter of surprise and most fascinating for anyone. The interaction between the current and magnetic field was due to a torque so as to rotate the wire—rotational effect of Ampère's force acting on the wire. This fact of perpetual rotations was appreciated and formulated by Ampère as the Newton of electricity [2].

Faraday might have received some idea from the mutual discussions between Davy and Wollaston. Faraday published his paper on his experiments without acknowledging the works of Davy and Wollaston. Davy was annoyed with this act of Faraday and kept him away from further research on electromagnetism by assigning him a heavy workload of optical measurement. Furthermore, Faraday was accused of plagiarizing the work of Wollaston. But, Faraday defended himself successfully to get out of the false accusation.

8.1.2 Invention of electromagnet

In the same year (1820) of Oersted's discovery of magnetic effect of electric current, Ampère showed that a piece of iron became magnetized while placed near a current. In 1824, British scientist William Sturgeon invented the electromagnet [3]. After few days, Arago observed that the iron filings clung to a wire while it carries current and drop when the circuit is broken. Then Arago and Ampère worked together and invented a modern electromagnet by winding a wire over an iron bar. They found that the iron bar behaved strongly magnetic while the current was on and lost its magnetism when current was cut off. These discoveries of Arago, Ampère and Oersted founded the laws governing electric circuits, establishing a link between magnetism and static electricity [4]. Ampère saw Oersted's demonstration, and after few days, he developed a mathematical expression to find the magnetic field and also magnetic forces.

8.2 Faraday's experiments of electromagnetic induction

Faraday conducted many experiments between 1820 and 1845. We will discuss his electromagnetic experiments inventing electric motor, electric generator and transformer etc.

8.2.1 Arago Rotations as an inspiration for mutual induction

In 1822, while measuring the magnetic force of earth near Greenwich, Dominique Francois Arago noticed the damping of non-magnetic metal near a compass needle. In 1824, Arago conducted a reverse experiment; he suspended a magnetic needle over a copper disc, and when he rotated the copper disc, the magnetic needle also rotated in the direction of rotation of the copper disc. This new kind of electro-magnetic action (Arago's rotation) attracted much attention because its cause was unknown to all scientists.

Faraday took it as a challenge to resolve the puzzle of Arago's rotation. Being motivated by it, Faraday conducted several experiments in vain (in his spare time) between 1824 and 1931 to explain Arago's rotation. As mentioned earlier, due to the difficult relation with Davy, Faraday was not allowed to conduct full-time research between 1825 and 1829, but was occupied with Davy's assignments. After the death of Davy in 1929, Faraday reinstated his full-time research on electromagnetism. On 29th August 1831, Faraday got his second breakthrough, that is, electromagnetic induction (the first invention was the homopolar motor in 1821). On the other hand, from the Oersted's discovery, Faraday was intending to get the reverse, that is, electricity from a magnetic field. With an experiment of two coils wrapped over a single iron ring, he could see a trace of transient current in a coil when the contact was made or broken in the other coil. This was actually the discovery of present-day *mutual* induction. Faraday noted as follows:

> A great length of copper wire, 1-20th of an inch in diameter, was wound round a cylinder of wood so as to compose two helices, the coils of which were inter-mixed, but prevented from touching each other by interposed threads of twine and calico. One helix was connected with a voltaic battery, and the other with a galvanometer. No effect was perceived on the latter, with a battery of 10 plates: a slight effect only with one of 100 plates; and a distinct deflection of the needle of the Galvanometer occurred when the contact was made with a battery of 120 plates. While the contact was preserved, the needle returned to its natural position, and was unaffected by the electric current passing through the wire connected with the battery; but on breaking the connexion, the needle of the galvanometer was again deflected, but in a direction contrary to that of its former deflection. Hence it is inferred that the electric current sent by the battery through one wire, induced a similar current through the other wire, but only at the moment the contact was made; and a current in the contrary direction when the passage of the electricity was suddenly interrupted.

In this first experiment, Faraday obtained a transient induced current. He was surprised to see that a building and dying (or decaying) current induced a transient current in the secondary coil, whereas the steady current could not induce any current at all!

8.2.2 Invention of the electric generator

In the very first day of his induction experiment (29th August 1831), Faraday asked himself, 'May not these transient effects be connected with causes of difference between power of metals in rest and in motion in Arogo's experiments' [2]?

He speculated that the force between the magnet and the rotating plate could be caused by the current induced in the plate. Then, two months later, in November 1831, to check his speculation he spun a copper disc with a constant angular velocity between the pole faces of the 'great magnet of the Royal Society' and could collect a constant current (instead of transient current) in a wire fitted with a sensitive galvanometer connected between the axis of rotation and periphery of the disc. He observed that the current was flowing radially in the disc perpendicular to its motion.

This was, in fact, the invention of the first electric generator converting mechanical energy *via* a magnetic field to electrical energy. On removing the driving force to the disc, he noticed that the motion of the disc dampened more quickly than that in the absence of the magnet. So, he concluded that the induced current in the copper disc behaved as a magnet and it interacted with the magnetic needle to rotate it. Thus, when all contemporary powerful mathematicians like Ampère, Weber, Gauss etc, could not resolve the cause of Arago's rotation by their mathematical skills, Faraday could resolve this puzzle experimentally.

So, Micheal Faraday discovered electromagnetic induction and explained Arago's rotation as an electromagnetic interaction between Foucault's current (induced eddy currents) in the copper disc and the magnetic needle. A deeper connection between electricity and magnetism was established [5].

In the second series of experiments, Faraday disturbed the magnetic environments of a coil fitted with a sensitive galvanometer by introducing or removing iron rods in to the coil. In this case he could behold transient current during insertion and withdrawal of the rod. In his words,

> Electric currents were also induced in a helix into which a soft iron cylinder was introduced, whenever that iron was rendered magnetically by induction from magnets applied to its ends. The sudden introduction or removal of a magnet, in the place of the iron cylinder, produced similar effects on the helix [2].

8.2.3 Faraday's law of induction

Faraday conducted several experiments with thin rectangular plates and loops passing between the poles of the magnets. To specify the direction of induced current was actually not so easy for Faraday because he was ignorant of vector product. So, on 4th November 1831, he coined the terms *magnetic lines of force* to state his law. He imagined the physical magnetic lines of force emanating from the North Pole and terminating on the South Pole outside the magnets (permanent and electromagnet). In January 1832, Faraday derived a common law from all his experiments regarding the condition or cause of induction as the cutting of magnetic lines or flux by a conductor. He states that: 'If a terminated wire moves so as to cut a magnetic curve, a power is called in to action which tends to urge an electric current through it' [2]. In other words, when an element of a conductor cuts the magnetic lines an emf is induced across it.

Initially, Faraday used this law for motion of the conductor or magnet or their relative motion. Later, he justified this law for time-varying currents in the loop, stating that a time-varying current generates a moving field or flux lines. We will discuss this in later sections.

8.3 Division of Faraday's experiments

Let us call the coil (loop) or conductor 'C' and the permanent magnet or electromagnet 'M'. In the last section we described the historical background of Faraday's experiments. However, in this section we need to divide these experiments disregarding the time sequence of Faraday's experiments based upon the two underlying principles involved to explain these experiments. These principles are *motional induced emf* and *transformer emf* (for time-varying currents). There was no concept of Einstein's relativity in Faraday's time. Motional induced emf is governed by Lorentz force (discovered by Lorentz in 1897), and transformer emf is governed by Faraday's law of induction. So, these are two separate principles, which was made clear after Einstein presented the special theory of relativity in 1905.

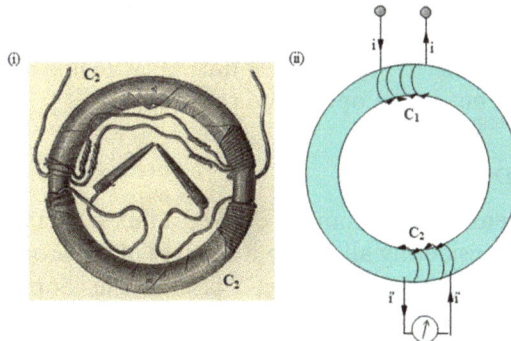

(i) Faraday discovered the mutual induction (mutual enduced emf) by his toroidal transformer in 1831;(ii) schematic diagram of the toroidal trasformer; a change in current in coil 1(C_1) induces a current in coil 2 (C_2) which can be detected by a sensitive galvanometer.

We can divide Faraday's experiments into the following three types based upon the aforementioned two principles according to the modern understanding.

8.3.1 First type of experiments

In the first type of experiments, the magnet is stationary and coil C is moved (rotated and translated) in all possible directions. We can find the electric current in the coil C except when coil C is rotated about the x-axis. While changing the area of the loop by squeezing it, we can find a deflection during the deformation of the coils.

We can notice that, in the above experiments, the total or part of the coil moves in the magnetic field of the stationary magnet (steady magnetic field).

8.3.2 Second type of experiments

In the second type of experiments coil C remains stationary and the magnet is moved (translated and rotated) in all possible directions. We can detect a current in the coil. No current flows in the coil when the magnet is rotated about the x-axis. In these experiments, the magnetic field moves because the magnet moves while the coil C is fixed.

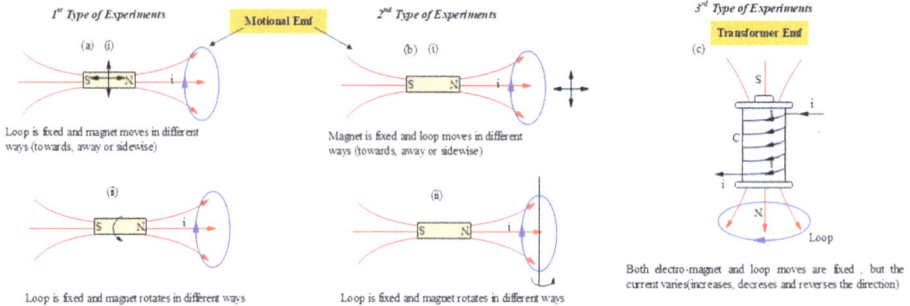

8.3.3 Third type of experiments

In our third type of experiments, the electromagnet is fixed relative to the coil. When we switch on the electromagnet, we can detect a brief current from a transient deflection in the galvanometer. Again at the time of switching off the electromagnet we can find a current, but in the opposite direction.

8.4 Faraday's flux formula

In 1845, a great mathematician Neumann derived a flux formula to measure the emf induced in a loop by considering all possible changes of the loop, such as area of the loop, orientation of the loop and magnetic field in the loop. In the third set of experiments, we can see that for any variation of the magnetic field, maximum current is induced in the coil when its plane is perpendicular to the magnetic field induction ($\theta = 90°$). However, for any time-varying magnetic field, no current flows when the plane of the coil is parallel to the magnetic field ($\theta = 90°$). Experimentally it is verified that

$$i \propto \cos \theta \tag{8.1}$$

Similarly, if the area of the coil changes, induced current i in the coil changes proportionately;

$$\text{or, } i \propto A \tag{8.2}$$

Likewise, the increment of the number of turns N of the coil increases the induced current proportionally.

$$\text{Then, } i \propto N \tag{8.3}$$

When a constant current flows in the electromagnet M, magnetic field remains constant. But, when we 'make' or/and 'break' the electromagnetic circuit, we find galvanometric deflections due to the variation of current that causes a variation of magnetic field. This means that the amount of induced current depends upon (the rate of) variation of magnetic field induction (rather than the magnitude of magnetic field induction). It is observed that, according to Lenz law, in all three cases, the induced current opposes the change in magnetic field. So, we can write

$$i \propto -\frac{\partial B}{\partial t} \tag{8.4}$$

The negative sign signifies the opposing nature of the induced current so as to change in magnetic field. Recapitulating the above expressions, we can write

$$i \propto -\frac{d}{dt}(NBA\cos\theta), \tag{8.5}$$

where $NBA\cos\theta = \phi$(flux).

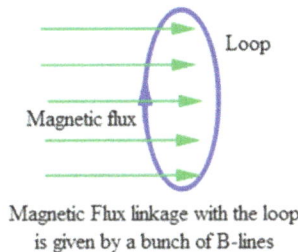

Magnetic Flux linkage with the loop
is given by a bunch of B-lines

Since $i = \dfrac{\varepsilon}{R}$, where, ε = induced electromotive force (emf) that causes the current and R = resistance of the coil C, we obtain $\varepsilon \propto -\frac{d\phi}{dt}$. For the sake of simplicity, assuming the constant of proportionality as unity, we obtain

$$\varepsilon = -\frac{d\phi}{dt},$$

where $\frac{d\phi}{dt}$ = rate of change of flux-linkage with the circuit. Please note that the above derivation of flux formula is a logical version of Neumann's flux formula. Please refer to section 8.8 for a more detailed derivation of the flux formula.

8.5 Motional induced emf

In this first set of experiments, let us observe the loop sitting on the magnet. Let the loop move in a steady magnetic field B which is not necessarily uniform. Since the loop is stationary relative to the observer (magnet), average velocity of the charge carrier is zero. Therefore, any charge carrier $+q$ does not experience magnetic force.

(i)

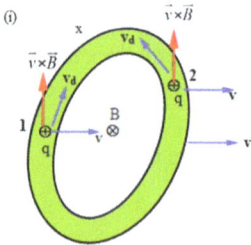

The work is done by the magnetic force $q\vec{v}\times\vec{B}$ is +ve at 1 and -ve at 2;so its total work done is zero

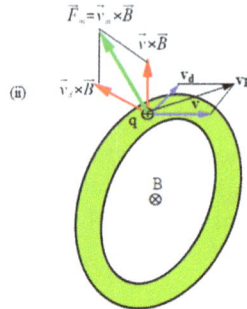

(ii)

No net work is done by the total magnetic force F_m because it is perpendicular to v_P .So, the external agent is ultimately respnsible for circulating a current

If we push the loop with a constant velocity v, each charge carrier of the loop begins to move with the same velocity v in the magnetic field. So, each charge carrier will experience magnetic force $q(\vec{v}\times\vec{B})$. The tangential component of the magnetic force pushes the charge carries (in reality, electrons in metals) inside the conductor. Eventually, each charge carrier moves with a constant drift velocity v_d producing an 'electric current'. But, the drift velocity v_d, causes an additional magnetic force $q(\vec{v_d}\times\vec{B})$ on each charged particle perpendicular to its motion ($\vec{v_d}$) inside the conductor; so. this force $q(\vec{v_d}\times\vec{B})$ cannot perform any work in drifting the charged particles. Then, who is really doing the work? Perhaps $q(\vec{v}\times\vec{B})$? Let us accept it for the time being. When a charge carrier circulates, sometimes the magnetic force $q(\vec{v}\times\vec{B})$ favours its motion v_d (as shown at point 1) performing positive work, whereas at some other points it hinders the motion v_d (as shown at point 2) performing negative work. However, the total work done by this magnetic force is

$$W_{\text{mag}} = q \oint (\vec{v}\times\vec{B}) \cdot \vec{dl}$$

in pushing a charged particle in an around trip. This must be positive so as to establish an induced current. Then, the work done by the magnetic force on a unit positive charge in an around trip is known as 'emf', given as

$$\varepsilon = \frac{W_{\text{mag}}}{q} = \oint (\vec{v}\times\vec{B}) \cdot \vec{dl}.$$

This emf is induced by the motion of the conductor; so, we call it *motional induced emf* denoted as

$$\varepsilon_m = \oint (\vec{v}\times\vec{B}) \cdot \vec{dl},$$

where v is the velocity of the conductor relative to the ground observer denoted as $\vec{v}_{co}$. Let us now try to find the root (ultimate) cause of motional induced emf. At the first look it appears as if the magnetic force $q\,(\vec{v}\times\vec{B})$ is responsible for motional induced emf, but the total magnetic force

$$\vec{F_m} = q\,(\vec{v}\times\vec{B}+\vec{v_d}\times\vec{B})$$

is perpendicular to the velocity

$$\vec{v_p} = (\vec{v_d} + \vec{v})$$

of the charged particle; so, it can perform positive work. On the other hand, the viscous force of the electron cloud, that is, 'f' being a resistive force cannot perform a net positive work on the charged particles. Then, what is solely responsible for producing a current in the conductor? Let us see the following example.

Example 1 (Cause of motional induced emf)

Let us write the formula of motional induced emf in a loop as

$$\varepsilon = \oint \vec{v}\times\vec{B}\left(=\frac{W_{mag}}{q}\right)$$

As you know, the work done by magnetic force is zero. If we put $W_{mag} = 0$, we get $\varepsilon = 0$. Then, where is the fallacy? What is actually the root cause of motional induced emf?

Solution

As mentioned earlier, a constant current is given by a drift velocity, which is caused by the tangential component of the magnetic force $q\,(\vec{v}\times\vec{B})$. But this force arises from of the (motion) velocity v of the conductor. Now, to find the ultimate cause of induction, we need to think what causes the motion of the conductor.

Will the conductor continue to move with the same velocity if the external agent withdraws the force F? Absolutely not. The conductor will slow down by the Ampère's force which is the backward component of the magnetic force F_m. This means that the external force F must counteract the backward component of F_m to pull the conductor with a uniform velocity. So, the external force F takes the credit of setting the induced current.

The work done by the external force while the rod undergoes a displacement ds during time dt is

$$\begin{aligned}
dW_{ext} &= F_{ext}\,ds = F_{ext}\,v\,dt \\
&= qv_d\,Bvdt \;(\because F_{ext} = qv_d B) \\
&= qv(v_d\,dt)B \\
&= qvB\,dl \;(\because v_d\,dt = dl)
\end{aligned}$$

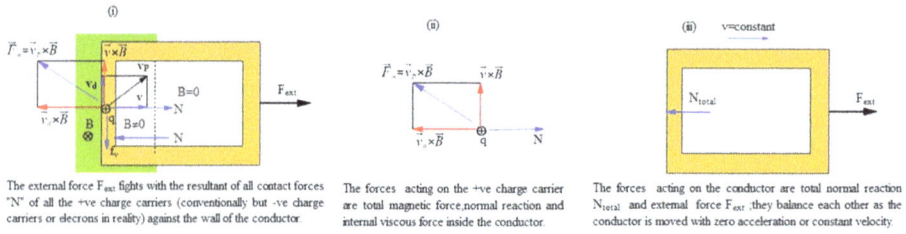

The external force F_{ext} fights with the resultant of all contact forces "N" of all the +ve charge carriers (conventionally but -ve charge carriers or elecrons in reality) against the wall of the conductor.

The forces acting on the +ve charge carrier are total magnetic force, normal reaction and internal viscous force inside the conductor.

The forces acting on the conductor are total normal reaction N_{total} and external force F_{ext}; they balance each other as the conductor is moved with zero acceleration or constant velocity.

Then, the total work done by the external agent in circulating the charge carrier q is

$$W_{ext} = \int dW_{ext} = q \oint vB \, dl = q \oint (\vec{v} \times \vec{B}) \cdot \overrightarrow{dl}$$

Hence, the motional induced emf is

$$\varepsilon_m = \frac{W_{ext}}{q} = \oint (\vec{v} \times \vec{B}) \cdot \overrightarrow{dl},$$

where v = velocity of conductor (observer) relative to the magnetic field.

This expression signifies that work done by the magnetic force in drifting the charge particles for a round trip in the loop is ultimately equal to the work performed by the external agent. Since the magnetic force as a whole does not perform any work, the external agent (force F) takes the full credit of pushing the charged particles around, constituting an induced current. Even though the external force F does not act on the charge carriers, it is able to do this job by developing a magnetic force F_m. Ans.

N.B: The tangential component of F_m performs positive work in drifting the charged particles, whereas its perpendicular component does equal negative work in pushing the charged particles back, ultimately summing them up, we will get zero work. Hence, the magnetic force serves as a mediator. It just redirects the power delivered by the external agent on the conductor, without performing any work, to the charged particles inside the conductor. In this way, the mechanical energy is converted into electrical energy via magnetic field. In this process, the magnetic force F_m acts as a converter and itself remains unaffected. So, you should not attribute the magnetic force F_m as the ultimate cause of induction. Then, the external agent that drives the conductor is ultimately responsible for induction.

Example 2 A smooth conducting bar of length l, slides with a constant velocity v on a conducting rail, fitted with a resistance R. Assuming inward magnetic field to be present throughout, find the
  (a) emf induced across the bar;
  (b) polarity of the bar;
  (c) electric field due to accumulated charges inside the bar.

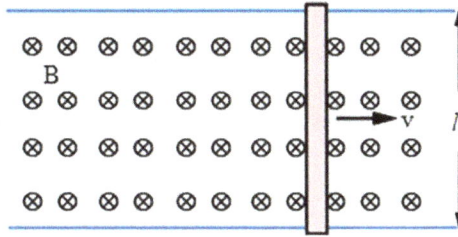

Solution

(a) Directly apply the formula,

$$\varepsilon_m = \oint (\vec{v} \times \vec{B}) \cdot \vec{dl}.$$

Since only the bar is moving as a part of the circuit and the circuit is open, emf is induced in the bar only. This is given as

$$\varepsilon_m = \int_0^l (v\,\hat{i}) \times (-B\,\hat{k}) \cdot dl\,\hat{j} = Bvl \text{ Ans.}$$

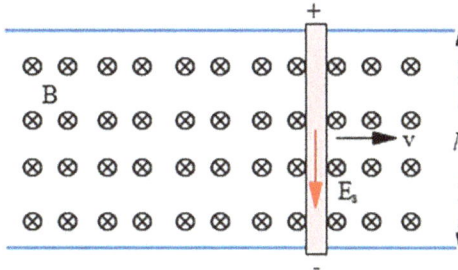

(b) As the positive charge carriers are pushed up, lower end A will be negatively charged and upper end B will accumulate the positive charges. Ans.

(c) The static charges accumulated at A and B will produce an electrostatic field $\vec{E_s}$ which is given as $\vec{E_s} = -vBj$. Ans.

8.6 Induced electric field

In Faraday's second type of experiments, the magnet moves and the coil is stationary relative to the observer. We should always imagine that the observer is fixed with the moving coil. Since, the velocity v of the loop (coil) relative to the observer is zero, the observer will not be able to detect a magnetic force $q\,(\vec{v} \times \vec{B})$. However, he will record the same current (as mentioned in the previous section) which signifies the same drift velocity v_d. Hence, in this case, the observer will report the presence of magnetic force q $(\vec{v_d} \times \vec{B})$ caused by the drift velocity v_d. This force cannot drive the electrons because it is perpendicular to the drift velocity v_d. Since the driving force is not magnetic, then what is responsible for this? To drive the charged particles, the only option left for us is electrical force but there seem to be no such electrical agents such as batteries present. Hence, we may call it an 'induced electric force' which is not directly associated with the charged particles. For this we need to define the corresponding electric field as *induced electric field*, given as

$$\vec{E} = \vec{v} \times \vec{B},$$

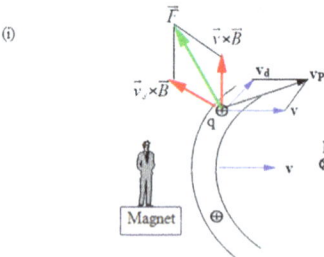

where v = velocity of the observer (loop) relative to the magnet (magnetic field).

Proof

In the first experiment, the observer is fixed with the magnet and the loop moves to the right relative to ground with velocity v, whereas in the second experiment the observer is fixed with the loop and the magnet moves to the left with the same speed. According to the theory of relativity, both the cases are equivalent. Hence, it is claimed that equal force acts on the charged particle $+q$ in both cases. Equating the forces on the charged particle,

$$\vec{F} = q\,(\vec{v} \times \vec{B} + \vec{v_d} \times \vec{B}) = q\,(\vec{E} + \vec{v_d} \times \vec{B})$$

So, the induced electric field E can be given as,

$$\vec{E} = \vec{v} \times \vec{B}$$

In Faraday's 1st series of experiments, observer is at rest and magnet moves. The net force acting on the charged particle of the (induced) current carrying conductor is $\vec{F}_n = \vec{v}_d \times \vec{B}$

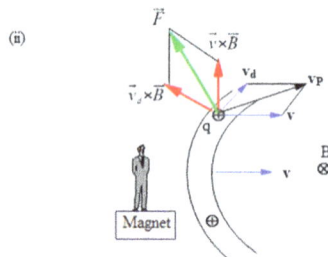

In Faraday's 2nd series of experiments, observer is at rest on the coil (conductor) and the magnet moves. The net force acting on the charged particle of the (induced) current carrying conductor is $F_n = q(\vec{v}_d \times \vec{B} + E)$

More generally, $\vec{E} = \vec{v}_{om} \times \vec{B}$,

where $\vec{v}_{om}$ = velocity of the observer relative to the magnetic field (or magnet). Ans.

Example 3 A rod of length l moves with a velocity $\vec{v} = v\hat{i}$ in a magnetic field $\vec{B} = -B\,\hat{k}$. (a) Find the force acting on the positive charge $+q$ and electron $-e$ when the observer is (i) fixed with the rod (ii) standing outside the rod (b). Show that $\vec{E}_{ind} + \vec{E}_{static} = 0$ for the observer fixed with the rod.

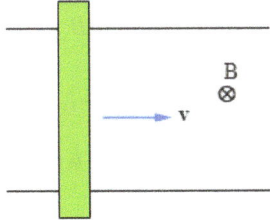

Solution

(a) When the observer is fixed with the rod he cannot observe Lorentz magnetic force because the velocity of charged particle q inside the rod relative to the observer is zero.

The forces acting on a positive charge q are given as follows:

Induced electric force is

$$\vec{F}_{ind} = (+q)\,(\vec{v} \times \vec{B})$$
$$= qvB\,\hat{j} \text{ Ans.}$$

The electrostatic force is

$$\vec{F}_s = -\vec{F}_{ind} = -qvB\,\hat{j} \text{ Ans.}$$

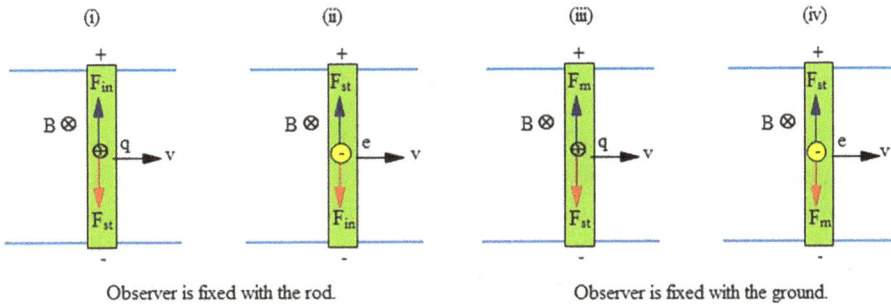

Observer is fixed with the rod.   Observer is fixed with the ground.

Magnetic force is

$$\vec{F}_m = q\,(\vec{v} \times \vec{B}) = qvB\,\hat{j} \text{ Ans.}$$

Static electric force is

$$\vec{F}_s = -\vec{F}_m = -qvB\,\hat{j} \text{ Ans.}$$

The forces acting on the electron are given as follows:
Induced electric force

$$\vec{F_i} = -e\,(\vec{v} \times \vec{B}) = -evB\,\hat{j} \text{ Ans.}$$

Electrostatic force

$$\vec{F_s} = -\vec{F_i} = +evB\,\hat{j} \text{ Ans.}$$

Magnetic force

$$\vec{F_m} = -evB\,\hat{j} \text{ Ans.}$$

Electrostatic force

$$\vec{F_s} = +evB\,\hat{j} \text{ Ans.}$$

(b) Since, $F_{\text{net}} = 0$ in both cases, we have

$$\vec{E_{\text{ind}}} + \vec{E_{\text{st}}} = 0 \text{ Proved.}$$

8.7 Concept of moving flux and induced electric field (optional)

In 1831, Michael Faraday introduced the concept of magnetic flux, defined as

$$\phi = \oint \vec{B} \cdot d\vec{A}$$

to explain electromagnetic induction. The concept of flux was borrowed from hydrodynamics. The liquid flux, that is, v-flux (time rate of flow of volume of liquid) is given as

$$\phi_v\left(=\frac{dV}{dt}\right) = \int \vec{v} \cdot d\vec{A},$$

where $\vec{v}$ is the flux density of $\vec{v}$-field (liquid flow). The difference between the liquid and magnetic flux is that the liquid particles move, whereas in magnetic flux nothing moves. When we put a bit of cork or straw on a liquid flux, its lines of motion are called streamlines (or less commonly v-lines). Likewise, if we put a small magnetic needle in a magnetic flux, the line of its orientation represents the lines of B-flux, but nothing moves in it like in hydrodynamics. However, we can at least imagine the invisible magnetic lines of force, which are really (physically) existing.

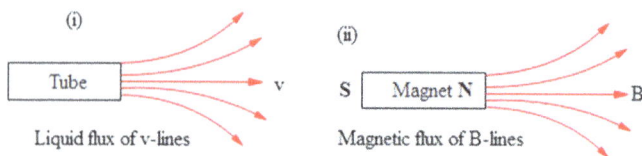

B-lines are analogous to v-lines

When a tube containing a flow of liquid moves, the flux of liquid will also move. When we move a torch, the flux (or beam) of light (energy per second) also moves. Similarly, when a magnet moves, the magnetic flux can also be imagined as a 'moving flux' that generates a changing magnetic field in space. We know that a current element behaves as a magnet, spreading its magnetic lines of force (flux). When a current element moves, we can imagine its B-flux moving in space. When a stationary conductor carries a time-varying current, the flux passing through a loop must change with time. Then, we can say that a time-varying current generates a moving flux. When a magnetic flux moves relative to an observer, we can detect the presence of an induced electric field.

The movement of the flux causes a variation of flux density B at any point in space. In other words, a time-varying magnetic field is associated with a moving flux. When we keep a conductor in a time-varying magnetic field, it is swept or cut by the moving magnetic flux lines. When the magnetic flux cuts (links with) the conductor, which is regarded as a physical path. In the absence of any conductor (or material body) we can imagine any line or contour (mathematical path) in the magnetic field and think that the imaginary lines are cut or swept by the magnetic lines of force, or B-lines.

The velocity of moving flux lines is equal to the velocity v_{mo} of the magnet relative to the observer who is fixed with the loop or standing on the loop. Then, the electric field induced in space can be given as

$$\vec{E} = \vec{v}_{om} \times \vec{B} = -\vec{v}_{mo} \times \vec{B},$$

where v_{mo} = velocity of magnetic flux relative to the observer.

Example 4 (Solenoid)

Let us consider a uniform inward magnetic field generated by a long solenoid fixed relative to ground. If you move inside the solenoid with a velocity v to the right, how do you perceive the induced E-field? Is the induced field conservative?

Solution

The induced electric field is

$$\vec{E} = \vec{v}_{om} \times \vec{B}, \text{ where } \vec{v}_{om} = v\,\hat{i} \text{ and } B = -B\,\hat{k}. \text{ So, we have, } \vec{E} = Bv\hat{j}$$

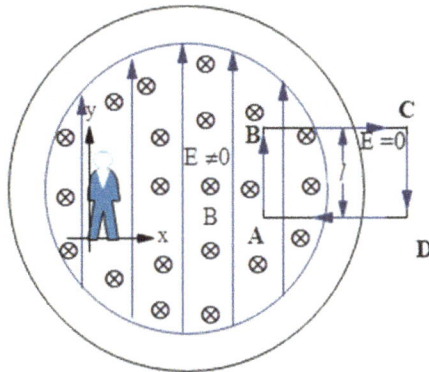

Both magnitude and direction of Induced E - field is governed by velocity v because $\vec{E} = \vec{v} \times \vec{B}$

Since $\vec{B}$ is constant and $\vec{v}$ is kept constant, obviously you will notice a uniform upward electric field of magnitude Bv. But you do not find any electric field outside the solenoid because outside a long solenoid no magnetic field exists; $(B = 0)$. If the electric field is present only inside the solenoid, its line integral in a closed path will not be equal to zero. So, it is not a conservative field. In this case, $\oint \vec{E} \cdot \vec{dl}$ will be equal to

$$\oint (\vec{v} \times \vec{B}) \cdot \vec{dl} = vBl,$$

which will be equal to zero if you take the path totally inside the solenoid. But, this does not mean that induced electric field $\vec{E}$ is conservative. For this, $\oint \vec{E} \cdot \vec{dl}$ must be zero for every closed path. *The electric field arising from the time-varying magnetic field is called the induced electric field, which is non-conservative.*

8.8 Derivation of Faraday's flux formula from the concept of moving flux (optional)

Let us sit on a conductor which moves in a time-varying (moving) magnetic flux (field). Let the velocity of the magnetic flux relative to the conductor (observer) be $v_{mc}(=v_{mo})$. Then, the induced electric field detected by the observer is

$$\vec{E} = -\vec{v}_{mo} \times \vec{B}$$

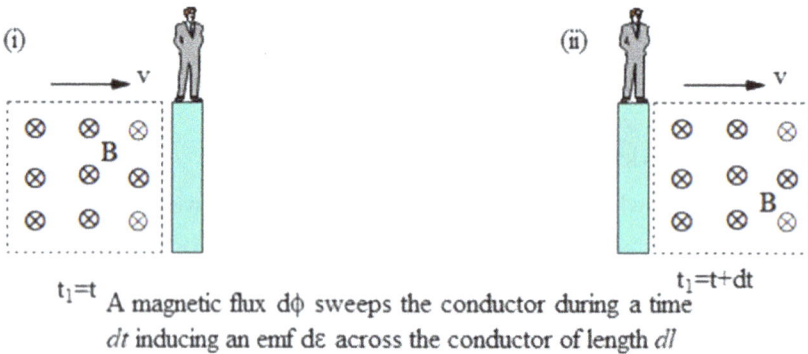

A magnetic flux $d\phi$ sweeps the conductor during a time dt inducing an emf $d\varepsilon$ across the conductor of length dl

The induced emf along the elementary length dl of the conductor (or any imaginary path) is

$$d\varepsilon = -\vec{E} \cdot \vec{dl}$$
$$= \vec{v}_{mo} \times \vec{B} \cdot \vec{dl}$$
$$= \frac{ds}{dt} \times \vec{B} \cdot \vec{dl} \left(\because \vec{v}_{mo} = \frac{ds}{dt} \right),$$

where ds = the elementary length covered by the conductor during time dt.

$$= -\frac{\vec{B} \cdot \vec{ds} \times \vec{dl}}{dt} = -\frac{\vec{B} \cdot d(\vec{\delta A})}{dt},$$

where $\vec{B} \cdot d(\vec{\delta A})$ = the magnetic flux swept by the conductor of length dl during time dt.
  Then, the total emf in the closed loop is

$$\varepsilon = -\int d\varepsilon = -\frac{d}{dt} \int \vec{B} \cdot \vec{dA},$$

where $\int \vec{B} \cdot \vec{dA}$ = the flux $d\phi$ swept by the closed loop during time dt.
  Then, we have

$$\varepsilon = -\frac{d\phi}{dt}$$

When the flux changes (moves) because of the relative motion of the electromagnet and the conductor, the resulting emf is termed as motional induced emf ε_m, whereas the emf caused by the moving flux owing to the variation of current in the electromagnet is known as transformer emf ε_t. When both the effects are present, the total emf induced can be written as

$$\varepsilon = \varepsilon_m + \varepsilon_t$$

These two emfs are based upon two different principles. We cannot derive one from the other. However, we are able to incorporate both the effects by a simple flux formula using Lorentz's force equation.
  From the above arguments, we recapitulate that the principle of motional induced emf is a manifestation of magnetic force given by Lorentz, whereas the transformer emf is dealt with the time-varying field which is the essence of Faraday's law of electromagnetic induction. Hence, we can state Faraday's law as

Whenever there is a time varying magnetic field (arising out of time varying current), an induced electric field is generated in space.

  Hence, Neumann's formula

$$\varepsilon = -\frac{d\phi}{dt}$$

cannot be truly stated as Faraday's law of electromagnetic induction. It is just a kind of 'flux formula' to tackle the problems. It incorporates the basic principles such as Lorentz's magnetic force and Faraday's law of electromagnetic induction.

Example 5 Faraday's law of electromagnetic induction states that 'A time-varying magnetic field gives rise to an induced electric field' in the space. Can you incorporate (generalize) this statement for all the experiments of Michael Faraday? Explain. Is there any other general statement fitting to all these three experiments? Narrate.

Solution

When you sit on the moving conductor of the loop, you can observe the change in magnetic field at any point in space or in the conductor referring to the first and second types of experiments. In the third types of experiments, a time-varying current will produce a time-varying magnetic field which will consequently induce an electric field. In this way, we can incorporate the statement for all three experiments.

In another way we can say that the magnetic flux cuts the conductor in the case of first and second experiments when there is a relative motion between the conductor and magnet and in the third experiment where a time-varying current produces a moving flux which sweeps the conductors to generate an induced electric field. Thus, both the statements are valid. Ans.

8.9 Properties of induced electric field (optional)

In a Betatron, the tangentially accelerating electrons reveal the existence of an induced electric field in a time-varying magnetic field. Instead of charged particles (electron beam) if we place a conducting loop in a time-varying magnetic field, the same current flows in the loop for uniform cross-section of the wire loop at a given instant. Hence, the current density $\vec{J}$ remains the same for all points in the loop at the given instant. This indicates that an equal magnitude of electric field intensity E' is experienced inside the conductor because

$$\vec{J} = \sigma \vec{E}',$$

where σ is the conductivity of the loop. The electric field E' inside the conductor need not be equal to the induced electric field E corresponding to any points in the space occupied by the conductor in its absence.

At each point of the loop, the tangential component of E starts driving the conduction electrons inside the conductor setting a net current in the conducting loop. The work done by the induced electric field E in displacing the electrons from that point through an elementary distance dl in the conducting loop is given as

$$dW = e \int \vec{E} \cdot \vec{dl}$$

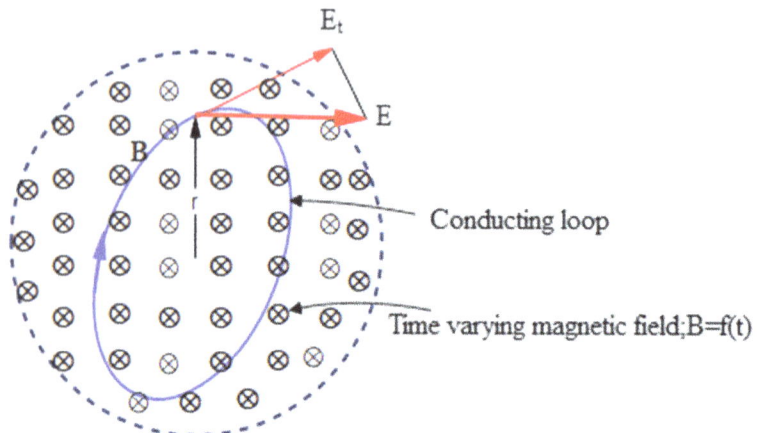

The varying magnetic field/flux induces an electric field whose tangential component would be helpful in driving the electrons and setting a current;ultimately the average electric field in the conductor is responsible for the flow of current

The work done for a round trip for a unit charge, i.e., the emf induced for the closed loop can be given as

$$\varepsilon = \oint \vec{E} \cdot \vec{dl}.$$

Since, $\varepsilon = iR$ according to Ohm's Law, where $i =$ induced current $= JA$ and

$$R = \frac{L}{\sigma A}$$

It gives

$$\frac{\oint \vec{E} \cdot \vec{dl}}{l} = \frac{Jl}{\sigma}.$$

Substituting

$$J = \sigma E'$$

to obtain

$$\oint \vec{E} \cdot \vec{dl} = \vec{E'} l.$$

Then

$$E' = \frac{\oint \vec{E} \cdot \vec{dl}}{l}.$$

This signifies that E' is not the induced electric field E at any point, rather it is its average value over the entire conducting loop. This is what we experience in practice manifested as an induced current. In fact, inside a uniform conducting loop carrying a constant current subjected to a time-varying magnetic field, one value of the electric field is experienced inside it, that is, the average induced electric field E' (not E).

Thus, Faraday's law of induction gives us the average induced electric field instead of actual induced electric field at any point in a conducting loop placed in a time-varying magnetic field.

The conductor may or may not disrupt the magnetic field pattern but according to the above discussions it is evident that the induced electric field pattern is disturbed. However, on the removal of the conductor the original electric field pattern will be restored.

8.9.1 Ionizing property of E_{ind}

When a conductor is placed in a time-varying magnetic field, the net electric field $\vec{E}_{net} = \vec{E}_{el} + \vec{E}_{ind} = 0$ inside a conductor. When we look outside the conductor, an electric field pattern is formed as the resultant of the electric field of induced charges and the external electric field. The field line must meet the conductor surface perpendicularly to satisfy the conductor property. The maximum electric field strength near the conductor must not exceed the dielectric strength of air 3×10^6 V m^{-1} to avoid breakdown. That is,

$$|\vec{E}_{\text{out}}| = |\vec{E}_{\text{ind}} + \vec{E}_{\text{el}}| \leqslant E_{\text{breakdown}}$$

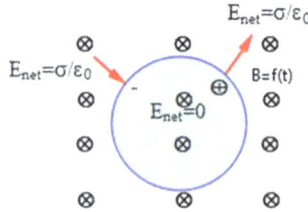

The varying magnetic induces an electric field which polarises the conductor to set up an electrostatic field. The sum of static and induced field inside the conductor is zero and just outside the conductor it is normal to the conductor surface.

It is worth noting the following.

The induced electric field has the ability to ionize a gas and make it conducting. At any point in the small gap (in the conducting loop placed in a time-varying magnetic field) we have a non-zero electric field

$$\vec{E} = \left(\vec{E}_{\text{el}} + \vec{E}_{\text{ind}} \right)$$

and inside the conductor $\vec{E} = 0$ as $\vec{E}_{\text{el}}$ and $\vec{E}_{\text{ind}}$ nullify each other because the static charges are induced. This leads to another induced field which is only associated with the static charges deposited, that is, an electrostatic field E_{el}. This counteracts the external (applied induced) electric field E_{ind}, completely inside the conductor.

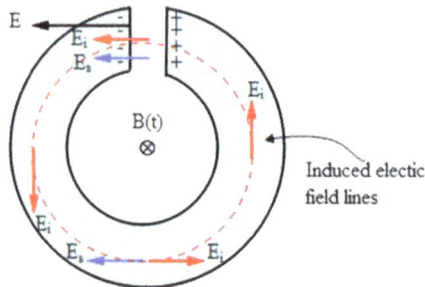

The net electric field in the air gap is $E = E_i + E_s$; it must be less than the breakdown strength of air to avoid the discharge in the air gap.

Hence, the total electric field vanishes inside the conductor and outside of it is associated with the combination of the applied induced electric field caused by the time-varying magnetic field and electric field of the static charges induced on the conductor.

In this way, the induced electric field pattern is distorted by the insertion of an open circuited conducting loop in a time-varying magnetic field but the magnetic field distribution remains unaffected. The electric field in the air (gas) should not exceed breakdown voltage to avoid sparking.

Here is an example to illustrate how an induced electric field can cause ionization of air.

Example 6 In the circular region of the magnetic field possessing axial symmetry let us assume that it varies with time. What is the minimum time in which a magnetic field of induction B can be safely established? E_b = breakdown potential gradient of the gas (or the medium) in which magnetic field varies.

Solution

Referring to the previous examples, if B changes at a rate of $\frac{dB}{dt}$, a distance r from the center, the induced electric field is given as

$$E = \frac{r}{2}\frac{dB}{dt},$$

since E is maximum at the periphery

$$E_{max} = \frac{R}{2}\frac{dB}{dt}.$$

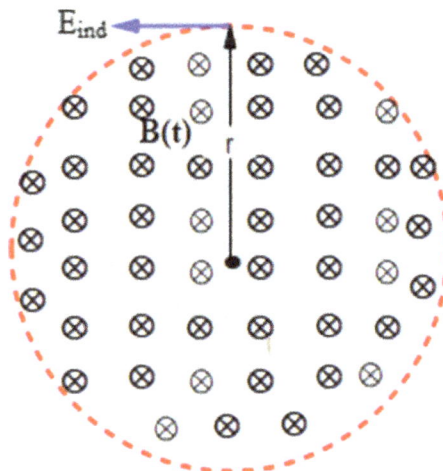

To avoid ionization at the periphery, $E_{max} \leqslant E_{breakdown}$.
Then,

$$\frac{r}{R}\frac{dB}{dt} < E_b.$$

It gives $dB \leqslant \frac{2F_b}{R}dt$. If during time T, B changes (increases) from 0 to B, we can write

$$\int_0^B dB \leqslant \frac{2E_b}{2}\int_0^T dt.$$

This yields, $T \geqslant \frac{RB}{2E_b}$. Then the minimum time is given as

$$T = \frac{RB}{2E_b} \text{ Ans.}$$

Example 7 A particle of charge q and mass m is constrained to move in a smooth insulated circular groove (path) coaxial to the axis of symmetry of the changing (time-varying) magnetic field. How does the current caused by the charged particle vary with time?

Solution

The induced electric field E is circular that pushes the charge q tangentially. So, the charged particle moves with a tangential acceleration

$$a = \frac{qE}{m}$$

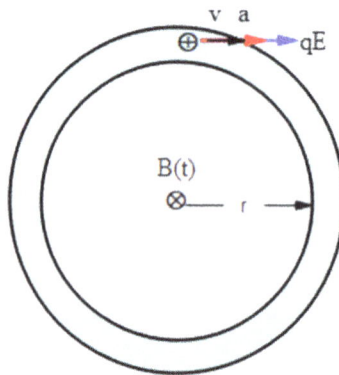

Thus, it increases its speed from zero to

$$v = at = \frac{qE}{m}t$$

during a time t constituting an electric current

$$i = \frac{q}{t}, \text{ where } T = \frac{2\pi}{v}.$$

This yields

$$i = \frac{e^2 E}{2\pi mr} t,$$

where

$$E = \frac{r}{2} \frac{dB}{dt}$$

as derived earlier. If we maintain $\frac{dB}{dt}$ constant E will remain constant.

This in turn goes on accelerating the particle to an incredible speed. Eventually a huge electric current will be built up after a very long time ($t \to \infty$). How do we reconcile the impractical infinite current? You can tackle this situation by adopting the principle of the special theory of relativity. Ans.

8.9.2 Non-conservativeness of induced electric field

As we have learnt, any time-varying magnetic field is associated with an induced electric field. This does not depend upon the presence or absence of any material medium; (conductor or dielectric etc) unlike ordinary cells. The induced electric field can polarize a dielectric material placed in it by exerting forces on bound charges. Furthermore, it can generate tremendous heat in conductors which is known as induction (or eddy current) heating. It can even ionize any medium when its strength exceeds the breakdown strength of that medium, as discussed earlier. When a dielectric is placed in a time-varying magnetic field, two types of fields exist in it. One is applied (induced) electric field and the other is the electrostatic field E_{el} of the bound charges. Then the effective electric field at any point can be written as

$$\vec{E} = \vec{E}_{el} + \vec{E}_{ind}.$$

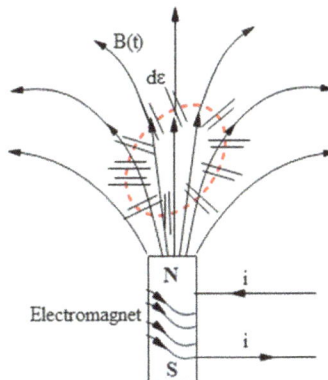

A reigon of time-varying magnetic field can be imagined as an array of elementary cells in any path.

A region of time-varying magnetic field (or induced electric field) can be imagined as a continuous distribution of hypothetical elementary (point-like) induced cells. These microscopic cells exist independently without any material medium so long as the magnetic field is varying. These are associated with induced electric field. As induced electric field is not directly linked with any charge separation (but depends on the rate of change in magnetic field), the induced cells are free from any charges (electrostatic field). When we take any hypothetical (mathematical) contour, the average induced electric field E_{av} along it gives us the total emf ε of the cells arranged in series, as shown in the figure, where ε is equal to the rate of change in magnetic flux enclosed by that contour, that is,

$$\varepsilon = \int d\varepsilon = \oint \vec{E} \cdot \vec{dl} = \frac{d\phi}{dt}.$$

Since $\frac{d\phi}{dt} \neq 0$, $\vec{E} \cdot \vec{dl} \neq 0$. Hence, E-field is non-conservative.

The electric field induced by the time-varying magnetic field is non-conservative, and we cannot use the concept of potential for this field. However, we can use Ohm's point form $\vec{J} = \propto \vec{E}$, where $\vec{E}$ is the resultant of static and induced electric fields.

Example 8 Discuss the effect of placing a conducting loop L outside the time-varying magnetic field, as shown in the figure.

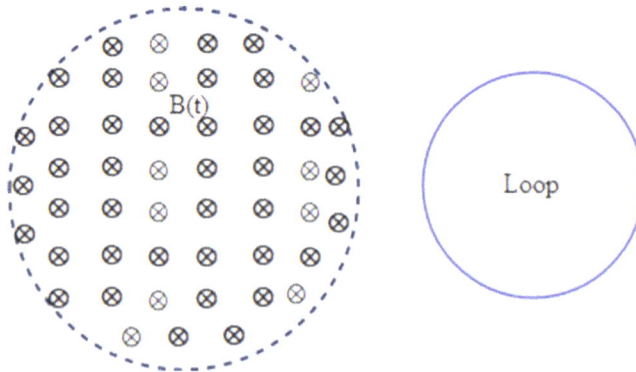

Solution

If the contour is taken outside the magnetic field, induced emf $\varepsilon = 0$. In these cases, $E_{av} = 0$, whereas E_{ind} may not be zero at any point in the contour. The emf $d\varepsilon$ of any elementary segment of the contour is given as $d\varepsilon = \vec{E_i} \cdot \vec{dl} = E_{av} dl$. If we sum up emfs of all the cells in the closed contour, we obtain $\varepsilon = \oint \vec{E} \cdot \vec{dl} = \oint E_{av}\, dl$.

Now, put $E_{av} = 0$ to obtain no net emf. In this case, partially if we place a conducting loop no current will flow. This is because of zero average induced electric field along the loop. It is inferred that the hypothetical elementary cells must be added algebraically. In this contour, in some cases, for some points $\overrightarrow{E}$ favours $\overrightarrow{dl}$ and in some other points $\overrightarrow{E}$ opposes $\overrightarrow{dl}$.

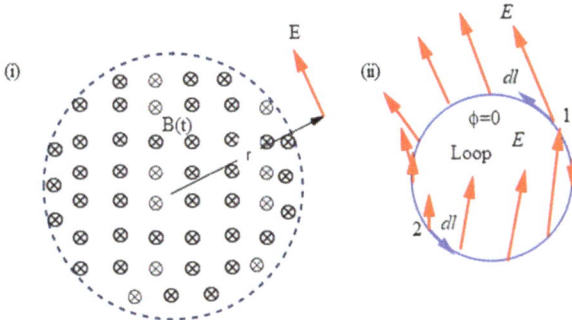

The closed line integral of the induced electric field E along the loop is zero because some places such as at point 1, E favours dl and some places like point 2, E opposes dl. In this case the nt work done by the induced electric field is zero.

Then, the closed line integral $\oint \overrightarrow{E} \cdot \overrightarrow{dl}$ becomes zero. Physically, this means that induced emf $\varepsilon = \oint \overrightarrow{E} \cdot \overrightarrow{dl}$, that is, numerically equal to the rate of change of flux passing through it is zero. This is possible only when $\phi = 0$ or constant; when $\frac{dB}{dt} = 0$ for constant magnetic induction, it is a magnetostatic case. In other words, a steady magnetic field cannot give rise to an induced emf, that is, $\varepsilon = \oint \overrightarrow{E} \cdot \overrightarrow{dl} = 0$. Please note that the induced electric field $E = 0$ for all points of a steady magnetic field (relative to an observer fixed with the field).

When we take a contour outside a time-varying magnetic field, $\varepsilon = \oint \overrightarrow{E} \cdot \overrightarrow{dl} = 0$ which does not mean that induced E-field is conservative. Ans.

8.10 Comparision of E_{ind} and B_{static} (optional)

Now we can see some sort of similarity between induced E-field and static B-field. We can compare the nature of the induced E-field with the static B-field. The induced E-field is not associated with any free charges directly. So, by putting $\rho = 0$ in Gauss's law we obtain $\oint \overrightarrow{E} \cdot \overrightarrow{dA} = 0$. Then, we compare it with a similar relation for magnetostatic field given as $\oint \overrightarrow{B} \cdot \overrightarrow{dA} = 0$. We find a clear resemblance between induced E-field and static B-field as both are solenoidal. This means that physically E-lines are closed (solenoidal) like B-lines. Mathematically, we can say that no net flux of E-lines passes through a closed surface like B-lines.

The induced electric field is solenoidal like the magnetic field.

(i)

Closed Gaussian
surface

B

(ii)

E

Closed Gaussian
surface

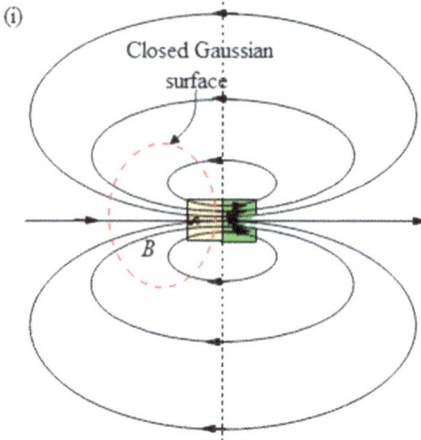

The total flux of B-lines passing through
the closed gaussian surface is $\phi_B = 0$

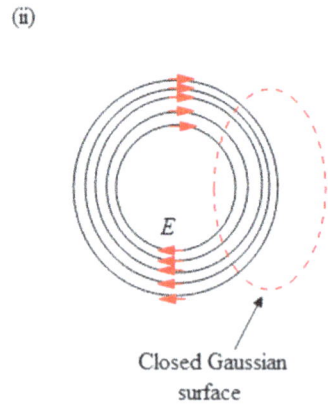

The total flux of B-lines passing through
the closed gaussian surface is $\phi_B = 0$

Finally, comparing Faraday's law

$$\oint \overrightarrow{E} \cdot \overrightarrow{dl} = \int \frac{\partial \overrightarrow{B}}{\partial t} \cdot \overrightarrow{dA} = \frac{\partial \phi}{dt} \text{ (for static magnetic field)}$$

with Ampère's circuital law

$$\oint \overrightarrow{B} \cdot \overrightarrow{dl} = \mu_0 i,$$

we can understand that, because of mathematical similarity between these two expressions, the nature of E-field is VORTEX (rotational) like B-field.

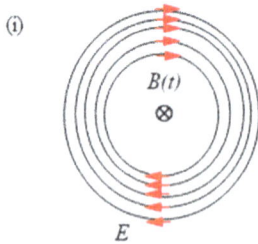

(i)

$B(t)$
$\otimes$

E

(ii)

$\otimes$ i

B

The inward decreasing B produces
vortex (clockwise) induced E-field;

$\oint \vec{E}.d\vec{l} = -d\phi/dt$

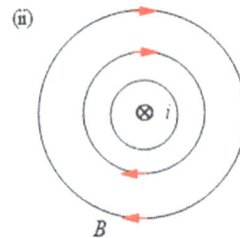

The inward current i produces
vortex (clockwise) B-field

$\oint \vec{B}.d\vec{l} = \mu_o i$

Physically, we can interpret the vortex field pattern of induced E-field and static B-field as follows. A closed line integral of induced electric field E in a contour is non-zero, that is, equal to rate of change of magnetic flux enclosed by the contour. Similarly, a closed line integral of magnetic field B in a contour is associated with a current enclosed by that contour. When i = current enclosed by a contour or loop = 0, $\oint \overrightarrow{B} \cdot \overrightarrow{dl} = 0$; then B-field is conservative (static). If the flux enclosed by a loop does not change with time, we can write $\phi = C$, $\oint \vec{E} \cdot \vec{dl} = 0$. Then, $\vec{E}$ – field is conservative (static).

The magnetic field can be conservative along a contour that does not enclose any current; but an induced E-field cannot be conservative along a contour although it does not enclose a net magnetic flux.

8.11 Difficulties in applying the flux formula

It is quite interesting to express or find the motional and transformer emf as the rate of change in the flux (ϕ) passing through the closed circuit (loop). The central idea behind the flux formula is to find the rate of change in the flux (ϕ) passing through the closed circuit (loop). The general flux formula is given as

$$\varepsilon = -\frac{d\phi}{dt}$$

which combines both motional emf ε_m and transformer emf ε_t,

$$\text{where } \phi = \int \vec{B} \cdot \vec{dA}.$$

The strength of this formula lies in its being handy and more useful to tackle the problems of electromagnetic induction without going through details of its micro-analysis (mechanism). In this way, we need not think about the types or reasons of induced emf (ε_m and ε_t). We just need to find the flux passing through or linked with the circuit or contour or loop as the function of time and then differentiate it. It will take care of both ε_m and ε_t simultaneously. We will present some illuminating problems in later sections and solve them only by using the flux formula conveniently by giving proper physical interpretation. Now we need to look into the weakness (limitations) of this formula. This formula is based on the concept of flux passing through or flux-linkage with the circuit. There are two terms based on the concept of flux passing through the circuit, such as 'flux' and 'circuit'. The 'flux' must pass through the 'circuit'. The flux which doesn't pass through the circuit is useless for our calculation. The flux which passes through the circuit is also termed as 'flux linked' or 'flux linkage'. We find it difficult to calculate the flux while handling some problems. Sometimes the flux linkage appears as a constant. The phrases 'flux passing', 'flux linking' and 'flux cutting' must be properly understood for calculation of flux and inductance.

In some problems we see no flux linkage but the circuit still carries an induced current. This is because of inadequate choice of the circuit (contour). Let us take the example of the sliding bar.

Example 9 The conducting bar in figure (i) or a conducting plate in figure (ii) moves with certain velocity $\vec{v} = v\hat{i}$ at the given position on the conducting rail. An upward magnetic field of induction $\vec{B} = B_0\hat{j}$ is present throughout. The rails are connected with a resistance R through vertical conductors. By using the flux formula how can we find the current passing through the resistor?

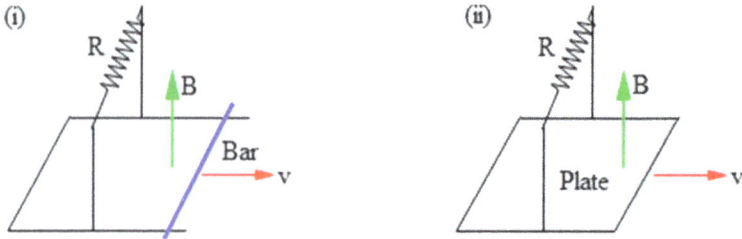

Solution

At the given position if we consider the bar just below the resistor the loop will be *BCDE* where we find no flux passing through the loop because the plane of the loop is parallel to the magnetic induction. Then the rate of change in flux $\frac{d\phi}{dt}$ is equal to zero yielding zero (no) induced current, which is contrary to the practical result. As we detect an induced current, the flux linkage changes with time. So, we need to choose the loop properly. While doing this, remember two points. Firstly, do not consider the loop with moving portion (bar) when it does not link any flux (or links a constant flux). We should consider the bar as part of the loop for any arbitrary position of the bar. Then, calculate the flux ϕ passing through this modified loop.

Here, the current loop is *BB′E′EDC* instead of *BEDC* which links a flux ϕ that comes out to be equal to *Blx*. So, the emf induced along the loop (bar) is $\varepsilon = \frac{d\phi}{dt} = Blv$ and the induced current is given as

$$i = \frac{\varepsilon}{R} = \frac{Blv}{R}.$$

Following the above principle, we should not take the closing side of the loop *BE* (dotted line drawn on the plate) just below the resistor because it does not enclose any flux. So, we can take the line *B′E′* at a distance x from *BE*. Now the flux passing through the loop is equal to *Blx*.

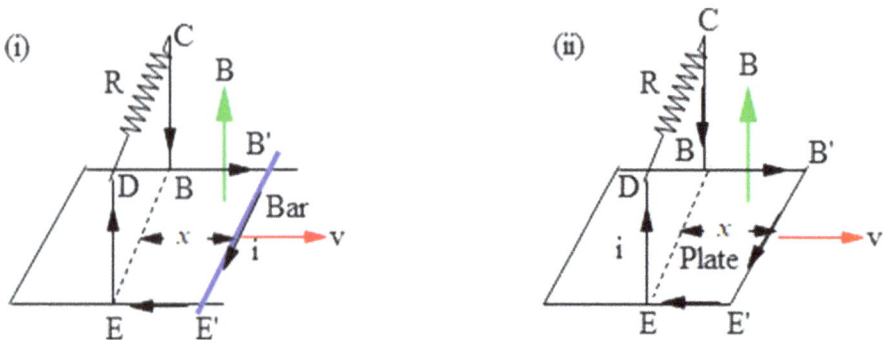

Then using the flux formula we can find the same result for this case of the moving plate. Ans.

Example 10 Now, consider the pure rotation of the bar about one of its ends in a vertically upward magnetic field as shown. The resistor R is connected by two vertical conducting bars between the fixed points A and B. The rotating bar occupies a position along AB at time $t = t_0$. Find the induced current in the resistor at that instant.

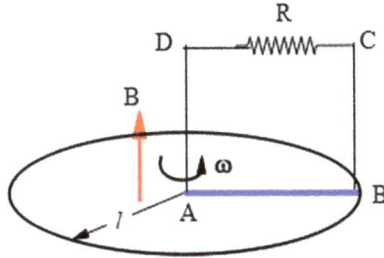

Solution

At the given position of the bar when it is lying just below the resistor, we consider $ABCD$ as a loop and we find no flux linkage through it. Following the argument used in the last example, let us consider the rotating bar at some other angular position where it does not lie just below the resistor. Let the bar occupy a position AB'. Now define the circuit as $AB'BCD$. The flux passing through it is given as

$$\phi = \frac{Bl^2}{2}\theta,$$

where $\theta = \omega t$. This yields

$$\phi = \frac{Bl^2}{2}\omega t.$$

Differentiation of both sides gives

$$\frac{d\phi}{dt} = \varepsilon = \frac{Bl^2\omega}{2},$$

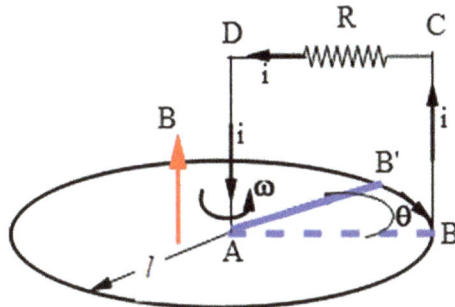

that is, the induced emf in the circuit. Then, the induced current

$$i = \frac{\varepsilon}{R} = \frac{Bl^2\omega}{2R} \quad \text{Ans.}$$

8.12 Lenz's law

Three years after the discovery of electromagnetic induction, in 1834, Friedrich Emil Lenz proposed a principle to find the direction of induced current. This is based on the following arguments.

Suppose a coil 1 carries a current i, which can be varied by a variable resistor R. For instance, current starts increasing in the coil 1 when R decreases. Let us assume that a clockwise current i_2 be induced in coil 2. When i_1 increases ($\frac{di_1}{dt} \neq 0$), i_2 must increase. Since, the flux produced by i_2 cuts coil 1, following the above assumption it will induce a current in coil 1 in the same sense. As a result, i_1 goes on increasing. Likewise, i_2 will go on increasing. This means the energy of the system of two coils increases infinitely without supplying any energy from outside, but this goes against the principle of energy conservation.

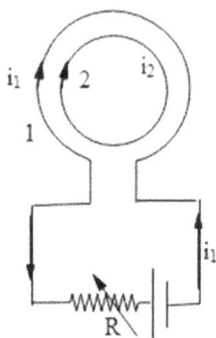

The current i_2 is induced so as to oppose the change in the current i_1

Hence, the current must be induced so as to nullify the change in flux. This suggests that the current is induced at the expense of the energy supplied by the applied emf. Therefore, the induced emf cannot nullify the applied emf completely. On the basis of conservation of energy, this can be stated as Lenz's law as follows.

'Current is induced so as to frustrate (oppose) the change in magnetic flux'. However, it doesn't necessarily oppose the magnetic flux. It should be noted that Lenz's law obeys Newton's third law as it obeys the principle of conservation of energy. *The induced magnetic flux may favour, or oppose the external or applied flux, but, it must oppose the variation or change in external magnetic flux.*

8.13 Application of flux formula $\varepsilon = -\dfrac{d\phi}{dt}$

While applying the flux formula, we must be aware of its practical difficulties. Only then can we solve many complicated problems easily with this formula without understanding many details of the mechanism of the induction principles. In this formula $\varepsilon = -\frac{d\phi}{dt}$, I assume the flux $\phi = \oint \overrightarrow{B} \cdot \overrightarrow{dA}$ as positive (+ve) for outward magnetic induction $\overrightarrow{B}$ and $\overrightarrow{dA}$ is always taken outward as per the convention used in this book. Hence, positive value of induced emf becomes anticlockwise current in a loop (circuit) placed in the plane of the paper or writing board or pad etc. The direction of induced emf can be calculated more conveniently by Lenz's law in which the induced current tries to oppose the variation of the magnetic flux. However, it cannot completely nullify the variation of flux because current is produced at the expense of the energy of the source of the magnetic flux. This is the essence of Lenz's law. I will explain the application of Lenz's law in the proceeding examples. As the flux formula is more mathematical, we need to find the flux passing through the suitable contour as the function of time. Then we just differentiate it. If $\frac{d\phi}{dt}$ is found +ve, the induced current is assumed anticlockwise, otherwise it is clockwise. Thus, this is a two-step procedure. Now, let us derive a suitable expression from the flux formula where the magnetic field varies with space (position) and time. If the flux $\phi = f(x, y, z)$; we can write the rate of its variation as $\frac{d\phi}{dt} = \frac{\partial\phi}{\partial t} + \frac{\partial x}{\partial t}\frac{\partial\phi}{\partial x} + \frac{\partial y}{\partial t}\frac{\partial\phi}{\partial y} + \frac{\partial z}{\partial t}\frac{\partial y}{\partial z}$.

The first term gives the induced emf because of variation of the magnetic field with time and the last three terms appear because of the motion of the circuit in a non-uniform magnetic field. Hence, the first term gives ε_t and the last three terms combiningly give ε_m. The partial variations $\frac{\partial\phi}{\partial x}, \frac{\partial\phi}{\partial y}, \frac{\partial\phi}{\partial z}$ are known as the gradients of the flux along the x-, y- and z-axis respectively. Then, $\vec{v} = \frac{dx}{dt}\hat{i} + \frac{dy}{dt}\hat{j} + \frac{dz}{dt}\hat{k}$; this formula is very useful in more complicated problems where magnetic field varies with time and space. It is quite suitable for the following examples.

Example 11 A square loop of side $l = 10$ cm with its sides parallel to the x and y-axes moves with velocity $v = -8\hat{i}$ m s^{-1}. An upward magnetic field is present which decreases along the negative x-axis with a gradient $\frac{\partial B}{\partial x} = 0.1$ T m^{-1} and increases with time at a rate of $\frac{\partial B}{\partial t} = 10^{-3}$ T s^{-1}. If the resistance of the loop is $2 \times 10^{-3}\ \Omega$, then find the magnitude and direction of induced current in the loop.

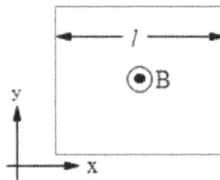

Solution
Direct application of the last formula yields $\varepsilon - \left(\frac{\partial\phi}{\partial t} + \frac{\partial\phi}{\partial x}\frac{dx}{dt}\right)$

$$\text{Put } \frac{\partial\phi}{\partial t} = \int \frac{\partial\vec{B}}{\partial t} \cdot \ \overrightarrow{dA} = \int \frac{\partial(B\hat{k})}{\partial t} \cdot \ dA\ \hat{k}$$
$$= \int \frac{\partial B}{\partial t}\ dA(\hat{k}\cdot\hat{k}) = \frac{\partial B}{\partial t}\int dA = \frac{\partial B}{\partial t}A = l^2\frac{\partial B}{\partial t}$$
$$= (0.1)^2 \times (0.1) = 10^{-3} \text{ weber s}^{-1} \text{ (or V) and}$$
$$\frac{\partial\phi}{\partial x}\cdot\frac{\partial x}{\partial t} = l^2\frac{\partial B}{\partial t}\frac{dx}{dt} = (0.1)^2(0.1)(-8) = -8 \times 10^{-3} \text{ V.}$$

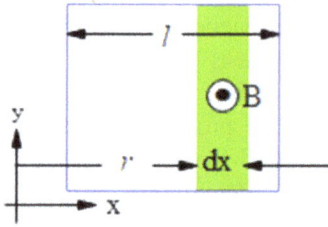

Summing up both the terms, obtain total induced emf

$$\varepsilon = -\frac{d\phi}{dt} = -(10^{-3} - 8 \times 10^{-3}) = +7 \times 10^{-3} \text{ V}$$

This causes an induced current

$$i = \frac{\varepsilon}{R} = \frac{+7 \times 10^{-3}}{2 \times 10^{-3}} = +3.5 \text{ A.}$$

This is positive as because magnetic induction decreases along the negative x-axis or increases along the positive x-axis. Ans.

Alternative method:

By following a general procedure, we can find the total flux ϕ passing through the square loop at any instant t as the sum of elementary flux $d\phi$ that passes through the elementary area ldr. So, $d\phi = (B\hat{k})(ldr\hat{k})$. Put $B = B_0 + 0.1r$, where $B_0=$ magnetic induction at the origin at that instant t which varies at a rate of $\frac{\partial B_0}{\partial t} = 10^{-3} \text{ T s}^{-1}$.

This gives $B_0 = 10^{-3}t + C$; where C is a constant. Thus, the total flux linked with the loop is given by summing up the elementary flux as $\phi = \int d\phi = \int_x^{x+l} (B_0 + 0.1r) \, ldr = B_0 l^2 + 0.05 l^2 (l + 2x)$, where $x = x_0$ distance covered by the loop during time t, that is equal to $x_0 + vt$ the loop moves towards the left and $B_0 = 10^{-3}t + C$. It gives $\phi = (10^{-3}t + C) + 0.05 l^2 \{1 + 2(x_0 - vt)\}$. Differentiating both sides with time we obtain $\frac{d\phi}{dt} = 10^{-3} + 0.05 l^2 (-2v)$ $= 10^{-3} + 0.05(0.1)^2 - (-2 \times 8) = -7 \times 10^{-3} \text{ V}$. Then the induced emf $\varepsilon = -\frac{d\phi}{dt} = 7 \times 10^{-3} \text{ V}$ is generated causing an anticlockwise current $i = \frac{\varepsilon}{R} = \frac{7 \times 10^{-3}}{2 \times 10^{-3}} = 3.5 \text{ A}$ in the loop. Ans.

Problem 1 For instance, when a magnet approaches a coil (loop), the number of magnetic lines of force passing through it increases. This means that we can see more and more magnetic lines added to the loop when viewing from the magnet. For this, 'we need to imagine the magnetic flux moving along with the magnet'. Then, how does the induced E-field pattern appear when observed from the loop?

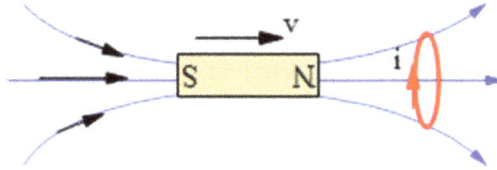

Solution

As shown in the figure below, moving B-field generates circular and coaxial induced electric field lines as seen by an observer fixed with the loop. We do not notice any induced current in the loop when the magnet rotates about its axis.

This reveals that no induced electric field is produced by mere axial rotation of the magnet. In fact, no flux cuts the loop during the rotation of the magnet. In other words, the flux linked with the loop does not change because there is no relative motion between the magnet and loop along their axes. However, when the magnet approaches the coil, we can imagine that the flux lines attached with the magnet also move with a velocity equal to that of the magnet.

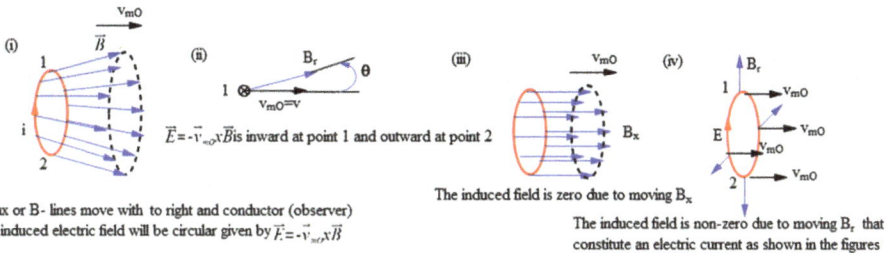

(i) v_{mO} $\vec{B}$

Magnetic flux or B- lines move with to right and conductor (observer) is fixed. The induced electric field will be circular given by $\vec{F} = -\vec{v}_{mO} x \vec{B}$

(ii) B_r θ $v_{mO} = v$ $\vec{E} = -\vec{v}_{mO} x \vec{B}$ is inward at point 1 and outward at point 2

(iii) v_{mO} B_x The induced field is zero due to moving B_x

(iv) B_r v_{mO} E The induced field is non-zero due to moving B_r that constitute an electric current as shown in the figures

In the figure you can see that B-flux lines move with a velocity v_{mo}. At any point on the loop, let θ be the angle between $\vec{B}$ and velocity of the observer relative to the magnetic field, that is, $\vec{v}_{om}$. Then, the induced E-field can be given as $\vec{E} = \vec{v}_{om} \times \vec{B}$. Its magnitude is equal to $v_{om} B \sin \theta$ and direction is tangential to the concentric circles spreading coaxially to the magnet and loop. In other words, we can experience an infinite series of concentric induced E-lines around the magnet. Ans.

Problem 2 Let us now try to apply the concept of induced electric field to find the force acting on a stationary particle of charge q and mass m kept on the perpendicular bisector of a straight conductor of length l meter carrying a current i. When the conductor moves with a velocity $\vec{v} = v_0 \hat{i}$, is there any force acting on the charge?

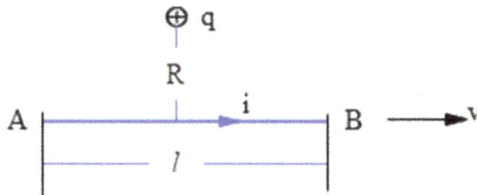

Solution

When we directly try to attack this problem, it seems as if there is no force exerted on the charged particle because it is at rest.

A stationary charge does not experience a magnetic force. When we look more carefully we can see that the magnetic field experienced by the charged particle changes at each point of space while the conductor (source of the magnetic field) moves.

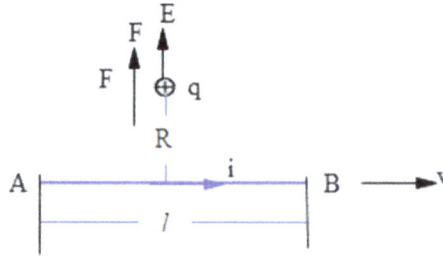

As we learnt, a changing magnetic field is associated with a moving magnetic flux which induces an electric field E. This can be given as $\vec{E} = \vec{v} \times \vec{B}$. Be careful about $\vec{v}$. This is not the velocity of the conductor (magnet). Instead, it is the velocity of the observer fixed with the charged particle 'relative to the conductor', that is $-v\hat{i}$. Substituting

$$B = \frac{\mu_0 i \, l \, \hat{k}}{2\pi R \sqrt{l^2 + 4R^2}}$$

due to the current-carrying conductor at the point, we find the particle accelerating up with an acceleration $\vec{a} = \frac{q\vec{E}}{m}$ since the charged particle seems to experience an induced electric field

$$\vec{E} = -\vec{v} \times \vec{B} = \frac{\mu_0 i v}{2\pi R \sqrt{l^2 + 4R^2}} \hat{j}$$

when viewed by the observer (at rest relative to the charged particle). Ans.

N.B: When we increase the length of the conductor to an incredibly large value, the magnetic field experienced by the charged particle remains constant, that is, equal to $\frac{\mu_0 i}{2\pi R}$. Therefore, it should not experience a magnetic force, but this is not true. So how do we explain this ambiguity? Apply the concept of relativity as described earlier to understand this fallacy.

The magnetic flux moves relative to the charged particle causing an induced electric field which pushes the charged particle.

Problem 3 You can apply the idea of induced electric field to this problem where a stationary conducting loop is subjected to a constant magnetic field of induction

$\vec{B} = B\,\hat{k}$, where the magnet moves due left with a speed v as shown in the following figure. The pole face of the magnet has dimensions $3a \times 3a$ and the dimensions of the loop is $a \times a$. Assuming no fringing of lines of force of magnetic field, draw the variation of current in the loop in function of time t.

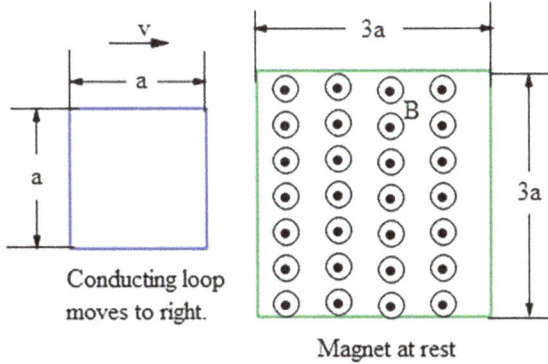

Conducting loop moves to right.

Magnet at rest

Solution

Look at the loop sitting on the magnet. The velocity of loop relative to the magnet $= \vec{v}_{lm} = \vec{v}_l - \vec{v}_m$. Substituting $\vec{v}_l = 0$ as the loop is stationary and $\vec{v}_m = -v\hat{i}$, we obtain $\vec{v}_{lm} = v\hat{i}$.

Consider a time $t(0 < t < \frac{a}{v})$ during which the entire loop comes into the magnetic field. Hence, $\vec{B}$ at left- and right side of the loop is zero and $+B\hat{k}$, respectively.

The portion of the loop outside the magnetic field does not experience any induced electric field since $B = 0$ for that region. The portion of the loop inside the magnetic field will experience an induced electric field as it experiences a magnetic field B. This is given as $\vec{E} = \vec{v}_{lm} \times \vec{B} = (v\hat{i}) \times (+B\hat{k}) = -vB\hat{j}$. Hence, it will induce an emf $\varepsilon = \oint \vec{E} \cdot \vec{dl} = \int_0^a \vec{E} \cdot \vec{dl} = vBa$ along AB. For upper and lower sides of the loop $\varepsilon = \oint \vec{E} \cdot \vec{dl} = 0$ because $\vec{E} \perp \vec{dl}$. As a whole, the induced emf $\varepsilon = vBa$ constitutes an anticlockwise current $i = \frac{\varepsilon}{R} = \frac{vBa}{R}$ in the loop which is valid for $0 < t < \frac{a}{v}$. During the time interval t; $(\frac{a}{v} < t < \frac{3a}{v})$, the loop remains completely inside the magnetic field which induces a downward electric field inside the region enclosed by the loop, as shown in the figure.

Then the induced emf $= \varepsilon = \oint \vec{E} \cdot \vec{dl} = \int_A^B \vec{E} \cdot \vec{dl} + \int_B^C \vec{E} \cdot \vec{dl} + \int_C^D \vec{E} \cdot \vec{dl} \int_D^E \vec{E} \cdot \vec{dl}$

$$Ea + 0 - Ea + 0 = 0$$

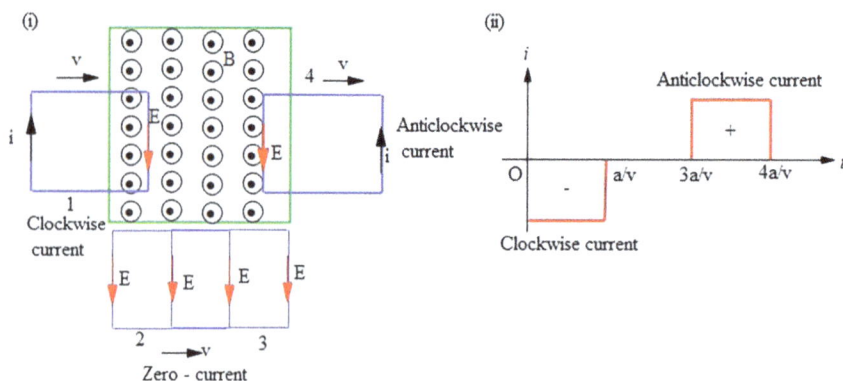

(i)

(ii)

Anticlockwise current

Clockwise current

Zero - current

Hence, no current flows during this time interval. It means, $i = 0$ for $(\frac{a}{v} < t < \frac{3a}{v})$. After this, again the loop remains partially in the magnetic field as the magnet moves towards the left. The upward induced electric field is present in the portion of the loop inside the magnetic field, whereas no induced electric field is observed in the portion of the loop lying outside the magnetic field.

Following the previous argument, we obtain the induced emf $\varepsilon = Ea = vBa$ along the loop which constitutes a current $i = \frac{vBa}{R}$ in clockwise direction during the time t; $\frac{3a}{v} < t < 4\frac{3a}{v}$. After the time $\frac{4a}{v}$, the loop will be completely free from the magnetic field. This means that zero (no) induced emf will appear causing zero (no) current after that time. Then $i = 0$ for $t > \frac{4a}{v}$. Ans.

Problem 4 A bar is rotated about one of its ends in the outward magnetic field of induction B with an angular speed ω. What is the motional induced emf across the bar? If we assume the centrifugal effect on the electrons of the bar, how will it affect the result?

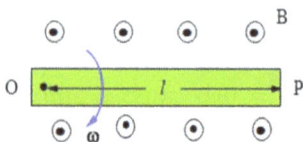

Solution

Let us take an elementary segment of length dl at a distance x from the centre of rotation O. The speed of the segment is $v = x\omega$. The motional emf induced across the small segment is $d\varepsilon_m = B(dx)v$. Substituting $v = x\omega$ and integrating over the total length of the bar, the total induced emf $\varepsilon_m = \int d\varepsilon_m = B\omega \int_0^l x\,dx = \frac{1}{2}Bl^2\omega$. Polarity of the induced emf is given by $\vec{v} \times \vec{B}$ as shown in the figure.

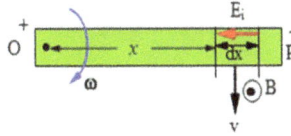

Due to the centrifugal effect, the loosely fitted electrons will be deposited at the free end P of the rod, making it more negative relative to the pivoted end O. So, there is an electric field induced due to the centrifugal effect which is conservative in nature. You can show that the net emf induced between the ends P and Q of the bar is given as

$$\varepsilon = \frac{1}{2}(Bl + \frac{m\omega l}{e})l\omega \text{ Ans.}$$

Problem 5 A conducting rod of length l is leaning against a vertical wall. An outward uniform magnetic field of induction B is applied. The bottom of the rod is pulled with a constant velocity v along x-direction. At any angular position $\theta = 60°$ of the rod, find the (a) induced emf between the ends of the rod (b) polarities of the ends of the rod.

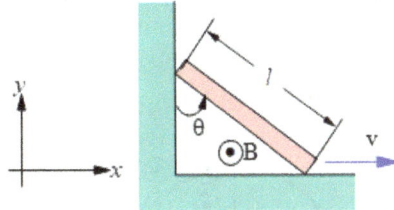

Solution
(a) Assume the triangle PQR as the loop at any instant t. The area enclosed by the loop is

$$A = \frac{xy}{2} = \frac{1}{2}(l \sin \theta)(l \cos \theta) = \frac{l^2}{4} \sin 2\theta$$

The flux ϕ through the loop is

$$\phi = \vec{B} \cdot \vec{A} = (B\hat{k}) \cdot \left(\frac{l^2}{4} \sin 2\theta \hat{k}\right) = \frac{Bl^2}{4} \sin 2\theta$$

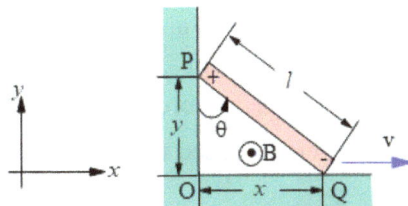

Differentiating ϕ with time t,

$$\varepsilon_m = -\frac{d\phi}{dt} = -\frac{Bl^2}{4}(2\cos 2\theta)\frac{d\theta}{dt},$$

where $\frac{d\theta}{dt} = \frac{v}{y} = \frac{v}{l\cos\theta}$.

$$\Rightarrow \varepsilon = -\frac{Blv}{2}\frac{\cos 2\theta}{\cos\theta}$$

Putting $\theta = 60°$, we have

$$\varepsilon = -\frac{Blv}{2}\frac{\cos 120°}{\cos 60°} = \frac{Blv}{2} \quad \text{Ans.}$$

(b) Positive sign of the *emf* signifies that the current will be anticlockwise if the path *PQR* is conducting. Then, the upper end of the rod is at negative potential relative to its lower end. Ans.

Problem 6 A semi-circular conducting loop of resistance R and loop area A is rotating about the x-axis in a uniform magnetic field of induction $\overrightarrow{B} = B\hat{j}$. At $t = 0$, the plane of the loop is parallel to the magnetic induction. If $B = 0$ in the left side of the axis of rotation of the loop, find the (a) expression for induced current in the loop as the function of time t, (b) total charge flowing through the conductor during $t = \frac{\pi}{\omega}$ from starting.

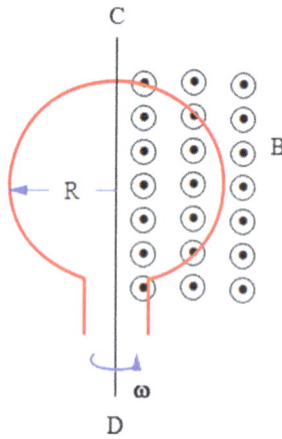

Solution

(a) In this case, each element of the curved portion of the loop has different speeds moving in different directions. Hence, the evaluation of the emf by the formula $\varepsilon_m = \oint \overrightarrow{v} \times \overrightarrow{B} \cdot \overrightarrow{dl}$ is more difficult. So, we have to use the flux formula $\varepsilon = -\frac{d\phi}{dt}$. For this, we need to find the flux ϕ passing through the loop, which is given as $\phi = \overrightarrow{B} \cdot \overrightarrow{A}' = BA' \cos \theta$, where $\theta = \frac{\pi}{2} - \alpha$; α is the angle rotated (swept) by the loop during time t.

This yields, $\phi = BA' \sin \omega t$. Then, the induced emf $|\varepsilon| = B\left(\frac{A}{2}\right)\omega \cos \omega t$ because $A' = $ maximum effective area under the magnetic field $=$ half of the loop area $= A/2$. This causes induced current $i = \frac{\varepsilon}{R} = \frac{BA\omega \cos \omega t}{2R}$ Ans.

(b) The charge flow during a time dt is

$$dq = idt = \frac{\varepsilon}{R}dt = \left(\frac{1}{R}\frac{d\phi}{dt}\right) \cdot dt = \frac{d\phi}{R}$$

Then, the total charge flown is

$$q = \frac{1}{R}\left|\int_{\phi_1}^{\phi_2} d\phi\right| = \frac{|\phi_2 - \phi_1|}{R}$$

Putting $\phi_1 = \phi|_{\theta=0}$ and $\phi_2 = \phi|_{\theta=90°}$, we have $q = \frac{BA}{(2R)}$ Ans.

Problem 7 A conducting coil comprising N loops is connected rigidly with an insulated bar that is fitted with a smooth pivot P. The coil is made to oscillate by the swinging bar so that its plane is perpendicular to the plane of the paper. The angle made by the bar with the vertical, changes with time according to the law $\theta = \theta_0 \sin \alpha t$, where α is a positive constant. If the coil has radius R and resistance R_0, find the expression for the induced current in the coil as the function of time. Assume that the coil is at the lowest position initially and a uniform horizontal magnetic induction B is applied, as shown in the figure.

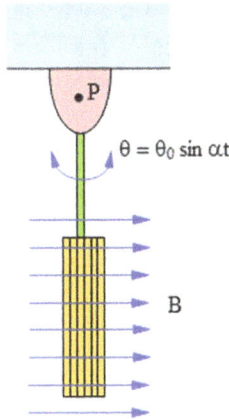

Solution
At any angular position θ, the coil encloses a flux $\phi = NBA \sin \theta$
The flux changes at a rate of $|\frac{d\phi}{dt}| = NBA(\frac{d\theta}{dt})\cos \theta$.

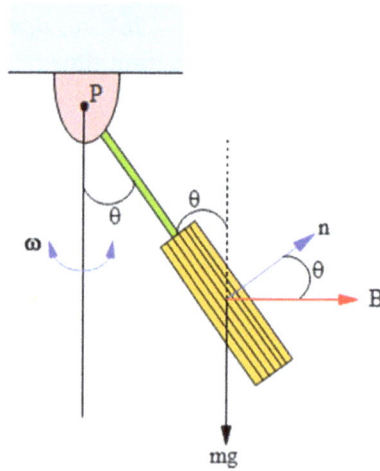

Then, the emf ε induced in the coil is given as $\varepsilon = NBA\omega \cos \theta$. This results in an induced current $i = \frac{\varepsilon}{R_0} = \frac{NBA}{R_0}\omega \cos \theta$ in the coil. Putting $A = \pi R^2$, $\theta = \theta_0 \sin \alpha t$ and $\omega = \frac{d\theta}{dt} = \theta_0 \alpha \cos \alpha t$, we have

$$i = \frac{NB\pi R^2 \theta_0}{R_0}\alpha \cos \alpha t \cos(\theta_0 \sin \alpha t) \text{ Ans.}$$

Problem 8 A conducting loop $ABCDA$ is arranged in space as shown in the figure below. It is subjected to a time-varying magnetic field of induction $B = (\hat{i} + \hat{j} + \hat{k})e^{-t}$. Let the resistance of the loop be R_0. Using the flux formula, find the magnitude and direction of induced emf and current.

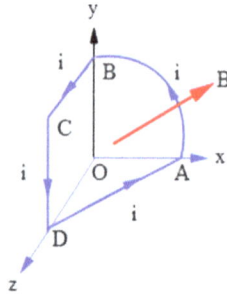

Solution

The area of the loop is given as $\vec{A} = a^2(\hat{i} + 0.5\hat{j} + \frac{\pi\hat{k}}{4})$.

Then, flux passing through the loop $=\phi = \vec{B} \cdot \vec{A} = (\hat{i} + \hat{j} + \hat{k})e^{-t} \cdot (\hat{i} + 0.5\hat{j} + \frac{\pi\hat{k}}{4})a^2$

$$\phi = \left(\frac{3}{2} + \pi/4\right)a^2 e^{-t}$$

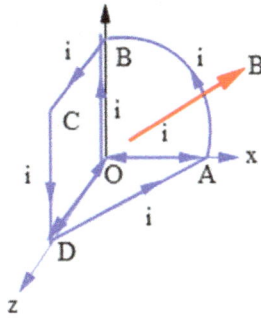

This varying flux induces an emf $\varepsilon = -\frac{d\phi}{dt} = \left(\frac{\pi+6}{4}\right)a^2 e^{-t}$

Substitute $t = 0$ to obtain initial emf induced $= \left(\frac{\pi+6}{4}\right)a^2$ Ans.

Problem 9 Shown in the figure is a simple network of three resistors $R_1 = 1\ \Omega$, $R_2 = 2\ \Omega$ and $R_3 = 3\ \Omega$. A battery of emf 0.3 V is connected with the loop 2. The dimensions of loops 1 and 2 are given. In loop 1, inward magnetic field varies with time as $B = e^{-2t}$ and in loop 2, outward magnetic field varies with time as $B = 2t$. Find the initial current flowing through the resistor R_3. Assume that $AB = 50$ cm, $BC = 100$ cm and $AE = 20$ cm.

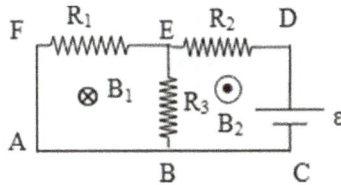

Solution

In loop 1, induced emf $\varepsilon = -\frac{d\phi}{dt}$, where $\phi = \vec{B} \cdot \vec{A}$.

Substituting $\vec{B} = -e^{-2t}\hat{k}$ and $\vec{A} = \frac{20}{100} \times \frac{50}{100}\hat{k}$, we have $\varepsilon_1 = -0.2e^{-2t}|_{t=0} = -0.2$ V. Negative sign signifies that the induced current in the loop 1 is clockwise. It is equivalent to putting a battery of emf $\varepsilon = 0.2$ V in loop 1 to set an anticlockwise current in the absence of other external agents (cells and circuits). Similarly, in loop 2 induced emf $\varepsilon = -\frac{d\phi}{dt}$, where $\phi = \vec{B} \cdot \vec{A}$. Here $\phi = 2t\hat{k}$ and $A = \frac{100}{100} \times \frac{20}{100}\hat{k}$. This gives $\varepsilon = -0.4$ V.

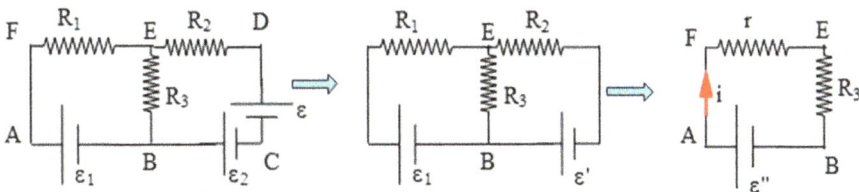

Negative sign signifies a clockwise current induced in both loops. Therefore, we can put the induced cells of emf 0.2 V and 0.4 V as shown in the figure. We should be careful while putting the induced cells. If we put them in the middle branch, it may lead to a different result; it is a matter of confusion where exactly we should put the induced cells with proper polarities. To avoid this complication, let us put the obtained induced cells (emfs) with the branches other than the middle branch. Remember that the polarity is made so as to produce a clockwise current in the loops. The right loop can be redrawn by substituting the combination of the induced cell and given cells by a single cell of emf $\varepsilon' = (0.4 - 0.3) = 0.1$ V. Now, the problem is reduced to a simple resistor-cell circuit.

We can now rearrange the circuit as a parallel combination of two cells between A and B whose equivalent cell emf can be given as $\varepsilon'' = \frac{0.2 \times 2 - 0.1 \times 1}{2 + 1} = 0.1$ V which is directed upward and its internal resistance $= r = \frac{1 \times 2}{1 + 2} = \frac{2}{3}$ Ω..

Now, the current i in 3 Ω resistor can be given as $i = \frac{0.1}{\frac{2}{3} + 3} = \frac{3}{110}$ A. Ans.

Problem 10 Shown in the figure is a cylindrical magnet of radius R. For the sake of simplicity, let the magnet have induction B throughout its volume. If you rotate the magnet with an angular speed ω, do you expect any current induced between the axis and the perimeter of the magnet, when connected by an external resistor R_0? If yes, find the current.

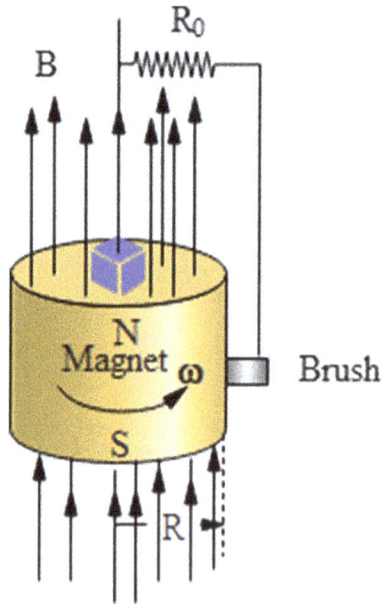

Solution

A magnet is associated with a magnetic field. We can see the magnet but not its magnetic field. When the magnet rotates about its axis with an angular speed ω, generally we think that the magnetic field of the magnet will rotate with the same angular velocity in the direction of rotation of the magnet. This gives an impression that the magnetic field would be stationary relative to the magnet. Then, erroneously we can think that the rotating magnetic (field) flux passing through any properly chosen loop $ABCDEF$ (as discussed earlier) remains unchanged as there is no relative velocity between the field flux and magnet. This would give a zero induced emf causing zero current through the resistor R_0. But, in this case, practically we find a current. So, where is the fallacy? Here we can see that the phrase 'moving magnetic field' has no true sense because the magnet (moves) rotates about its axis, whereas its magnetic field does not change. This is because the magnetic field is symmetrical about the axis of rotation. So, the problem can be reduced to a rotating metallic cylinder (or disc) of radius R in a static magnetic field of induction B. Following the previous procedure, you can find an induced current $i = \frac{BR^2 \omega}{2R_0}$. Here the correct loop is *ABCDEFGHA* as shown in the figure. Ans.

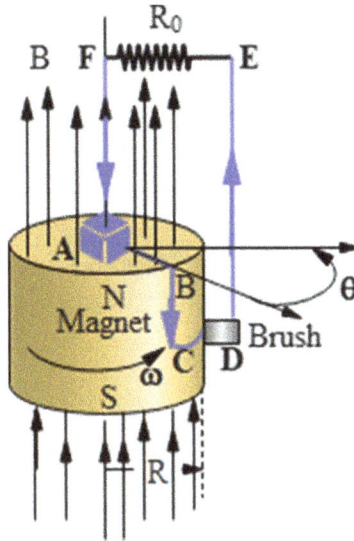

Problem 11 In a long solenoid, let the current vary with time causing a time varying magnetic induction inside the solenoid. If the rate of variation of magnetic induction is $\frac{dB}{dt}$, find the variation of induced electric field as the function of radial distance from the axis of the solenoid.

Solution

For a long solenoid, at any radial distance $r < R$ ($R =$ radius of the solenoid), the magnetic induction is $B = \mu_0 ni$.

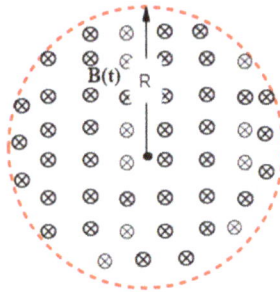

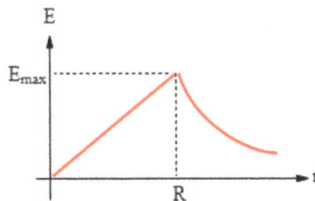

At any $r > R$, $B = 0$. This means that flux $\phi = \pi r^2 B (r \leqslant R)$ is linked with the loop of radius $r \leqslant R$. Then the emf induced in this loop is given as $\varepsilon = \frac{d\phi}{dt}$. Putting $\phi = \pi r^2 B$ we obtain $\varepsilon = \pi r^2 \frac{dB}{dt}$. Since $\varepsilon = \oint \vec{E} \cdot \vec{dl} = E(2\pi r)$ we get $E = \frac{r}{2} \frac{dB}{dt}$ for $r \leqslant R$. That means E varies linearly inside the solenoid (similarly, for any loop of radius $r \geqslant R$). $\varepsilon = \oint \vec{E} \cdot \vec{dl} = E(2\pi r) = 2\pi r E = \pi r^2 \frac{dB}{dt}$ and $E = \frac{R^2}{2r} \frac{dB}{dt}$ for $(r \geqslant R)$. Hence, E varies inversely with r outside the solenoid. Ans.

Problem 12 In the above example, let us place two concentric conducting loops of radius $a(<R)$ and $b(>R)$ one inside and the other outside the region of magnetic field of the previous problem.

(a) Putting $R = b$, find the ratio of:

(i) power dissipated;

(ii) electric fields in the two concentric circular conducting loops;

(b) Draw the induced electric field pattern.

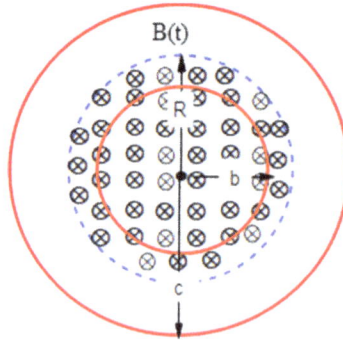

Solution

(a) (i) The ratio of the power dissipated $= \frac{P_1}{P_2} = \frac{\frac{\varepsilon_1^2}{R_1}}{\frac{\varepsilon_2^2}{R_2}}$, where $\varepsilon_1 = \pi a^2 \frac{dB}{dt}$,

$\varepsilon_2 = \pi b^2 \frac{dB}{dt}$, $R_1 = \frac{\rho(2\pi a)}{S}$ and $R_2 = \frac{\rho}{S}(2\pi b)$, where S = area of cross-section of the wires.

Then, $\frac{P_1}{P_2} = \left(\frac{a}{b}\right)^3$ Ans.

(ii) The ratio of electric fields $\frac{E_1}{E_2} = \frac{\left(\frac{a}{2}\frac{dB}{dt}\right)}{\frac{b^2}{2c}\frac{dB}{dt}}$ obtained by referring to the derived expressions $\frac{E_1}{E_2} = \frac{ac}{b^2}$ Ans.

(b) Field pattern: $E \propto r$ for $r \leqslant b$ and $E \propto \frac{1}{r}$ for $r > b$.

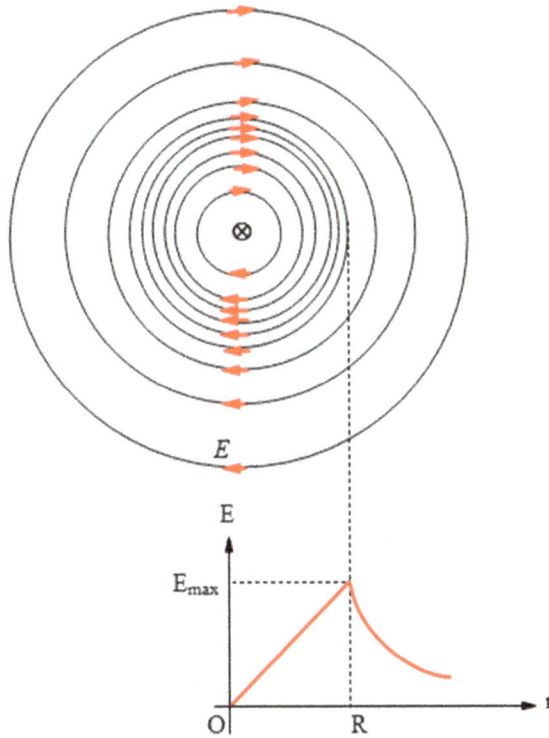

We have shown the direction of induced electric field by assuming that inward magnetic field B is decreasing. Hence, electric field lines become denser towards the periphery of the solenoid and again decrease from the periphery, as shown in the figure. The direction of $\vec{E}$ depends on the algebraic value of $\frac{dB}{dt}$. (+ve for increase in B and -ve for decrease in B) Ans.

N.B: In the previous problem, we observed that induced electric field E, increases linearly from the centre (axis) to the periphery of the circular region of radius b in which a magnetic field is present and decreases hyperbolically from the periphery to infinity. Since $i \propto E$, we experienced different currents in the two coaxial loops.

Problem 13 In the last example, let us remove the loops and place a spiral coil of resistance R_1 of total number of turns N made of insulated thin conducting wires in the circular portions of radii ranging from $r = a$ to $r = c$, as shown in the figure below. Assuming the rate of variation of inward magnetic induction as $\frac{dB}{dt}$, find the corresponding induced current in the coil between the ends (at $r = a$) and B (at $r = b$). Assume that the spiral is placed coaxial with the solenoid that produces a magnetic field. Put the radius of the solenoid as $R = b$.

Coil

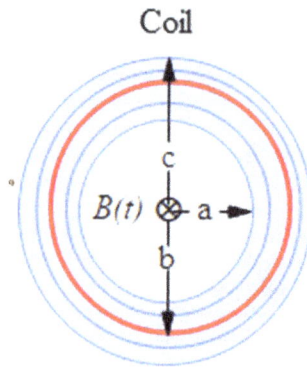

Solution

As, the spiral is comprised of a number of closely spaced circular loops, we can assume each loop as an approximately concentric circle for the sake of mathematical simplicity. Let us take an elementary ring of thickness dr which encloses dN number of loops turns. The flux linked with the thin ring $= d\phi = (dN)BA$; A = area enclosed by the shaded portion $= \pi r^2$. Then we have $d\phi = (B)(\pi r^2)(dN)$. The total flux linked with the coil or spiral $= \int d\phi = \int_a^b B\,\pi r^2 dN$, where $dN = \frac{N}{c-a}dr$.

We take the upper limit b instead of c because a magnetic field is absent in the region between $r = b$ to $r = c$ causing zero flux passing through that region. Substituting dN we obtain $\phi = \frac{\pi NB}{c-a}\int_a^b r^2 dr = \frac{\pi NB(b^3 - a^3)}{3(c-a)}$.

If B varies, ϕ will also vary with a rate $\frac{d\phi}{dt} = \frac{\pi(b^3 - a^3)}{3(c-a)}\frac{dB}{dt}$. This numerically equals the emf induced between points A and B. This results an induced current $i = \frac{\varepsilon}{R_1} = \frac{\pi N(b^3 - a^3)}{3R_1(c-a)}\left(\frac{dB}{dt}\right)$. Ans.

Problem 14 Let us consider the time varying uniform magnetic field in the circular region. Fix an insulated ring concentric with the circular region. Inside the ring, place a smooth bead of mass m and charge q. Then increase the magnetic field of induction from B to $2B$. If we assume that the bead was at rest initially, (a) find the angular momentum attained by the charged particle. Compute the magnetic moment of the current caused by the motion of the charged particle due to the change in magnetic field.

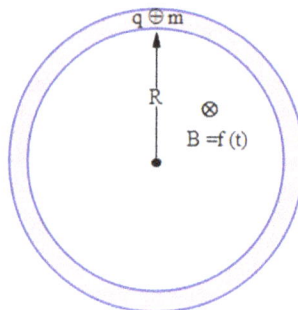

Solution

(a) When the magnetic field increases by dB during time dt, let the magnetic flux change by $d\phi$. As discussed earlier, we know that vortex electric field

$$E = \frac{\pi R^2}{2\pi R}\frac{dB}{dt} = \frac{RdB}{2dt}.$$

This electric field tangentially accelerates the charged particle producing an angular acceleration

$$\alpha = \frac{(qE)R}{mR^2}.$$

That increases the angular speed by $d\omega$ during time dt. Then,

$$d\omega = \alpha dt = \frac{qE}{mR}dt$$

Putting the value of $E = (R/2)(dB/dt)$, we have

$$d\omega = \frac{q}{2mR}(RdB) = q\left(\frac{dB}{2m}\right).$$

When B increases from B to $2B$. ω increases from 0 to ω, so, integrating both sides,

$$\omega = \frac{qRB}{2mR} = \frac{qB}{2m}.$$

Hence the change in angular momentum of the charged particle is given as

$$\Delta L = mR^2\omega = \frac{qBR^2}{2} \quad \text{Ans.}$$

(b) The resulting magnetic dipole moment of the current constituted by the charged particle

$$\mu = \pi R^2 i \quad \text{where} \quad i = \frac{q}{T} \quad \text{and} \quad T = \frac{2\pi}{\omega}.$$

Substitution of all the above values yields

$$\mu = \frac{q^2 BR^2}{4m} \quad \text{Ans.}$$

N.B: *The angular momentum of the charged particle is not conserved because the radiated electromagnetic wave will also carry equal and opposite angular momentum.*

Problem 15 A charged disc of thickness y, uniform surface charge density σ and density ρ is free to rotate about the vertical axis. If $\frac{dB}{dt}$ is the rate of variation of the vertical magnetic field that is present uniformly in the circular region as described in the last problem, find the angular acceleration of the disc. Assume that the axis of rotation of the disc coincides with the center of the circular region of the magnetic field.

Solution

Referring to the previous problem, the circular induced electric field at any distance r from the centre is $E = \frac{r}{2}\frac{dB}{dt}$. The torque produced $= d\tau = \int_0^{q_{ring}} (dq)\, E \cdot r$ due to the charged ring $= q_{ring} \cdot Er = (q_{ring})\frac{r^2}{2}\frac{dB}{dt}$.

Putting $q_{ring} = \sigma \cdot 2\pi r dr$, we have $d\tau = \pi\sigma r^3 \cdot dr\left(\dfrac{dB}{dt}\right)$

Integrating, the net torque acting on the disc is

$$\tau = \pi\sigma\frac{dB}{dt}\frac{r^4}{4}\bigg|_0^R.$$

Then, we have $\tau = \frac{\pi\sigma R^4}{4}\frac{dB}{dt} = \frac{mR^2}{2}\alpha$

Putting $m =$ mass of the disc $= \rho(\pi R^2 y)$, we have $\alpha = \frac{\sigma}{2\rho y}R^2\left(\frac{dB}{dt}\right)$. Ans.

Problem 16 A very small loop of radius r moves coaxially with a constant velocity v, away from bigger loop of radius R which is carrying a steady current i. It is assumed that the smaller loop begins to move from the centre of the bigger loop along their common axis. (a) Find the flux passing through the smaller loop in the function of time. (b) Find the expression for emf induced in the smaller loop in the function of time.

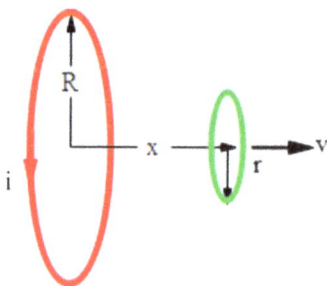

Solution

(a) At any distance of separation x between the loops, the magnetic field of the bigger loop $= \vec{B} = \frac{-\mu_0 i r^2 \hat{i}}{2(R^2 + x^2)^{3/2}}$. Since loop 2 is very small compared to loop 1, $\vec{B}$ can be assumed uniform at any axial position. Therefore, the flux linked with the loop $2 = \phi = \vec{B} \cdot \vec{A}$, where A is area of the loop $2 = -\pi r^2 \hat{i}$. This gives

$$\phi = -\frac{\mu_0 \pi r^2 i}{2(R^2 + x^2)^{3/2}}, \quad x = vt \text{ Ans.}$$

(b) This equation shows how the flux linkage with the smaller loop varies with time. The time rate of change in flux, that is,

$$\text{emf } \varepsilon = -\frac{d\phi}{dt} = \frac{-\mu_0 \pi r^2 i}{2}\frac{d}{dt}(R^2 + v^2 t^2)^{-3/2} = \frac{3\mu_0 v^2 r^2 \pi^2 i t}{2(R^2 + v^2 t^2)^{5/2}} \text{ Ans.}$$

Problem 17 At $t = 0$, a conducting bar of length l is given a velocity v_0 to the left and simultaneously pulled by a constant horizontal force F. The conducting bar can slide on a smooth conducting rail which is fitted with a battery of emf ε and a resistor of resistance R. An outward uniform magnetic field of induction B is applied as shown. (a) Find the velocity of the bar as a function of time 't'. (b) Analyze the motion of the bar and draw the v–t graph.

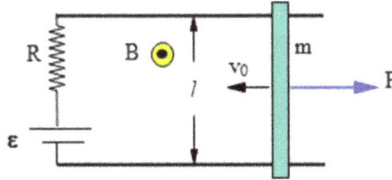

Solution

(a) For the sake of simplicity of understanding, the following steps are given in detail to solve the problem of a sliding bar on a conducting rail.

Step 1: Assume the direction of motion, that is given by the velocity vector of the bar. After some time t, suppose the bar moves to the right with a velocity v.

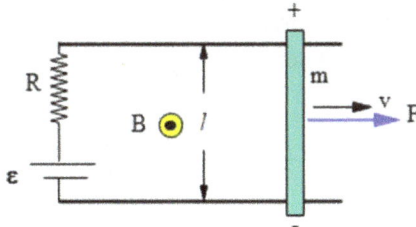

Step 2: Then, find the magnitude and polarity of the emf induced across the bar. The magnitude of induced emf is equal to 'Blv' (as discussed earlier). Its polarity is given according to Fleming's right-hand rule as shown in the figure below.

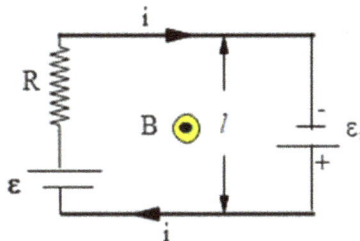

Step 3. Replace the bar mentally by a battery of emf $\varepsilon_i = Blv$ with its polarity as shown below and draw the equivalent circuit diagram.

Step 4. Generally, we do not know the exact direction of current in the circuit, which is the superposition of all currents. One is due to the applied emf ε and the other is due to the emf induced across the bar. In this case the actual direction of current is clockwise because both the emfs are favouring each other. But, in general, we have to assume any arbitrary direction of the (resultant) current in the bar when the indued emf opposes the applied (external) emf.

Step 5. Then, apply Kirchhoff's circuital law to obtain the instantaneous current in terms of the velocity of the bar. The corresponding equation is $\varepsilon - iR + \varepsilon_i = 0$ (where $\varepsilon_i = Blv$) which yields the current as $i = +\frac{\varepsilon + Blv}{R}$

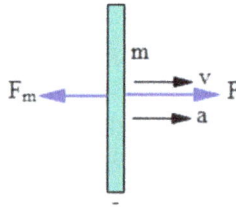

Step 6. Calculate the magnitude and direction of the magnetic force F_m on the bar, applying Ampère's force law and Fleming's left-hand rule. In this arrangement, $F_m = ilB$, where $i = \frac{\varepsilon + Blv}{R}$ and it is directed leftward.

$$\text{So, } F_m = \left(\frac{\varepsilon + Blv}{R}\right)lB$$

Step 7. Now apply Newton's second law of motion for the bar to obtain $F\hat{i} - F_m\hat{i} = m\vec{a}$ to obtain $\vec{a} = \left\{\frac{(F - F_m)}{m}\right\}\hat{i}$.

Step 8. Then you can write the differential equation and solve it to obtain displacement and velocity of the bar as the function of time.

Putting $\vec{a} = \frac{dv}{dt}\hat{i}$ and $F_m = \left(\frac{\varepsilon + Blv}{R}\right)lB$ in the last expression, we obtain the differential equation $\frac{dv}{dt} = \frac{FR - (\varepsilon + Blv)lB}{mR}$. Separating the variables, $\frac{dv}{FR - (\varepsilon + Blv)lb} = \frac{dt}{mR}$.

As during time t velocity $\vec{v}$ of the bar changes from $-v_0\hat{i}$ to $v\hat{i}$. Integrating both sides, we have $\int_{(-v_0)}^{v} \frac{dv}{FR - \varepsilon - B^2l^2v} = \frac{1}{mR}\int_{0}^{t} dt$

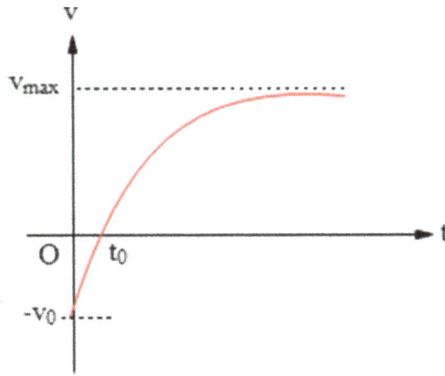

Evaluating the integral we obtain the vector expression of velocity as

$$\vec{v} -= \left\{ \frac{(FR - \varepsilon lB) - (FR - \varepsilon lB + B^2 l^2 v_0)e^{-\frac{B^2 l^2}{mR}t}}{B^2 l^2} \right\} \hat{i} \text{ Ans.}$$

(b) At $t = 0$, the bar moves with $\vec{v} = -v_0 \hat{i}$.

After a long time ($t \to \infty$) the bar attains a constant velocity $v_t = \frac{FR - \varepsilon lB}{B^2 l^2}$, where v_t is termed as terminal velocity. The above expression tells us that the bar stops instantaneously ($v = 0$) at time $t = t_0 = \frac{mR}{B^2 l^2} \ln \frac{(FR - \varepsilon lB + B^2 l^2 v_0)}{FR - \varepsilon lB}$.

Hence, the bar reverses its motion from left to right at $t = t_0$ and moves with a constant velocity v_t after a long time, as shown in the above graph.

We can observe that, the magnetic force F_m varies linearly with velocity v. If the speed of the bar is greater, it will experience greaterAmpère's force. As a result, the net force acting on the bar decreases. Hence, the acceleration of the bar gradually decreases. In other words, the bar speeds up at a slower rate. After a long time its acceleration becomes zero as the magnetic (Ampère) force F_m nullifies the applied force F. Eventually, the bar continues to move with a constant velocity known as terminal velocity v_t. Then the rod carries a constant current ($i = \frac{F}{lB}$) and experiences no net force. Ans.

N.B: We can discuss some possible cases of the set of sliding bar when,
  (a) neither battery nor external force is present;
  (b) only external force is present (generator);
  (c) only battery is present (motor);
  (d) battery is substituted by a constant (i) voltage (ii) current generator;
  (e) neither force nor battery is present.

Problem 18 In the last problem, let us remove the battery and withdraw the external force F from the previous set. Then project the bar with a speed v_0 to the right at $t = 0$
  (i) Find the variation of distance x covered by the bar with time t and calculate the maximum distance covered by it.
  (ii) What is the heat dissipated in the circuit during the time t?

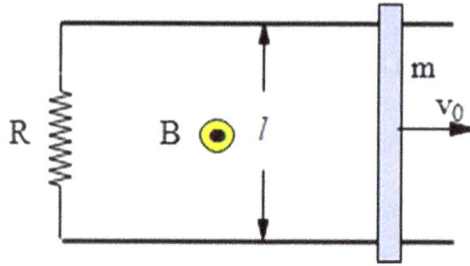

Solution

Referring to the general expression obtained in the last problem,

(i) Put $F = 0$ and $\varepsilon = 0$ to obtain the velocity of the bar as the function of time; $v = f(t)$. Then, integrating v with time, the distance moved by the bar in time t is

$$x = \int_0^t v\,dt = \frac{mv_0 R}{B^2 l^2}\left(1 - e^{-\frac{B^2 l^2}{mR}t}\right) \text{ Ans.}$$

Substituting $t = \infty$, the total distance covered is

$$x_{\max} = \int_0^{t=\infty} v\,dt = \frac{mv_0 R}{B^2 l^2} \text{ Ans.}$$

(ii) Integrating the instantaneous power dissipated in the circuit during time t we obtain

$$Q = \int_0^t P\,dt = \int_0^t i^2 R\,dt, \quad \text{where } i = \frac{Blv}{R} \text{ and } v = v_0 e^{-\frac{B^2 l^2}{mR}t}$$

Evaluating the integral and simplifying the factors, we have, $Q = \frac{1}{2}mv_0^2$ Ans.
N.B: 1. By applying the work–energy theorem, $Q = $ -change in kinetic energy. 2. By applying the work–kinetic energy theorem, you can show that velocity versus distance is linear.

Problem 19 Shown in the figure is a smooth conducting rail fitted with a resistor R which is situated at a distance x_0 from a long straight conductor carrying a current i_0. The conducting bar of mass m and length l is pushed with a velocity v_0. Derive an expression for its velocity as a function of distance covered x.

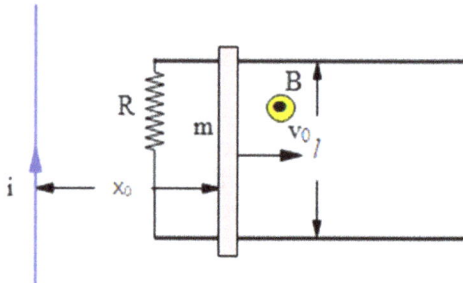

Solution

As discussed earlier, magnetic force F_m acting on the bar $= iBl$, where $i =$ induced current $= \frac{Blv}{R}$. Here, $B =$ magnetic field induction at any point of the moving bar due to the long straight current conductor $= \frac{\mu_0 i_0}{2\pi(x + x_0)}$.

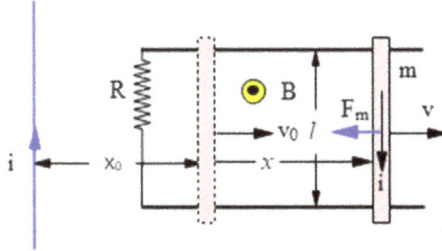

After evaluation, we have

$$F_m = \frac{\mu_0^2 i_0^2 v}{4\pi R(x + x_0)^2}$$

Applying Newton's second law, the net force acting on the bar is

$$F_m = -\frac{mdv}{dt} = -\frac{mvdv}{dx}$$

Using the last two equations and integrating both sides after cancelling 'v' from both sides, we obtain

$$\int_0^{v_0} v = \frac{\mu_0^2 i_0^2 l^2}{4\pi^2 Rm} \int_0^x \frac{dx}{(x + x_0)^2}.$$

After evaluating the integration and simplifying the terms,

$$v = v_0 - \frac{\mu_0^2 i_0^2 l^2}{4\pi^2 mR}\left(\frac{1}{x_0} - \frac{1}{x + x_0}\right) \text{ Ans.}$$

Problem 20 (Linear generator)

A conducting bar of length l is dragged on a smooth conducting rail with constant velocity as shown in the figure below. An outward magnetic field of induction B is applied. If a resistor R is connected to the rail, find the:

(a) Ampère's force acting on the bar;
(b) power developed by the external force;
(c) electrical power delivered by the external agent;
(d) thermal power generated in the resistor;
(e) power delivered by the magnetic field;
(f) power delivered by the tangential component of the magnetic force.

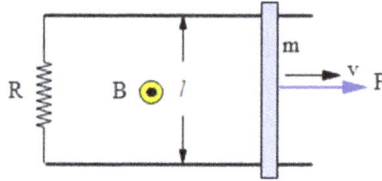

Solution

(a) When the bar moves with a velocity v in the magnetic field of induction B, the induced emf $\varepsilon_i = Blv$. This generates an induced current $i = \frac{Blv}{R}$ in the bar. Now the current-carrying bar moving in the magnetic field experiences a magnetic force $F_m = ilB = \frac{B^2l^2v}{R}$. Ans.

(b) Power developed by the external agent $= P_{ext} = \vec{F} \cdot \vec{v}$ where F can be obtained by applying Newton's second law on the bar. As the bar moves with constant velocity v, it experiences no net force, $F - F_m = \frac{mdv}{dt} = 0$.

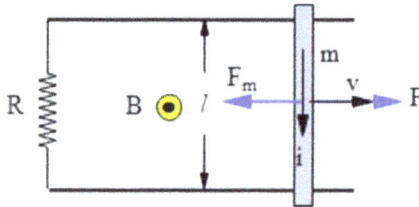

Thus, substituting $F = F_m = \frac{B^2l^2v}{R}$ in the previous expression, we obtain $P_{ext} = Fv = \frac{B^2l^2v^2}{R}$. Ans.

(c) Thermal power generated in the resistor $= P_{ther.} = i^2R$. Substitution of i in this equation yields $P_{ther.} = \frac{B^2l^2v^2}{R}$. Ans.

(d) Electrical power generated $= P_{el} = \varepsilon_i i = (Blv)(\frac{Blv}{R}) = \frac{B^2l^2v^2}{R}$ Ans.

(e) Power imparted by the magnetic field can be expressed as $\vec{F'}_m \cdot \vec{v_R}$ where v_R is the resultant of v and v_d of the charge carrier inside the bar. Since $\vec{v_R} \perp \vec{F'}_m$ power imparted by magnetic field is zero; $v_d = $ drift velocity of the charged particle. Ans.

(f) Power delivered by the tangential component of $\vec{F_m} = \vec{F_m} \cdot \vec{v} = -F_m v = \frac{B^2l^2v^2}{R}$ is obtained after substituting the value of F_m. We can see that this is numerically equal to the heat loss per second. The sum of work done by all the forces acting on the bar $= W_{ext} + W_{mag.} + W_{battery} + W_f = 0$. This is equal to its change in its kinetic energy, which is equal to zero as it moves with a constant velocity. As the magnetic force does not perform work, the sum of work done by the battery and that of the external agent is zero if the bar is moved with a constant velocity. Ans.

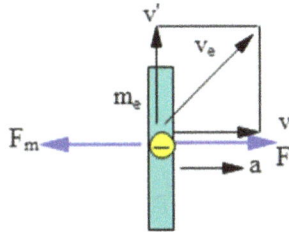

N.B: Ampère's force is a component of the total magnetic force perpendicular to the bar. The work done by magnetic force is zero, but the work done by Ampère's force is non-zero.

Problem 21 A rectangular loop of dimensions $b \times l$ is released from a certain height h into a region of horizontal magnetic field of induction $\overrightarrow{B} = -B\,\hat{k}$.

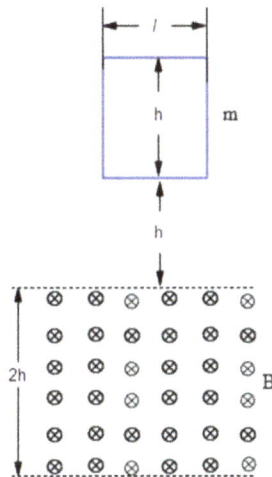

What must be the height of fall h so that the loop will move with a constant velocity in the magnetic field?

Solution

For the loop to move with constant velocity v in the magnetic field the net force acting on it must be zero. Hence, magnetic force nullifies the effect of gravity.

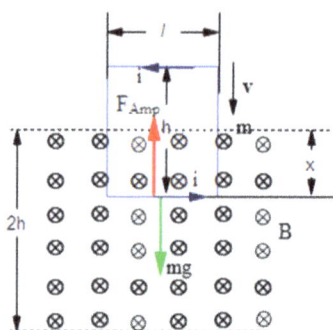

Then, $F_m = mg$. Substituting $F_m = \frac{B^2 l^2 v}{R}$, we obtain $v = \frac{mgR}{B^2 l^2}$.

This is the terminal speed. This velocity is attained by the loop after falling through a height $h = \frac{v^2}{2g}$ in gravity. Putting the value of v, we have $h = \frac{m^2 g R^2}{2 B^4 l^4}$. Ans.

Problem 22 (Linear motor)

Let us withdraw the external force F from the original set of problem 17 and then analyse the motion of the bar.

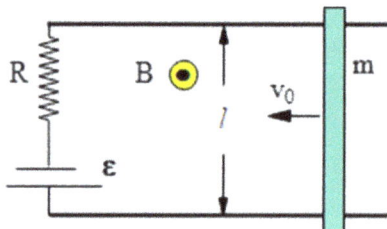

Solution

Directly putting $F = 0$ in the velocity–time equation of problem 17 we obtain the velocity of the bar at any instant t as

$$v = \frac{\varepsilon}{Bl}\left(1 - e^{-\frac{B^2 l^2}{mR}t}\right) - v_0 e^{-\frac{B^2 l^2}{mR}t}$$

If we assume that the bar is at rest initially, put $v_0 = 0$.

Then, its velocity after a time t can be given as

$$v = \frac{\varepsilon}{Bl}\left(1 - e^{-\frac{B^2 l^2}{mR}t}\right)$$

When $t \to \infty$, the bar moves with a terminal speed $v_t \to \frac{\varepsilon}{Bl}$

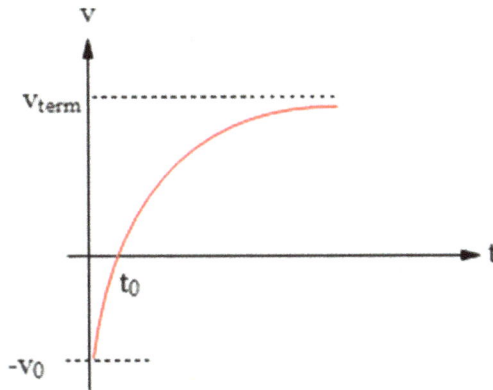

The speed of the bar is limited to $v_t = \varepsilon/Bl$ because of electromagnetic induction. Ans.

Problem 23 (Conducting rail fitted with a capacitor)
Discuss the effect of an external force F on a smooth bar placed on a conducting horizontal rail fitted with a capacitor and subjected to a steady magnetic field.

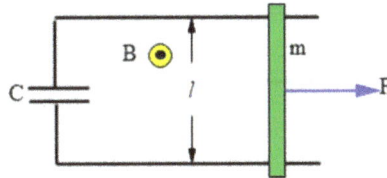

Solution
The constant force F tries to accelerate the bar; as a result, the bar picks up a speed v in the magnetic field, so, emf $\varepsilon_i = Blv$ is induced across it. As the capacitor is connected to the loop, it will be charged to the same potential difference $\varepsilon_i = Blv$. So, one of the plates of the capacitor continuously receives a charge $q = CV_c = C\varepsilon_i$ by the induced emf (rod moving in the magnetic field), whereas the other plate loses equal charge at the same rate. Thus, a current flows in the loop by the induced emf developed across the rod which can be given as $i = \frac{dq}{dt} = \frac{d}{dt}(C\varepsilon) = C\frac{d}{dt}(Blv) = BlC\frac{dv}{dt} = BlCa$. Now the current-carrying bar experiences a magnetic force $F_m = ilB = B^2l^2Ca$. This force F_m resists the external force F acting on the bar. So, the net force $F_{net} = F - F_m$ acting on the bar accelerates it with an acceleration a. Applying Newton's second law on the bar, we have $F - F_m = ma$. Substitution of F_m, the acceleration of the bar is

$$a = \frac{F}{(B^2l^2C + m)}$$

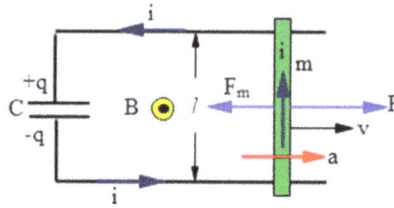

It tells us that the bar will move with a constant acceleration in the direction of applied force F. Ans.

Problem 24 Now remove the capacitor and connect an inductor of inductance L in the previous example. Ignore the resistance for sake of simplicity. Let us suppose that the initial velocity of the bar is zero. If a constant external force F acts on the bar, describe the nature of motion of the bar. You can refer to the next chapter to understand the concept of inductance.

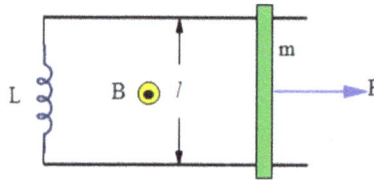

Solution

When the bar attains a rightward velocity v after a time t by the applied force F, an anticlockwise induced current i is set up. This current generates an Ampère's force F_m which acts to left on the bar. Ampère's law gives $F_m = ilB$. Applying Kirchhoff's circuital law for the loop, we have $\varepsilon_L + \varepsilon_{ind} = 0$. Or, $-L\frac{di}{dt} + Blv = 0$ or $Ldi = Blvdt$.

When the bar moves through a distance x, current in the loop increases from 0 to i during time t. Hence, $\int_0^i di = Bl \int_0^t vdt = Bl \int_0^x dx$. This gives the current $i = \frac{Bl}{L}x$. It shows that the current in the loop varies linearly with the displacement of the bar. This current develops the magnetic force $F_m = ilB = \frac{B^2l^2}{L}x$.

Then, the net force acting on the bar is $F_{net} = F - F_m$

$$\text{or, } m\frac{d^2x}{dt^2} = F - \frac{B^2l^2}{L}x$$

$$\text{or, } \frac{d^2x}{dt^2} + \frac{B^2l^2}{mL}x = \frac{F}{m}$$

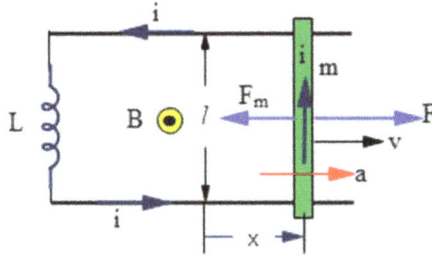

This differential equation tells us that, the bar oscillates simple harmonically with an angular frequency, $\omega = \frac{Bl}{\sqrt{mL}}$ and amplitude $A = \frac{FL}{B^2l^2}$. The equation of motion is

$x = \frac{FL}{B^2l^2}(1 - \cos\frac{Bl}{\sqrt{mL}}t)$. Ans.

N.B: If instead of applying the force the bar is given an initial velocity v_0, the nature of motion of the bar will be simple harmonic with an angular frequency

$$\omega = \frac{Bl}{\sqrt{ml}} \text{ and amplitude } A = \frac{\sqrt{ml}}{BL}v_0 \text{ Ans.}$$

Problem 25 In the previous example, if we substitute the inductor by a resistor of resistance R,

(a) How does the velocity of the rod change with time?

(b) Find the

    (i) speed of the rod as the function of distance x.

    (ii) distance as the function of time and the total distance covered by the rod.

    (iii) Draw v–t, x–t and v–x graphs.

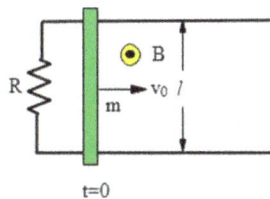

Solution

(a) Let the rod acquire a velocity v after covering a distance x at time t, The flux swept by the rod is

$$\phi = Blx$$

The rate of change in flux, that is, induced emf is

$$\varepsilon = \left|\frac{d\phi}{dt}\right| = \frac{Bldx}{dt} = Blv$$

Then, the induced current in the rod is

$$i = \frac{\varepsilon}{R} = \frac{Blv}{R}$$

Hence, the Ampère's force acting on the rod is

$$F_{\text{om}} = ilB = \frac{B^2l^2v}{R}$$

Since, this force opposes the velocity, we can write

$$m\frac{dv}{dt} = -\frac{B^2l^2v}{R}$$

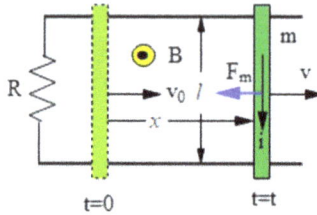

Then, $\frac{dv}{v} = -\frac{B^2l^2}{mR}dt$

Since, during time t, the speed of the rod changes from v_0 to v, integrating, we have

$$\int_0^v \frac{dv}{v} = -\frac{B^2l^2}{mR}\int_0^t dt$$

This gives

$$v = v_0 e^{-\frac{B^2l^2}{mR}t} \text{ Ans.}$$

(b)

(i) We can also write $\frac{dv}{dt} = v\frac{dv}{dx} = \frac{B^2l^2v}{mR}$

$$\Rightarrow v = v_0 - \frac{B^2l^2x}{mR} \text{ Ans.}$$

(ii) Putting $v = v_0 e^{-\frac{B^2l^2}{mR}t}$ in the last equation, we have

$$v_0 e^{-\frac{B^2l^2}{mR}t} = v_0 - \frac{B^2l^2x}{mR}$$

$$x = \frac{mRv_0}{B^2l^2}\left(1 - e^{-\frac{B^2l^2}{mR}t}\right) \text{ Ans.}$$

When t tends to infinite or v tends to zero, the rod undergoes a maximum displacement

$$x_{max} = \frac{mRv_0}{B^2l^2} \text{ Ans.}$$

(iii) The graphs of v–t, x–t and v–x are given as follows:

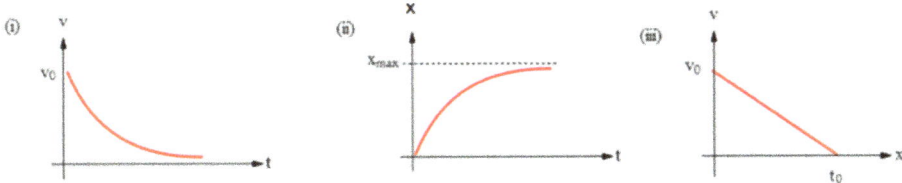

In the above graphs we can see that velocity decreases with time exponentially to zero in graph (i); the displacement increases with time exponentially from zero to a maximum value (maximum distance covered) in graph (ii); the velocity of the bar decreases linearly with distance (but exponentially with time) to zero in graph (iii). Ans.

Problem 26 (Magnetohydrodynamics)

A parallel plate air capacitor of square plate of length l is placed completely inside a flowing liquid dielectric of resistivity ρ and relative permittivity ε_r. The liquid flows with a velocity v between the plates horizontally, while a vertical uniform magnetic field of induction B is present. If the capacitor is connected to an external resistor R, find the (a) current, (b) Ampère's force per unit volume, (c) induced electric field, (d) voltage dropped across the capacitor, (e) static electric field, (f) difference between induced and static field, (g) charge induced.

Solution

(a) Here the capacitor behaves as a battery as the flowing liquid dielectric moving in the upward magnetic field between the plates generates a motional induced emf given as

$$\varepsilon = Bvd$$

The polarity of the induced emf is given as shown in the figure. The effective resistance of the liquid between the capacitor is

$$R' = \frac{\rho d}{A} = \frac{\rho d}{l^2}$$

So, the total resistance of the circuit is

$$R_{eff} = R + R' = R + \frac{\rho d}{l^2}$$

Then the current flowing through the circuit is

$$i = \frac{\varepsilon}{R_{eff}} = \frac{Bvd}{R + \frac{\rho d}{l^2}}$$

$$\Rightarrow i = \frac{Bl^2vd}{Rl^2 + \rho d} \quad \text{Ans.}$$

(b) The Ampère's force that opposes the velocity is

$$F = idB$$

Putting the obtained value of current, we have

$$F = \frac{B^2l^3dv}{Rl^2 + \rho d}$$

The Ampère's force per unit volume is

$$\frac{F}{vol} = \frac{B^2l^3dv}{\left(Rl^2 + \rho d\right)\left(l^2d\right)} = \frac{B^2lv}{\left(Rl^2 + \rho d\right)} \quad \text{Ans.}$$

(c) The induced electric field is

$$E_{ind} = Bv \quad \text{Ans.}$$

(d) The voltage drops across the capacitor is

$$V = iR_{dieltric}$$

$$= \left(\frac{Bl^2vd}{Rl^2 + \rho d}\right)\left(\frac{\rho d}{l^2}\right)$$

$$= \frac{\rho d^2 Bv}{Rl^2 + \rho d} \quad \text{Ans.}$$

(e) The static electric field is

$$E_{static} = \frac{V}{d}$$

$$= \frac{\rho d^2 Bv}{\left(Rl^2 + \rho d\right)d} = \frac{\rho Bdv}{Rl^2 + \rho d} \quad \text{Ans.}$$

(f) The difference between the induced and static electric field is

$$E_{ind} - E_{static} = Bv - \frac{\rho Bdv}{Rl^2 + \rho d} = \frac{Rl^2 Bv}{Rl^2 + \rho d} \quad \text{Ans.}$$

(g) The charge induced in the capacitor is

$$q = CV = \left(\frac{\varepsilon_0\varepsilon_r l^2}{d}\right)\left(\frac{\rho B d^2 v}{Rl^2 + \rho d}\right)$$

$$= \frac{\varepsilon_0\varepsilon_r l^2 \rho B d v}{Rl^2 + \rho d} \quad \text{Ans.}$$

N.B: In this case the induced electric field is greater than the static electric field due to the resistance of the liquid. However, for a good conducting liquid, both fields are nearly equal.

Problem 27 A metallic thin long cylinder of mass M and radius R is smoothly pivoted at its centre. It has a uniform charge distribution of density σ. A string is wrapped over the ring and it is connected to a bob of mass m. If the bob is released from rest, assuming that the string does not slip, find the angular acceleration of the cylinder.

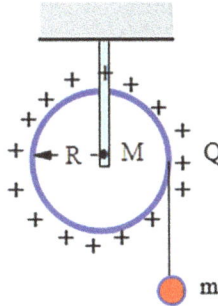

Solution
The weight of the bob generates a tension that rotates the charged thin cylinder with an angular acceleration α, say. The rotation of charged ring produces a magnetic field. At an angular speed ω of the cylinder, the magnetic field is given by Ampère's circuital law as follows:
For a long strip of width δl,

$$B\delta l = \mu_0 i_{enclosed},$$

If dq = charge contained on the strip, the current associated with the moving charged strip is

$$i = \frac{dq}{dt} = \frac{\sigma\delta l(Rd\theta)}{dt} = \sigma\delta lR\omega$$

Using the last two equations, we have

$$B\delta l = \mu_0\left(\sigma \delta l R\omega\right)$$

So, the axial uniform magnetic field produced by the spinning thin charged cylinder is

$$B = \mu_0 \sigma R\omega \tag{8.6}$$

Since the angular speed increases, the magnetic field also increases at a rate

$$\frac{dB}{dt} = \mu_0 \sigma R\frac{d\omega}{dt} = \mu_0 \sigma R\alpha \tag{8.7}$$

So, the rate of change of flux linked with the rotating cylinder is equal to emf induced, given as

$$-\pi R^2\frac{dB}{dt} = \varepsilon_{\text{ind}} = E_{\text{ind}}2\pi R$$

$$\Rightarrow E_{\text{ind}} = -\frac{R}{2}\frac{dB}{dt} \tag{8.8}$$

The negative sign tells us that the induced electric field is circular pointing in a clockwise sense. Putting the obtained value of dB/dt, we have

$$E_{\text{ind}} = -\frac{R\mu_0\sigma R\alpha}{2} \tag{8.9}$$

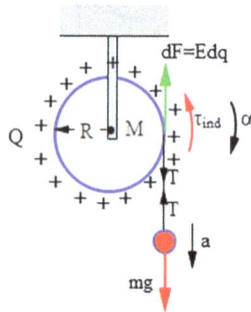

This induced electric field produces an anticlockwise torque on the elementary strip of charge dq, which can be given as

$$d\tau = \left(E_{\text{ind}}dq\right)R = \left(\frac{R\mu_0\sigma R\alpha}{2}dq\right)R$$

The induced torque acting on the cylinder is

$$\tau = \int d\tau = \int \left(\frac{R\mu_0\sigma R\alpha}{2}dq \right)R = \frac{\mu_0\sigma R^3\alpha}{2}\int dq$$

$$\Rightarrow \tau = \frac{\mu_0\sigma R^3 Q\alpha}{2}$$

The charge Q of the cylinder is

$$Q = (2\pi Rl)\sigma$$

Using the last two equations,

$$\tau = \frac{\mu_0\sigma R^3(2\pi Rl)\sigma\alpha}{2} = \pi\sigma^2 R^4\mu_0\alpha l \tag{8.10}$$

This torque opposes the anticlockwise torque TR produced by the tension about the axis of rotation of the cylinder. So, the net anticlockwise torque is

$$\tau_{\text{net}} = TR - \tau = I\alpha \tag{8.11}$$

Using the last two equations, we have

$$TR - \pi\sigma^2 R^4\mu_0\alpha l = I\alpha$$

$$\Rightarrow TR = \left(\pi\sigma^2 R^4\mu_0 l + I\right)\alpha$$

$$\Rightarrow TR = \left(\pi\sigma^2 R^4\mu_0 l + MR^2\right)\alpha \left(\because I = MR^2\right) \tag{8.12}$$

Writing the force equation on the bob m, we have

$$\Rightarrow mg - T = ma = mR\alpha$$

$$\Rightarrow T = m(g - R\alpha) \tag{8.13}$$

Using last two equations (8.12) and (8.13), we have

$$\Rightarrow m(g - R\alpha)R = \left(\pi\sigma^2 R^4\mu_0 l + MR^2\right)\alpha$$

$$\Rightarrow \alpha = \frac{mg}{\pi\sigma^2 R^3\mu_0 l + (M + m)R} \quad \text{Ans.}$$

Problem 28 A circular loop of radius R carries a current i. A small loop of radius r placed coaxially is moved with a constant velocity from a bigger loop with a constant velocity v along the common axis. Find the force of interaction between these two loops at a distance of separation.

Solution

With reference to problem 16, the emf induced in the small loop is

$$\varepsilon = -\frac{d\phi}{dt} = -A_{small}\frac{dB}{dt}$$
$$= -A_{small}\frac{dB}{dx}\frac{dx}{dt} = -A_{small}\frac{dB}{dx}v$$

(8.14)

The induced current is

$$i' = \frac{\varepsilon}{R_0} = -\frac{A_{small}}{R_0}\frac{dB}{dx}v$$

(8.15)

The magnetic moment of the small loop is

$$\mu_m = i'A_{small}$$

(8.16)

Using the last two equations, we have

$$\mu_m = i'A_{small} = -\frac{\left(A_{small}\right)^2}{R_0}\frac{dB}{dx}v$$

(8.17)

The force of interaction between two current loops is given as

$$\Rightarrow F = \mu_m\frac{dB}{dx}$$

(8.18)

Using the last two equations,

$$\Rightarrow F = -\frac{\left(A_{small}\right)^2}{R_0}\left(\frac{dB}{dx}\right)^2 v$$

(8.19)

The magnetic induction due to the bigger loop is

$$B = \frac{\mu_0 i R^2}{2(R^2 + x^2)^{5/2}}$$

(8.20)

Then differentiating with x, the gradient of magnetic field along the x-axis is

$$\frac{dB}{dx} = \frac{d}{dx}\left(\frac{\mu_0 i R^2}{2(R^2 + x^2)^{3/2}}\right) = \frac{3\mu_0 i R^2 x}{2(R^2 + x^2)^{3/2}}$$

(8.21)

Using equations (8.19) and (8.21), we have

$$\Rightarrow F = -\frac{\left(\pi r^2\right)^2}{R_0}\left\{\frac{3\mu_0 i R^2 x}{2(R^2 + x^2)^{3/2}}\right\}^2 v$$

$$\Rightarrow F = -\frac{9}{4}\frac{\pi^2 \mu_0^2 i^2 r^4 R^4 v^2}{R_0(R^2 + x^2)^5} \quad \text{Ans.}$$

References

[1] Al-Khalili J 2015 The birth of the electric machines: a commentary on Faraday (1832) 'Experimental researches in electricity' *Phil.l Trans. R. Soc. A: Math. Phys. Eng. Sci.* **373** 20140208

[2] Darrigol O 2003 *Electrodynamics from Ampere to Einstein* (Oxford University Press)

[3] Sturgeon W, Christie S H, Gregory O and Barlow P 1824 No. III. Improved electro-magnetic apparatus *Trans. Soc. Inst. Lond. Encourage. Arts, Manuf. Commerce* **43** 37–52

[4] Magnusson C E 1920 The centennial of the discoveries of Oersted, Arago, and Ampère *J. Am. Inst. Electr. Eng.* **39** 1031–3

[5] Thompson S P 1898 *Michael Faraday: his life and work* (Cassell)

IOP Publishing

Problems and Solutions in Electricity and Magnetism

Pradeep Kumar Sharma

Chapter 9

Inductance

9.1 Introduction

Inductance was discovered by Micheal Faraday with his first experiment on 29th August 1831. In the same year, inductance was also discovered independently by Joseph Henry in the United States of America. The unit of inductance is the Henry named after Joseph Henry.

The very familiar example of inductors is a choke coil that chokes or blocks unwanted electrical noise (high-frequency signals). A choke coil is used in tube light to arrest the large frequencies of alternating current and it generates a high voltage to just start the fluorescence in the tube. It has the ability of storing magnetic energy when a current flows through it. It is also known as induction coil or inductor. Inductance (L) is the property of a conductor by the virtue of which it stores magnetic energy when it carries a current. It is analogous to capacitance (C) of a piece of a conductor which is defined as the ability of storing electrostatic energy when it carries a charge. Inductors store magnetic energy and capacitors store electrostatic energy. However, resistance (R) is another property of a conductor by the virtue of which it dissipates heat when it carries a current. These three circuit elements R, L and C are called passive because they need permanent external energy source to work, each of them can drop a voltage across them. On the other hand, any sources of electricity such as a battery or generators are called active circuit elements because they have the ability of generating electrical energy. An inductor is used as a coil in an LC circuit for tuning a radio set. Inductors are used to avoid power loss. Inductors have a wide range of applications in choking, blocking high-frequency noise in a filter circuit and storing and transferring energy in power converters (dc–dc or ac–dc). A transformer is a great example of a step up of electrical power for power transmission and a step down of the electrical power for our day-to-day use in our electrical appliances. A transformer is a combination of primary and secondary coils which are basically inductors. By the property of the mutual induction of the coils, a transformer can transform the voltage by transferring a changing magnetic field from primary to secondary coil.

doi:10.1088/978-0-7503-6477-5ch9

9.2 Self-inductance

9.2.1 Definition of self-inductance in a steady magnetic field

A current-carrying conductor produces a magnetic field inside and also outside the conductor. A fraction of this self-magnetic flux that links with the conductor itself is known as its internal flux ϕ_{int}. The remaining flux which is linked with other circuits outside the conductor is known as external flux ϕ_{ext}. So, the total flux linkage ϕ of the current i flowing in a conductor is equal to the sum of internal and external flux, given as

$$\phi = \phi_{int} + \phi_{ext}$$

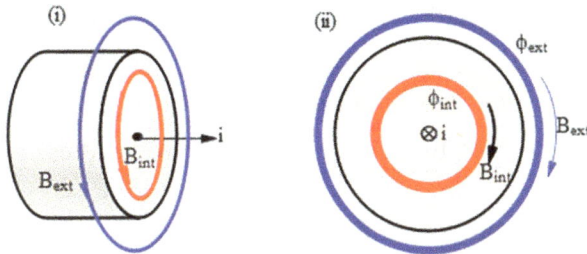

Internal and external flux linkage

Since, the self-flux ϕ of the conductor is directly proportional to the current i flowing through it, the ratio ϕ/i is a constant quantity which is known as (self) inductance L of the conductor.

$$L = \frac{\phi}{i} = \frac{\phi_{int}}{i} + \frac{\phi_{ext}}{i}$$

We call the first term internal inductance L_{int} and the second term is known as external inductance L_{ext} of the conductor. Then, the total inductance L can be written as

$$L = L_{int} + L_{ext}$$

Putting $i = 1$ A, in the forumula $L = \phi/i$, we have

$$L = \phi$$

This tells us that *whenever an electrical system carries a current of 1 A, the self-inductance of the system is numerically equal to the total flux linked with it.*

Example 1 A closely-spaced coil having $N = 100$ turns carries a current of 1.5 A. If the self-inductance of the coil is equal to 2 mH, what is (a) the total flux linked with the coil, (b) the total flux linked with each loop (turn) of the coil, (c) the flux passing through the coil, and (d) total flux passing through an infinite plane perpendicular to the length of the coil.

Solution

(a) The magnetic flux Φ linked with a coil can be calculated using the formula:
$$\Phi = L \cdot i,$$
where, $L = 2$ mH $= 2 \times 10^{-3}$ H and $i = 1.5$ A. So, we have
$$\Phi = 2 \times 10^{-3} \times 1.5 = 3 \times 10^{-3} \text{ Weber}.$$
Since the loops of the coil are closely placed, the total flux linkage is equal to flux linkage per turn multiplied with number of turns (loops). Ans.

(b) Since there are 100 turns in the coil, the flux linked per turn is given as
$$\Phi/N = 3 \times 10^{-3}/100 = 3 \times 10^{-5} \text{ Weber/turn Ans}.$$

(c) Total flux passing through the coil is
$$\Phi = 2 \times 10^{-3} \times 1.5 = 3 \times 10^{-3} \text{ Weber Ans}.$$

(d) Total flux passing through the coil is coming out of the coil to form closed loops. This means that total incoming and outgoing flux passing through an infinite plane perpendicular to the coil is always zero. Ans.

9.2.2 Definition of inductance in a time-varying field

If a magnetic field varies with time, its source must vary with time. We understood that a time-varying current produces moving magnetic flux. When the moving flux cuts a conductor it produces an electromotive force (emf). This means that each element of a conductor will behave as a tiny cell. So, the entire piece of a conductor (coil) can be treated as the series combination of tiny cells. So, the total coil behaves as a battery.

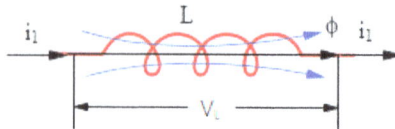

Time varying flux ϕ induces a volatage V_L across the inductor

In other words, when a current i in a coil (or an inductor) varies with time, it produces moving flux lines that cut the conductor. As a result, an induced electric field appears at each and every point of the coil. This induces an emf in the coil. This self-induced emf opposes the change in current in the coil. Then, we can say that a voltage is dropped across the inductor in the external circuit but not inside the conductor because $E = 0$ inside the conductor.

The induced voltage V_L can be given as

$$V_L = -\frac{d\phi}{dt},$$

where $\phi =$ flux passing through the inductor.

$$\Rightarrow V_L = -\frac{d}{dt}(Li) \ (\because \phi = Li)$$

$$\Rightarrow V_L = -L\frac{di}{dt}$$

So, the self-inductance of the coil or conductor can be given as

$$L = \frac{|V_L|}{\left|\frac{di}{dt}\right|}$$

If the current changes at a rate $\frac{di}{dt} = 1 \text{ A s}^{-1}$, we have

$$L = V_L$$

This means that self-inductance of a system is numerically equal to the voltage induced in it when current flowing through it varies at a rate of one ampere per second.

Example 2 Find the self-inductance L of the coil whose resistance $R = 3\,\Omega$, a current $i = 2.5$ A, which is increasing at a rate of 10^3 A s^{-1}. Put $\varepsilon = 10$ V.

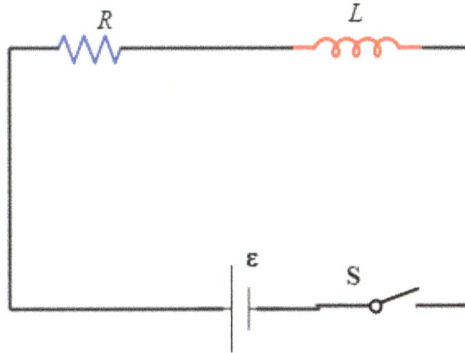

Solution
Applying of Kirchoff's voltage law in the circuit,

$$\varepsilon = V_L + V_R,$$

$$\Rightarrow \varepsilon = L\frac{di}{dt} + iR$$

$$\Rightarrow L = \frac{\varepsilon - iR}{\frac{di}{dt}} \text{ H}$$

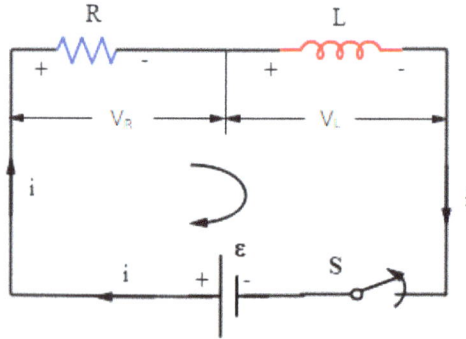

$$= \frac{10 - 2.5 \times 3}{10^3} = 2.5 \times 10^{-3} \text{ H Ans.}$$

9.3 Inertial properties of an inductor; growth and decay of current

9.3.1 Growth of current

In the previous sections you have learnt that whenever a current varies in an inductor, a voltage V_L is induced across the inductor which is given as

$$V_L = L\frac{di}{dt}$$

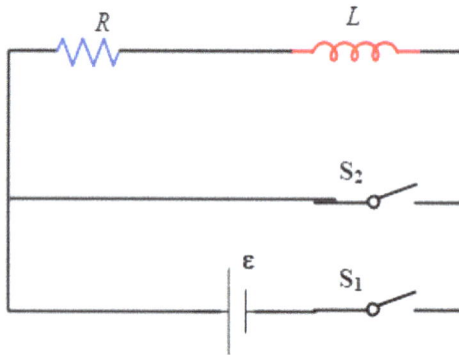

Let us assume that the current in the inductor varies suddenly. This means, with no (zero) time ($\delta t = 0$), current increases from zero to i. Hence, di/dt is infinite. If this happens, an incredible voltage must be induced in the inductor. Hence, to set up a current i, the applied emf must be stronger than the induced voltage so as to oppose it. Thus, a source of an infinite emf is needed to produce a current in the coil but this is not possible. So, we can reject the idea of a sudden or abrupt growth of current in an inductor and accept the fact that the current grows gradually. Let us see, exactly in which way current builds up in the RL circuit after closing the key at time $t = 0$. After a time t, as the current i increases at a rate di/dt, there results a voltage drop $V_L = L(di/dt)$ across the inductor and the current i drops a voltage $V_R = iR$ across the resistor R.

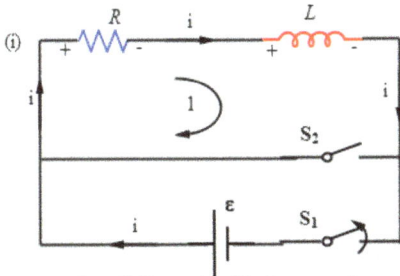

In an R-L curcuit, while the current is building, induced voltage opposes the current

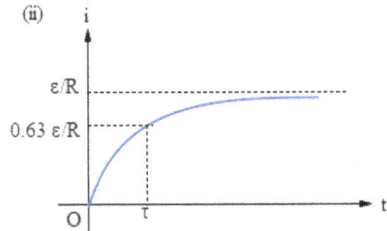

In an R-L circuit current builds up exponentially

9.3.2 Time constant

Applying Kirchoff's second (voltage) law,

$$-V_L - V_R + \varepsilon = 0$$

$$\Rightarrow -L\frac{di}{dt} - iR + \varepsilon = 0$$

$$\Rightarrow L\frac{di}{dt} + iR = \varepsilon$$

$$\Rightarrow L\frac{di}{dt} = \varepsilon - iR$$

$$\Rightarrow \frac{di}{\varepsilon - iR} = \frac{dt}{L}$$

Since current increases from zero to i during a time t,

$$\int_0^i \frac{di}{\varepsilon - iR} = \frac{1}{L}\int_0^t dt$$

$$\Rightarrow i = \frac{\varepsilon}{R}(1 - e^{-\frac{t}{\tau}}), \quad \text{where } \tau = \frac{L}{R}.$$

This equation tells that the current builds up exponentially to a steady value of magnitude $\frac{\varepsilon}{R}$ after a long time ($t \to \infty$). Practically, current becomes steady after a few (practically five to six) time constants (τ) of an RL circuit. If we put $t = \infty$ and $\frac{\varepsilon}{R} = i_0$ (maximum or steady-state current) in the previous equation we have

$$i = i_0\left(1 - \frac{1}{e}\right) = 0.63i_0$$

So, the time constant is a measure of the rate or speed of building the current in an *RL* circuit. During a time equal to a time constant, the current builds up to 63% of the maximum or steady-state current. Then, logically, you can say that, after five to six time constants, the current becomes nearly constant or steady. So, we can conclude that,

> An inductor does not allow any current in it to grow abruptly. The growth of current happens exponentially in an inductor.

9.3.3 Decay of current

After closing the key K_2 at $t = 0$, let the current fall at a rate of di/dt. Hence, the inductor acts as a battery that continues to send the current in the same direction with the polarities as shown. Now, applying Kirchoff's circuital law (KCL) we have

$$V_L + V_R = 0,$$

$$\Rightarrow L\frac{di}{dt} + iR = 0$$

$$\Rightarrow \frac{L}{i}\frac{di}{} = -\frac{R}{L}dt$$

If the current drops from i_0 to i during a time t after closing key K_2,

$$\int_{i_0}^{i} \frac{di}{i} = -\frac{R}{L}\int_0^t dt$$

$$\Rightarrow i = i_0 e^{-\frac{t}{\tau}},$$

where $i_0 = \frac{\varepsilon}{R}$ and $\tau = \frac{L}{R}$.

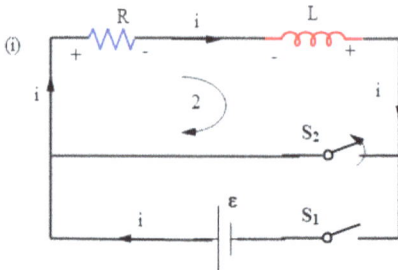

(i) When the current decays, the induced voltage favours the current

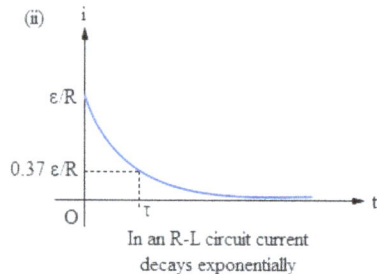

(ii) In an R-L circuit current decays exponentially

This shows that the current decays exponentially. During $t = \tau$, it falls to 37% of the initial current i_0. This means, practically it takes roughly four or five time constants to reduce the current to zero even though theoretically it takes infinite time ($t \to \infty$) for a complete decay of the current.

Recapitulating, *an inductor in a closed circuit does not allow a sudden variation (increase or decrease) of current flowing through it. This property by virtue of which an inductor opposes the variation of current flowing in it, is known as 'inductance'.*

9.3.4 Analogy between self-induction and inertia

Following the above arguments, we compare an 'inductor' with 'mass' (inertia). In mechanics, you cannot increase or decrease the velocity of an object with no time. So, sudden change in velocity is impossible in mechanics. This means that it takes a finite (non-zero) time to change the velocity of a body. Then, it is equivalent to say that, 'matter resists the change in its state (velocity)'. This is Newton's first law or law of inertia. The property of matter by the virtue of which it opposes the change in its state is termed as inertia. 'Mass' of an object is the measure of its inertia. If the mass is greater, it will be more difficult to set it in motion (or to stop it).

Likewise, any electrical system (circuit) opposes the change in its own current by inducing an emf in it due to the action of induced electric field. This is known as self-induction and this opposing (inertial) nature of the electric circuit to any change in its current is known as self-inductance. *The portion of the circuit responsible for it is called 'inductor'.*

Let us demonstrate the inertial nature of inductors through the following example.

Example 3 Shown in the figure is a system of two identical bulbs B_1 and B_2 connected in parallel with a battery of emf ε. B_2 is connected in series with an inductor L. Describe the variation of brightness of the bulbs after closing the switch S. Is there any difference when we open the switch S after a long time? Draw an i–t graph by using the logic and avoiding mathematical calculations.

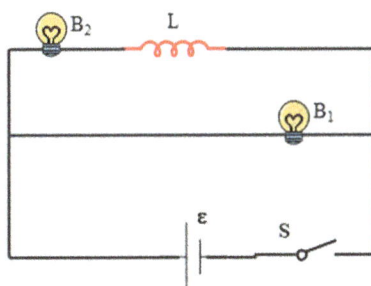

Solution

Practically the bulbs are simple resistors of resistor R, say, without a noticeable inductive effect. As the bulbs are joined in parallel with the battery, we should expect them to glow with equal brightness after closing the switch. But the presence of the inductor induces a voltage V_L across it. As the current increases, the induced voltage opposes the applied voltage ε for all the times. This makes B_1 glow brighter than B_2 till a steady-state current flows through B_2. It will glow with equal brightness of the other bulb in the steady state after some time.

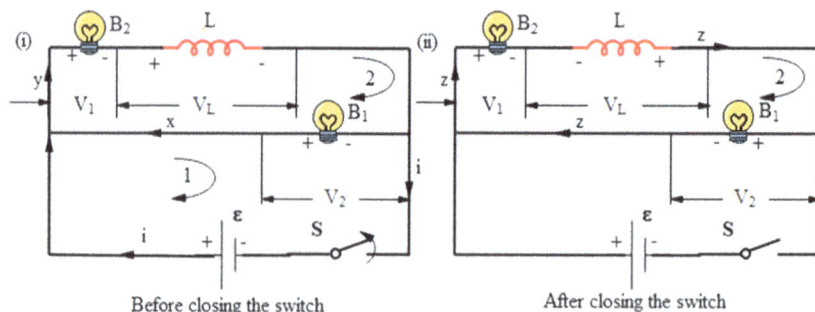

Before closing the switch   After closing the switch

After a long time, let us reopen the switch. This tries to destroy the magnetic field (current) rapidly. As a result, a huge voltage is induced so as to retain the current. So, the polarity of the inductor reverses in order to maintain the same equal current through both B_1 and B_2. However, current flows in a direction opposite to the initial current of B_1. The currents in both the bulbs that measure their brightness fall gradually to zero after some time. Variation of currents in B_1 and B_2 are shown in the diagrams. Ans.

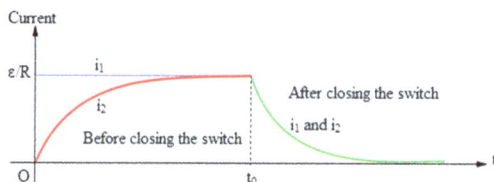

9.4 Self-energy stored in an inductor

As learnt in the previous section, while a current builds up at a rate of di/dt in an inductor, self-induced voltage dropped across the coil is

$$V_L = L\frac{di}{dt}$$

As the induced voltage opposes the applied emf, the source (battery or generator) performs positive work in pushing the charges against the induced voltage. If an elementary charge dq is pushed by the source (applied electric field E_a) from a and b against the induced voltage V_L (or induced electric field E_i) during a time dt, constituting an electric current i, the work dW performed by the source during the transportation of the elementary charge dq between the terminals of the inductor is given as

$$dW = V_L dq$$

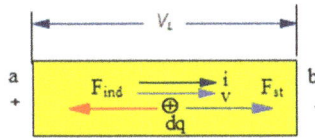

The battery or genertor does a positive work on the conductor in setting a current i which is stored in the form of magnetic energy associated with this curremt.

Using the last two equations, we have

$$dW = Lidi$$

Integrating both sides, the total work done by the source in establishing a current i is

$$W = \int dW = L \int_0^i i \, dl = \frac{Li^2}{2}.$$

You know that current in a conductor is a stream of free electrons with certain drift speed. Hence, an increase in current is followed by an increment of kinetic energy of the electrons. So, in this process, positive work is done by the source in increasing the current in an inductor. If we ignore heat loss etc, the work done by the source is stored in the form of magnetic energy, that is, the excess kinetic energy of the electrons inside the conductor.

The induced current generates a magnetic field that can apply forces and torque to other current elements (magnets) according to Ampère's law. Now you can say that a magnetic field (force) has the ability of performing work. But Lorentz's force law tells us that a magnetic force cannot perform any work because F_m is perpendicular to the velocity v of the charged particles. To resolve this ambiguity, you should note that, 'Ampère's force is a component of the total magnetic force F_m' and it has the ability of doing mechanical work by displacing current elements. Needless to mention, the other component of the magnetic force will do equal negative work to produce no net work done. So, a magnetic field possesses certain energy. Since a magnetic field is produced by an electric current, we can say that a current-carrying inductor can store (magnetic) energy U by the virtue of its magnetic field established by the current. Then we can write

$$U = \frac{Li^2}{2}$$

If you connect the two terminals of an inductor by a resistor R (bulb), the stored magnetic energy will be released in other forms such as heat, sound and light.

Example 4 Find the total heat developed in an RL circuit of a decaying current.
   Solution
   As explained earlier, in an RL circuit, current decays from i_0 to i according to the relation

$$i = i_0 e^{-\frac{R}{L}t}$$

The energy dissipated in the resistor can be given as

$$Q = \int_0^\infty i^2 R \, dt$$

Substituting i and integrating we have

$$Q = \frac{L i_0^2}{2}$$

which is equal to the initial energy U stored in the inductor. This confirms that the total energy stored in the inductor is liberated in the form of heat, sound and light to the surroundings.

Now we can conclude that whenever an inductor L carries a current i, it establishes its 'self-flux ϕ' and stores 'a magnetic energy U' that has three mathematical forms given as

$$U = \frac{L i^2}{2} = \frac{\phi^2}{2L} = \frac{i\phi}{2}$$

Since the stored energy is due to its 'self-flux', this is known as 'self-energy'. Hence, any portion of an electrical circuit is capable of storing some amount of 'self-magnetic energy' (due to its own flux). This phenomenon is known as 'self-induction' of the circuit. It may be localized (lumped) or pervaded (distributed) in the circuit. Ans.

9.5 Conservation of magnetic flux

Let us bring an inductor coil of resistance R to an external magnetic field that may increase with time. This time-varying magnetic field will induce a current in the coil. When this current builds up or increases with time, the coil will induce a voltage called self-induced voltage due to its self inductance. The self-induced emf will oppose the applied or external emf. So, the net emf will be dropped due to the resistance of the coil. By applying KCL and Ohms law, we can write

$$iR = \varepsilon_{app} - \varepsilon_{self}$$

This means that, the self-induced emf cannot cancel out the applied emf completely due to the presence of resistance of the circuit. If the resistance R of the circuit is zero or negligible,

$$\varepsilon_{net}(= \varepsilon_{app} - \varepsilon_{self}) = iR = 0 \quad (R = 0).$$

We can see that the induced emf cancels the applied emf. Eventually, Lenz's law is most clearly exhibited in perfect and super conductors where the resistance is zero or nearly zero.

Then how much current flows in a superconducting circuit or a pure inductor with zero or negligible resistance? If we put $R = 0$ in the expression

$$i = \frac{\varepsilon_{net}}{R},$$

we obtain an infinite current which itself is non-physical. We can solve this fallacy by the following logic: when we decrease the resistance to zero ($R \to 0$), the net emf will approach zero ($\varepsilon_{net} \to 0$). Then we have the current

$$i = \lim_{R \to 0} \frac{\varepsilon_{net}}{R} = \text{a finite quantity.}$$

This tells us that no net emf is required to maintain the flow of a finite current in a resistanceless circuit. So, there is no net change in the flux linkage to the circuit it has zero resistance.

Using the expressions

$$\varepsilon_{app} = \frac{d\phi_{ext}}{dt} \text{ and } \varepsilon_{self} = \frac{d\phi_{self}}{dt}$$

we obtain

$$\varepsilon_{net} = \frac{d}{dt}(\phi_{ext} - \phi_{self}).$$

If you prefer to call ($\phi_{ext} - \phi_{self}$) 'the net flux ϕ' linked with the superconducting circuit, we have

$$\varepsilon_{net} = \frac{d\phi}{dt}.$$

So, in the flux formula

$$|\varepsilon| = \frac{d\phi}{dt},$$

the flux ϕ is neither the external flux nor the self-flux; rather it is the 'net (total) flux', i.e., algebraic sum of these two fluxes.

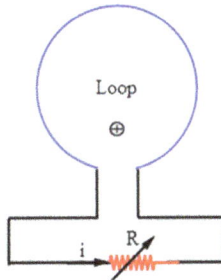

If the resistance of a circuit or loop tends to zero, the total flux linked with the circuit remains conserved.

From the above discussions we learnt that,

$$\varepsilon_{net} = \frac{d\phi}{dt} = 0,$$

when resistance of the loop (circuit) is zero; $R = 0$. Hence, the 'net flux' $\phi(= \phi_{ext} - \phi_{self})$ linked with the circuit remains constant for a resistanceless circuit, which is the essence of 'conservation of total flux' in a superconducting circuit.

The conservation of magnetic flux is an important principle in electromagnetism, specifically related to Faraday's law of electromagnetic induction. It states that the magnetic flux through a closed surface is constant, meaning that any change in the magnetic field or area will lead to a corresponding change in the induced emf in a circuit.

Example 5 An inductor L and zero-resistance circuit is connected with a battery of emf ε through a key switch S. After closing the switch S, at time $t = 0$, let us find the variation of current.

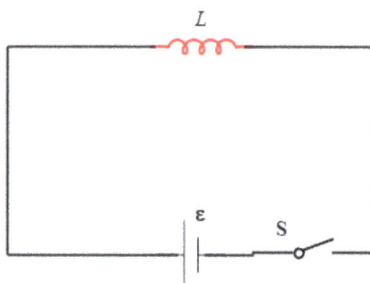

Solution

In this case, the emf of the battery is the external emf. If a current i flows after a time t, this current is the source of self-flux of the inductor (coil) and mutual flux between the coil and the other parts of the circuit. Then the algebraic summation of these two fluxes gives us the total flux which remains constant. According to the law of conservation of flux, the net flux is the algebraic sum of self and (external or mutual) flux given as

$$\phi = \phi_m + \phi_s$$

Differentiating both sides with time,

$$\frac{d\phi}{dt} = \frac{d\phi_s}{dt} + \frac{d\phi_{ext}}{dt}$$

Since the total or net flux is a constant,

$$\frac{d\phi}{dt} = 0$$

Then, we have

$$0 = \frac{d\phi_s}{dt} + \frac{d\phi_{ext}}{dt}$$

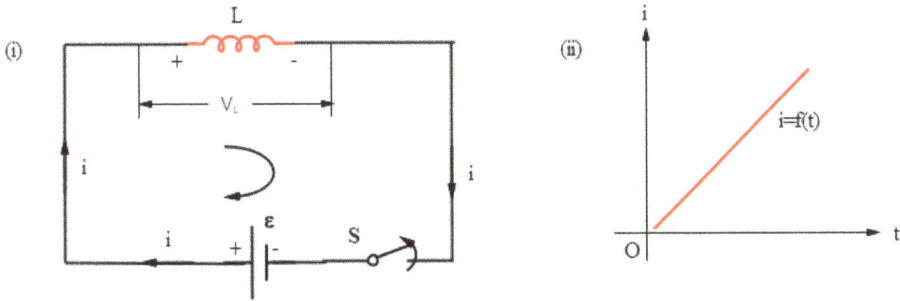

Current increases linearly with time in an ideal inductor when fitted with an ideal cell

Putting $\frac{d\phi_{ext}}{dt} = \varepsilon$ and $\frac{d\phi_s}{dt} = -L\frac{di}{dt}$, we have

$$0 = -L\frac{di}{dt} + \varepsilon$$

$$\Rightarrow L\frac{di}{dt} = \varepsilon.$$

Then current i flowing in the circuit can be given as

$$i = \frac{\varepsilon}{L} \int_0^t dt = \frac{\varepsilon}{L}t.$$

(Alternatively, by using KCL we can write $\varepsilon_a - \varepsilon_s = 0$; substituting $\varepsilon_a = \varepsilon$ and $\varepsilon_s = L\frac{di}{dt}$ we will obtain the same expression.)

This tells us that, the current in a pure inductive or zero-resistance circuit cannot be incredibly large in a finite time. It will take a long time to build a huge current in a zero-resistance circuit. Due to the property of self-inductance, current does not rise abruptly. In this case, the current increases linearly with time. However, in an RL circuit the current increases exponentially after a long time theoretically, but practically it is four to five time constants. Ans.

9.6 Calculation of self-inductance

You have already learnt that when a current flows through any circuit, it produces a magnetic field in the surrounding space. The flux ϕ of the magnetic field passing through any contour (loop) of area A can be given as

$$\phi = \int \overrightarrow{B} . d\overrightarrow{A} = \int B \, dA \cos\theta$$

as discussed earlier. Biot–Savart's law tells us that magnetic field due to an elementary segment of a current-carrying circuit at any point is directly proportional to the current i flowing through it. Hence, flux ϕ linked with the loop is directly proportional to the current i causing this flux. So, the flux is directly proportional to current; $\phi \propto i$. Then, the ratio of flux and current, that is, $\frac{\phi}{i}$ = constant. This constant of proportionality is known as 'inductance' of the circuit. Hence, the ratio of flux and the current causing this flux is known as inductance of the electrical system (inductor).

Example 6 (Solenoid)

    (a) A long solenoid has n turns per unit length. What is the self-inductance of the solenoid. Assume that the solenoid has closely spaced loops or turns, A = area of each loop of the solenoid and l = length of the solenoid.

    (b) If the linear dimensions of a long solenoid are doubled, what happens to its self-inductance?

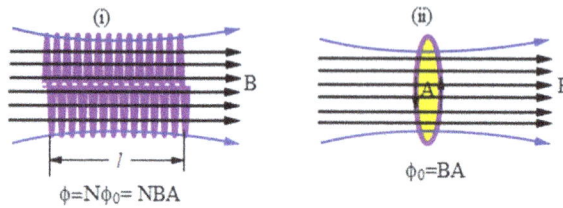

$\phi = N\phi_0 = NBA$

$\phi_0 = BA$

Solution

    (a) For a long solenoid carrying a current i its magnetic induction is given as

$$B = \mu_0 ni$$

In the figure it is evident that each loop of the solenoid is linked with its own magnetic flux

$$\phi_0 = BA$$

The total flux passing through a part of the solenoid of length l = (number of turns) flux passing through each turn $=(nl)\phi_0$. Then, the total flux of the portion of the solenoid is linked with itself (solenoid). In other words we can say that the total flux produced by the solenoid passes through it. Therefore, this flux is termed as its 'self-flux'. Substituting the value of ϕ in the previous expression we obtain the total flux

$$\phi = \mu_0 n^2 \, Ali$$

passing through the portion. Then the ratio of self-flux and the current causing that flux, which we call self-inductance is given as

$$L = \frac{\phi}{i} = \mu_0 n^2 \, Al$$

This is known as self-inductance of the solenoid denoted by the letter L. Then the self-inductance per unit length is given as

$$L/l = \mu_0 n^2 A \text{ Ans.}$$

(b) The self-inductance of the solenoid is

$$L = \mu_0 n^2 \, Al = \mu_0 n^2 \text{ (volume)}$$

As the linear dimensions (length l and radius r) of the solenoid are doubled, the area $A = \pi r^2$ will be quadrupled. So, the volume occupied by the solenoid will be eight times. Thus, the self-inductance becomes eight times. Ans.

9.7 Mutual inductance

9.7.1 Mutual inductance in a steady magnetic field

By this time, you have understood the concept by virtue of which an emf is induced in an inductor by the variation of its own flux linkage. Now we will talk about the mutual flux—the flux linked with one coil (or circuit) caused by the current flowing in the other. So, we take two coils C_1 and C_2. When coil C_1 carries a current i_1, a fraction of its own flux, i.e., ϕ_{21} is linked with coil C_2. For any steady current, ϕ_{21} is directly proportional to the current i_1 flowing in coil C_1; $\phi_{21} \propto i_1$. The constant of proportionality is known as mutual inductance M between the coils, which is given as

$$M_{21} = \frac{\phi_{21}}{i_1}$$

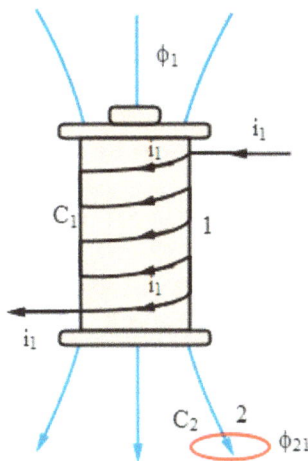

ϕ_{21} = The portion of the total flux ϕ_1 produced by the current i_1 in the coil 1 links with the coil 2

This coefficient (M_{21}) couples the flux ϕ_{21} with i_1. Therefore, this is also known as 'coefficient of coupling' between the flux linked with one coil due to the current flowing in the other. In other words, we can say that the ratio of flux ϕ_{21} linked with the coil C_2 and the current i_1 flowing in C_1 is termed as M_{21}. Similarly, if a current i_2 flows in coil C_2, a flux ϕ_{12} links with the coil C_1. Then, ϕ_{12}/i_2 is given as M_{12};

$$\phi_{12}/i_2 = M_{12}$$

These two coefficients M_{12} and M_{21} need not be equal to each other; they can be equal when the loops or coils are placed very close to each other.

The mutual inductance between two coils is defined as the flux linked with one coil if a current of one ampere flows in the other coil.

9.7.2 Definition of mutual inductance M in a time-varying field

In the previous section, the mutual inductance was derived from a static magnetic field as the ratio of static magnetic flux and the constant current causing that flux. You can also define the mutual inductance for a time-varying field. For this purpose, let the current i_1 in the coil C_1 vary *slowly* (quasistatic). So, the flux ϕ_{21} at any instant can still be treated linear with the current i_1 causing that flux. Putting a constant of proportionality M_{21}, we have

$$\phi_{21} = M_{21}i.$$

Differentiating both sides with respect to time,

$$\frac{d\phi_{21}}{dt} = M_{21}\frac{di_1}{dt},$$

where $\frac{d\phi_{21}}{dt}$ is numerically equal to the voltage induced V_2 in the coil C_2 because of variation of flux ϕ_{21} due to coil C_1 linked with coil C_2.

Then, the mutual inductance between two coils is given as

$$M_{21} = \frac{V_2}{\frac{di_1}{dt}}$$

It tells us that, whenever the current in one circuit changes, the flux caused by this current linked with the other coil will also change. In consequence, a voltage is induced in the latter so as to oppose the variation of current in the former coil.

Following the above argument, we can write

$$M_{12} = \frac{V_1}{\frac{di_2}{dt}}.$$

This means that when the current i_2 in coil C_2 changes at a rate of $\frac{di_2}{dt} = 1$ A s^{-1}, the mutual inductance M_{12} is numerically equal to the voltage induced in coil C_1.

In general, mutual inductance M between any two coils (in the absence of ferromagnetic) can be defined as the amount of voltage induced in one coil when the current in the other coil changes at a rate of 1 A s^{-1}.

9.7.3 Coupled circuit and mutual induction

You have seen in the previous section that, the variation of current in coil C_1 induces a current in coil C_2. Then how is it accomplished as they are not connected by any (material) medium? Let us discuss this in terms of wave phenomena.

In the previous section, we observed that variation of current in any circuit (conductor) generates an electromagnetic wave (emw) that travels with the speed of light c in free space. Following this principle, the time-varying current in C_1 generates an emw which can propagate in space from C_1 to C_2. This takes a time $t = \frac{l}{c}$, where l = distance between the coils. As the emw interacts with C_2, it generates (induces) an electric current, as described earlier. For small distance l, the time lag is so small that it seems to induce a current in C_2 as soon as the current starts varying in C_1. In fact, it takes a finite time. Let us illustrate the above facts in the following examples.

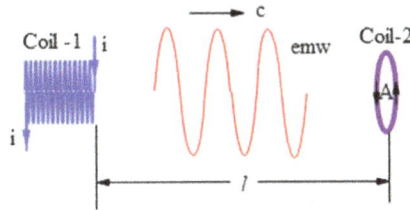

Example 7 Describe the interaction between a radio set and a cloud.

Solution

The interaction of cloud and radio set is very familiar. When we observe a flash of lightning in the clouds, our radio set gets disturbed producing a crackling sound. It happens because of the sudden increase in current in the circuit of the radio set. In the lightning, a huge variable current flows between the clouds. This generates a time-varying electromagnetic field that travels in space in the form of an emw that interacts with the circuit of the running radio set inducing an extra current in it. So, we can say that the radio set and clouds are electromagnetically coupled. Although the distance between the radio set and clouds is large, the lightning and crackling of a radio set occurs almost instantaneously. However, after viewing the lightning it takes a considerable time to hear the crackling sound between the clouds (thunderstorm). This happens because velocity of the emw (light) is much greater than that of sound (thunderstorm). Ans.

Example 8 Describe the mutual induction between a radio set and a transmitter.

Solution

The transmitting section of a radio station and receiving section (radio set) are electromagnetically coupled in the process of mutual induction. Nevertheless, practically the change in currents in the millions of radio sets cannot cause any disturbance in the transmitter as the latter is electromagnetically shielded and equipped with necessary devices to prevent the mutual electromagnetic interaction. In other words, 'the radio set is coupled with the transmitter (but not the reverse)'. This is known as 'one-sided coupling' under the afore-mentioned conditions.

So, any two circuits carrying variable currents are electromagnetically coupled. As the current in one circuit varies, it induces a voltage in the other. Hence, we can call this phenomenon 'mutual (electromagnetic) induction or mutual induction' Then the circuits are said to be 'mutually coupled' electromagnetically. The coupling may be one directional or it may happen from both the circuits as happens in a transformer. Ans.

Example 9 Explain the mutual induction in a transformer. Is a transformer an active element?

Solution

In transformers, primary and secondary coils are mutually coupled by the process of mutual induction. The alternating voltage across the primary coil generates a voltage across the secondary coil via emw generated by the variable current flowing in the primary coil. In this way, energy transfer takes place from one circuit to the other via the emw. We can conclude that one circuit energizes the other by the principle of electromagnetic mutual induction, in a broader sense.

Transformers are not the active element although they can increase the voltage or current. It is because a transformer acts obeying the principle of energy conservation; the product of current and voltage must be a constant in an ideal transformer. It has no ability of increasing the energy or power (both current and voltage simultaneously). If one increases, the other has to decrease so as to maintain the transfer of power ($= iV$) constant; so, transformers are passive elements. Ans.

9.7.4 Relation between M_{12} and M_{21} (reciprocity theorem)

With reference to the previous section, whenever any current-carrying loop (maintaining a constant current i) is shifted slowly in an external magnetic field so as that the change in the flux linkage with the loop is ϕ, the work done by the external agent can be given as

$$W = i\phi.$$

Let us use this expression in the present case of two conducting loops 1 and 2 carrying currents i_1 and i_2, respectively. We need to find the coefficients of mutual induction M_{12} and M_{21} between the loops.

It is evident from the diagrams that loop 2 links with a flux ϕ_{21} due to the current i_1 flowing in the loop 1. When we shift loop 2 slowly to infinity maintaining its current i_2 constant, it will link no (zero) flux at infinity. Then, the flux passing through the loop decreases by an amount ϕ_{21}. Following the last expression, the external work done in shifting the coil is given as

$$W_1 = i_2\phi_{21}.$$

Similarly, if we shift loop 1 slowly to infinity maintaining the current constant, the corresponding external work is

$$W_2 = i_1 \phi_{12}.$$

Since the initial arrangements (configurations) are the same in both cases, from a symmetrical point of view, equal work must be performed. So, we can write

$$W_1 = W_2.$$

Substituting the obtained values of W_1 and W_2 we have

$$i_2 \phi_{21} = i_1 \phi_{12}$$

$$\Rightarrow \phi_{21}/i_1 = \phi_{12}/i_2.$$

$$\Rightarrow M_{12} = M_{21};$$

This tells us that both the coefficients of mutual induction are equal (in the absence of any ferromagnetic materials). This relation is known as the reciprocity theorem.

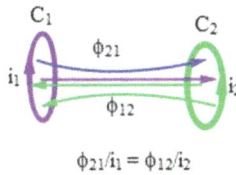

$$\phi_{21}/i_1 = \phi_{12}/i_2$$

9.7.5 Significance and application of the reciprocity theorem

According to the reciprocity theorem we can write

$$M_{12} = M_{21},$$

$$\Rightarrow \frac{\phi_{12}}{i_2} = \frac{\phi_{21}}{i_1}.$$

If equal current flows in the loops, putting $i_1 = i_2 = i$, we have

$$\phi_{12} = \phi_{21}.$$

This relation tells us that when equal current flows in the loops, they link equal fluxes due to each other's currents—mutual fluxes ϕ_{12} and ϕ_{21} are equal.

This 'significant result' helps us to solve some interesting problems regarding the calculation of 'mutual inductance' between two circuits.

Example 10 When the centres of two coaxial conducting loops 1 and 2 of radii r and $R(r < <R)$, respectively, are separated by a distance x, find the mutual inductance between them.

Solution

If we give a current i in loop 1 to find ϕ_{21}, the magnetic field distribution due to loop 1, i.e., B_1, is not uniform over the area enclosed by loop 2. Hence, it is difficult to calculate ϕ_{21}. Then, we take the help of the reciprocity theorem to equate ϕ_{21} with ϕ_{12}. For this, we have to set same current i in loop 2. Since area of loop 2 is much smaller than that of loop 2 ($r \ll R$), magnetic field distribution B_2 due to loop 2 can be assumed uniform over the area $A_1 = \pi r^2$ of loop 1. Substituting

$$B_2 = \frac{\mu_0 R^2 i}{2(R^2 + x^2)^{3/2}}$$

in $\phi_{12} = B_2 A_1$, we have

$$\phi_{12} = \frac{\pi \mu_0 i R^2 r^2}{2(R^2 + x^2)^{3/2}}.$$

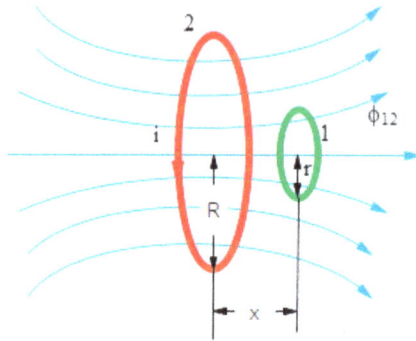

Then, the mutual inductance is given as

$$M = (M_{12} = M_{21}) = \frac{\pi \mu_0 R^2 r^2}{2(R^2 + x^2)^{3/2}} \text{ Ans.}$$

9.7.6 Calculation of mutual inductance between two coils from their self-inductance

Suppose two coils carry currents i_1 and i_2, respectively. The self-inductance of the first coil is given as

$$L_1 = \frac{\phi_{11}}{i_1}$$

The self-inductance of the second coil is given as

$$L_2 = \frac{\phi_{22}}{i_2}$$

The mutual inductance between the coils is given as

$$M = \frac{\phi_{12}}{i_2} = \frac{\phi_{21}}{i_1}.$$

When the identical coils are placed very close side by side, they can interlink a maximum flux. In that case, there is no difference between self-flux and mutual flux; then

$$\phi_{11} = \phi_{12} = \phi_{21} = \phi_{22}$$

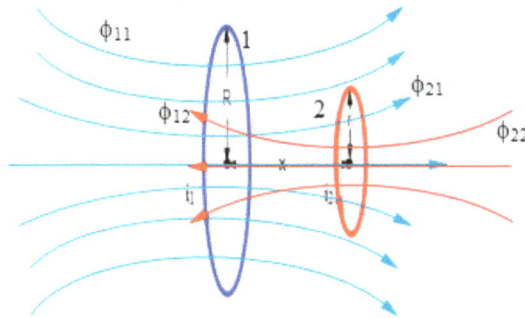

The self and mutual flux of two coaxial current carrying coils

Hence, $L_1 L_2$ will be equal to M^2. In other words, maximum mutual inductance is equal to the geometric mean of the self-inductances of the coils, that is,

$$M = \sqrt{L_1 L_2}$$

But, more generally,

$$M \leqslant \sqrt{L_1 L_2}$$

in most practical cases because a fraction of the (self-) flux produced by one coil is linked with the other. We call this fraction K 'coefficient of flux linkage,' which is given as

$$K = \frac{\text{mutual flux}}{\text{self-flux}}; \ 0 \leqslant K \leqslant 1.$$

Then we have

$$M = K\sqrt{L_1 L_2},$$

where K can also be termed as 'coefficient of coupling' of magnetic flux between any two circuits.

9.8 Combination of inductors

9.8.1 Series combination

9.8.1.1 Non-interacting coils
Let us take two coils connected in series and electromagnetically isolated. If a current i increases at a rate of di/dt, the total voltage drop V_{ab} is equal to the summation of induced voltages due to each coil.

$$V_{ab} = V_1 + V_2.$$

If we choose a coil of inductance L so as to induce the same voltage V_{ab} for the same variation of current, we can call it equivalent inductance.

Substituting, $V_1 = L_1\frac{di}{dt}$ and $V_2 = L_2\frac{di}{dt}$, we obtain

$$V = V_1 + V_2 = L_1\frac{di}{dt} + L_2\frac{di}{dt} = (L_1 + L_2)\frac{di}{dt}$$

Let us replace the combination by a single inductor of inductance L that drops the same voltage V for the same rate of change of current, then it is called an equivalent inductance.

Putting $V = L\frac{di}{dt}$ in the last expression, we have

$$L = L_1 + L_2$$

If n non-interacting inductors are connected in series, the equivalent inductance can be given as

$$L = L_1 + L_2 + \cdots + L_n$$

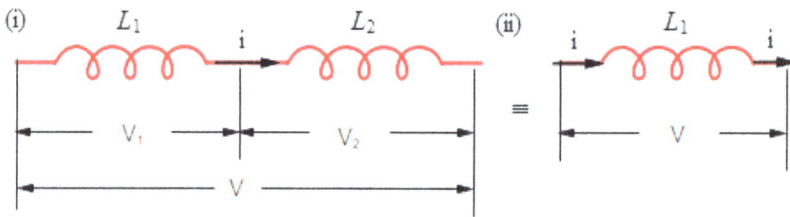

9.8.1.2 Interacting coils

Favouring fluxes
If the coils are electromagnetically coupled, mutually induced voltage will come into play in addition to self-induced voltage. If the flux of one coil favours the flux of the other, as shown in figure (i), the total flux is equal to the simple addition of self- and mutual flux in each coil.

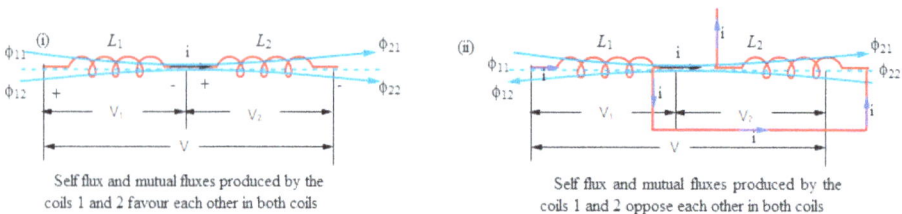

Self flux and mutual fluxes produced by the coils 1 and 2 favour each other in both coils

Self flux and mutual fluxes produced by the coils 1 and 2 oppose each other in both coils

Then the voltage induced across the coil of inductance L_1 is given as

$$V_1 = \frac{d\phi_1}{dt} = \frac{d(\phi_{11} + \phi_{12})}{dt} = L_1 \frac{di}{dt} + M \frac{di}{dt}$$

$$\Rightarrow V_1 = L_1 \frac{di}{dt} + M \frac{di}{dt}$$

Similarly, the voltage induced across the coil of inductance L_2 is given as

$$V_2 = L_2 \frac{di}{dt} + M \frac{di}{dt}$$

The total induced voltage across the series combination of the coils is given as

$$V = V_1 + V_2 = (L_1 + L_2 + 2M) \frac{di}{dt}$$

Then, the equivalent inductance is

$$L = \frac{V}{\frac{di}{dt}} = L_1 + L_2 + 2M.$$

Opposing fluxes

When the flux of one coil opposes the flux of the other coil, as shown in figure (ii), the total flux is equal to the difference of self-flux and mutual flux in each coil. Then, the total voltage drop across the coil of inductance L_1 is given as

$$V_1 = \frac{d\phi_1}{dt} = \frac{d(\phi_{11} - \phi_{12})}{dt} = L_1 \frac{di}{dt} - M \frac{di}{dt}$$

$$\Rightarrow V_1 = L_1 \frac{di}{dt} - M \frac{di}{dt}$$

Similarly, the total induced voltage across the coil of inductance L_2 is given as

$$V_2 = L_2 \frac{di}{dt} - M \frac{di}{dt}$$

The total induced voltage across the series combination of the coils is given as

$$V = V_1 + V_2 = (L_1 + L_2 - 2M) \frac{di}{dt}$$

Then, the equivalent inductance is

$$L = \frac{V}{\frac{di}{dt}} = L_1 + L_2 - 2M.$$

Generalizing the above results, we can write the equivalent inductance of two interacting coils connected in series as,

$$L = L_1 + L_2 \pm 2M,$$

where $+$ and $-$ sign are for favouring and opposing fluxes, respectively. When there is no interaction between the coils, M is equal to zero. So, the equivalent inductance is given as,

$$L = L_1 + L_2.$$

9.8.2 Parallel combination

For parallel combination of the non-interacting inductors, voltage between the inductors is given as

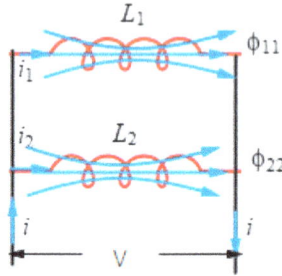

$$V = \frac{d\phi_{11}}{dt} = \frac{d\phi_{22}}{dt}$$

$$\Rightarrow V = L_1 \frac{di_1}{dt} = L_2 \frac{di_2}{dt} \tag{9.1}$$

Applying Kirchoff's first law,

$$i = i_1 + i_2.$$

Differentiating both sides with respect to time, we obtain

$$\frac{di}{dt} = \frac{di_1}{dt} + \frac{di_2}{dt} \tag{9.2}$$

Using equations (9.1) and (9.2), we have

$$V\left(\frac{1}{L_1} + \frac{1}{L_2}\right) = \frac{di}{dt}.$$

If L is the equivalent inductance connected between the terminals, for the same variation $\frac{di}{dt}$, it induces the same voltage V, we can write

$$\frac{1}{L} = \frac{di/dt}{V} = \frac{1}{L_1} + \frac{1}{L_2}.$$

For n non-interacting inductors in parallel, the equivalent inductance is given as

$$\frac{1}{L} = \sum_{i=1}^{i=n} \frac{1}{L_1}.$$

Example 11 If a parallel combination of inductors as shown in the last figure draws a current $i = at$, find the (a) currents in each inductor and (b) voltage induced.
Solution

(a) For two inductors connected in parallel, referring to the last figure, we have

$$V = L_1 \frac{di_1}{dt} = L_2 \frac{di_2}{dt} \tag{9.3}$$

and integrating both sides we obtain

$$L_1 i_1 = L_2 i_2 \tag{9.4}$$

Applyng of Kirchoff's first law,

$$i = i_1 + i_2, \tag{9.5}$$

we have $i_1 = \frac{L_2 i}{L_1 + L_2}$ and $i_2 = \frac{L_1 i}{L_1 + L_2}$, where $i = at$. Ans.

(b) Putting the obtained value of i_1 in equation (9.3), we have

$$V = \frac{L_1 L_2}{L_1 + L_2} \frac{di}{dt} = \frac{L_1 L_2}{L_1 + L_2} \frac{d(at)}{dt} = \frac{L_1 L_2 a}{L_1 + L_2} \text{ Ans.}$$

9.9 Magnetic energy stored in the system of two interacting coils

9.9.1 Interacting coils

In the last section we talked about the isolated coil. If we take two isolated coils and try to set up individual currents i_1 and i_2 by directly connecting them to external emfs, the work done by them in doing so is equal to the sum of $\frac{1}{2}L_1 i_1^2$ and $\frac{1}{2}L_2 i_2^2$; this is what we call 'self-energy' stored in the magnetic field of the coils. It is strictly valid for 'electromagnetically isolated coils'. Even though the coils are kept near, we can isolate them by putting them in 'suitable metallic boxes'. Now we remove the boxes and allow them to interact electromagnetically. This means that a portion of the total flux ϕ of each current-carrying coil passes through itself; these are known as self-fluxes ϕ_{11} and ϕ_{22}, respectively. The rest of the flux of one coil is linked with the other coil. These are known as mutual fluxes ϕ_{12} and ϕ_{21}, respectively; (ϕ_{21} is the flux of coil 1 that links with coil 2). Since the current flows in the same sense in each coil, as shown in figure (i), the total fluxes favour each other in both coils; so the total flux linked with the coils can be written as

$$\phi_1 = \phi_{11} + \phi_{21} \text{ and } \phi_2 = \phi_{22} + \phi_{12}.$$

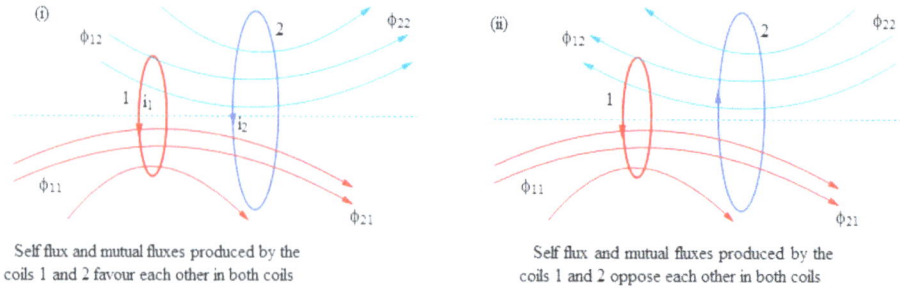

(i) Self flux and mutual fluxes produced by the coils 1 and 2 favour each other in both coils

(ii) Self flux and mutual fluxes produced by the coils 1 and 2 oppose each other in both coils

Since the current flows in the opposite sense in coil 2, as shown in figure (ii), the total fluxes oppose each other in both coils; so the total flux linked with the coils can be written as

$$\phi_1 = \phi_{11} - \phi_{21} \text{ and } \phi_2 = \phi_{22} - \phi_{12}.$$

In general, we can write

$$\phi_1 = \phi_{11} \pm \phi_{21} \text{ and } \phi_2 = \phi_{22} \pm \phi_{12},$$

where $\phi_{11} = L_1 i_1$, $\phi_{22} = L_2 i_2$, $\phi_{21} = M i_1$ and $\phi_{12} = M i_2$ (since $M_{21} = M_{12}$) for direct current or slowly varying current (quasistatic approximation). Then the net flux-linkage of the interacting coils are given as follows:

$$\phi_1 = L_1 i_1 \pm M i_1, \quad \phi_2 = L_2 i_2 \pm M i_2.$$

9.9.2 Work done in setting currents in the electromagnetically coupled coils

Since the coils are electromagnetically coupled, any attempt in changing the current or flux in one coil results in a change in current and flux of the other coil. When the total flux ϕ_1 and ϕ_2 change, the coils induce emfs ε_1 and ε_2 which can be given as

$$\varepsilon_1 = -\frac{d\phi_1}{dt} \text{ and } \varepsilon_2 = -\frac{d\phi_2}{dt}.$$

Then the batteries connected to the coils (not shown in the coils) will have to fight against these induced emfs to increase (or decrease) the currents. Let the charge passing through the coils be dq_1 and dq_2 during time dt, respectively. Then, the total work done against the induced emfs during time dt is given as

$$dW = -\varepsilon_1 dq_1 - \varepsilon_2 dq_2.$$

Substituting ε_1 and ε_2 we obtain

$$dW = d\phi_1 \frac{dq_1}{dt} + d\phi_2 \frac{dq_2}{dt} = i_1 d\phi_1 + i_2 d\phi_2.$$

Now substituting ϕ_1 and ϕ_2 and rearranging the terms we have

$$dW = L_1 i_1 di_1 + L_2 i_2 di_2 \pm M d(i_1 i_2).$$

Finally, integrating both sides, we have the total work done (by the external cells) in establishing currents i_1 and i_2, which is given as

$$W = \frac{L_1 i_1^2}{2} + \frac{L_2 i_2^2}{2} \pm M i_1 i_2.$$

This amount of energy is stored in the magnetic field of the coils.

9.9.3 Self energy and mutual energy

Hence, the energy stored in the magnetic field of two coils carrying currents i_1 and i_2 and having self-inductance L_1 and L_2 and mutual inductance M can be given as

$$U = \frac{L_1 i_1^2}{2} + \frac{L_2 i_2^2}{2} \pm M i_1 i_2.$$

Note that the extra term $M i_1 i_2$ appears due to the mutual coupling between the coils. This is the extra work done by the external cells to fight against the mutual induced emfs in the coils. Therefore, this is known as 'mutual energy'. The first two terms are termed emfs as 'self- (or intrinsic) energy' stored.

When the mutual flux and self-flux favour each other, we can take $+M i_1 i_2$, and when they oppose each other, we can use $-M i_1 i_2$. This tells us that the mutual energy stored, i.e., $\pm M i_1 i_2$, is algebraic quantity. As the mutual energy depends on the mutual inductance M, we can conclude that M is also an algebraic quantity. The mutual inductance depends upon the direction (sense) of flow of current, orientation of the coils and distance between the coils.

The physical meaning of 'self-energy' is that it is the work done by the cell in setting the currents i_1 and i_2 in electromagnetically isolated coils. The 'mutual energy' is equal to the extra work done by the cells in maintaining the same currents i_1 and i_2 in the coils when they are allowed to interact while they are slowly brought from infinity to the given positions. This extra work is done against the (mutual induced) emf in one coil because of the variation of current in the other. *The total energy of a system of interacting current-carrying coils is equal to the self-energy plus mutual energy of the system; self-energy is always positive, and mutual energy can be positive and negative.*

9.10 Field expression for energy

9.10.1 Magnetic energy density

As derived earlier, the magnetic energy stored in a coil of self-inductance L is equal to $\frac{1}{2}L i^2$. For the sake of simplicity, let us take a very long solenoid that carries a current i. The flux passing through each loop is $\phi_1 = BA$, where $B = \mu_0 n i$, $n =$ number of turns per unit length and A is the area of cross-section of the solenoid.

Then $\phi_1 = \mu_0 n i A$. The total flux linked with the portion of length l of the solenoid is $\phi = N\phi_1$, where $N =$ number of turns of solenoid in the length l. Since the number of turns per unit length $= n = \frac{N}{l}$, the total flux linkage is $\phi = nl\phi_1$. Then substituting ϕ_1, the total flux linkage is

$$\phi = \mu_0 n^2 i A l.$$

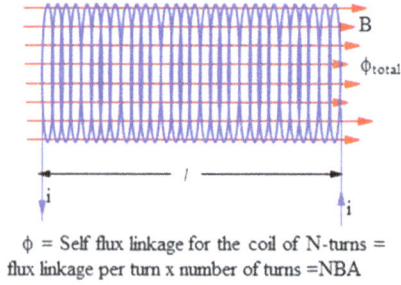

ϕ = Self flux linkage for the coil of N-turns =
flux linkage per turn x number of turns =NBA

The energy stored in the magnetic field of the segment of length l is

$$\Delta U = \frac{\phi i}{2}$$

Substituting ϕ and $i\left(=\dfrac{B}{\mu_0 n}\right)$, we have $\Delta U = \dfrac{\mu_0 n^2 A l i}{2}$, where $Al = \Delta V$

Then, the energy density is

$$\frac{dU}{dV}\left(=\frac{\Delta U}{\Delta V}\right) = \frac{B^2}{2\mu_0}.$$

This is known as a 'magnetic energy' density

$$u_v = \frac{B^2}{2\mu_0}.$$

In general, it can be written as

$$U_v = \frac{1}{2}\overrightarrow{B}.\overrightarrow{H},$$

where $\overrightarrow{B} = \mu_0 \mu_r \overrightarrow{H}$; μ_r = relative permeability and $\overrightarrow{B}$ is the resultant magnetic field density.

Taking the volume integration of energy density, the total magnetic energy is

$$U = \int u_v dv = \int \frac{B^2}{2\mu_0}dv.$$

9.11 Interpretation of field energy when two magnetic fields are simultaneously present

9.11.1 Total magnetic field energy

Suppose there are two magnetic fields $\overrightarrow{B_1}$ and $\overrightarrow{B_2}$ present simultaneously. Then the resultant field induction at any point in the space is given as

$$B = \left| \overrightarrow{B_1} + \overrightarrow{B_2} \right| = \sqrt{B_1^2 + B_2^2 + 2\overrightarrow{B_1} \cdot \overrightarrow{B_2}}.$$

Substituting B in the previous formula of energy density, we have

$$u_v = \frac{B_1^2}{2\mu_0} + \frac{B_2^2}{2\mu_0} + \frac{\overrightarrow{B_1} \cdot \overrightarrow{B_2}}{\mu_0}.$$

Then, the total field energy stored in any region of space can be given as a combination of self-energy and mutual energy in the following expression.

$$U = \int u_v dv = \int \frac{B_1^2}{2\mu_0} dv + \int \frac{B_2^2}{2\mu_0} dv + \int \frac{\overrightarrow{B_1} \cdot \overrightarrow{B_2}}{\mu_0} dv.$$

Comparing this with the previous expression, we have

$$U = \frac{L_1 i_1^2}{2} + \frac{L_2 i_2^2}{2} \pm M i_1 i_2$$

In the last expression, the first two terms are called intrinsic or self-energy of B_1 and B_2, respectively, and the third term is known as (mutual energy of interaction) between the magnetic fields $\overrightarrow{B_1}$ and $\overrightarrow{B_2}$.

For a single coil of self-inductance L carrying a current i, energy stored U can be written in terms of self-inductance as

$$U = \frac{1}{2} L i^2.$$

It can also be given in terms of the field vector $\overrightarrow{B}$ as

$$U = \int \frac{B^2}{2\mu_0} dv.$$

Equating these two equations, the self-inductance can be given as

$$L = \frac{1}{2\mu_0 i^2} \int B^2 dv.$$

This is the field expression for self-inductance $\overrightarrow{B}$. Let us demonstrate the use of the above formula through the following example.

Example 12 Using the field expression of self-inductance, find the self-inductance of a long straight conductor of circular cross-section.
   Solution
   The magnetic field B inside the conductor can be given as

$$B = \frac{\mu_0 i r}{2\pi R^2}, \quad r \leqslant R.$$

Let us consider a thin cylindrical shell of radius r, length l and thickness dr. Let the energy du be stored in the elementary region of volume $dv(=2\pi r l dr)$. Then substituting B and dv in the expression

$$L = \frac{1}{2\mu_0 i^2} \int B^2 dv,$$

we have

$$L = \frac{1}{\mu_0 i^2} \int_0^R \left(\frac{\mu_0 i r}{2\pi R^2} \right)^2 \cdot 2\pi r l \ dr.$$

Integrating it from $r = 0$ to $r = R$ for the entire cylindrical volume of length l, the inductance per unit length can be given as

$$L = \frac{\mu_0}{8\pi} H/m \text{ Ans.}$$

9.12 RL circuits

An RL circuit is comprised of resistors and inductors connected in series and parallel. The resistor does not allow the current to go to infinite and the inductor resists the change in current by inducing voltage in it. Inductors can also store magnetic energy. The response of RL circuits to both AC and DC will be different.

The RL circuits are used in electronics and electrical devices as delay circuits in DC that control how fast current changes. RL circuits can act as low pass filter circuits that can block certain frequencies and allow the others to pass through them. They are used as an inductive load control that controls the build-up and decay of currents in electrical machines. They can prevent the voltage spikes that can damage the switches and electronics of electrical systems. Thus an RL circuit can ensure safe and efficient operation of electrical machines such as motors, solenoids and trasformers etc. They can be used for signal tuning in RF (radio frequency) and communication circuits. RL circuits are used in signal processing in shaping and delaying the signals.

Let us discuss the behaviour of a simple RL circuit when connected to a DC source in the following examples.

Example 13 Shown in the figure below is a simple RL circuit fitted with a seat of emf ε. When we close the switch K_1, current flows in the circuit. Find the:
  (a) initial current and current flowing in the circuit after a long time,
  (b) potential drop across the resistor and inductor at $t = 0$ and after a long time,
  (c) power dissipated in the resistor at any time t,
  (d) heat dissipated in the resistor during time t,
  (e) charge flown through the circuit during time t, and
  (f) energy stored in the inductor at time t.

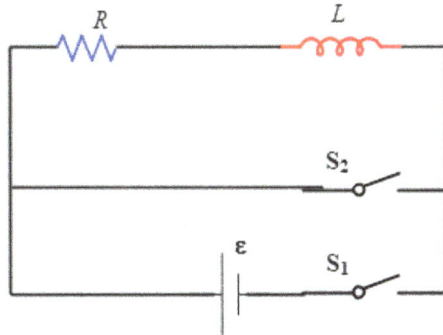

Solution

(a) When key K_1 is closed, let the current i flow in the circuit at any instant t. Applying KCL through the loop, we have

$$-iR - L\frac{di}{dt} + \varepsilon = 0$$

$$\Rightarrow L\frac{di}{\varepsilon - iR} = dt.$$

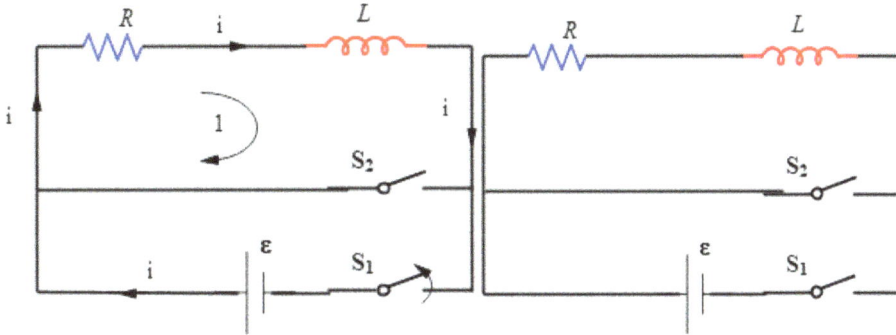

Since current i increases from zero to i during time t we write

$$\int_0^i \frac{di}{\varepsilon - iR} = \int_0^t \frac{dt}{L}$$

Evaluating the integration, we obtain

$$i = \frac{\varepsilon}{R}(1 - e^{-\frac{R}{L}t}) \text{ Ans.}$$

Simply put $t = 0$ and $t = \infty$, in the above expression to obtain

$$i|_{t=0} = 0 \text{ and } i|_{t=\infty} = \frac{\varepsilon}{R} \text{ Ans.}$$

(b) Substitution of i in the first equation yields,

$$V_R = iR = \varepsilon(1 - e^{-\frac{R}{L}t}) \text{ Ans.}$$

The voltage across the inductor is

$$|V_L| = \varepsilon - iR = \varepsilon e^{-\frac{R}{L}t}; \text{ putting } t = 0 \text{ and } t = \infty,$$

the initial and final voltage across the resistor and inductor are given as follows:

$$V_R|_{t=0} = 0 \text{ and } V_L|_{t=0} = \varepsilon - V_R|_{t=0} = \varepsilon$$

$$V_R|_{t=\infty} = \varepsilon \text{ and } V_L|_{t=\infty} = \varepsilon - V_R|_{t=\infty} = 0 \text{ Ans.}$$

(c) The instantaneous power dissipation in the resistor is

$$P_R = \frac{V^2}{R} = \frac{\varepsilon^2}{R}(1 - e^{-\frac{R}{L}t})^2 \text{ Ans.}$$

(d) The heat dissipated in the resistor is

$$Q = \int_0^t P_R dt = \frac{\varepsilon^2}{R} \int_0^t (1 - e^{-\frac{R}{L}t})^2 dt$$

$$= \frac{\varepsilon^2}{R} \times \left\{ t + \frac{L}{2R}(1 - e^{-\frac{2R}{L}t}) - \frac{2L}{R}(1 - e^{-\frac{R}{L}t}) \right\} \text{ Ans.}$$

(e) The charge flowing in the circuit during time t is

$$q = \int_0^t i \, dt$$

By substituting i and evaluating the integration, we have

$$q = \frac{\varepsilon}{R} \left\{ t - (L/R)\left(- e^{-\frac{R}{L}t} \right) \right\} \text{ Ans.}$$

(f) Energy stored in the inductor at time t is

$$V_L = \frac{1}{2}Li^2 = \frac{L\varepsilon^2}{2R^2}\left(1 - e^{-\frac{R}{L}t} \right)^2$$

obtained after putting the substituting value of i. Ans.

9.13 Magnetic forces from an energy point of view

Let us consider n number of current-carrying conductors (loops, say) fitted with batteries. All the conductors are kept fixed in their positions. We want to find the mechanical force (Ampère's force) on the kth conductor, in the x-direction, say. For this purpose, if we shift the kth conductor, say, by a distance dx in the configuration of the conductors, the magnetic field distribution will change. So, obeying Faraday's law of induction, the induced electric field appears in a space that interacts with the electrons of all the conductors. In consequence, a transient current flows in each conductor. In other words, the motion of the kth conductor will cause induced emfs in the other conductors. This will disturb the initial currents by setting transient induced currents. Hence, the applied emfs (cells) will have to do positive or negative work by sending extra charges to compensate (oppose) the variation of current. When the extra charges flow against the applied emfs (cells), they do negative work and when these charges flow in favour of emf of the cells, they do positive work. The flow of this extra (transient current) charge happens for a very short time. Then, each conductor (loop) will retain their initial currents and variation of flux (associated with variation of current) will disappear exponentially. Thus, each conductor tries (tends) to maintain its current by its own external cells fitted with it. If we sum up all the positive and negative works of the external cells, we obtain the total work done by the cells. Let the total work performed by the cells be dW during a time dt. This work (energy supplied by the cells) is partly utilized in increasing the magnetic energy of the system. Let the magnetic energy of the system change by dU during time dt due to the change in current and fluxes in the system. The surplus energy is utilized in changing (increasing) the kinetic energy (KE) of the moving kth conductor or loop. That is what we call 'work done by the mechanical force (Ampère's force) which is responsible for displacing the kth conductor (loop), denoted as dW_{mech}. Disregarding the heat loss and change in electrostatic energy (capacitance) of the system, applying the work–energy theorem, the net elementary work done by all cells is given as

$$dW_{cells} = dU + dW_{mech},$$

where

$$dW_m = F_k \, dx_k.$$

So, the general expression for the force on the kth conductor is

$$F_k = \frac{dW_{cell} - dU}{dx_k}$$

(i) If the flux is kept constant, $\phi = C$, there is no change in flux in each conducting loop during the displacement of the kth loop; so, we have $d\phi = 0$. Hence.

$$dW_{cell} = \sum i \, d\phi = 0.$$

Then, force acting on the kth conductor can be given as

$$F_k = \frac{-dU}{dx_k}\bigg|_{\phi=C}.$$

(ii) If the current is kept constant in each loop (conductors), $i = C$,

$$dW_{\text{cell}} = \sum \varepsilon_i \cdot dq_i = \sum \frac{d\phi_i}{dt} dq_i = \sum i_i d\phi_i$$

As derived earlier, the elementary change in stored magnetic energy is

$$dU = \frac{1}{2} \sum i_i \cdot d\phi_i$$

Using the last two equations, we have

$$dW_{\text{cell}} = 2dU.$$

Then, we have

$$F_k = + \left. \frac{dU}{dx_k} \right|_{i=C}$$

In both ways (using the above formulae) we can find the force experienced by any conductor.

In a nutshell,

$$dW_{\text{mech}} = -dU|_{\phi=c} = -dU|_{i=c}$$

We will see that both the formulae are self-sufficient and yield the same result in example 14.

Example 14 A smooth conducting rod is sliding on a conducting rail fitted with a battery of emf ε. Find the magnetic force experienced by the rod if the current i in the circuit is maintained constant. Assume that the inductance of the circuit varies with x.

Solution
Method 1
At any distance x from the given position, if L is the inductance of the system and i the current flowing in the conductors, the magnetic potential energy stored is

$$U_m = \frac{Li^2}{2} = \frac{\phi^2}{2L},$$

where $\phi=$ flux enclosed by the loop.
For $\phi = $ constant,

$$F_m = -\frac{dU_m}{dx} = \frac{\phi^2}{2L} \frac{dL}{dx}.$$

Substituting $\frac{\phi}{L} = i$ we have

$$F_m = \frac{i^2}{2} \frac{dL}{dx} \quad \text{Ans.}$$

Alternative method:
Again, using the other expression

$$F_m = \frac{dU_m}{dx}; \text{ for } i = \text{constant},$$

where $U_m = \frac{Li^2}{2}$, we have

$$F_m = \frac{i^2}{2}\frac{dL}{dx} \text{ Ans.}$$

N.B: Now we conclude that both formulae give the same answer.

Then it is just the matter of mathematical operation in the process of constant flux or constant current ($\phi = c$ or $i = c$). The rod experiences the same magnitude of the force in the direction of increasing x.

Problem 1 A conducting rod of mass m and length l is placed on a smooth conducting rail which is attached to the inductor L. If we drag the rod with a constant velocity v on the rail, how does the current vary in the inductor?

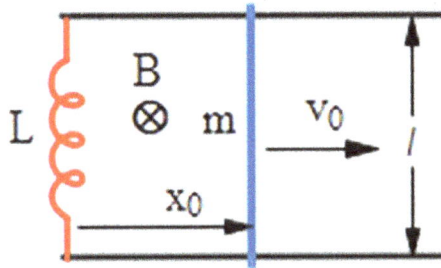

Solution

As the resistance of the circuit is zero, we can apply the concept of conservation of magnetic flux. The net flux lined with the circuit is given as

$$\phi = \phi_{\text{ext}} - \phi_{\text{self}}$$

While the rod moves a distance x, the change in flux will induce a current i, say, in it. So, its self-flux that opposes the external flux is equal to Li. Then, the net flux linked with the loop is given as

$$\phi = Bl(x + x_0) - Li$$

which must remain constant. Then equating it with the initial flux

$$\phi_0 (= Blx_0),$$

we have the current

$$i = \frac{Bl}{L}x.$$

When the rod moves with a constant velocity v, substituting $x = vt$, we have

$$i = \frac{Blv}{L}t \text{ Ans.}$$

Alternative method:
As the rod moves with a velocity v, the motional induced emf is

$$\varepsilon_{app} = Blv$$

The self-induced emf is

$$\varepsilon_{self} = Ldi/dt$$

Applying Kirchoff's voltage law, we have

$$\varepsilon_{app} = \varepsilon$$

Using the last three equations,

$$Ldi/dt = Blv$$

$$\Rightarrow Ldi = Blvdt$$

Integrating both sides, we have

$$\int_0^i Ldi = \int_0^t Blvdt$$

$$\Rightarrow i = \frac{Blvt}{L} \text{ Ans.}$$

N.B: If we keep on pushing the rod with constant velocity v, we have to continuously apply a force which must nullify the Ampère's force

$$F_{amp}\left(= ilB = \frac{B^2l^2v}{L}t\right).$$

This means that we have to apply an ever increasing force to maintain a constant current which is quite non-physical for a long time.

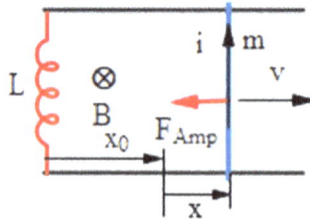

Problem 2 Referring to the previous example, instead of trying to push the rod with a constant velocity, if we give an initial velocity v_0 to the rod, how does (a) the rod move and (b) the current inside the inductor/rod vary with time?

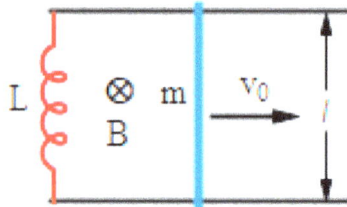

9-37

Solution

(a) Since the rod carries a current

$$i = \frac{Bl}{L}x$$

after moving through a distance x (referring to the last example), it experiences a magnetic force F_m in a backward direction. Substituting $i = \frac{Bl}{L}x$ in $F_m = ilB$, we have

$$F_m = \frac{B^2l^2}{L}x,$$

which decelerates the rod. Then, substituting the value of F_m in

$$F = -\frac{md^2x}{dt^2},$$

we obtain the differential equation

$$\frac{d^2x}{dt^2} + \frac{B^2l^2}{mL}x = 0$$

This represents a simple harmonic motion whose solution is

$$x = A \sin \omega t,$$

where $\omega = \frac{Bl}{\sqrt{mL}}$ and $A = \frac{v_0}{\omega}$ Ans.

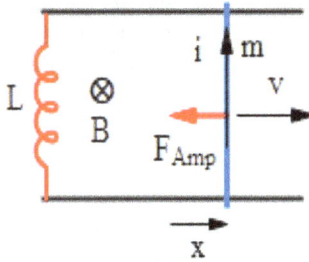

(b) Then substituting x in the equation

$$i = \frac{Bl}{L}x,$$

we have $i = \frac{(\sqrt{mL})v_0}{Bl} \sin \frac{Bl}{\sqrt{mL}}t$.

This tells us that, the rod will execute simple harmonic motion. Ans.

Problem 3 A small circular superconducting loop of mass m, inductance L and radius r is free to move along the x-axis. If its plane is perpendicular to the x-axis and it is given a velocity v_0 along the x-axis, find (a) the variation of current in the loop as the function of its displacement x. Assume that magnetic field varies according to the relation $B = B_0 + \alpha x$, where α is a positive constant. (b) Prove that the motion of the loop is simple harmonic and find the angular frequency of oscillation.

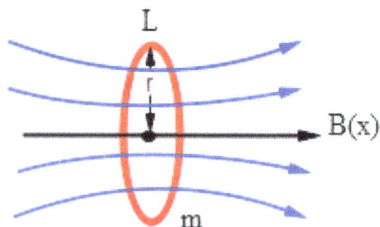

Solution

(a) Let us assume that after time t, the loop undergoes a displacement x. During this time, the external flux increases from $B_0 A$ to BA and the self-flux increases from zero to Li. As the self-flux opposes the external flux in this case, to conserve the net flux linked with the loop, we can write

$$B_0 A = BA - Li$$

Then substituting $A = \pi r^2$ and $B = B_0 + \alpha x$, we have

$$i = \frac{\pi r^2 \alpha}{L} x$$

We can assume a uniform magnetic field B in the loop because its area A is small. Please note that, in this problem, the applied emf comes from the motion of the loop in the non-uniform magnetic field. So, in this case, the applied emf is 'motional induced emf ε_m'. Ans.

(b) The magnetic energy possessed by the loop is

$$U = \frac{1}{2} Li^2$$

The force acting on the loop upon a small displacement along the x-axis is equal to the gradient of magnetic energy along the x-direction, which is given as

$$F = -\frac{\partial U}{\partial x} = -\frac{\partial}{\partial x}\left(\frac{1}{2} Li^2\right) = -Li\frac{\partial i}{\partial x},$$

where $i = \frac{\pi r^2 \alpha}{L} x$.

$$F = -Li\frac{\partial}{\partial x}\left(\frac{\pi r^2 \alpha}{L} x\right) = -L\left(\frac{\pi r^2 \alpha}{L} x\right)\frac{\pi r^2 \alpha}{L}$$

$$\Rightarrow F = -\frac{\pi^2 r^4 \alpha^2}{L} x$$

So, the loop oscillates along the x-axis with an effective spring constant

$$k = \frac{\pi^2 r^4 \alpha^2}{L}$$

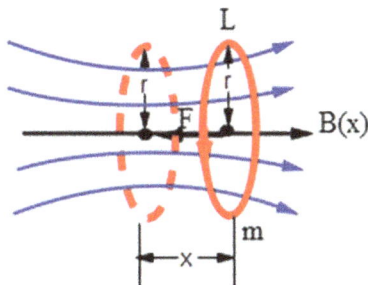

So, the frequency of oscillation is

$$\omega = \sqrt{k/m} = \sqrt{\frac{\pi^2 r^4 \alpha^2}{mL}} = \frac{\pi r^2 \alpha}{\sqrt{mL}} \quad \text{Ans.}$$

Problem 4 A superconducting ring of radius R and self-inductance L is kept coaxial with a long solenoid of radius $r(r < R)$. When the current i_0 flowing in the solenoid is reduced to $\frac{i_0}{2}$, find the current induced in the ring. Assume $n =$ number of turns of the solenoid per unit length.

Solution

Initial flux passing through the ring $\phi_1 = \pi r^2 \cdot B$, where $B = \mu_0 n i_0$ because there is zero self-flux as the ring carries no initial current. Then,

$$\phi_1 = \mu_0 n r^2 i_0.$$

When the current in the solenoid decreases to $\frac{i_0}{2}$, the external flux passing through the ring decreases to $\frac{\mu_0 n r^2 i_0}{2}$.

To conserve the total flux, the ring will induce a current i. It gives a self-flux Li. When added with the external flux, it yields

$$\phi_2 = Li + \frac{\mu_0 \pi r^2 i_0}{2}.$$

Since, $\phi_1 = \phi_2$, by substituting the values of ϕ_1 and ϕ_2, we have

$$Li + \frac{\mu_0 \pi r^2 i_0}{2} = \mu_0 \pi r^2 i_0.$$

This gives

$$i = \frac{\mu_0 \pi r^2 i_0}{2L} \quad \text{Ans.}$$

N.B: Please note that, here the applied emf is generated because of a time-varying magnetic field. In other words, the transformer emf is the applied emf in this case. The self-flux favours the external flux because external flux decreases. When calculating $\phi_{\text{ext}} = BA$ we used $A = \pi r^2$ (but not πR^2) because outside the solenoid $B = 0$.

Problem 5 A uniform magnetic field B is established along the axis of the conducting ring of radius r and self-inductance L as shown. Find (a) the induced current in the ring, (b) the magnetic field at the centre of the ring.

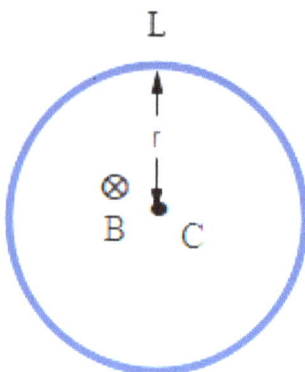

Solution

(a) Initial flux passing through the ring $\phi_1 = 0$ because initially the current and magnetic field are absent. And the ring carries no initial current. When a uniform magnetic field B is established, variation of magnetic field induces a current i, say, in the loop. So, the induced opposing flux is

$$\phi_{\text{ind}} = Li$$

The net flux is equal to the algebraic sum of the external or applied and the induced or self-flux, which is given as

$$\phi_2 = \phi_{\text{ind}} + \phi_{\text{ext}} = -Li + BA$$

To conserve the total flux,

$$\phi_1 = \phi_2$$

$$\phi_2 = -Li + BA = \phi_1 = 0$$

This gives

$$i = \frac{BA}{L} = \frac{B\pi r^2}{L} \text{ Ans.}$$

(b) The magnetic field due to the induced current i at the centre of the loop is

$$B' = \frac{\mu_0 i}{2r} = \frac{\mu_0}{2r}\left(\frac{B\pi r^2}{L}\right) = \frac{\mu_0}{2r}\frac{B\pi r^2}{L}$$

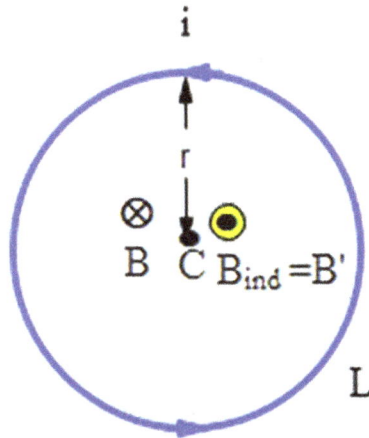

Then, the net magnetic field at the centre of the loop is

$$B = B - B' = B\left[1 - \frac{\mu_0 \pi r^2}{2L}\right] \text{ Ans.}$$

Problem 6 (Straight conductor)

A long straight long current-carrying conductor carries current i. Find the magnetic flux that passes through the conductor.

Solution

If we assume an outward current flowing through the conductor, an anticlockwise magnetic field induction B is produced inside and outside the conductor. We know that B varies linearly inside the conductor given as

$$B = \frac{\mu_0 i r}{2\pi R^2}.$$

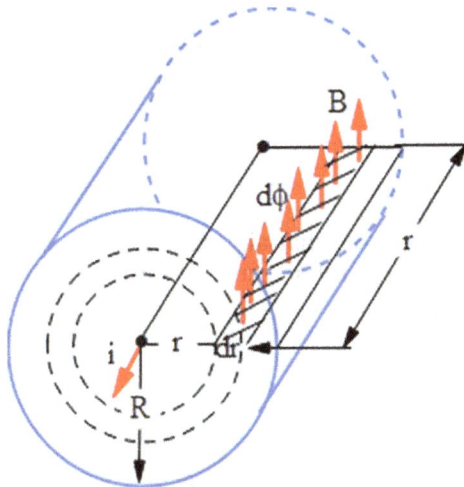

Then, inside the conductor the magnetic flux $d\phi$ passing through an elementary strip of thickness dx and length l can be given as $d\phi = Bldr$. Substitute $B = \frac{\mu i r}{2\pi R^2}$ to obtain

$$d\phi = \frac{\mu_0 i l}{2\pi R^2} r\, dr$$

Integrating the elementary flux passing through the conductor it can be written as

$$\phi = \int d\phi = \frac{\mu_0 i l}{2\pi R^2} \int_0^R r\, dr = \frac{\mu_0 i l}{4\pi}.$$

Hence, the self-inductance of the conductor per unit length should be written as

$$L = \frac{\phi}{i} = \frac{\mu_0}{4\pi} \text{ Ans.}$$

but this is double that of the actual value. Then, where is the mistake?

N.B: The obtained expression for L is incorrect by a factor ½. To avoid this ambiguity, we can calculate the self-inductance of the conductor by energy method. Please refer to example 12 that uses expression $L = \frac{1}{\mu i^2} \int B^2 dv$ followed by the energy method.

Problem 7 Find the inductance of a long straight conductor of circular cross-section by flux-linkage method which is different from flux passing.

Solution

Let us suppose that the conductor carries an outward current i_0. Then, the current i enclosed by an Amperian loop of radius r is given as $i = \frac{i_0}{\pi R^2} \cdot \pi r^2 = \frac{i_0 r^2}{R^2}$. Then, applying Ampère's circuital law we have $B 2\pi r = \mu_0 i$. Substituting i, we have

$$B = \frac{\mu_0 i_0 r}{2\pi R^2}; \ r \leqslant R$$

Now take a differential strip of width dr and length l parallel to the axis of the conductor. The magnetic flux passing through this elementary strip is $d\phi = B \cdot dA$ where dA = area of the strip $= l\, dr$. Substituting B we have.

$$d\phi = \frac{\mu_0 i_0 l}{2\pi R^2} r\, dr$$

Since, the Amperian loop encloses $\frac{r^2}{R^2}$ of the total current, the differential flux linked with the current will be equal to the $\frac{r^2}{R^2}$ of the total differential flux passing through the elementary strip. Hence, note that any elementary flux $d\phi$ passing through the conductor cannot link with the total current flowing through it. Then, the flux $d\phi$ linked with the current is given by

$$d\phi' = \frac{r^2}{R^2} d\phi$$

Substituting $d\phi$, we have

$$d\phi' = \frac{\mu_0 i_0 l}{2\pi R^4} r^3\, dr$$

Now integrating both sides, the total flux (linked with the current i_0) is

$$\phi' = \frac{\mu_0 i_0 l}{2\pi R^4} \int_0^R r^3 dr$$

This gives the inductance

$$L = \frac{\mu_0 l}{8\pi}$$

Hence, inductance per metre of the conductor is

$$L/l = \frac{\mu_0}{8\pi} = \frac{1}{2} \times 10^{-7}\,\text{H m}^{-1}\,\text{Ans.}$$

N.B: Here, you can make an important point that, in the formula $L = \frac{\phi}{i}$, ϕ is not simply the flux passing through the area of the loop; rather it is the flux linked with the current flowing through the conductor.

Problem 8 (Coaxial cable)

In this age of communication, coaxial cables are very popular. Coaxial cable consists of two conductors. There is an inner solid conductor of circular cross-section of radius a. It is surrounded by a coaxial hollow conductor of inner radius b and outer radius c. This is known as the sheath. If an inward current flows through the inner conductor, it comes out through the outer conductor. Therefore, the outer conductor provides a return path to the current. Assuming free space between the conductors, find the inductance per unit length of the coaxial cable.

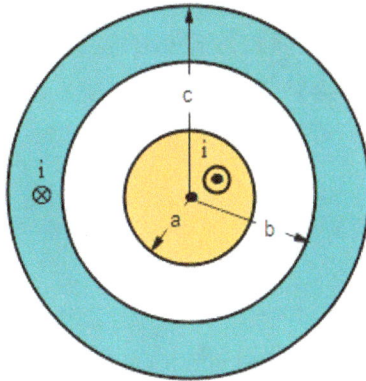

Solution

Referring to chapter 6, the expressions for the magnetic field can be directly written as

$$B = \frac{\mu_0 i}{2\pi a^2} r \text{ for } r \leqslant a,$$

$$B = \frac{\mu_0 i}{2\pi r} \text{ for } a \leqslant r \leqslant b,$$

and, $B = \dfrac{\mu_0 i(c^2 - r^2)}{2\pi r(c^2 - b^2)}$ for $b \leqslant r \leqslant c$ and $B = 0$ for $r \geqslant c$

Considering the complication involved with the flux method we prefer to solve this problem by energy method. As we know, magnetic energy stored per unit volume is equal to $\dfrac{B^2}{2\mu_0}$, the total energy stored in the magnetic field in the space occupied by each metre of the coaxial cable is given as

$$U_B = \int \frac{B^2}{2\mu_0} dv.$$

Since, $U_B = \frac{1}{2}Li^2$ as derived earlier, using these two equations, we obtain

$$L = \frac{1}{\mu_0 i^2} \int B^2 dv$$

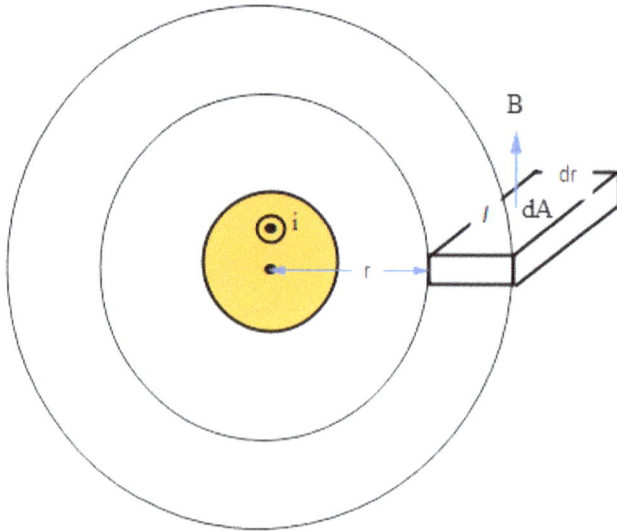

Considering the space between the cylindrical region from 0 to a, a to b and b to c, we have,

$$L = \frac{1}{\mu_0 i^2} \left[\int_0^a B^2 dV + \int_a^b B^2 dV + \int_b^c B^2 dV \right]$$

Putting the corresponding values of magnetic induction and $dV = 2\pi r\, dr$ (because we assume a segment of coaxial cable of length $l = 1$ m) we obtain

$$L = \frac{\mu_0}{2\pi} \left[\frac{1}{a^4} \int_0^a r^3 dr + \int_a^b \frac{1}{r} dr + \frac{1}{(c^2 - b^2)^2} \int_b^c \left(\frac{c^2 - r^2}{r} \right)^2 dr \right].$$

Evaluation of integration yields

$$L = \frac{\mu_0}{2\pi}\left[\frac{1}{4} + \ln\frac{b}{a} + \frac{1}{(c^2+b^2)}\left\{\frac{c^4\ln\frac{c}{b}}{c^2-b^2} - \frac{3c^2}{4} + \frac{b^2}{4}\right\}\right] \text{H m}^{-1}\text{Ans.}$$

N.B: If the sheath is very thin, assuming ($b \gg a$) you can show that the inductance per unit length of the inner core is

$$\frac{\mu_0}{2\pi}\left[\frac{1}{4} + \ln\frac{b}{a}\right]\text{H m}^{-1}.$$

Problem 9 Two thin long conducting tapes of width a are separated by a distance h, as shown in the figure below. Current is uniformly distributed over the surface area of each tape. Assuming linear density of the current over its width as i_0, find the inductance of the tape per unit length.

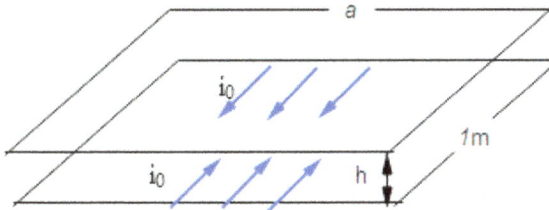

Solution

When $h \ll a$, we can assume uniform magnetic field in the space between the tapes. We know that $B = \mu_0 i_0$ between two tapes when it carries a current of linear density i_0. Zero (no) magnetic field exists outside the tapes because magnetic field of one tape is cancelled by the magnetic field of the other.

Assuming 1 m length of each tape, the flux ϕ passing through the system, that is, area $S = (1)(h)$ between them, can be given as $\phi = B \cdot S = Bh$.

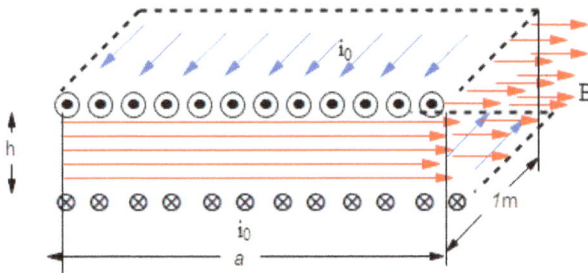

Putting $B = \mu_0 i_0$, we obtain $\phi = \mu_0 i_0 h$; since the current $i = i_0 a$ passing through each tape, the inductance of the system (tapes) per unit length $= L = \frac{\phi}{i} = \frac{1}{a}\frac{\phi}{i_0}$. Putting the values of $\phi = \mu_0 i_0 h$ and $i = i_0 a$, we obtain $L = \frac{\mu_0 h}{a}$. Ans.

Problem 10 This is a system comprising a rectangular conducting loop of dimensions $b \times c$ and a long straight conductor. (a) If a current i flows in the rectangular loop, find the flux generated by the loop passing through the semi-infinite plane located at the left side of the long straight conductor. (b) Find the mutual inductance between the straight conductor and the rectangular loop.

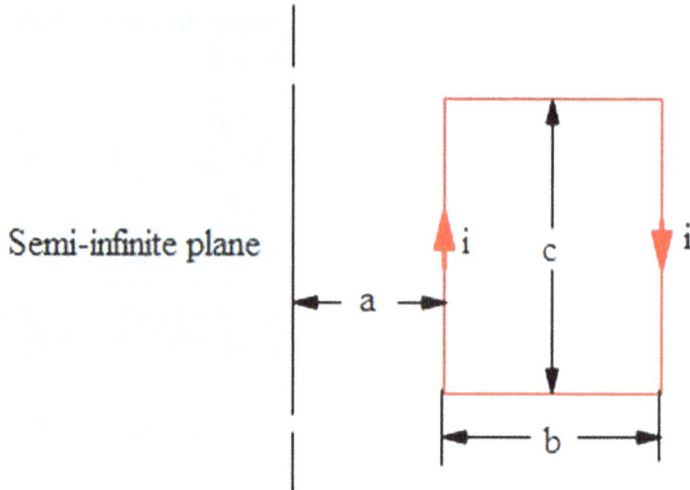

Solution

(a) If we intend to compute ϕ_{12} we have to set up a current i in loop 2. Since the magnetic field distribution due to the current loop 2 is very complex, it is very difficult to calculate the flux ϕ_{12} linked with the infinite loop (containing an infinitely long conductor). Then, we will have to take the help of the reciprocity theorem to find ϕ_{21}. As $\phi_{21} = \phi_{12}$ for equal current flowing through the loops (circuits), we establish the same current i in the infinite conductor (loop 1) and then try to find the flux ϕ_{21} linked with the rectangular loop (loop 2) which is much easier to calculate.

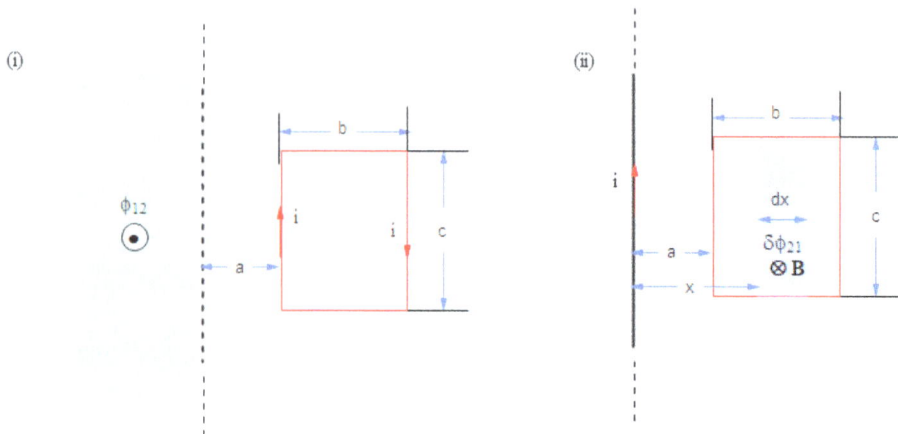

We know the magnetic field distribution of a long straight conductor carrying a current i as $B = \frac{\mu_0 i}{2\pi r}$ at a distance r from it. By assuming a vertical elementary strip of width dr at a distance r from the long straight conductor, we have the elementary flux passing through the strip $= d\phi_{21} = BdA$, where $dA = cdr$; substituting B, we have $d\phi_{21} = BdA$, where $dA = cdr$; substituting B, we have

$$d\phi_{21} = \frac{\mu_0 i c}{2\pi} \cdot \frac{dr}{r}.$$

Integrating the elementary fluxes, the total flux is

$$\phi_{21} = \int d\phi_{21} = \frac{\mu_0 i c}{2\pi} \int_a^{a+b} \frac{dr}{r} = \frac{\mu_0 i c}{2\pi} \ln \frac{a+b}{a} \text{ Ans.}$$

(b) This gives mutual inductance

$$M = \frac{\phi_{21}}{i} = \frac{\mu_0 c}{2\pi} \ln \frac{a+b}{a} \text{ (between circuits 1 and 2) Ans.}$$

Problem 11 (Ballistic galvanometer)

After putting the tiny magnet having dipole moment μ at the centre of the conducting loop of radius R, pull the magnet completely out of the loop along its axis. What is the charge flowing through the ballistic galvanometer G fitted with the loop?

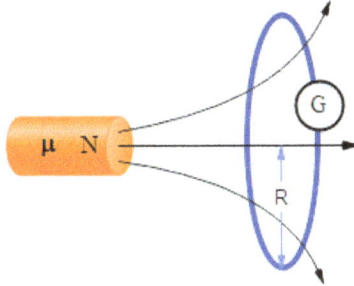

Solution

As mentioned earlier, the magnet can be imagined as a current loop 1 of area A enclosed by it. Furthermore, we can imagine a current i flowing in loop 1 which will behave as the magnet of a given dipole moment μ. Hence, $\mu = iA$. Removal of the magnet is equivalent to removal of the current loop. In this process, the flux ϕ_0 linked with the bigger loop 2 due to the magnet (smaller loop 1), decreases to zero. In consequence, a current i is induced in loop 2. Suppose, after a time t from the beginning of removal of the magnet, the flux decreases at a rate of $\frac{d\phi}{dt}$ which is numerically equal to the emf ε induced in loop 2. It results in a current $i = \frac{\varepsilon}{R}$ in loop 2. Then the charge q flowing through the ballistic galvanometer can be given as

$$q = \left| \int_0^t idt \right| = \left| \int_0^t \frac{\varepsilon}{R_0} dt \right| = \left| \frac{1}{R} \int_{\phi_0}^0 \frac{d\phi}{dt} \cdot dt \right| = \frac{1}{R_0} \left| \int_{\phi_0}^0 d\phi \right| = \frac{\phi_0}{R_0}.$$

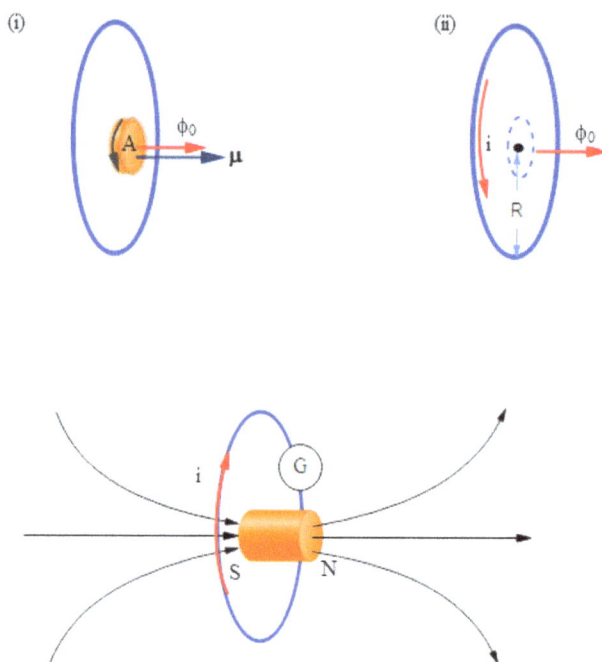

Direct calculation of ϕ_0 is cumbersome. However, by using reciprocity theorem it can be measured easily. For this reason we assume the same current i flowing through loop 2 so as to cause a flux linkage ϕ_0 with loop 1 (magnet). Now $\phi_0 = BA_1$. Since loop 1 is much smaller than loop 2, we can assume a uniform (axial) magnetic field B in it. Then putting $B = \frac{\mu_0 i}{2R}$, we have

$$\phi_0 = \frac{\mu_0 iA}{2R}.$$

Finally, substituting $iA = \mu$, we have

$$q = \frac{\phi_0}{R_0} = \frac{\mu_0 \mu}{2RR_0} \text{ Ans.}$$

Problem 12 (Modified RL circuit)

Let us connect another resistor R_0 between two terminals of the battery in a single-loop RL circuit and see the difference after closing the key. (a) Find the current through R as the function of time. (b) What is the heat dissipated in the resistor R during the time until the current across R and R_0 becomes equal, assuming $R_0 = 2R$? (c) Find the ratio of initial ($t = 0$) and final ($t = \infty$) current passing through the battery. (d) When the key is disconnected, find the total heat dissipated in the resistor R_0 from the instant of disconnecting

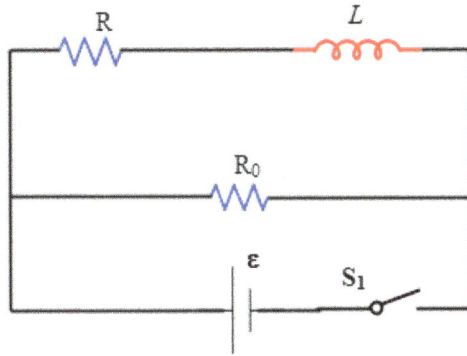

Solution

(a) Applying KCL for loops 1 and 2 we obtain

$$-i_1 R - L\frac{di_1}{dt} + (i - i_1)R_0 = 0$$

$$-(i - i_1)R_0 + \varepsilon = 0$$

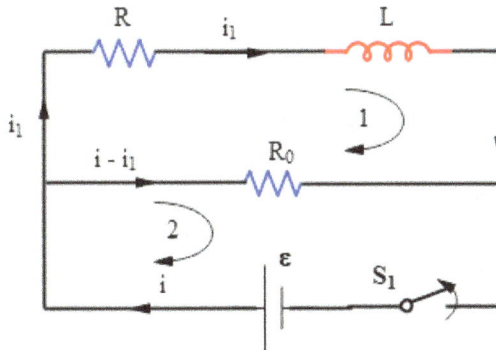

Eliminating i from the above two equations we obtain the differential equation

$$L\frac{di_1}{dt} + i_1 R = \varepsilon$$

whose solution is given as

$$i_1 = \frac{\varepsilon}{R}\left(1 - e^{-\frac{R}{L}t}\right).$$

Substitution of i_1 and the second equation yields

$$i = \frac{\varepsilon}{R_0} + \frac{\varepsilon}{R}\left(1 - e^{-\frac{R}{L}t}\right) \text{ Ans.}$$

(b) When the currents in R and R_0 are equal, $(i - i_1) = i_1$. Then $i_1 = \frac{i}{2}$.
   Substitute i and i_1 and put $R_0 = 2R$ to obtain

$$\frac{\varepsilon}{R}\left(1 - e^{-\frac{R}{L}t}\right) = \frac{\varepsilon}{2R}$$

It gives $e^{-\frac{R}{L}t} = \frac{1}{2}$, then $t = \frac{L}{R}\ln 2$.
   Then the heat dissipated in the resistor R during the obtained time t, is given as

$$Q_R = \int_0^t i_1^2\, R\, dt = \frac{\varepsilon^2}{R}\int_0^t \left(1 - e^{-\frac{R}{L}t}\right)^2 dt.$$

Evaluation of integration yields

$$Q = \frac{\varepsilon^2 L}{R^2}(\ln 2 - \frac{5}{8})\ \text{Ans.}$$

(c) After a long time, the current flows through both resistors which is given as

$$i_\infty = \frac{\varepsilon}{R_{eq}},$$

where

$$R_{eq} = \frac{R_0 R}{R + R_0}$$

$$\Rightarrow i_\infty = \frac{\varepsilon(R + R_0)}{R R_0}$$

The initial current flows through resistor R_0 because the inductor behaves as an open circuit initially as it does not allow the current to grow abruptly. So, the initial current is

$$i_0 = \frac{\varepsilon}{R_0}$$

After a long time, the entire current flows through the inductor and no current flows through the resistor. This is because the inductor has zero resistance. It is given as

$$i_\infty = \frac{\varepsilon}{R}$$

Then the ratio of initial ($t = 0$) to final ($t = \infty$) current through R_1 is

$$\frac{i_0}{i_\infty} = \frac{R}{R + R_0}\ \text{Ans.}$$

(d) The energy stored in the inductor in steady state is

$$U = \frac{1}{2}L i_\infty^2$$

When we again open the key, the total heat dissipated is equal to the energy stored in the inductor. So, we have

$$Q = \frac{1}{2}Li_\infty^2$$

Putting $i_\infty = \frac{\varepsilon(R + R_0)}{RR_0}$ in the last expression, we have

$$Q = \frac{1}{2}L\left(\frac{\varepsilon(R + R_0)}{RR_0}\right)^2 \text{ Ans.}$$

N.B: We can also find the initial and final (steady-state) current by putting $t = 0$ and $t = \infty$ in the expression $i = f(t)$. We can also find the heat dissipated in each resistor by taking the proportionality relation between heat and resistance.

Problem 13 Slightly modifying the previous problem, we can get another interesting *RL* circuit as shown in the following figure. (a) If you close the key at time $t = 0$, find the variation of current with time in the inductor L, resistors R_1 and R_2. (b) Plot the current versus time in (a). (c) Find the heat dissipated in R_2. (d) After a long time, again open the key and find the heat dissipated in R_2. (e) Find the ratio of initial ($t = 0$) to final ($t = \infty$) current through R_1.

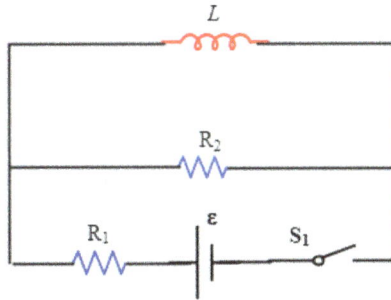

Solution

(a) For the assumed current distribution applying Kirchoff's voltage law for loops 1 and 2 we obtain,

$$\varepsilon - (i - i_1)R_2 - iR_1 = 0 \tag{9.6}$$

$$-L\frac{di}{dt} + (i - i_1)R_2 = 0 \tag{9.7}$$

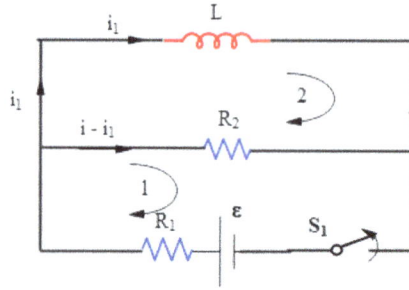

Eliminating i between the last two equations we obtain a differential equation for i_1 given as

$$\frac{di_1}{dt} = \frac{(\varepsilon - i_1 R_1)R_2}{(R_1 + R_2)L}.$$

$$\Rightarrow i_1 = \frac{\varepsilon}{R_1}\left(1 - e^{-\frac{R_1 R_2}{(R_1 + R_2)L}t}\right) \text{ Ans.}$$

Substitute i_1 in equation (9.7) to obtain

$$i = \frac{\varepsilon}{R_1} - \frac{\varepsilon R_2}{R_1(R_1 + R_2)} e^{-\frac{R_1 R_2}{(R_1 + R_2)L}t} \text{ Ans.}$$

Then current through the resistor

$$i' = i - i_1 = \frac{\varepsilon}{R_1 + R_2} e^{-\frac{R_1 R_2 t}{(R_1 + R_2)L}} \text{ Ans.}$$

(b) You can see that the total current i increases from $\frac{\varepsilon}{R_1 + R_2}$ to $\frac{\varepsilon}{R_1}$ exponentially. The current i_1 increases from zero to $\frac{\varepsilon}{R_1}$ exponentially. Ans.

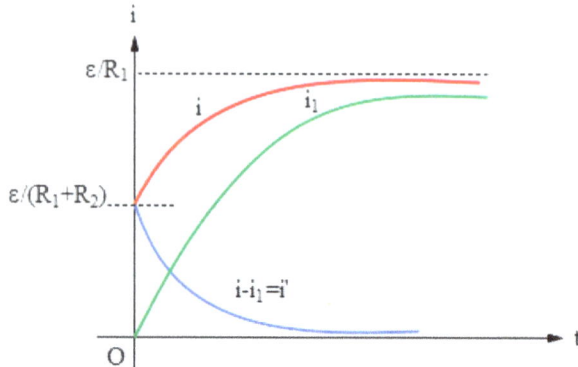

(c) Heat dissipated in

$$R_2 = Q \int_0^\infty i'^2 R_2 \, dt$$

Substitute i_1 to obtain

$$Q = R_2 \left(\frac{\varepsilon}{R_1 + R_2} \right)^2 \int_0^\infty e^{-\frac{2R_1 R_2 t}{(R_1 + R_2)L}} dt.$$

Evaluation of the integration yields

$$Q = \frac{\varepsilon^2 L}{2(R_1 + R_2)R_1} \quad \text{Ans.}$$

(d) After a long time, the current flows through the inductor because it has zero resistance. It is given as

$$i_\infty = \frac{\varepsilon}{R_1}$$

Th energy stored in the inductor in steady state is

$$U = \frac{1}{2} L i_\infty^2$$

When we again open the key, the heat dissipated in R_2 is equal to the energy stored in the inductor. So, we have

$$Q = \frac{1}{2} L i_\infty^2$$

Putting $i = \varepsilon/R_1$, we have

$$Q = \frac{1}{2} L (\varepsilon/R_1)^2 = \frac{L\varepsilon^2}{2R_1^2} \quad \text{Ans.}$$

(e) The initial current flows through resistor R_2 because the inductor behaves as an open circuit initially as it does not allow the current to grow abruptly. So, the initial current is

$$i_0 = \frac{\varepsilon}{R_1 + R_2}$$

Then the ratio of initial ($t = 0$) to final ($t = \infty$) current through R_1 is

$$\frac{i_0}{i_\infty} = \frac{R_1}{R_1 + R_2} \quad \text{Ans.}$$

Problem 14 (Coaxial loops)

Two coaxial circular single-turn loops of radii R and $r(r < < R)$ are carrying constant currents i_1 and i_2, respectively. The distance of separation between the loops is x as shown in the figure below. Find the force of interaction between the loops.

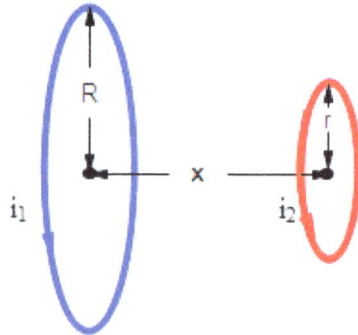

Solution

Let us solve the problem by assuming a constant current in the circuits (loops). The magnetic force of interaction is given as

$$F_{\text{mag}} = \frac{dU_{\text{mag}}}{dx},$$

where U_{mag} = the magnetic energy of interaction given as

$$U_{\text{mag}} = \frac{L_i i_1^2}{2} + M i_1 i_2 + \frac{1}{2} L_2 i_2^2;$$

$+ M i_1 i_2$ is written because fluxes of the loops favour each other. Here, everything is constant except the mutual inductance M because it is the function of distance x. However, $M i_1$ can be written as ϕ_{21}. Then we can write

$$F_{\text{mag}} = \frac{dU_{\text{mag}}}{dx} = i_2 \frac{d\phi_{21}}{dx}.$$

Since, $r \ll R$, we can assume a uniform magnetic induction in the smaller loop. Then, the flux passing through the smaller loop due to the current i_1 in the bigger loop is given as $\phi_{21} = B_{21} \cdot \pi r^2$, where

$$B_{21} = \frac{\mu_0 i_1 R^2}{2(R^2 + x^2)^{3/2}}.$$

This gives

$$\phi_{21} = \frac{\pi \mu_0 i_1 r^2 R^2}{2(R^2 + x^2)^{3/2}}.$$

Then substituting ϕ_{21} in the above expression of force, we have

$$
\begin{aligned}
F_{\text{mag}} &= i_2 \frac{d}{dx}\left(\frac{\pi \mu_0 i_1 r^2 R^2}{2(R^2 + x^2)^{3/2}} \right) \\
&= \frac{\pi \mu_0 i_1 i_2 r^2 R^2}{2} \frac{d}{dx}\left(\frac{1}{(R^2 + x^2)^{3/2}} \right) \\
&= -\frac{3\pi \mu_0 i_1 i_2 r^2 R^2 x}{2(R^2 + x^2)^{5/2}} \quad \text{Ans.}
\end{aligned}
$$

The negative sign signifies that F_{mag} points in the direction of decreasing x. In other words, the loops are attracted towards each other along their common axis.

Problem 15 If a thin long cylinder of radius R carries a current i, find the magnetic pressure exerted on the wall of the cylinder.

Solution

Let us consider an elementary cross-sectional area dA of the tube. It carries a current

$$di = \frac{i}{A} \cdot dA = \frac{idx}{2\pi R}.$$

This thin current filament of length l (say) interacts with the magnetic field B_2 of the other portion of the tube except the current filament under consideration, because it cannot interact with its own magnetic field B_1.

The force experienced by the filament is $dF = (di)lB_2$. Substituting di, we have

$$dF = \frac{i(ldx)B_2}{2\pi R}.$$

Then the magnetic pressure on the surface is

$$P = \frac{dF}{dA} = \frac{dF}{ldx} = \frac{iB_2(ldx)}{ldx(2\pi R)} = \frac{iB_2}{2\pi R} \tag{9.8}$$

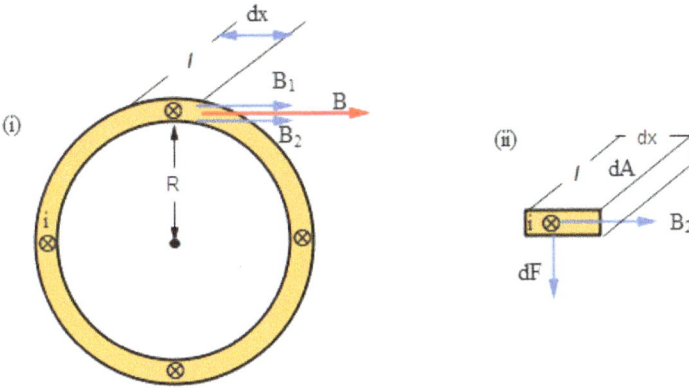

Let us find B_2; since zero (no) current is enclosed by a coaxial Amperian loop inside the tube, the net magnetic field inside the tube is $B = |\overrightarrow{B_1} + \overrightarrow{B_2}| = 0$. This gives $\overrightarrow{B_1} = -\overrightarrow{B_2}$. That means $B_1 = B_2$. Since, $B = \frac{\mu_0 i}{2\pi R}$ just outside the tube and $\overrightarrow{B} = \overrightarrow{B_1} + \overrightarrow{B_2}$, we can understand that $\overrightarrow{B_1}$ and $\overrightarrow{B_2}$ are parallel and equal in magnitude at the surface of the tube. Hence, we have

$$B_1 = B_2 = \frac{B}{2} = \frac{\mu_0 i}{4\pi R} \tag{9.9}$$

Substituting B_2 from equation (9.9) in equation (9.8) we have

$$P = \frac{\mu_0 i^2}{8\pi^2 R^2} \text{ Ans.}$$

Alternative method:

Since $P = \frac{B^2}{2\mu_0}$ as derived in the previous example, substituting $B = \frac{\mu_0 i}{2\pi R}$ we will get the same answers but remember, in the formula $P = \frac{B^2}{2\mu_0}$, B is the total magnetic field induction (neither B_1 nor B_2).

Problem 16 Let us now apply the concept of mutual inductance between two coils A and B in a varying field produced by the current i flowing in coil A, which increases at a rate of 10^3 A s^{-1}. Coil B is connected with a resistor $R = \frac{1}{2}\ \Omega$ and a capacitor $C = 10^{-6}$ farad. We find a current of magnitude 2 A induced in coil B. In ideal conditions (neglecting the effects of self-induction), if the mutual inductance between the coils is equal to 3×10^{-3} H, find the electrostatic energy stored in the capacitor.

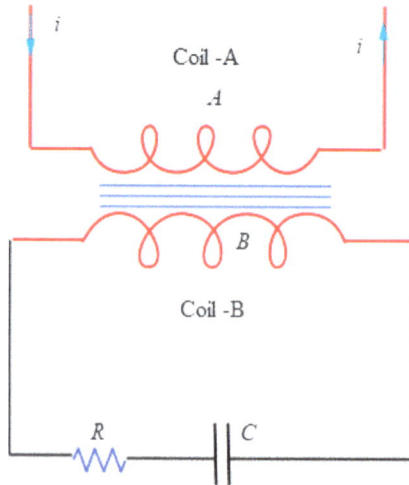

Solution

Let the mutual inductance between the coils be M. Since the current in coil A changes at a rate of $\frac{di}{dt}(=10^3$ A s$^{-1})$, a mutually induced voltage V_B is induced across the terminals of coil B which is given as $V_B = M\frac{di}{dt}$; as $\frac{di}{dt}$ is constant, coil B acts as a source of constant voltage. This induced voltage (V_B) sends a current $i = 2$ A in coil B. In this process, the capacitor is charged to 'q' dropping a voltage q/C across it.

When coil B carries a current i', a voltage $i'R$ is dropped across the resistor R. Then, by applying KCL in the circuit containing coil B, we have

$$V_B - V_R - V_C = 0,$$

where $V_R = i'R$, $V_B = M\frac{di}{dt}$ and $V_C = \frac{q}{C}$.

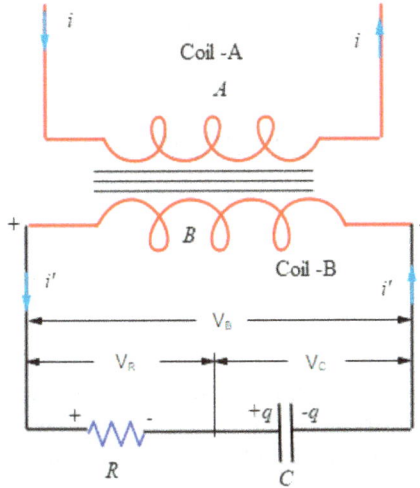

It gives

$$q = \left(M\frac{di}{dt} - i'R\right)C.$$

Putting $\frac{di}{dt} = 1000$ A s^{-1}, $i' = 2$ A, $M = 0.003$ H, $C = 1$ micro farad and $R = 0.5$ ohms, in the last equation, we have

$$q = 2 \times 10^{-6}C.$$

Then the desired electrostatic energy stored in the capacitor is

$$U = \frac{q^2}{2C} = \frac{(2 \times 10^{-6})^2}{2 \times 10^{-6}} = 2 \times 10^{-6}\text{J Ans.}$$

IOP Publishing

Problems and Solutions in Electricity and Magnetism

Pradeep Kumar Sharma

Chapter 10

Alternating current circuits

10.1 Introduction

An alternating current (AC) circuit is one where the electric current reverses its direction periodically, contrary to that in DC circuits where current flows in one direction. AC circuits are the backbone of modern electrical systems, from household electrical machines (appliances) to the industrial power generators, motors and transformers. In an AC circuit, current and voltage vary with time sinusoidally. The common values of frequency of alternations of current or voltage can be 50 Hz in most of the countries (India, UK, Japan etc) and 60 Hz (in USA, Japan etc). Most of the countries adopt a standard rating of 230 V, 50Hz supply for single-phase AC supply for household applications. In 1831, Micheal Faraday invented the first prototype AC generator. In 1832, Hippolyte Pixii, a French instrument maker, built the first practical AC generator using Faraday's principle. Later, Nikola Tesla made a significant contribution in advancing the technology of AC generators. Let us now talk about the AC generator, which is the prime source of power supply in AC circuits. In the last chapter we talked about the AC generator which generates sinusoidally varying voltage (emf) as the magnetic flux passing through the loop changes as

$$\phi = NBA \cos \omega t,$$

where N = number of turns of the coil, B = magnetic field and A = loop area.

doi:10.1088/978-0-7503-6477-5ch10

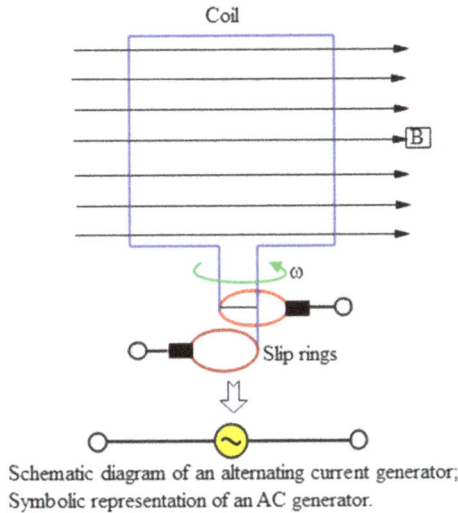

Schematic diagram of an alternating current generator;
Symbolic representation of an AC generator.

Following Faraday's law of induction, the variation of flux induces an emf $\varepsilon(=V) = -\frac{d\phi}{dt} = NAB\omega \sin \omega t$ or, $V = V_0 \sin \omega t$, where $V_0 = NAB\omega =$ maximum terminal voltage.

The generator voltage cannot be controlled from outside because it depends on its internal structures and properties characterized by ω, B, A and B. Hence we call it 'active circuit element' as it activates the others (R, L and C). When the generator is connected with a load (R, L and C), the electrons of the conductors are forced to oscillate with the frequency of the AC voltage. This is known as forced oscillation. As the electrons oscillate back and forth by the applied driving sinusoidal (alternating) voltage (or electric field) of the AC generator, the drift velocities alternate their magnitude and directions. In other words, an alternating current flows in the load. Since the electrons cannot promptly respond to the applied voltage, the current and voltage of the circuit will not have their peaks (hills) or crests (valleys) at a time. Then, you can say that the current i will not generally follow the voltage V in phase. Assuming that current lags the voltage by an angle ϕ, we can write

$$i = i_0 \sin(\omega t - \phi),$$

where $i_0 =$ maximum current or current amplitude.

In this chapter, we will explain how to find the current i_0 and phase difference ϕ between V and i for resistors, capacitors and inductors, which will form the basis of AC circuits.

We will adopt different techniques like phasor, complex algebra to find i_0 and ϕ. This will ultimately help us to analyse how resistors, capacitors and inductors respond to the applied (or external) voltage (or electric field) in the AC electric machine (motor, generator and transformer) fitted in modern day-to-day electrical appliances like washing machines, freezers, TVs, etc.

10.2 Phasors

As we know, in simple harmonic motion (SHM), the displacement of an oscillating particle (point) can be given as the (y or x) projection of the rotating radius vector whose tip undergoes a uniform circular motion.

10.2.1 Definition

The angular frequency of oscillation of the particle (or displacement) is equal to angular frequency of rotation of the radius vector $\vec{r}$.

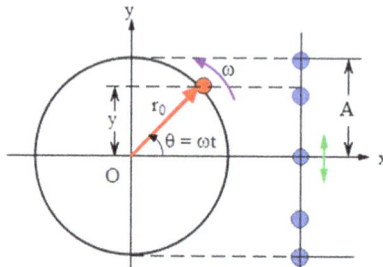

As the radius vector rotates with an angular frequency ω, its x- and y- projections oscillate with same angular frequency ω.

The displacement of the oscillating particle is $y = A \sin \omega t$ which can be represented as the y-projection of the rotating radius vector $\vec{r}$ as $\vec{r} = r(\cos \omega t \hat{i} + \sin \omega t \hat{j})$.

The projection of $\vec{r}$ along the y-axis gives us $y = A \sin \omega t$ and that along x-axis gives us $x = A \cos \omega t$, where $A = r$.

This means that any rotating vector is known as a phasor, which can represent any oscillating (alternating) quantity such as displacement, velocity, current, voltage, etc. The phasor $\vec{A}$ can be given as

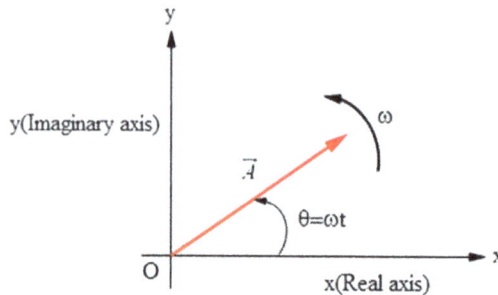

A rotating vector is a phasor that represents a sinusoidally varying quantity; the length A of the vector refers to the maximum value or amplitude of that quantity.

$\vec{A} = A \cos \omega t \hat{i} + A \sin \omega t \hat{j}$, where the x- and y-projections (components) of the phasor such as $A \cos \omega t$ or $A \sin \omega t$ represents an oscillating quantity. So, an oscillating quantity can be treated as a rotating vector or phasor, whose frequency of rotation is equal to the frequency of variation of the oscillating quantity. *So, a phasor is a rotating vector (in the complex plane) that represents a sinusoidal varying quantity.* This is a standard mathematical tool or process to ease the calculations in AC circuits.

10.2.2 Phase angle

The phase angle of an oscillating quantity $\vec{A}$ is equal to the angle θ rotated by the phasor $\vec{A}$ relative to the +ve x-axis during time t. If the initial phase is non-zero, the phase angle at time t can be given as

$$\theta = \omega t + \phi_0, \quad \text{where} \quad \phi_0 = \text{initial phase.}$$

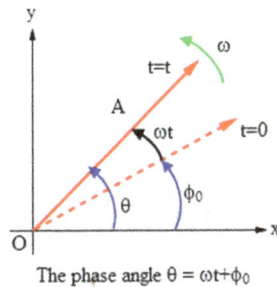

The phase angle $\theta = \omega t + \phi_0$

10.2.3 Addition of two oscillating (alternating quantities)

Let us assume that the alternating quantities a_1 and a_2 maintaining a phase difference ϕ are added to give a new alternating quantity a. So, we can write

$$a = a_1 + a_2 = A_1 \sin \omega t + A_2 \sin(\omega t + \phi)$$
$$= A_1 \sin \omega t + A_2 \sin \omega t \cos \phi + A_2 \cos \omega t \sin \phi$$
$$= (A_1 + A_2 \cos \phi)\sin \omega t + A_2 \sin \phi \cos \omega t$$
$$= A \sin(\omega t + \beta),$$

where $A = \sqrt{A_1^2 + A_2^2 + 2A_1 A_2 \cos \phi}$ and $\beta = \tan^{-1}\left(\dfrac{A_2 \sin \phi}{A_1 + A_2 \cos \phi}\right)$.

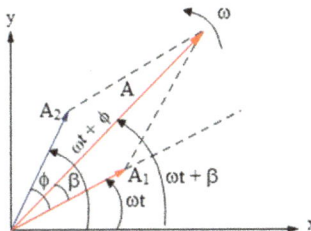

Two phasors A_1 and A_2 having a phase difference ϕ are added to yield a resultant phasor A obeying parallelogram law of vectors.

This tells us that two alternating quantities having the same angular frequency and the same or different amplitudes, can be added by adding their phasors using the parallelogram law of vectors.

10.3 Resistor, capacitor and inductor connected to an AC supply separately

10.3.1 Resistor connected to an AC supply

Let us connect a pure resistive load of resistance R across the terminals of an AC generator whose voltage (or emf) varies as

$$V = V_0 \sin \omega t \qquad (10.1)$$

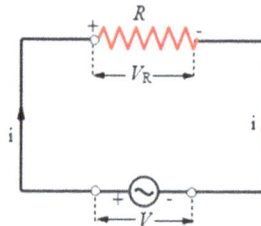

A resistor R drops a voltage V_R which is equal to the applied alternating voltage $V = V_0 \sin \omega t$.

At any instant, assuming the polarities of the generator and resistor as shown in the figure, applying Kirchhoff's circuital law (KCL), we have

$$V + V_R = 0$$

$$\Rightarrow V - iR = 0$$

$$\Rightarrow V = iR$$

$$\Rightarrow i = \frac{V}{R} = \frac{V_0 \sin \omega t}{R}$$

$$\Rightarrow i = i_0 \sin \omega t, \qquad (10.2)$$

where $i_0 = \frac{V_0}{R}$.

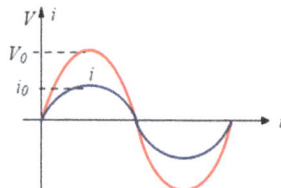

The current flowing in the resistor is in phase with the voltage dropped across the resistor (= applied voltage).

This tells us that the voltage dropped across the resistor and current are in phase.

10-5

Equations (10.1) and (10.2) indicate that the current and voltage attain their peak (hill) or crest (valley) at the same time. This means that the phase difference between current i flowing in the resistor and voltage V dropped across the resistor is zero.

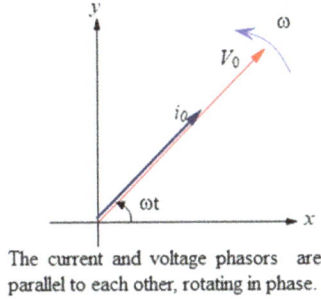

The current and voltage phasors are parallel to each other, rotating in phase.

In other words, the phasors i_0 and V_0 are parallel and rotate in the same plane having the same phase $\theta = \omega t$.

Recapitulating, the voltage across the resistor is in phase with the current passing through it.

10.3.2 Capacitor connected to an AC supply

Let us connect a pure capacitive load of capacitance C to the AC supply of voltage

$$V = V_0 \sin \omega t \tag{10.3}$$

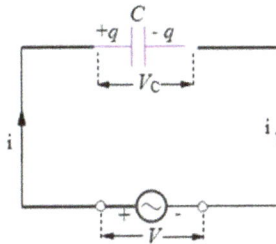

An capacitor C drops a voltage V_C which is equal to the applied alternating voltage V when connected to an AC supply.

Assuming the polarities of the applied voltage source and the capacitor and then applying KCL, we have

$$V + V_C = 0$$

$$\Rightarrow V - \frac{q}{C} = 0$$

$$\Rightarrow q = CV$$

$$\Rightarrow q = CV_0 \sin \omega t$$

$$\Rightarrow i = \frac{dq}{dt} = CV_0 \frac{d}{dt} \sin \omega t$$

Differentiating the charge q of the capacitor, the current flowing in the circuit is

$$\Rightarrow i = i_0 \cos \omega t, \quad \text{where} \quad i_0 = CV_0 \omega$$

$$i = i_0 \sin\left(\omega t + \frac{\pi}{2}\right) \tag{10.4}$$

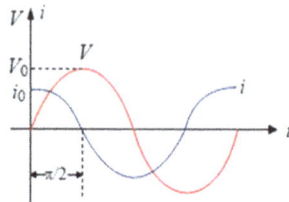

Current flowing in the pure capacitor leads the voltage dropped across the capacitor or applied voltage by 90^0 when connected to an alternating voltage source.

Comparing equations (10.3) and (10.4), we understand that when the voltage starts from zero, the current attains its maximum. This means, current leads the voltage by an angle of $\frac{\pi}{2}$ radian (or 90°).

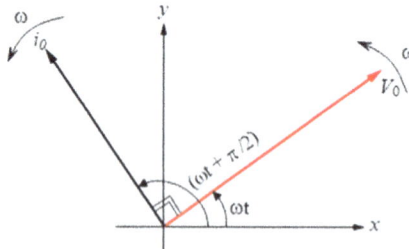

Current phasor leads the voltage phasor by 90^0 when a capacitor is connected to an ideal alternating voltage.

In terms of phasors, we can show that the current phasor i_0 rotates ahead of voltage phasor V_0 (with the same angular velocity ω) by an angle of $\frac{\pi}{2}$ radian.

The current across a capacitor leads the voltage across the capacitor by $\frac{\pi}{2}$ radian.

10.3.3 Inductor connected to an AC supply

Let us now connect an inductive load, an ideal coil of inductance L to an AC supply of voltage

$$V = V_0 \sin \omega t \tag{10.5}$$

Assuming the polarities across the generator and coil and applying KCL, we have

$$V_L + V = 0$$

$$\Rightarrow -\frac{Ldi}{dt} + V = 0$$

Or, $di = Vdt/L$

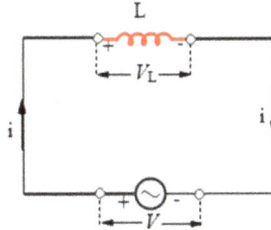

An inductor L drops a voltage V_L which is equal to the applied alternating voltage V when connected to an AC supply.

Integrating di, the current flowing in the circuit is

$$i = \int \frac{V}{L} dt$$

$$\Rightarrow i = \frac{1}{L} \int V_0 \sin \omega t \, dt \ (\because V = V_0 \sin \omega t)$$

$$\Rightarrow i = -\frac{V_0}{\omega L} \cos \omega t$$

$$\Rightarrow i = i_0 \sin\left(\omega t - \frac{\pi}{2}\right), \tag{10.6}$$

where $i_0 = \frac{V_0}{\omega L}$.

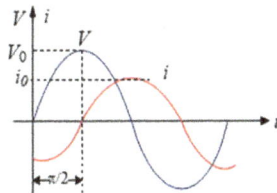

Voltage dropped across a pure inductor which is equal to the applied alternating voltage, leads the current flowing in it by 90 degree.

Comparing equations (10.5) and (10.6) we can state the following.

When the voltage is maximum, the current in the inductor is zero. This means that the current lags behind the voltage by 90° or $\frac{\pi}{2}$ radian.

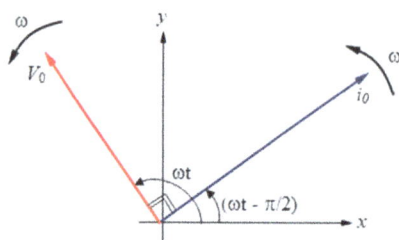

Voltage phasor leads the current phasor by 90^0 when a inductor is connected to an ideal alternating voltage.

The phasor diagrams show that the current and voltage phasors i_0 and V_0 rotate with the same angular speeds ω, i_0 lags the voltage by $\frac{\pi}{2}$ radian.

The current in the inductor lags the voltage across the inductor by $\frac{\pi}{2}$ radian.

10.4 Impedance

10.4.1 Definition

Impedance of a circuit elements R, L and C is the property of that circuit element by the virtue of which a potential (or more precisely a voltage) is dropped across it when a current flows through it. Thus, the circuit elements resist the flow of current. The principle behind this property may be different for different circuit elements. It is given as the ratio of maximum voltage dropped and maximum current flowing through the element.

$$Z = V_0/i_0.$$

10.4.2 Resistive impedance

In resistors, by the process of collision with the electron cloud and metallic kernels as discussed in current electricity, it opposes the flow of current and drops a voltage given as $V_0 = i_0 R$.

The ratio of voltage and current amplitudes is $\frac{V_0}{i_0} = Z_R = R$, which does not depend on the frequency of the AC supply.

Hence, impedance in a resistor is equal to its resistance which is also called resistive impedance.

10.4.3 Capacitive impedance

In a capacitor, no conducting path is there. Hence, no conduction current can flow through a capacitor. In steady state of DC supply, a capacitor offers infinite resistance as it carries zero current in it. However, in AC supply, the charge of the capacitor plates will change sinusoidally.

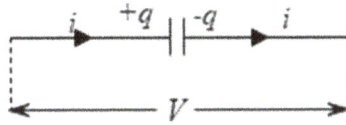

A capacitor resists the flow of current by dropping a voltage across it by inducing equal and opposite charges on its plates.

$$q = CV_0 \sin \omega t = q_0 \omega \cos \omega t, \quad \text{where } q_0 = CV_0.$$

Hence, the conduction current outside the capacitor is

$$i = \frac{dq}{dt} = CV_0 \omega \cos \omega t$$

which is numerically equal to the displacement current inside the capacitor. Please note that displacement current, in essence, is a time-varying electric field in the capacitor then the ratio of maximum voltage and maximum current is called capacitive impedance $\frac{V_0}{i_0} = Z_C = \frac{1}{\omega C}$.

This is also called capacitive reactance X_C because the capacitor reacts by dropping a voltage V_C across the capacitor connected to an AC supply. Then, we can write

$$X_C = \frac{1}{\omega C}$$

We can see that X_C increases hyperbolically with the frequency of an AC supply and vice versa as shown in the graph in example 1.

Example 1 Explain why the effective resistance of a capacitor decreases with frequency.

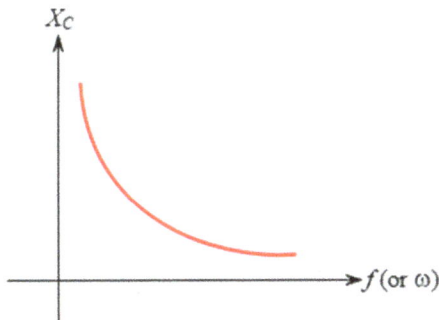

Solution

When we increase the frequency of the generator, the frequency of changing the polarity of the capacitor increases. It means that the building and decaying of charges happen more rapidly by the generator of higher frequency. Eventually, for a given maximum voltage, the maximum current increases. Thus, the effective resistance (capacitive reactance) decreases. When $\omega \to 0$ (constant current source), the capacitor offers infinite resistance as $X_C \to \infty$. When $\omega \to \infty$, the capacitive current will be incredibly high and it behaves as a zero resistor ($X_C \to 0$). Ans.

10.4.4 Inductive impedance

The voltage is dropped across the inductor when a back emf is induced in it due to change in current and flux linked with it obeying Faraday's law of electromagnetic induction.

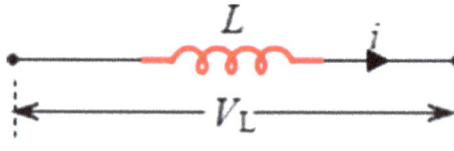

An inductor resists the flow of current by inducing a voltage across it.

The voltage across the inductor is $V = L\frac{di}{dt}$.
Assuming $i = i_0 \sin(\omega t - \phi)$,

$$V = L\frac{d}{dt}i_0 \sin(\omega t - \phi)$$
$$= Li_0\omega \cos(\omega t - \phi)$$

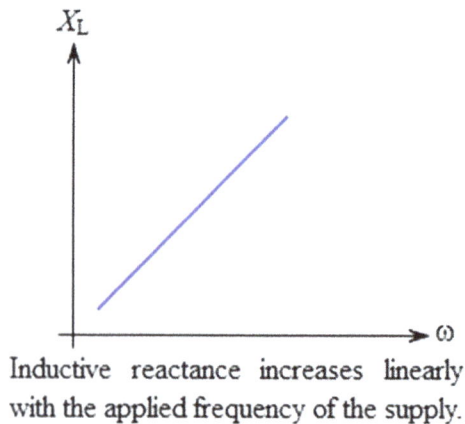

Inductive reactance increases linearly with the applied frequency of the supply.

Then, $\frac{V_0}{i_0}(Z_L)$ is called an inductive reactance X_L.

$$\Rightarrow X_L = \frac{V_0}{i_0} = \frac{\omega Li_0}{i_0}$$

$$\Rightarrow X_L = \omega L$$

The above equation tells us that $X_L \propto \omega$

Example 2 Find the maximum current flowing in a resistanceless coil of inductance $L = 0.1$ H. connected to an AC supply of peak voltage $V_0 = 314$ V and frequency $f = 50$ Hz.

Solution

The inductive reactance $X_L = \omega L = 2\pi f L = 2 \times 3.14 \times 50 \times 0.1 = 31.4\ \Omega$

So, the maximum current $i_0 = \frac{V_0}{X_L} = \frac{314}{31.4} = 10$ A Ans.

Example 3 Interpret the physical significance of inductive reactance and explain why it is directly proportional to the frequency of the applied voltage.

Solution

If ω increases, the current changes by more rapidly. Since $\frac{di}{dt}$ is high, the rate of change of flux, that is, induced voltage V_L across the inductor will be higher. So, the ratio of voltage and current amplitude, that is, impedance will be higher. It physically means that greater voltage will be required to produce a given maximum current in the inductor, at higher frequency of the supply voltage of the generator. In other words, an inductor opposes the flow of a current more vigorously at higher frequency. Thus, the effective impedance or reactance of an inductor increases with the applied frequency. When $\omega = 0$ (constant current), $V_L = 0$. This means that the inductor does not oppose the flow of a steady current as discussed in the last chapter (*RL* circuit in DC). Ans.

Example 4 A capacitor of capacitance $C = (\frac{1}{3.14})$ μF is connected to an alternating emf given by $V = V_0 \sin \omega t$, where $V_0 = 50$ V and $\omega = 314$ and 3.14×10^6 rad s^{-1}.

Solution

The maximum (peak) currents for the reactance is

$$X_C = \frac{1}{\omega C} = \frac{3.14}{314 \times 10^{-6}} = 10^4\ \Omega \text{ at } \omega = 314 \text{ rad s}^{-1}.$$

The corresponding peak current is $i_0 = \frac{V_0}{X_C} = \frac{50}{10^4} = 0.5 \times 10^{-2}$ A

Similarly, at $\omega = 3.14 \times 10^6$ rad s^{-1}, $X_C = \frac{1}{\omega C} = \frac{1}{3.14 \times 10^6 \times \left(\frac{10^{-6}}{3.14}\right)} = 1\ \Omega$.

Then, $i_0 = \frac{V_0}{i_0} = \frac{50}{1} = 50$ A.

A capacitor acts as an open switch (circuit) at $\omega = 0$ and an inductor acts as an open switch (circuit) at $\omega \to \infty$.

N.B: Summarizing the above facts, the maximum current is

$$i_0 = \frac{V}{Z},$$

where

$Z_R = R =$ resistive impedance; i_0 in phase with V_0

$Z_L = \omega L = X_L =$ inductive reactance; i_0 lags V_0 by 90°

$Z_C = \frac{1}{\omega C} = X_C =$ capacitive reactance; i_0 leads V_0 by 90°.

10.5 Use of complex numbers in AC circuits

10.5.1 Definition

As we know, imaginary (complex) numbers are the real numbers multiplied by $j = \sqrt{-1}$. As shown in the figure below, the number C in a complex plane can be given as the vector sum of two directed numbers, one is real number A and the other is an imaginary number jB. Then the phasor C can be given as $C = A + jB$.

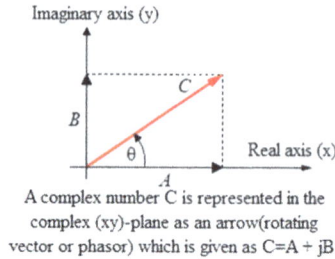

A complex number C is represented in the
complex (xy)-plane as an arrow(rotating
vector or phasor) which is given as C=A + jB.

Another way of representing the complex number C is $C = |C|\, e^{j\theta}$, because $e^{j\theta} = \cos\theta + j\sin\theta$, where $|C| =$ magnitude (or modulus) of the complex number C, and $\theta =$ angle made by the phasor with the real +ve axis.

$|C|$ and θ describe the magnitude and direction of a phasor C in a complex plane: $|C| = \sqrt{A^2 + B^2}$ and $\theta = \tan^{-1}\frac{B}{A}$. This means that, C is a rotating vector in the complex plane.

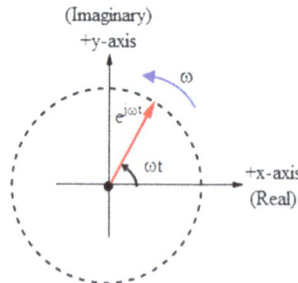

The unit vector or phasor rotates in the complex
(xy)-plane with an angular frequency ω.

In an AC circuit the rotating vectors, that is, amplitudes of current and voltage can be represented in the complex plane by the help of a rotating unit vector or a phasor that rotates in a complex (xy-) plane with an angular frequency equal to the angular frequency of alternating voltage or current. We can use the Euler formula

$$e^{\theta} = \cos\theta + j\sin\theta, \quad \text{where } \theta = \omega t;\ \omega = \text{angular velocity of the unit vector.}$$

The two terms $\cos\omega t$ and $j\sin\omega t$ give real and imaginary components of the rotating vector. Either one can be used to represent a sinusoidal function of time.

The instantaneous value of any oscillating quantity a (current, voltage etc) can be given as $a = Ae^{j\omega t} = A(\cos\omega t + j\sin\omega t)$, where $A =$ amplitude of the oscillating quantity a.

10.5.2 Addition of two oscillating quantities

Let us add the oscillating quantities a_1 and a_2 having a phase difference ϕ.

$$\text{If } a_1 = A_1 e^{j\omega t}, \text{ we can write } a_2 = A_2 e^{J(\omega t + \phi)}.$$

Then, the sum of a_1 and a_2 is

$$
\begin{aligned}
a &= a_1 + a_2 \\
&= A_1 e^{j\omega t} + A_2 e^{j(\omega t + \phi)} \\
&= A_1 e^{j\omega t} + A_2 e^{j\omega t} e^{\phi} \\
&= (A_1 + A_2 e^{j\phi}) e^{j\omega t}
\end{aligned}
$$

N.B: By using the identity $e^{j\phi} = \cos \phi + j \sin \phi$, we can show that $a = A e^{j(\omega t + \beta)}$, where $A = \sqrt{A_1^2 + A_2^2 + 2A_1 A_2 \cos \phi}$ and $\beta = \tan^{-1} \frac{A_2 \sin \beta}{A_1 + A_2 \cos \beta}$ making with A_1.

1. Any complex variable Z (or C) can represent two physical quantities by two variables x (or A) and y (or B) as

$$Z = x + jy, \quad \text{where } j = \sqrt{-1},$$

$$x = \text{Re}(Z) \text{ or coefficient of } j^0$$

$$\text{and } y = \text{Im}(Z) \text{ or coefficient of } j^1$$

The higher powers of j can be reduced by using the relations $j^2 = -1$, $j^3 = -j$.
The phasor Z (or C) is more algebraic than pictorial (diagrammatic).

2. Since $e^{i\theta} = \cos \theta + j \sin \theta$ using the above relation and putting $\theta = \omega t$, any alternating voltage can be written as

$$V_0 \sin \omega t = \text{Im}(V_0 e^{i\omega t})$$

$$\text{and } V_0 \cos \omega t = \text{Re}(V_0 e^{j\omega t}),$$

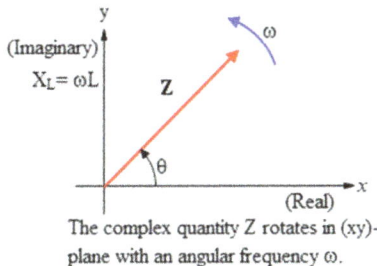

The complex quantity Z rotates in (xy)-plane with an angular frequency ω.

Then, $Z = x + jy = C e^{j\theta}$, where $C = \sqrt{x^2 + y^2}$ and $\theta = \tan^{-1} \frac{y}{x}$.

3. The advantages of complex analysis procedure are based on the following facts:

(i) Derivative and integration of exponential remain exponential;

(ii) The exponential terms get cancelled to reduce the differential equation to simple algebraic equations.

4. The imaginary axis is not that which does not exist. In reality, it is the y-axis representing the imaginary number j (or i) but not to be compared as representing the unit vector $\hat{j}$ or $\hat{i}$.

Let the applied voltage be V which can be given as $V = V_0 \sin \omega t$.

In a complex variable (notation) the voltage can be given as $V = \text{Im}\,(V)$ or $V = \text{Im}\,(V_0 e^{j\omega t})$.

The above expression can simply be written as $V = V_0 e^{j\omega t}$.

10.5.3 Resistor

Resistor current in the resistor is $i = \dfrac{V}{R} = \dfrac{V_0}{R} e^{j\omega t}$ or $i = i_0 e^{j\omega t}$, where $i_0 = \dfrac{V_0}{R}$.

For a resistor R connected with an ideal AC supply, the current and voltage phasors rotate in phase in the complex (xy)-plane with an angular frequency ω.

From the equation of i and V, we can show that V and i are in phase.

10.5.4 Capacitor

The current in the capacitor is

$$i = \frac{dq}{dt} = C\frac{dV}{dt} = C\frac{d}{dt}V_0 e^{j\omega t} = JCV_0\omega e^{j\omega t} \text{ or } i = ji_0 e^{j\omega t}, \quad \text{where } i_0 = CV\omega$$

Comparing the equations of V and i we understand that, since $i_0 = CV_0\omega$ is real, ji_0 will be pointed in the +ve j-axis direction. It means, ji_0 leads V_0 by 90° which turns out to be the same as proved by the phasor analysis.

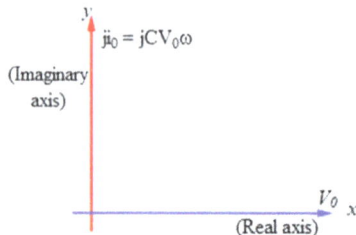

The current amplitude in the complex plane , that is, ji₀ = jV₀ω leads the voltage amplitude by 90 degree for a capacitor connected to an ideal AC voltge source.

10.5.5 Inductor

The current in the inductor is

$$i = \int \frac{V\,dt}{L}$$

$$= \frac{1}{L} \int V_0 e^{j\omega t}\,dt$$

$$= \frac{1}{j\omega L} e^{j\omega t}$$

$$= \frac{i_0}{j} e^{j\omega t},$$

where $i_0 = \omega L$

$$\Rightarrow i = -j i_0 e^{j\omega t} \left(\because \frac{1}{j} = \frac{j}{j^2} = \frac{j}{-1} = -j \right).$$

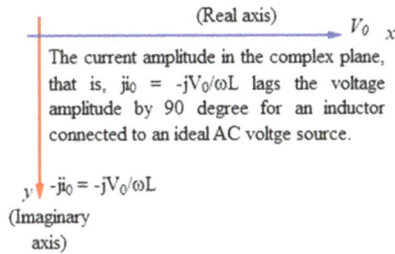

(Real axis)

V_0 x

The current amplitude in the complex plane, that is, $ji_0 = -jV_0/\omega L$ lags the voltage amplitude by 90 degree for an inductor connected to an ideal AC voltge source.

y $-ji_0 = -jV_0/\omega L$

(Imaginary axis)

Comparing the equations of V and i, we understand that, since $-j$ is multiplied with the real quantity i_0, the current phasor must be directed in the $-$ve imaginary axis. In other words, current lags the voltage by $\frac{\pi}{2}$ radian.

10.5.6 Impedance in complex form

As discussed earlier, the current in a resistor is

$$i = \frac{V_0 e^{j\omega t}}{R}.$$

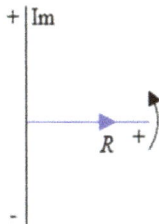

$+$ Im

R $+$

$-$

Resistnce points in positive real axis.

Then, the resistive impedance, that is, resistance in AC circuit is real which is given as

$$Z_R = \frac{V_0}{i_0}$$

Or,

$$Z_R = R$$

In capacitors the current can be given as

$$i = \frac{-V_0 e^{j\omega t}}{j/\omega C}.$$

Then, the capacitive reactance is

$$X_C = \frac{V_0}{i_0}$$

$$\text{Or, } Z_C = X_C = -\frac{j}{\omega C}$$

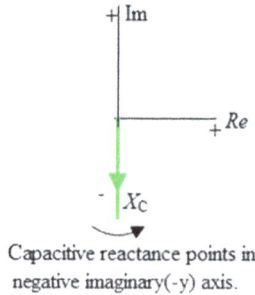

Capacitive reactance points in
negative imaginary(-y) axis.

This tells us that $-\dfrac{j}{\omega C}$ is the complex form of capacitive reactance.

In inductors, the current is $i = \dfrac{V_0 e^{j\omega t}}{j\omega L}$.

Thus, the inductive reactance is $Z_L = X_L = \dfrac{V_0}{j_0} = j\omega L$

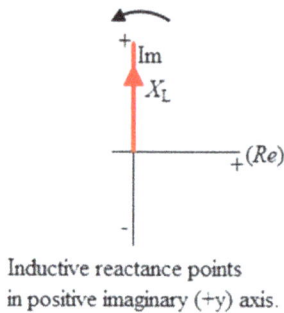

Inductive reactance points
in positive imaginary (+y) axis.

This tells us that X_L can be given as an imaginary number like X_C, but directed in $+j$ (+ve imaginary axis).

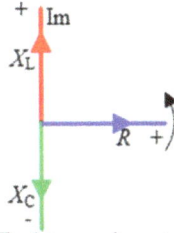

Orientation of Resistance along +x axis; inductive and capacitive reactances oppose each other lying on y-axis.

The sum of all impedances in series connection of R, L and C can be given as

$$Z = Z_R + Z_C + Z_L$$

$$= R - \frac{j}{\omega C} + j\omega L$$

$$\Rightarrow Z = R + j\left(\omega L - \frac{1}{\omega C}\right)$$

The last expression tells us that the total impedance is the sum of resistive, capacitive and inductive impedances in series circuits; using complex numbers (algebra) we can write

$$Z_R = R$$

$$Z_L = j\omega L = X_L \text{ and } Z_C = \frac{1}{j\omega C} = -\frac{j}{\omega C} = X_C.$$

10.6 Series R–L–C circuit with AC supply

Let us connect three circuit elements resistor R, capacitor C and inductor L in series with an alternating voltage

$$V = V_0 \sin \omega t$$

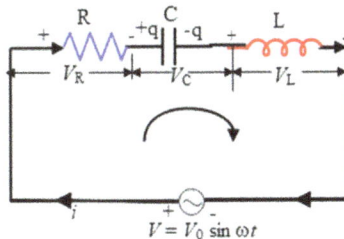

Series R-L-C circuit with an AC supply.

Let the current i pass through each circuit element lagging a phase ϕ relative to the applied voltage. Then, we can write

$$i = i_0 \sin(\omega t - \phi)$$

We need to find i_0 and ϕ in three different methods.

10.6.1 Method 1 (differential equation)

Following the conventions of polarities at any instant and applying KCL, we have

$$V + V_R + V_C + V_L = 0$$

$$\Rightarrow V_0 \sin \omega t - iR - \frac{q}{C} - L\frac{di}{dt} = 0$$

$$\Rightarrow L\frac{di}{dt} + \frac{q}{C} + iR = V_0 \sin \omega t$$

$$\Rightarrow \frac{d^2q}{dt^2} + \frac{R}{L}\frac{dq}{dt} + \frac{q}{LC} = V_0 \sin \omega t$$

The solution of the above differential equation is

$$q = -\frac{V_0}{\omega Z} \cos(\omega t - \phi), \quad \text{where } Z = \sqrt{|X_C - X_L|^2 + R^2} \text{ and } \phi = \tan^{-1}\frac{|X_C - X_L|}{R}.$$

Then, $i = \frac{dq}{dt}$

$$\Rightarrow i = \frac{V_0}{Z} \sin(\omega t - \phi), \quad \text{where } Z = \sqrt{R^2 + X^2} \text{ and } \phi = \tan^{-1}\frac{X}{R},$$

$$X = |X_C - X_L| = \left|\frac{1}{\omega C} - \omega L\right|.$$

10.6.2 Method 2 (trigonometric approach)

As derived, $V_R + V_C + V_L = V$

$$\Rightarrow iR + \frac{q}{C} + L\frac{di}{dt} = V \tag{10.7}$$

$$\text{Let, } i = i_0 \sin(\omega t - \phi) \tag{10.8}$$

$$\text{Then, } \frac{di}{dt} = i_0\omega \cos(\omega t - \phi) \tag{10.9}$$

By integrating i with time, we have

$$q = \int i \, dt = i_0 \int \sin(\omega t - \phi) \, dt$$

$$\Rightarrow q = -\frac{i_0}{\omega} \cos(\omega t - \phi) \tag{10.10}$$

Substituting i from equation (10.8), $\frac{di}{dt}$ from equation (10.9) and q from equation (10.10) in equation (10.7), we have

$$i_0 R \sin(\omega t - \phi) - \frac{i_0}{\omega C} \cos(\omega t - \phi) + \omega L i_0 \cos(\omega t - \phi) = V_0 \sin \omega t$$

$$\Rightarrow i_0 \left[R \sin(\omega t - \phi) + \left(\omega L - \frac{1}{\omega C} \right) \cos(\omega t - \phi) \right] = V_0 \sin \omega t$$

Let, $\omega L - \frac{1}{\omega C} = X$ and $\sqrt{R^2 + X^2} = Z$.

Then, we have

$$i_0 \left[\frac{R}{\sqrt{R^2 + X^2}} \sin(\omega t - \phi) + \frac{X}{\sqrt{R^2 + X^2}} \cos(\omega t - \phi) \right] \sqrt{R^2 + X^2} = V_0 \sin \omega t$$

Let us assume that

$$\frac{R}{\sqrt{R^2 + X^2}} = \cos \beta; \quad \text{so,} \quad \frac{x}{\sqrt{R^2 + X^2}} = \sin \beta, \quad \text{where } \beta = \tan^{-1} \frac{X}{R}$$

Then, the last expression can be written as

$$i_0 Z [\cos \beta \sin(\omega t - \phi) + \sin \beta \cos(\omega t - \phi)] = V_0 \sin \omega t$$

$$\Rightarrow i_0 Z \sin(\omega t - \phi + \beta) = V_0 \sin \omega t$$

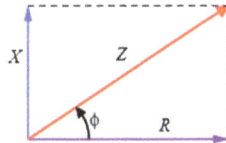

In a series R-L-C circuit with an AC supply, the impedance is the resultant of resistance R and reactance X ;it is given as $Z = (R^2+X^2)^{1/2}$;the phase angle is given as $\phi = \cos^{-1}(R/Z) = \sin^{-1}(X/Z) = \tan^{-1}(X/R)$

Comparing both sides of the last expression, we have two outcomes; the first one is $i_0 Z = V_0$. This gives us the current amplitude

$$i_0 = \frac{V_0}{Z},$$

where $Z = \sqrt{R^2 + X^2}$; $X = \left| \omega L - \frac{1}{\omega C} \right|$

The second one is given as

$$\sin(\omega t - \phi + \beta) = \sin \omega t$$

This gives us the phase angle between the supply voltage and circuit current as

$$\phi = \beta = \tan^{-1} \left(\frac{X}{R} \right)$$

10-20

$$\Rightarrow \phi = \tan^{-1}\left(\frac{X}{R}\right).$$

10.6.3 Method 3 (phasor diagram)

It is more convenient to solve an AC circuit by drawing a phasor diagram. Let the applied voltage amplitude V_0 rotate through an angle $\theta = \omega t$ from the $+x$-axis. Dropping a projection onto the y-axis we can write

$$V = V_0 \sin \omega t$$

Since the current is assumed to be lagging the applied voltage by ϕ radian, the phase of the current amplitude i_0 is equal to $(\omega t - \phi)$. Taking its projection along the y-axis we can get $i = i_0 \sin(\omega t - \phi)$.

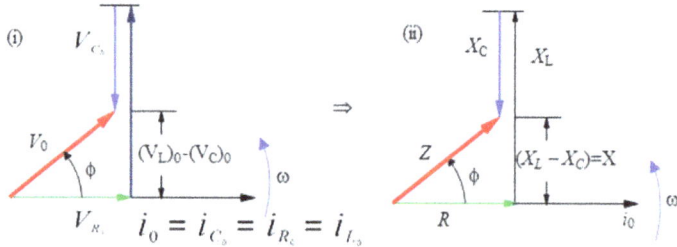

The applied voltage phasor (amplitude) is given interms of capacitive, inductive and resistive voltage phasors as $V_a = \sqrt{|V_L - V_C|^2 + V_R^2}$;dividng both sides by the current amplitude, we can get the impedance Z as follows.

If the inductive voltage (reactance) is greater than capacitive voltage(reactance), the resulting circuit is inductive with a net reactance $X = X_L - X_C$ and resistance R;so, the impedance Z is the resultant of R and X, given as $Z = (R^2 + X^2)^{1/2}$ and the phase angle is given as $\phi = \cos^{-1}(R/Z) = \tan^{-1}(X/R) = \sin^{-1}(X/Z)$.

Since the same current i flows in each circuit element, potential difference (or potential drop or voltage drop) across them will be different in phase as discussed earlier. In the resistor, i_0 and V_{R_0} are in phase; so, we have shown the phasors i_0 and V_{R_0} in parallel both having phase angle $\omega t - \phi$; $V_{R_0} = i_0 R$. Since, the current in the inductor lags the voltage V_L dropped across the inductor by 90°, the phasor V_{L_0} will lead V_{R_0} or i_0 by 90°. On the other hand, the capacitive voltage V_C lags the capacitive i current by 90°. So, the phasor V_{C_0} lags i_0 by 90° as shown in the diagram. Then, the net voltage (or applied voltage) is given as

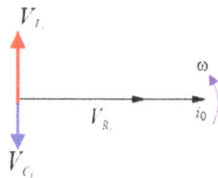

In series R-L-C circuit, the voltage dropped in resistor R is in phase with current and the voltage dropped in the capacitor C lags the current by 90° and the voltage dropped in the inductor L leads the current by 90°.

$$V_0 = \sqrt{|V_{L_0} - V_{C_0}|^2 + V_{R_0}^2}$$

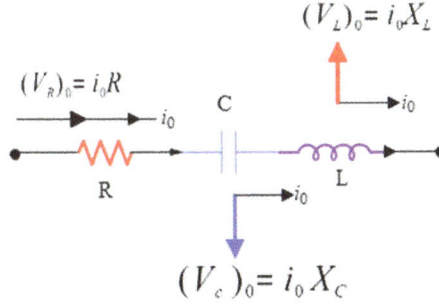

In series R-L-C circuit, the phasor representation of voltage dropped in each circuit element and current are given relative to the circuit current phasor. The voltage dropped in resistor is in phase with circuit current. The voltage dropped in the capacitor leads and that in the inductor lags the circuit current by 90^0.

$\vec{V_0} = \vec{V_{R_0}} + \vec{V_{L_0}} + \vec{V_{C_0}}$, where V_{R_0} is in phase with i_0; V_{L_0} leads i_0 by $\frac{\pi}{2}$ rad; V_{C_0} lags i_0 by $\frac{\pi}{2}$ rad.

$$= \sqrt{|i_0 X_L - i_0 X_C|^2 + (i_0 R)^2}$$
$$= i\sqrt{(X_L - X_C)^2 + R^2}$$

Then, the current amplitude is

$$i_0 = \frac{V_0}{\sqrt{|X_L - X_C|^2 + R^2}} = \frac{V_0}{Z}, \quad \text{where } Z = \sqrt{|X_L - X_C|^2 + R^2}$$

The phase difference ϕ between V and i can be given as

$$\tan \phi = \frac{|V_{L_0} - V_{C_0}|}{V_{R_0}}$$
$$= \frac{|i_0 X_L - i_0 X_C|}{i_0 R}$$
$$= \frac{|X_L - X_C|}{R}$$

$$\Rightarrow \phi = \tan^{-1} \frac{|X_L - X_C|}{R}$$

1. Dropping the perpendicular from the tip of the phasors V_0, V_{R_0}, V_{C_0} and V_{L_0} onto the y-axis (or taking the projections of the phasors (rotating vectors), we can get their instantaneous values such as

$$V = V_0 \sin \omega t$$

$$i = i_0 \sin(\omega t - \phi); \quad i_0 = \frac{V_0}{Z}$$

$$V_R = V_{R_0} \sin(\omega t - \phi); \quad V_{R_0} = i_0 R$$

$$V_L = V_{L_0} \sin\left(\omega t - \phi + \frac{\pi}{2}\right) = V_{L_0} \cos(\omega t - \phi); \quad V_{L_0} = i_0 X_L$$

$$V_C = V_{C_0} \sin\left(\omega t - \phi - \frac{\pi}{2}\right) = -V_{C_0} \cos(\omega t - \phi); \quad V_{C_0} = i_0 X_C$$

2. Hence, in the equation $V = V_R + V_L + V_C$, put the instantaneous values, V, V_R, V_L and V_C to obtain

$$V_0 \sin \omega t = i_0 R \sin(\omega t - \phi) + i_0 X_L \cos(\omega t - \phi) - i_0 X_C \cos(\omega t - \phi)$$
$$= i_0[R \sin(\omega t - \phi) + (X_L - X_C)\cos(\omega t - \phi)]$$
$$= i_0 Z \sin(\omega t - \phi + \beta)$$

Comparing both sides, we have $i_0 = V_0/Z$ and $\beta = \phi$, where $Z = \sqrt{R^2 + |X_L - X_C|^2}$ and $\beta = \phi \tan^{-1} \frac{|X_L - X_C|}{R}$

3. Please remember that $\vec{V_0} = \vec{V_{R_0}} + \vec{V_{L_0}} + \vec{V_{C_0}}$, but not $\vec{V} = \vec{V_R} + \vec{V_L} + \vec{V_C}$.
   If you write $V_0 = iZ$, $V_R = iR$, $V_L = iX_L$ and $V_C = iX_C$ you will get $Z = R + X_L + X_C$ which is a wrong answer.

4. *Phasor diagram:*
   The applied voltage amplitude (phasor) must be equal to the resultant of voltage amplitude across the resistor, capacitor and inductor as shown in the phasor diagram. So, the applied voltage amplitude is given as

$$V_0 = \sqrt{V_{R_0}^2 + |V_{L_0} - V_{C_0}|^2}, \quad i_0 = \frac{V_0}{Z} \text{ and } \phi = \tan^{-1}\frac{X}{R},$$

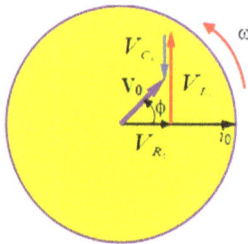

In series R-L-C circuit, the phasor diagram is igven; The voltage phasor in resistor R is in phase with current phasor and the voltage phasor in the capacitor C lags the current phasor by 90^0 and the voltage phasor in the inductor L leads the current phasor by 90^0.

The net reactance can be given as

$$X = |X_L - X_C|; \quad X_L = \omega L \text{ and } X_C = \frac{1}{\omega C}$$

10.6.4 Method 4 (complex analysis)

The sum of voltage phasors is equal to the applied voltage phasor.

$$\Rightarrow \vec{V_0} = \vec{V_{R_0}} + \vec{V_{L_0}} + \vec{V_{C_0}}$$

where

$$V_{R_0} = i_0 R$$
$$V_{L_0} = ji_0. \quad X_L = j\omega L i_0$$

$$V_{C_0} = -ji_0 X_C = -\frac{j}{\omega C} i_0$$

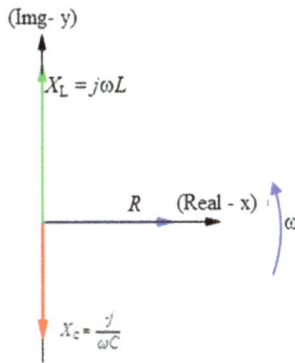

Phasor diagram of resistance and reactances; inductive reactance is ahead of resistance by 90^0 and ahead of capacitive reactances by 180^0; so these two reactances oppose each other in series R-L-C circuit.

Then, the net voltage phasor is

$$V_0 = i_0 \left(R + j\omega L - \frac{j}{\omega C} \right)$$

$$\text{So, } \frac{V_0}{i_0}(=Z) = R + j\left(\omega L - \frac{1}{\omega C} \right)$$

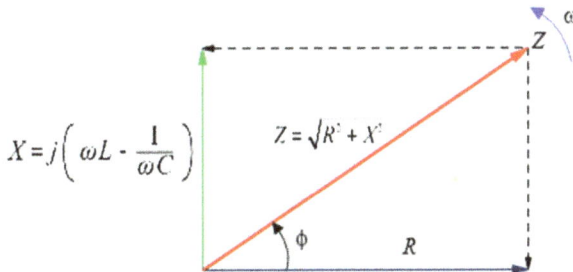

If the inductive reactance is greater than capacitive reactance, the resulting circuit is inductive with a net reactance $X=X_L-X_C$ and resistance R; so, the impedance Z is the resultant of R and X, given as $Z = (R^2+X^2)^{1/2}$ and the phase angle is given as $\phi=\cos^{-1}(R/Z) = \tan^{-1}(X/R) = \sin^{-1}(X/Z)$.

Hence, the complex notation of impedance is

$$Z = R + j\left(\omega L - \frac{1}{\omega C}\right)$$

Its magnitude is given as

$$\Rightarrow Z = \sqrt{R^2 + \left(\omega L - \frac{1}{\omega C}\right)^2}$$

and the phase angle between supply voltage and circuit current is given as $\phi = \tan^{-1}\frac{X}{R}$, where $X = |\omega L - \frac{1}{\omega C}|$

Then, the instantaneous current in the R—L—C circuit is $i = \frac{V}{Z}$. Putting the complex impedance, we have,

$$i = \frac{V_0 e^{j\omega t}}{R + j\left(\omega L - \frac{1}{\omega C}\right)}$$

After simplification, we have

$$i = \frac{V_0}{\sqrt{R^2 + \left|\omega L - \frac{1}{\omega C}\right|^2}} e^{j(\omega t - \phi)},$$

where $\frac{V_0}{R^2 + |\omega L - \frac{1}{\omega C}|^2} = \frac{V_0}{Z_0} = i_0$ (peak current) and $\phi = \tan^{-1}\frac{X}{R} = \cos^{-1}\frac{R}{Z}$.

10.7 Parallel R–L–C circuit with AC supply

Let us connect the resistor, capacitor and inductor in parallel with the supply (AC generator) voltage $V = V_0 \sin \omega t$.

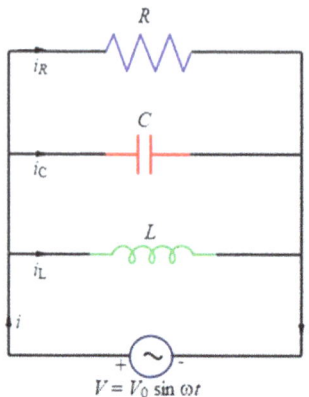

Parallel R-L-C circuit is connected with an alternating voltage. The current in C leads that in R by 90^0 and the current in R leads that in L by 90^0.

10.7.1 Method 1

Each circuit element will drop the same voltage, which is equal to the applied voltage.

Hence, $V_R = V_C = V_L = V$ at any instant. Then, the current passing through each circuit element will be different in phase.

Since i_R and V_R are in phase, we have

$$i_R = \frac{V_R}{R} = \frac{V}{R} = \frac{V_0 \sin \omega t}{R}$$

Since, i_C leads $V_C(=V)$ by 90°, we have

$$i_C = i_{C_0} \sin\left(\omega t + \frac{\pi}{2}\right) = \frac{V_0}{X_C} \cos \omega t$$

Since, i_L lags $V_L(=V)$ by 90°, we have

$$i_L = i_{L_0} \sin\left(\omega t - \frac{\pi}{2}\right) = -\frac{V_0}{X_L} \cos \omega t$$

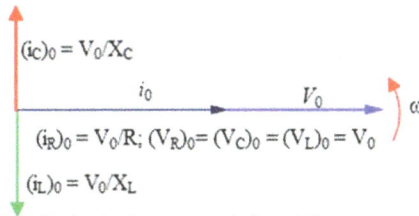

$(i_C)_0 = V_0/X_C$

i_0 V_0 ω

$(i_R)_0 = V_0/R;\ (V_R)_0 = (V_C)_0 = (V_L)_0 = V_0$

$(i_L)_0 = V_0/X_L$

In parallel R-L-C circuit, the currents in L and C oppose each other because the current in C is ahead of that in R by 90^0 and the current in R is ahead of that in L;so, the total circuit-current is given as $i_0 = [\{(i_C)_0 - (i_L)_0\}^2 + \{(i_R)_0\}^2]^{1/2}$

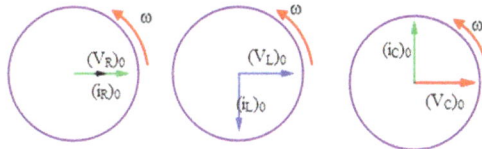

$(V_R)_0$ $(V_L)_0$ $(i_C)_0$

$(i_R)_0$ $(i_L)_0$ $(V_C)_0$

In parallel R-L-C circuit, the voltage dropped in R ,L and C are equal to the applied voltage and they are in phase with each other; however, the current in R is in phase with the applied voltage;the current in C is ahead of that in R by 90^0. The current in R leads the current in L by 90^0.

Then, the total current is

$$i = i_R + i_C + i_L$$

$$= \frac{V_0}{R} \sin \omega t + \frac{V_0}{X_C} \cos \omega t - \frac{V_0}{X_L} \cos \omega t$$

$$= V_0 \left[\frac{1}{R} \sin \omega t + \left(\frac{1}{X_C} - \frac{1}{X_L} \right) \cos \omega t \right]$$

$\Rightarrow i = \frac{V_0}{Z} \sin(\omega t + \phi)$, where

$$\frac{1}{Z} = \sqrt{\left(\frac{1}{R}\right)^2 + \left(\frac{1}{X_C} - \frac{1}{X_L}\right)^2} \quad \text{and} \quad \phi = \tan^{-1}\left\{\frac{\left(\frac{1}{X_C} - \frac{1}{X_L}\right)}{\frac{1}{R}}\right\};$$

$$\text{Put } X_C = \frac{1}{\omega C} \quad \text{and} \quad X_L = \omega L.$$

10.7.2 Method 2

Since, $\vec{i_0} = \vec{i}_{R_0} + \vec{i}_{C_0} + \vec{i}_{L_0}$, where i_{C_0} and i_{L_0} are opposite in phase and both are perpendicular to i_{R_0} as shown in the phasor diagram, we can write

$$i_0 = \sqrt{i_{R_0}^2 + (i_{C_0} - i_{L_0})^2}$$

$$\Rightarrow \frac{V_0}{Z} = \sqrt{\left(\frac{V_0}{R}\right)^2 + \left(\frac{V_0}{X_C} - \frac{V_0}{X_L}\right)^2}$$

$$\Rightarrow \frac{1}{Z} = \sqrt{\frac{1}{R^2} + \left(\frac{1}{X_C} - \frac{1}{X_L}\right)^2}$$

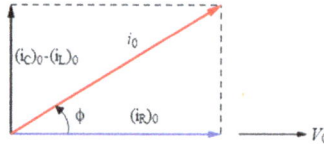

In parallel R-L-C circuit, the currents in L and C oppose each other (as the current in C is ahead of that in L by 180°); so, the total circuit-current is given as $i_0 = [\{(i_C)_0 - (i_L)_0\}^2 + \{(i_R)_0\}^2]^{1/2}$

Then, the phase angle ϕ can be given as

$$\phi = \tan^{-1}\left(\frac{i_{C_0} - i_{L_0}}{i_{R_0}}\right)$$

$$= \tan^{-1}\left(\frac{\frac{V_0}{X_C} - \frac{V_0}{X_L}}{\frac{V_0}{R}}\right)$$

$$= \tan^{-1}\frac{\frac{1}{X_C} - \frac{1}{X_L}}{\frac{1}{R}}$$

Substituting $X_C = \frac{1}{\omega C}$ and $X_L = \omega L$, we have

$$\phi = \tan^{-1}\left(\frac{\omega C - \frac{1}{\omega L}}{\frac{1}{R}}\right)$$

10.7.3 Method 3

The complex form of the impedance terms can be given as $Z_R = R$

$$Z_L = X_L = j\omega L \text{ and } Z_C = X_C = \frac{1}{j\omega C}$$

Then, the current amplitude is

$$i_0 = \frac{V_0}{Z}$$

$$= V_0\left(\frac{1}{Z_R} + \frac{1}{Z_L} + \frac{1}{Z_C}\right)$$

$$= V_0\left(\frac{1}{R} + \frac{1}{j\omega L} + \frac{1}{1/j\omega C}\right)$$

$$= V_0\left(\frac{1}{R} + \frac{1}{j\omega L} + j\omega C\right)$$

$$\Rightarrow \frac{i_0}{V_0} = \frac{1}{Z} = \frac{1}{R} + \frac{1}{j\omega L} + j\omega C$$

$$\Rightarrow \frac{1}{Z} = \frac{1}{R} - \frac{j}{\omega L} + j\omega C$$

$$\Rightarrow \frac{1}{Z} = \frac{1}{R} - j\left(\omega C - \frac{1}{\omega L}\right)$$

$$\Rightarrow i(t) = \left[\frac{1}{R} - j\left(\omega C - \frac{1}{\omega L}\right)\right]V_0 e^{j\omega t}$$

$$= \frac{V_0}{Z} e^{j(\omega t + \phi)},$$

where,

$$\frac{1}{Z} = \sqrt{\left(\frac{1}{R}\right)^2 + \left(\omega C - \frac{1}{\omega L}\right)^2} \text{ and } \phi = \tan^{-1}\left(\frac{\omega C - \frac{1}{\omega L}}{\frac{1}{R}}\right)$$

10.8 Power in AC circuits

Let us consider any electrical device (freeze, washing machine or a simple heater coil etc). It is called load; let its impedance be Z, which may be a combination of R, L and C. Let an AC generator send a current i, say, in it. While doing so, let a voltage dropped between the terminals of the equipment or circuit be V, which is known as the input or supply or external or applied voltage to the circuit. The power delivered by the supply voltage (or source of this voltage) is $P = Vi$, where $V =$ potential drop across the circuit and $i =$ current drawn by the circuit from the supply (or applied voltage).

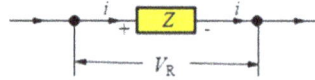

As the circuit element (impedance) drops a voltage V while carrying a current i, the instantaneous power absorbed by it is P = iV.

10.8.1 Power dissipated in a resistor

Let the electrical appliance be a heater coil which is a pure resistor of resistance R, say.

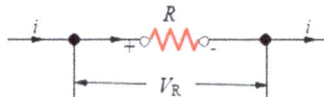

As the resistor drops a voltage V_R while carrying a current i, the instantaneous power absorbed by the resistor is $P_R = iV_R$.

If the coil draws a current $i = i_0 \sin(\omega t - \phi)$, the voltage drop V across the resistor which is 'in phase' with the current can be given as $V_R = V_{R_0} \sin(\omega t - \phi)$.

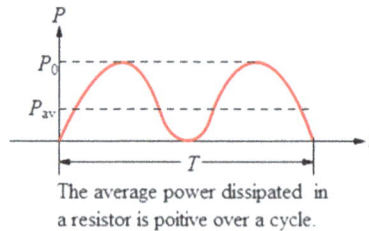

The average power dissipated in a resistor is poitive over a cycle.

Then, the instantaneous power delivered to the resistor or absorbed by the resistor is

$$P_R = Vi = V_R i$$
$$= V_{R_0} \sin(\omega t - \phi) i_0 \sin(\omega t - \phi)$$
$$= i_0^2 R \sin^2(\omega t - \phi) \; (\because V_{R_0} = i_0 R)$$

For a time interval which is much greater than the time period of the AC ($t \gg T = \frac{2\pi}{\omega}$), the average power delivered by the generator to the resistor is $P_{av} = i_0^2 R \langle \sin^2(\omega t - \phi) \rangle$.

Since, $\langle \sin^2(\omega t - \phi) \rangle = \frac{1}{2}$, we have $P_{av} = \frac{1}{2} i_0^2 R$.

Since the power delivered by the generator averaged over a long time is non-zero, that is, $\frac{i_0^2 R}{2}$, we can conclude that this power is lost in the dorm of heat, sound and light in the resistor as discussed in DC circuits. This tells us that, *the average power dissipated by a resistor is equal to $i_0^2 R/2$.*

10.8.2 RMS current

In a DC circuit, let the constant current i_{eff} dissipate same average power in the same resistor which is given as

$$P = i_{\text{eff}}^2 R$$

Comparing the above expression with

$$P = \frac{i_0^2}{2} R$$

we can write, $i_{\text{eff}} = \sqrt{\frac{i_0^2}{2}}$ which is called 'root mean square (rms)' current

$$\Rightarrow i_{\text{eff}} = i_{\text{rms}} = \frac{i_0}{\sqrt{2}}.$$

Then, we can define i_{rms} as the effective current in AC which can be made equivalent to DC (direct current) to dissipate same average power.

Similarly, the root mean square (RMS) voltage is given as

$$V_{\text{rms}}\text{(rms voltage)} = \frac{V_0}{\sqrt{2}}$$

For instance, the peak AC voltage is $V_0 = 250$ V. Then, the effective (or rms) voltage to produce same power output averaged over a long time is $V_{\text{rms}} = \frac{V_0}{\sqrt{2}} = \frac{250}{\sqrt{2}}$ V.

In an AC circuit, most of the time the current and voltage are less than their peak (maximum) values. Hence, the average power dissipation in the resistor will be less than $V_0 i_0$ or $i_0^2 R$, rather it is $\frac{1}{2} V_0 i_0 = \frac{V_0^2}{2R} = \frac{i_0^2}{2} R = i_{\text{rms}}^2 R = \frac{V_{\text{rms}}^2}{R} = V_{\text{rms}} i_{\text{rms}}.$

10.8.3 Power in an inductor

Let us consider an inductor coil of reactance X_L. The power delivered to the inductor by the generator in fighting against the induced voltage V_L dropped across the inductor while injecting a current i is

$$P = V_L i,$$

where $i = i_0 \sin(\omega t - \phi)$ and $V_L = i_0 X_L \sin(\omega t - \phi + \frac{\pi}{2}) = i_0 X_L \cos(\omega t - \phi)$ ($\because V_L$ leads i_L by 90°).

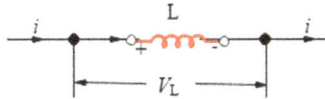

As the inductor drops a voltage V_L while carrying a current i, the instantaneous power absorbed by the inductor is $P_L = iV_L$.

Then, the instantaneous power absorbed by the inductor or delivered by the supply is

$$P = i_0^2 X_L \sin(\omega t - \phi)\cos(\omega t - \phi)$$

$$= \frac{i_0^2}{2} X_L \sin 2(\omega t - \phi)$$

As you know, the average value of sine function over a time much greater than $T = \frac{2\pi}{\omega}$ is zero; $[\sin 2(\omega t - \phi)]_{\text{av}}$, so we can write $P_{\text{av}} = 0$ for an inductor.

This tells us that, no net power is delivered by the generator over a cycle or over a time much larger than the time period of rotation of the turbine of the generator. This is because, during building up the current in the inductor, the generator does positive work or the inductor in pushing the current (+ve charges) against the induced emf across the inductor for time $\frac{T}{2}$. For the second half of the time period (half-cylce), the stored magnetic energy is released by the inductor back to the supply. While doing so, the inductor does equal positive work on the generator when the current goes to zero from maximum. As a whole, no net work is done over a time period T or full cycle. Hence, the average power $(= \frac{W_{\text{total}}}{\text{total time}})$ for a pure inductor is zero.

An inductor does not dissipate a net energy from the supply during one cycle of the AC.

10.8.4 Power in a capacitor

The power supplied by the generator to send a current i against the voltage V_C induced across the capacitor is $P = iV_C$, where $i = i_0 \sin(\omega t - \phi)$ and $V_C = i_0 X_C \sin(\omega t - \phi - \frac{\pi}{2})$ ($\because V_C$ lags i_C by 90°).

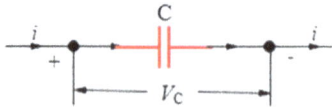

As the capacitor drops a voltage V_C while carrying a current i,
the instantaneous power absorbed by the capacitor is $P_C = iV_C$.

Then, the instantaneous power absorbed by the capacitor or delivered by the supply is

$$P = -i_0^2 X_C \sin(\omega t - \phi)\cos(\omega t - \phi)$$

$$= -\frac{i_0^2}{2} X_C \sin 2(\omega t - \phi)$$

Since $[\sin 2(\omega t - \phi)]_{\text{av}} = 0$, $P_{\text{av}} = 0$ for a capacitor.

This tells us that the average power delivered by the source to the capacitor is zero for one cycle (or practically for a time $t \gg T$). This is because the generator does positive work to charge the capacitor during the building of the charge for $t = \frac{T}{2}$ (first half cycle). For the second half cycle, the charged capacitor does equal positive work on the generator in sending the charges back by releasing all the stored electrostatic energy. So, no net work is done in the process of charging and discharging a pure capacitor by an AC supply in a complete cycle. In other words, the average power = net work done/time $(T) = 0$.

Thus, an ideal capacitor does not absorb a net energy from the supply during one cycle of AC.

The average power dissipated or absorbed by a pure capacitor is zero.

10.8.5 Power in a series R–L–C circuit

Let us now consider a load (series R–L–C) that draws a current i when the supply voltage is V. The instantaneous power delivered by the supply to the external load (comprises a resistor R, inductor L and capacitor C connected in series) of impedance Z is $P = iV$, where $V = V_0 \sin \omega t$ and $i = i_0 \sin(\omega t - \phi)$ as discussed earlier.

Then, the instantaneous power absorbed by the load or delivered by the supply is

$$P = V_0 i_0 \sin \omega t \cdot \sin(\omega t - \phi)$$
$$= V_0 i_0 \sin \omega t (\sin \omega t \cos \phi - \cos \omega t \sin \phi)$$
$$= V_0 i_0 \sin^2 \omega t \cos \phi - V_0 i_0 \sin \omega t \cos \omega t \sin \phi$$
$$= V_0 i_0 \cos \phi \sin^2 \omega t - \frac{V_0 i_0}{2} \sin \phi \sin 2\omega t$$

The average value of $\sin^2 \omega t = \frac{1}{2}$ and the average value of $\sin 2\omega t = 0$ over a cycle; so, the average power delivered by the source to the load is

$$P_{av} = \frac{1}{2} V_0 i_0 \cos \phi = V_{rms} i_{rms} \cos \phi.$$

So, the average power absorbed by a circuit depends upon the factor $\cos\phi$.

10.8.6 Power factor

Since the factor $\cos \phi$ in the expression of power decides the average power dissipation, we can call it power factor (*pf*). The average power P_{av} is maximum, when *pf* = 1 or $\phi = 0$ for a pure resistive load.

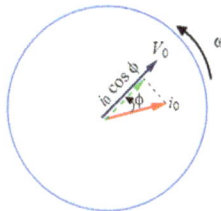

If the phase angle between the applied voltage phasor V_0 and circuit current phasor i_0 is ϕ, the power delivered by the generator to the circuit is given as $P = V_0 i_0 \cos\phi$.

P_{av} is zero, when $pf = 0$ or $\phi = 90°$ in the case of a pure inductive or capacitive load, as described earlier.

Using phasor notation, the average power can be given as half of the product of applied voltage phasor V_0 and the projection $i_0 \cos \phi$ of the current phasor i_0 onto V_0; $P_{av} = \frac{V_0}{2}(i_0 \cos \phi)$, where

$$\cos \phi = \frac{R}{Z}, \; Z = \sqrt{R^2 + X^2} \text{ and } X = |X_L - X_C|.$$

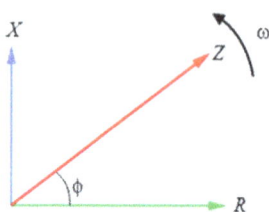

If the phase angle between the applied voltage phasor V_0 and circuit current phasor i_0 is ϕ, the resistance and reactance of a circuit is given as X=Zsinϕ and R = Zcosϕ.

Then, the average power is

$$P_{av} = V_{rms} i_{rms} \frac{R}{Z}$$
$$= \left(\frac{V_{rms}}{Z}\right) i_{rms} R = i_{rms}^2 R \left(\because \frac{V_{rms}}{Z} = i_{rms} \right)$$

In an R–L–C circuit, the average power (or energy) supplied by the source will be dissipated in the resistance of the external load, whereas the inductance and capacitance of the load do not consume any net power in a time much greater than the period of the oscillation of the electrical wave. Hence, capacitors are connected in the inductive circuit of motors, compressors, etc, to improve the power factor to maximize the power absorption. For the consumption of power in the household electronics and electrical appliances, we are charged for the average power (not instantaneous power) consumed. At high frequency, a fraction of the energy supplied is lost in radiation of electromagnetic waves (emw) generated by the AC circuit and voltage oscillations in inductors and capacitors. The rest of the supplied energy will be used in doing mechanical work for moving the parts of the electrical machines.

Example 5 Find the average power drawn by the *RL* circuit from the AC supply.

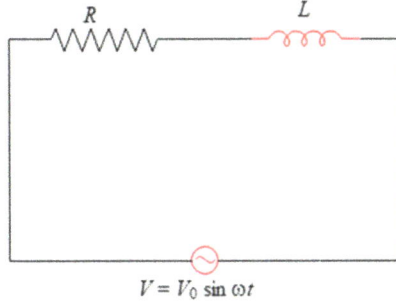

$$V = V_0 \sin \omega t$$

Solution

The average power drawn by the *R–L* circuit is

$$P_{av} = \frac{1}{2} V_0 i_0 \cos \phi, \quad \text{where} \ \cos \phi = \frac{R}{Z}$$

$$\Rightarrow P_{av} = V_0 i_0 \frac{R}{2Z}$$

$$= \frac{V_0 R}{2Z} \left(\frac{V_0}{Z} \right) \left(\because i_0 = \frac{V_0}{Z} \right)$$

Putting $Z^2 = R^2 + \omega^2 L^2$, we have

$$P_{av} = \frac{V_0^2 R}{2(R^2 + \omega^2 L^2)} \ \text{Ans.}$$

10.9 Resonance

10.9.1 Series *R–L–C* circuit

As derived earlier, in *R–L–C* series circuit the current flowing is $i = i_0 \sin (\omega t - \phi)$ where $i_0 = \frac{V_0}{Z}$

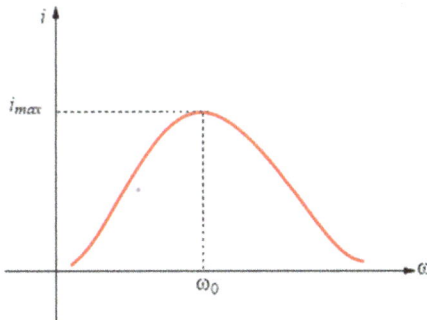

In series R-L-C circuit, the current amplitude first incresaes, reaches a maximum value at resonat freqency and then decreases to zero, with increasing frequency of the supply.

So, the rms current is given as

$$i_{\mathrm{rms}} = \frac{V_{\mathrm{rms}}}{Z}$$

$$= \frac{V_{\mathrm{rms}}}{\sqrt{R^2 + X^2}}$$

$$= \frac{V_{\mathrm{rms}}}{\sqrt{R^2 + |X_L - X_C|^2}}$$

$$\Rightarrow i_{\mathrm{rms}} = \frac{V_{\mathrm{rms}}}{\sqrt{R^2 + \left|\omega L - \frac{1}{\omega C}\right|^2}}$$

When $\omega \to 0$, $i_{\mathrm{rms}} \to 0$; when $\omega \to \infty$, $i_{\mathrm{rms}} \to 0$.

The value of i_{rms} is maximum when $X = 0$

$$\Rightarrow X_L = X_C$$

$$\Rightarrow \omega L - \frac{1}{\omega C} = 0$$

$$\Rightarrow \omega = \omega_0 = \frac{1}{\sqrt{LC}}$$

For the above frequency, the inductive and capacitive reactance are equal and opposite. Hence, the phase angle between the applied voltage V and circuit current i is zero or power factor is maximum, which is equal to one; so, the total impedance is resistive. This condition of the circuit is called 'resonance', at which the circuit draws maximum current given as

$$i_{\max} = \frac{V_0}{R}$$

At resonance, the voltages across the inductor and capacitor are equal and opposite. This means, at $\omega = \omega_0$, $V_L = V_C$. As a result, no net current is drawn due to the combined effect of inductor and capacitor at resonance. Then, the resonant current i_0 is drawn due to the resistor only, because the impedance of the circuit is totally resistive. Since the current is maximum, power dissipation will be maximum at resonance. As the circuit draws maximum current, it is called an 'acceptor circuit' at resonance. Tuning of radio is a familiar example of electromagnetic resonance.

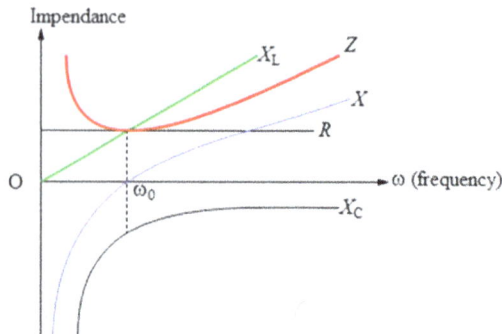

Variation of impedances (resistive, capacitive and inductive) with frequency of the supply; at the resonance of series R-L-C circuit, the impedance is minimum, net reactance is zero; the net impedance =R; so the circuit is purely resistive.

10-35

Recapitulating, at resonance in series R—L—C circuit,
1. *i is maximum*
2. *Z is minimum; Z = R*
3. *X is zero*
4. *$V_L + V_C = 0$*
5. *$\phi = 0$ (i and V are in phase)*
6. *pf = 1*
7. *The maximum average power is drawn and dissipated in the circuit.*

10.9.2 Parallel R–L–C circuit

In a parallel R–L–C circuit, current is

$$i = \frac{V_0}{Z} \sin(\omega t - \phi)$$

Then, the current amplitude and rms currents are given as follows:

$$i_0 = \frac{V_0}{Z}, \quad i_{rms} = \frac{V_{rms}}{Z}$$

Since, $\frac{1}{Z} = \sqrt{\frac{1}{R^2} + \left(\frac{1}{X_C} - \frac{1}{X_L}\right)^2}$, where $\frac{1}{X_C} = \frac{1}{X_L}$, the parallel R–L–C circuit has minimum value of $\frac{1}{Z}$ or maximum value of impedance. Hence, it will draw minimum current at $\omega_0 = \frac{1}{\sqrt{LC}}$ ($\because X_C = X_L$). At resonance, the currents flowing through the inductor and capacitor are equal in magnitude and oppositely directed. Hence, the current flowing through L and C will cancel. So, at resonance, the total current in parallel R–L–C circuit is drawn due to the resistor only. Since the circuit draws minimum current at resonance, this is called a 'rejector circuit'.

1. *Recapitulating, at resonance in parallel R–L–C, $\omega_0 = \frac{1}{\sqrt{LC}}$.*
2. *Z is maximum.*
3. *i is minimum.*
4. *Currents in inductors and capacitors are equal and opposite.*
5. *The circuit is resistive.*

Example 6 A 200 km long telegraph wire has capacitance of 0.014 μF km^{-1}. If it carries an emw of frequency 5 kHz, find the inductance required to be connected in series for maximum current.

Solution

The impedance is minimum when *i* is maximum.

Then, $X_L = X_C$

$$\Rightarrow \omega L = \frac{1}{\omega C}$$

$$\Rightarrow L = \frac{1}{\omega^2 C}$$

$$= \frac{1}{4\pi^2 f^2 C}; \ (C = \text{capacitance per unit length} \times \text{total length of the wire})$$

$$= \frac{1}{4(9.8)(5 \times 10^3)^2(0.014 \times 10^{-6} \times 200 \times 10^3)}$$

$$= 0.362 \text{ mH Ans.}$$

10.10 Transformer

10.10.1 Introduction

It is evident from our daily experience that household electrical equipments like TV, freezer, washing machine, etc, operate at a high voltage of 250 V, while we operate electric toaster and radio, etc, at lower voltages. This means that we need different voltages to operate different equipment. We need low voltage to handle electrical items safely. On the other hand, we cannot operate any machine at any voltage we like. For instance, power generators operate at 22 kV without any insulation breakdown and corona discharge. The generated electrical power is transmitted at a voltage 500 kV, reducing the line current to $\frac{22}{500}$ times. This reduces the copper (joule heat) loss by $(\frac{22}{500})^2$ times.

Thus, purposefully we need to transmit electrical power at high voltage to reduce copper loss and use the electric power at low voltage to reduce electrical breakdown for easily handling the equipment without any electrical hazard.

10.10.2 Definition

Hence, we need a device which can change the voltage of an AC supply according to our requirement without any appreciable change in its power, which is called a 'transformer'. It works obeying the principle of mutual induction with an efficiency between 90% and 99%.

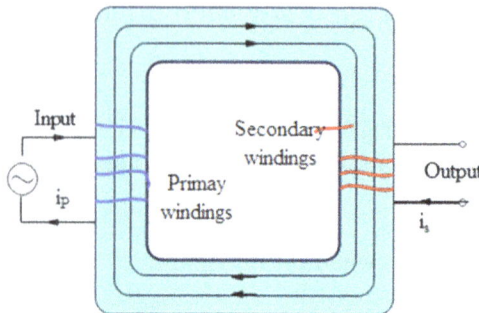

In an ideal trasformer, energy conservation is given as
$i_p V_p = i_s V_s$, where $V_p = -N_p(d\phi/dt)$ and $V_s = -N_s(d\phi/dt)$; $\phi = $fux linked with each loop or turn of the windings.

There are two coils wrapped over the same iron core. The coil which is connected to the AC supply is called primary (P) and the other coil connected to the external load (R, L and C) is called secondary (S) coil. The AC supply causes an alternating current in the P-coil. This generates an alternating magnetic field B in the iron core. As the iron core has high permeability, almost all B-lines pass through it. As a result, an equal flux ϕ, say, passes (links) through each loop of both primary and secondary coils. If we assume N_P and N_S as the number of turns in P and S coils, respectively, the net flux passing through them can be given as

$$\phi_P = N_P\phi \text{ and } \phi_S = N_S\phi$$

As i_P changes sinusoidally, ϕ_P will also change sinusoidally. The time-varying flux ϕ_P induces an emf V_P across the primary coil given as

$$V_P = -\frac{d\phi_P}{dt} = -N_P\frac{d\phi}{dt} \tag{10.11}$$

Simultaneously, the change in flux ϕ_S in the secondary, induces an emf (voltage) V_S across the secondary which is given as

$$V_S = -\frac{d\phi_S}{dt} = -N_S\frac{d\phi}{dt} \tag{10.12}$$

Comparing equations (10.11) and (10.12), we have $\frac{V_P}{V_S} = \frac{N_P}{N_S}$, where $\frac{N_P}{N_S}$ = turn-ratio and $V_P = V$ (applied voltage) neglecting the resistance of the primary coil.

If $N_S > N_P$, $V_P > V_S$, then the transformer is said to be a step-up transformer. If $N_S < N_P$, $V_S < V_P$, then the transformer is called a step-down transformer. Depending on our requirement, by adjusting the turn ratio, we can increase or decrease the voltage in the AC circuit. This particular advantage of AC over DC makes AC machines more popular and practical than DC machines.

10.10.3 Impedance matching

When a pure resistive load R_L is connected with the secondary winding, the induced voltage across S-coil sends some alternating current through it. We call it load current i_S. Since i_S is alternating, it generates a back emf across the primary coil, so as to obey Lenz's law. As the applied voltage is kept constant by the generator, the primary coil draws more current to nullify (compensate) the back emf. Neglecting the losses (copper and hysteresis etc), the power drawn by the P-coil from the source (generator) exactly matches the power that the S-coil delivers to the load.

$$\Rightarrow V_P i_P = V_S i_S \tag{10.13}$$

Due to the load resistance R_L, the secondary current is

$$i_S = \frac{V_S}{R_L} \tag{10.14}$$

Using equations (10.13) and (10.14), we get the current in the primary as

$$i_P = \frac{V_S i_S}{V_P}$$

$$= \frac{V_S \left(\frac{V_S}{R_L}\right)}{V_P}$$

$$= \frac{V_S^2}{V_P^2} \cdot \frac{V_P}{R_L}$$

$$= \left(\frac{N_S}{N_P}\right)^2 \cdot \frac{V_P}{R_L} \left(\because \frac{V_S}{V_P} = \frac{N_S}{N_P}\right)$$

$$\Rightarrow i_P = \frac{V_P}{\left(\frac{N_P}{N_S}\right)^2 R_L} = \frac{V_P}{R_{eq}}, \text{ say.}$$

Then, it is equivalent to say that the same current i_P can be caused by the P-coil by connecting an equivalent (or effective) resistance R_{eq} with the primary coil given as

$$R_{eq} = \left(\frac{N_P}{N}\right)^2 R_2$$

This significant property of the transformer is called impedance matching.

Hence, in addition to transformations of current and voltage, a transformer matches the impedance between the supply and the consumer (load) circuit to draw maximum power.

Maximum power transfer can occur between the power source and load by coupling them with a transformer of suitable turn ratio.

For instance, a transformer is connected between 1 k Ω output audio amplifier with 8 Ω speaker that ensures maximum possible transfer of audio signal into the speaker from the amplifier.

Example 7 An ideal step-down transformer has turn $\frac{N_P}{N_S} = 25$. If the input voltage is 400 V, find the (i) output voltage and (ii) ratio of currents in primary and secondary.
   Solution

(i) It is given that

$$\frac{V_i}{V_O}(=\frac{V_P}{V_S}) = \frac{N_P}{N_S} = 25$$

$$\Rightarrow V_O = \frac{400}{25} = 16 \text{ V Ans.}$$

(ii) The ratio of currents is

$$\frac{i_P}{i_S} = \frac{V_S}{V_P} = \frac{N_S}{N_P} = \frac{1}{25} \text{ Ans.}$$

Problem 1 The rms (root mean square) value of the current for the given sinusoidal voltage variation is equal to 5 A.

$$V = V_0 \,|\sin \omega t|$$

Find the average power absorbed by the heater coil. Put $f = 50$ Hz and average voltage is 146 V and rms current $= 5$ A. The power factor is equal to 0.99.

Solution

The average value of the sinusoidal voltage is

$$V_{av} = \frac{\int_0^T |V|\,dt}{T} = \frac{\int_0^T \sin\frac{2\pi}{T}\,dt}{T}$$

$$\text{Or, } 146 = \frac{2V_0}{\pi}$$

So, the peak voltage $= V_0 = 230$ V; the peak current $= (5)(1.414) = 7.07$ A. So, the average power loss in the coil $=$ (peak current)(peak voltage)(power factor)/2. Or, average power $= (7.07)(230)(0.98)(0.5) = 731.745$ W. Ans.

Problem 2 A resistance $R = 20\ \Omega$ drops a voltage of 80 V across it when connected to an AC supply of $V_0 = 100$ V and $\omega = 314$ rad s^{-1}. If the inductor L is connected in series with the resistor, find (i) the potential dropped across the inductor, (ii) inductive reactance, and (iii) inductance L.

Solution

Let V_L, V_R be the voltage amplitudes across the resistor and inductor, respectively.

(i) By using phasor algebra, $V_L^2 + V_R^2 = V_0^2$

$$V_L = \sqrt{V_0^2 - V_R^2}$$
$$= \sqrt{100^2 - 80^2}$$
$$= 60 \text{ V Ans.}$$

(ii) $iX_L = V_L$ and $i_0 = \frac{(V_R)_0}{R} = \frac{80}{20} = 4$ A. Since the current is 4 A, the inductive reactance is

$$X_L = \frac{60}{4} = 15 \text{ ohm Ans.}$$

(iii) The inductance is

$$L = \frac{15}{314} = 0.0477 \text{ H Ans.}$$

Problem 3 (Series R–L circuit in AC)

A series R–L circuit is subjected to an alternating voltage $V = V_0 \sin \omega t$. Find $i = f(t)$ in the circuit.

Solution

If $V = V_0 \sin \omega t$, referring phasor diagram in section 10.6, we have

$$i = i_0 \sin(\omega t - \phi),$$

where

$$\phi = \cos^{-1} \frac{V_{R_0}}{V_0} = \tan^{-1} \frac{V_{L_0}}{V_{R_0}}$$

$$= \cos^{-1} \frac{i_0 R}{i_0 Z} = \tan^{-1} \frac{i_0 X_L}{i_0 R} = \cos^{-1} \frac{R}{Z}$$

$$= \tan^{-1} \frac{X_L}{R} \text{ and } i_0 = \frac{V_0}{Z} = \frac{V_0}{\sqrt{R^2 + X_L^2}}$$

Then, the current in the circuit is

$$i = \frac{V_0}{\sqrt{R^2 + X_L^2}} \sin\left(\omega t - \tan^{-1} \frac{X_L}{R}\right)$$

Substituting $X_L = \omega L$, we get

$$i = \frac{V_0}{\sqrt{R^2 + \omega^2 L^2}} \sin(\omega t - \tan^{-1} \frac{\omega L}{R}) \text{ Ans.}$$

N.B: 1. In a series RL circuit, current lags the applied voltage by an angle $\phi = \cos^{-1} \frac{R}{Z} = \tan^{-1} \frac{X_L}{R}$, where $Z = \sqrt{R^2 + X_L^2}$ and $X_L = \omega L$. The maximum current $i_0 = \frac{V_0}{Z} = \frac{V_0}{\sqrt{R^2 + \omega^2 L^2}}$.

2. In complex form, $Z = R + j\omega L$ and $i = \frac{V_0 e^{j\omega t}}{R + j\omega L} = \frac{V_0}{\sqrt{R^2 + \omega^2 L^2}} e^{j(\omega t - \phi)}$, where $\phi = \cos^{-1} \frac{R}{X_L}$.

Problem 4 In the previous problem discuss how the impedance and current vary with the frequency of AC supply. Draw the variation of impedance and peak current with frequency f.

Solution

The impedance of the RL circuit is

$$Z = \sqrt{R^2 + X_L^2}$$

$$= \sqrt{R^2 + \omega^2 L^2} \ (\because X_L = \omega L)$$

$$= \sqrt{R^2 + 4\pi^2 L^2 f^2} \ (\because \omega = 2\pi f)$$

When $f = 0$, $Z = R$ and when $f \to \infty$, $Z \to \infty$.

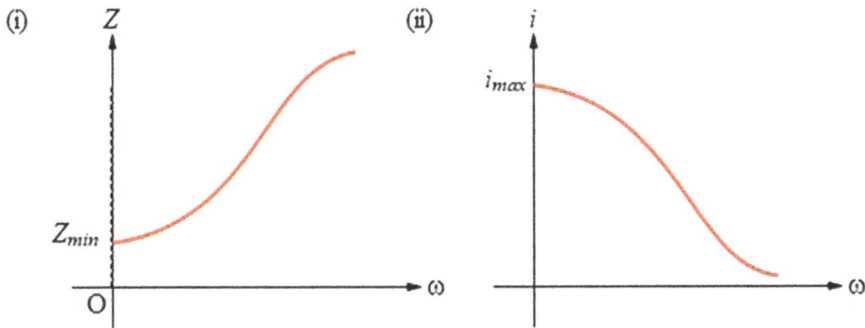

The peak current is

$$i_0 = \frac{V_0}{Z}$$

$$= \frac{V_0}{\sqrt{R^2 + 4\pi^2 L^2 f^2}}$$

When $f = 0$, $i_0 = \frac{V_0}{R}$ and when $f \to \infty$, $i_0 \to 0$ Ans.

N.B: 1. $X_L = \omega L$ and $Z = \sqrt{R^2 + \omega^2 L^2}$

2. $i = \frac{V_0}{\sqrt{R^2 + \omega^2 L^2}}$

3. When ω increases Z increases from R to ∞ and hence current decreases from maximum $\frac{V_0}{R}$ to zero.

Problem 5 (Series RC circuit in AC)

A series RC circuit is subjected to an alternating voltage given as $V = V_0 \sin \omega t$. Find the (a) impedance and (b) phase difference between current and applied voltage, (c) current flowing in the circuit as the function of time.

Solution

(a) The phasor diagram shows that in the resistor, V_{R_0} is in phase with i_0 and capacitor V_{C_0} lags the current by $90°$.

Then using the concept of phasor, we have

$$V_0 = \sqrt{V_{X_0}^2 + V_{X_0}^2}$$

$$= \sqrt{i_0^2 R^2 + i_0^2 X_C^2}$$

Then, the impedance of the circuit is

$$Z = \frac{V_0}{i_0}$$

$$= \sqrt{R^2 + X_C^2}$$

$$= \sqrt{R^2 + \frac{1}{\omega^2 C^2}} \text{ Ans.}$$

(b) Then, the current i_0 leads the voltage V_0 by an angle ϕ given as

$$\phi = \tan^{-1}\frac{V_{C_0}}{V_{R_0}} = \cos^{-1}\frac{V_{R_0}}{V_0}$$

$$= \tan^{-1}\frac{i_0 X_C}{i_0 R} = \cos^{-1}\frac{i_0 R}{i_0 Z}$$

$$= \tan^{-1}\frac{X_C}{R} = \cos^{-1}\frac{R}{Z}$$

$$= \tan^{-1}\frac{1/\omega C}{R} = \cos^{-1}\frac{R}{\sqrt{R^2 + 1/\omega^2 C^2}}$$

$$= \tan^{-1}\frac{1}{\omega RC} = \cos^{-1}\left(\frac{\omega RC}{\sqrt{R^2\omega^2 C^2 + 1}}\right) \text{ Ans.}$$

(c) The current in the circuit as the function of time is

$$i = \frac{V_0}{\sqrt{R^2 + X_C^2}}\sin(\omega t - \phi),$$

where $\phi = \tan^{-1}\frac{1}{\omega RC}$ and $X_C = \frac{1}{\omega C}$ Ans.

N.B: 1. In a series RC circuit, $Z = \sqrt{R^2 + X_C^2}$, where $X_C = \frac{1}{\omega C}$ and $\phi = \tan^{-1}\frac{1}{R\omega C}$.

2. The current $i = i_0 \sin(\omega t - \phi)$, where $i_0 = \frac{V_0}{Z}$.

3. In complex form, $Z = R - \frac{1}{\omega C}$ and $i = \frac{V_0}{R - \frac{j}{\omega C}}e^{j\omega t} = \frac{V_0}{\sqrt{R^2 + \frac{1}{\omega^2 C^2}}}e^{j(\omega t - \phi)}$, where

$\phi = \tan^{-1}\frac{1}{\omega RC}$.

4. i_0 leads V_0 by $\phi = \tan^{-1}\frac{1}{\omega RC}$ in a series RC circuit.

Problem 6 In the last example, draw the variation of impedance and current in the circuit with the varying angular frequency ω.

Solution

Since, $Z = \sqrt{R^2 + \frac{1}{\omega^2 C^2}}$ if $\omega \to 0$, $Z \to \infty$

If $\omega \to \infty$, $Z \to R$.

Then, the current amplitude is given as

$$i_0 = \frac{V_0}{Z}$$

$$= \frac{V_0}{\sqrt{R^2 + \frac{1}{\omega^2 C^2}}}$$

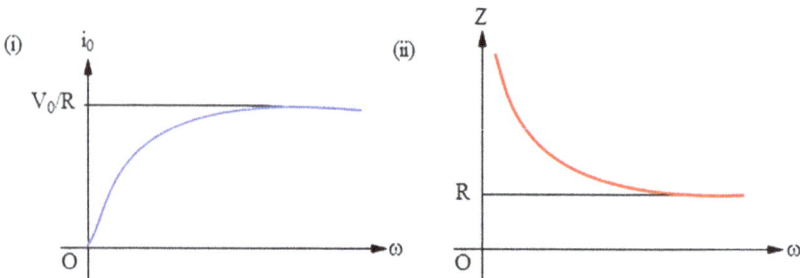

When $\omega \to 0$, $i_0 \to 0$; when $\omega \to \infty$, $i_0 \to \frac{V_0}{R}$; when ω increases, X_C decreases. Hence Z decreases increasing the peak current from zero to $\frac{V_0}{R}$. Ans.

Problem 7 (AC applied to an LC circuit)

A series LC circuit is connected to an AC supply (voltage) $V = V_0 \sin \omega t$. Find the peak current and phase angle ϕ between current and applied voltage.

Solution

Method 1 (Trigonometry):

Applying KCL, we have

$$V = V_C + V_L,$$

where $V = V_0 \sin \omega t$, $V_C = i \sin(\omega t - \phi - \frac{\pi}{2})$ and $V_L = i_0 X_L \sin(\omega t - \phi + \frac{\pi}{2})$

Then,

$$V_0 \sin \omega t = -i_0 X_C \cos(\omega t - \phi) + i_0 X_L \cos(\omega t - \phi)$$

$$V_0 \sin \omega t = i_0(X_L - X_C)\cos(\omega t - \phi)$$

Comparing both sides, the current amplitude is

$$i_0 = \frac{V_0}{X_L - X_C} = \frac{V_0}{\omega L - \frac{1}{\omega C}} \text{ Ans.}$$

The phase angle is

$$\phi = \frac{\pi}{2} \text{ Ans.}$$

Method 2 (Phasor):

Since, $R = 0$,

$$Z = \sqrt{R^2 + X^2} = X = |X_L - X_C|$$

Then, the phase angle is equal to

$$\cos^{-1} \frac{R}{Z} = \cos^{-1} 0 = \frac{\pi}{2}$$

and the current amplitude is

$$i_0 = \frac{V_0}{Z} = \frac{V_0}{X}$$

$$= \frac{V_0}{X_L - X_C}$$

$$\Rightarrow i_0 = \frac{V_0}{\omega L - \frac{1}{\omega C}} \text{ Ans.}$$

Method 3 (Complex number):

$$Z = \frac{j}{\omega C} + j\omega L = j\left(\omega L - \frac{1}{\omega C}\right)$$

Then, the current amplitude is

$$i_0 = \frac{V_0}{Z} = \frac{V_0}{|j(\omega L - \frac{1}{\omega C})|} = \frac{V_0}{|\omega L - \frac{1}{\omega C}|} \text{ and } \phi = \frac{\pi}{2} \text{ Ans.}$$

N.B: In a series LC circuit subjected to AC supply,

1. Current i_0 lags if $(X_L > X_C)$ and leads if $(X_C > X_L)$ the voltage V_0 by 90°.
2. $Z = |X_L - X_C| = |\omega L - \frac{1}{\omega C}|$
3. $i = i_0 \sin(\omega t - \frac{\pi}{2}) = i_0 \cos \omega t; i_0 = \frac{V_0}{Z}$.
4. In complex form, $i = \frac{V_0}{j(\omega L - \frac{1}{\omega C})}e^{j\omega t} = \frac{V_0}{\omega L - \frac{1}{\omega C}}e^{j(\omega t - \frac{\pi}{2})}$
5. At $\omega = \frac{1}{\sqrt{LC}}, i_0 \to \infty$.

At the resonant frequency $\omega = \frac{1}{LC}$, $X = 0$ which is called resonance at which the circuit current

$$i \to \infty \text{ and } Z = 0 \text{ as shown in the figure in section 10.9.1.}$$

Problem 8 An alternating emf of frequency $f = 50$ Hz, peak voltage $V_0 = 21$ V is applied to a series circuit of resistance $R = 20\ \Omega$, an inductance $L = 100$ mH and a capacitor of $C = 30\ \mu F$. Find (a) maximum current i_0, (b) phase difference ϕ between current and applied voltage, and (c) $i = f(t)$.

Solution

(a) The impedance of the circuit is

$$Z = R + j\left(\omega L - \frac{1}{\omega C}\right)$$

$$= 20 + j\left[314 \times 100 \times 10^{-3} - \frac{1}{314 \times 30 \times 10^{-6}}\right]$$

$$\Rightarrow Z = 20 - j74.76$$

$$\Rightarrow Z = 77.38\ \Omega.$$

Hence, $i_0 = \frac{V_0}{Z} = \frac{21}{77.38} = 0.27$ A Ans.

(b) Since Im (Z) is $-$ve, the circuit is capacitive. Hence, the current leads the voltage by an angle ϕ given as

$$\phi = \tan^{-1}\frac{X}{R} = \tan^{-1}\left(\frac{-74.76}{20}\right) = -75° \text{ Ans.}$$

(c) Hence, $i = i_0 \sin(\omega t - \phi) = i_0 \sin\{\omega t - (-75°)\}$

$$\text{or, } i = 2.73 \sin(314t + 75°) \text{ Ans.}$$

Problem 9 A choke coil is needed to operate an arc lamp at $V_{rms} = 130$ V and frequency $f = 50$ Hz. If the rms current in the arc lamp is 10 A and effective resistance of the arc lamp is 5 Ω, find the inductance of the choke coil.

Solution

From the phasor diagram, we have

$$V_{L_0}^2 + V_{R_0}^2 = V_0^2$$

Or,

$$i_0^2 X_L^2 + i_0^2 R^2 = V_0^2$$

Or,

$$X_L = \sqrt{\frac{V_0^2 - i_0^2 R^2}{i_0^2}}$$

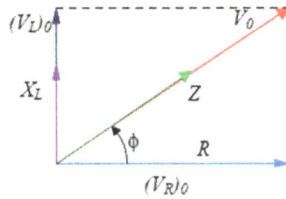

Or,

$$\omega L = \sqrt{\left(\frac{V_0}{i_0}\right)^2 - R^2}$$

Or,

$$L = \frac{1}{2\pi f}\sqrt{\left(\frac{V_{rms}}{i_{rms}}\right)^2 - R^2}$$

$$= \frac{1}{314}\sqrt{\left(\frac{130}{10}\right)^2 - 5^2}$$

$$= 3.82 \times 10^{-2} \text{ H Ans.}$$

Problem 10 An alternating voltage $V_0 = 100$ V with angular frequency ω is connected across the capacitor and inductor having $X_L = 5\,\Omega$ and $X_C = 10\,\Omega$. Find the current in each branch and total current at time t.

Solution

The applied voltage is

$$V = V_0 \sin \omega t$$

Since i_C leads $V_C = V$ by $\frac{\pi}{2}$ rad, we have

$$i_C = \frac{V_0}{X_L} \sin\left(\omega t + \frac{\pi}{2}\right) = \frac{100}{10} \cos \omega t$$
$$= 10 \cos \omega t \text{ Ans.}$$

Since i_L lags $V_L = V$ by $\frac{\pi}{2}$ rad, we have

$$i_L = \frac{V_0}{X_L} \sin\left(\omega t - \frac{\pi}{2}\right) = -\frac{100}{5} \cos \omega t = -20 \cos \omega t \text{ Ans.}$$

Then, the total current is

$$i = i_L + i_C$$
$$= 10 \cos \omega t + (-20 \cos \omega t)$$
$$= -10 \cos \omega t \text{ Ans.}$$

Problem 11 Find the current drawn by the given circuit from the AC supply of voltage $V = V_0 \sin \omega t$.

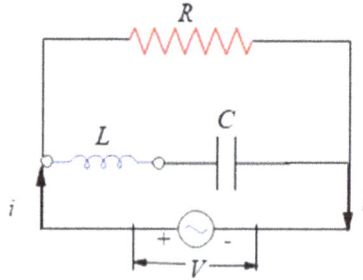

Solution

The impedance is given as

$$\frac{1}{Z} = \frac{1}{R} + \frac{1}{j\omega L - \frac{j}{\omega C}}$$
$$= \frac{1}{R} + \frac{\omega C}{j(\omega^2 LC - 1)}$$
$$= \frac{1}{R} - \frac{j\omega C}{(\omega^2 LC - 1)}$$

Then,

$$i_0 = \frac{V_0}{Z} = \frac{V_0}{R} - j\frac{V_0 \omega C}{\omega^2 LC - 1}$$

Then,

$$i = i_0 e^{j\omega t}$$

$$= V_0 \left(\frac{1}{R} - \frac{j\omega C}{\omega^2 LC - 1} \right) e^{j\omega t}$$

$$= i_0 e^{j(\omega t + \phi)} = i_0 \sin(\omega t + \phi),$$

where

$$\phi = \tan^{-1} \frac{\dfrac{\omega C}{\omega^2 LC - 1}}{\dfrac{1}{R}} = \tan^{-1} \left(\frac{\omega CR}{\omega^2 LC - 1} \right) \text{ and } i_0 = V_0 \sqrt{\frac{1}{R^2} + \left(\frac{V_0 \omega C}{\omega^2 LC - 1} \right)^2} \quad \text{Ans.}$$

N.B: The currents in the other branches are given by

$$i_R = \frac{V_0}{R} \sin \omega t, \quad i_L = \frac{V_0 \omega C}{(\omega^2 LC - 1)} \sin(\omega t + \phi).$$

Problem 12 A resistor R is connected in series with a coil. The system is subjected to an AC supply of peak voltage V_0. If the peak voltages dropped across the resistor R and the coil are V_1 and V_2, respectively, find the power dissipated in the coil.

Solution

The power dissipated in the coil is

$$P = iV_2 \cos \phi \tag{10.15}$$

From the phasor diagram,

$$V_1^2 + V_2^2 + 2V_1 V_2 \cos \phi = V_0^2 \tag{10.16}$$

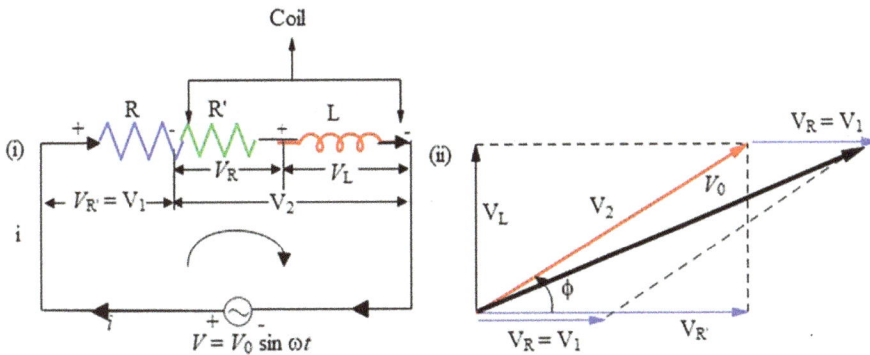

Substituting $V_2 \cos \phi$ from equation (10.16) in equation (10.15),

$$P = \frac{i(V_0^2 - V_1^2 - V_2^2)}{2V_1}, \quad \text{where } i = \frac{V_1}{R}.$$

Then,

$$P = \frac{V_0^2 - V_1^2 - V_2^2}{2R} \text{ Ans.}$$

N.B: You can find that the total power dissipated in the series combination is given as

$$P_{\text{total}} = \frac{V_0^2 - V_2^2}{2R}.$$

Problem 13 In a series R–L–C circuit, we have the same current at angular frequencies ω_1 and ω_2. Find the resonant frequency.
   Solution
   The current in a series R–L–C circuit is

$$i = \frac{V_0}{\sqrt{R^2 + \left|\omega L - \frac{1}{\omega C}\right|^2}}$$

Since, $i_1 = i_2$ at ω_1 and ω_2, we have

$$\text{or, } \frac{V_0}{\sqrt{R^2 + \left(\omega_1 L - \frac{1}{\omega_1 C}\right)^2}} = \frac{V_0}{\sqrt{R^2 + \left(\omega_2 L - \frac{1}{\omega_2 C}\right)^2}}$$

$$\text{or, } \omega_1 L - \frac{1}{\omega_1 C} = \omega_2 L - \frac{1}{\omega_2 C}$$

$$\text{or, } L(\omega_1 - \omega_2) = \frac{1}{C}\left(\frac{1}{\omega_1} - \frac{1}{\omega_2}\right)$$

$$\text{or, } \omega_1 \omega_2 = \frac{1}{LC}$$

Substituting $\frac{1}{LC} = \omega_0^2$, we have $\omega_0 = \sqrt{\omega_1 \omega_2}$ Ans.

Problem 14 A parallel plate air capacitor C is corrected to an AC supply of angular frequency ω with a resistor R connected in series. If the capacitor is filled completely with a dielectric slab, the power dissipation in the resistor increases four-fold. What is the relative permittivity of the dielectric?
   Solution
   The power dropped across the resistor is proportional to the square of the current flowing through it. Since the power dissipation doubles, the current increases by a factor 4. Since $I = V/Z$, for the same peak voltage, the impedance Z will decrease by a factor 2. So, we can write the ratio of impedances as

$$Z_2/Z_1 = 1/2$$

$$\Rightarrow Z_2 = Z_1/2 \tag{10.17}$$

So, the initial value of impedances is

$$Z_1 = \sqrt{R^2 + \frac{1}{\omega^2 C_1^2}} \tag{10.18}$$

So, the final value of impedances is

$$Z_2 = \sqrt{R^2 + \frac{1}{\omega^2 C_2^2}} \tag{10.19}$$

Using the last three equations, we have

$$2\sqrt{R^2 + \frac{1}{\omega^2 C_2^2}} = \sqrt{R^2 + \frac{1}{\omega^2 C_1^2}}$$

$$\Rightarrow \frac{1}{\omega^2 C_1^2}\left(1 - \frac{4C_1^2}{C_2^2}\right) = 3R^2$$

$$\Rightarrow 1 - \frac{4C_1^2}{C_2^2} = 3C_1^2 \omega^2 R^2$$

$$\Rightarrow 1 - \frac{4}{\varepsilon_r^2} = 3C_1^2 \omega^2 R^2 (\because C_2/C_1 = \varepsilon_r)$$

$$\Rightarrow \varepsilon_r = \frac{2}{\sqrt{1 - 3C^2\omega^2 R^2}}(\because C_1 = C) \quad \text{Ans.}$$